DISCRETE MATHEMATICAL MODELS

with applications to social, biological, and environmental problems

FRED S. ROBERTS

Rutgers University

Prentice-Hall, Inc.

Englewood Cliffs, New Jersey

Library of Congress Cataloging in Publication Data

ROBERTS, FRED S.
  Discrete mathematical models, with applications to
social, biological, and environmental problems.

  Includes bibliographies and index.
  1. Biology—Mathematical models. 2. Social sciences
—Mathematical models. I. Title.
QH3235.R6   511′.8′0243   75-20153
ISBN   0-13-214171-X

To My Family

10 9 8 7 6 5 4 3

Printed in the United States of America

PRENTICE-HALL INTERNATIONAL, INC., *London*
PRENTICE-HALL OF AUSTRALIA, PTY. LTD., *Sydney*
PRENTICE-HALL OF CANADA, LTD., *Toronto*
PRENTICE-HALL OF INDIA PRIVATE LIMITED, *New Delhi*
PRENTICE-HALL OF JAPAN, INC., *Tokyo*
PRENTICE-HALL OF SOUTHEAST ASIA (PTE.) LTD., *Singapore*

Contents

iii

Preface

This book is intended for a variety of audiences, including mathematicians, social scientists, biologists, environmental scientists, etc. I will argue in the Introduction that there are two directions of interaction between mathematics and any applied field. First, mathematics can be applied to that field; second, that field can stimulate the development of new mathematics. These interactions have been exhibited between mathematics and physical problems for a long time. As the interactions between mathematics and such newer areas of application as social, biological, and environmental problems become more serious, there is need to educate both mathematicians and non-mathematicians in the mathematics which is playing a role in this interaction. The by-now-common "Finite Math" books do this on an elementary level for a certain type of finite or "discrete" mathematics.[1] This book is a more advanced treatment of the discrete mathematical tools which are being used in these newer areas of application. It illustrates both the applications of mathematics to these various applied subjects and the impact of these applied subjects on the development of new mathematics.

Although this book can be used for reference, it is primarily a textbook. It can be used for a variety of courses. I have used a preliminary version of the book several times in a sophomore through senior level course at

[1]For further discussion of the nature of "discrete" mathematics, the reader is referred to the Introduction.

Rutgers University on Mathematical Models in the Social and Biological Sciences and in a junior-senior course in Graphs, Games, and their Applications. The enrollment in these courses was about 50% math majors and the rest from a variety of areas in social and biological science; it included several graduate students from disciplines outside mathematics. I have also used the book in mathematics graduate courses in Mathematical Models, in Applied Graph Theory, and in Measurement and Decisionmaking—these courses made much greater use of the proofs and theoretical material presented. The material from the book has also been used in note form in similar courses at a number of other institutions. Finally, I was fortunate enough to use the material in two Summer Institutes for college and junior college teachers of mathematics and related fields such as Operations Research, Engineering, Envrionmental Science, etc.

The 1-semester Math Models course uses material on modelling (Chapter 1), graph theory (a brief treatment[2] of Chapter 2), signed graphs and balance (Sec. 3.1), weighted digraphs and pulse processes (a brief treatment of Chapter 4), and Markov chains (a brief treatment of Chapter 5). It closes with either *n*-person games (a brief treatment of Chapter 6), group decisionmaking (a brief treatment of Chapter 7), or measurement and utility (a brief treatment of Chapter 8). In general, in the undergraduate version of this course, proofs are de-emphasized, and building and evaluation of models is emphasized. At present, the undergraduate version of this course is taught at Rutgers as a follow-up to a course in "Finite Mathematics," which covers the language of sets, elementary topics in linear algebra, counting techniques, elementary probability, etc. (The Finite Math course has some calculus as a prerequisite)

A 1-year course in Mathematical Models in the Social and Biological Sciences could cover parts of each chapter in the book. It should start with all of Chapters 1 and 2, then cover most of Chapter 3 (perhaps up through Sec. 3.5.1), and conclude with a brief treatment of the remaining chapters. Similar 1-semester or 1-year courses could emphasize environmental problems. The 1-semester graphs and games course covered most of Chapters 1, 2, 3, and 6. A 1-semester applied graph theory course covers much of Chapters 1, 2, 3, and 4. Finally, a measurement and decisionmaking course covers Chapters 1, 7, and 8, with careful treatment of exercises and supplementary readings from the references.

I have tried to keep the interdependencies in this book to a minimum, so that the book can be used in a variety of ways. The following sections and subsections in Chapter 2 are essential for Chapters 2 to 5: 2.1, 2.2.1, 2.2.4, 2.2.6, 2.3.1, 2.3.2 (not the proofs), and 2.4.1. Much of Chapter 6 can be

---

[2]See the table of dependencies which follows the preface for a detailed description of the "brief treatments" referred to.

read after reading Sec. 2.1 and parts of Sec. 2.2.1, and Chapters 7 and 8 have no prerequisites. Chapters 3–8 are essentially independent, though occasional reference to earlier chapters is made. If much of Chapter 3 is covered, I recommend covering all of Chapter 2 first, or beginning with the brief treatment of Chapter 2 and then returning to additional topics as they are needed. The dependencies within chapters are diagrammed at the end of the preface. Material which could be covered in a brief treatment is also described. Further guidance for what can be omitted is contained in the text.

There are a number of mathematical prerequisites for this book. The language of sets is used throughout. So are elementary logical symbols and arguments. The reader should be familiar with such terms as necessary condition, sufficient condition, converse, contrapositive, and so on. Also basic to much of the book is elementary linear algebra. However, beyond assuming certain ability to manipulate matrices and vectors (and in one place determinants), I have tried to make the book self-contained as far as linear algebra is concerned. The development also uses some probability theory, but essentially only in Chapter 5 (Markov Chains) and briefly in Sec. 8.5.2 (The Expected Utility Hypothesis). The reader who has not been exposed to the elementary theory of probability, say at the level of a book in Finite Math, will have trouble with that material. He should understand how to calculate probabilities, how to use tree diagrams, and what it means to find conditional probabilities and expected values. Counting techniques, again at the level of a Finite Math course, are also used in places. Used throughout are simple terms about functions, for example, domain, range, one-to-one, onto, etc. Some ideas from the calculus are also used in places, in particular, the idea of limit. The student without at least one semester of calculus will have trouble reading these parts.

These are the formal mathematical prerequisites for much of the book. It is my experience that the book can be used, if almost all proofs are omitted, by students with no more background than a good finite math course and a good one-semester calculus course. However, a year of calculus is strongly recommended and the student who has the added mathematical sophistication of a full course in probability or linear algebra will get much more out of this material.

Some of the subsections or proofs use more advanced mathematical tools, for example group theory and analytic or topological arguments. Other subsections simply require a fair amount of mathematical maturity. These subsections or proofs are starred or moved to the end of a section, and can be omitted without loss of continuity. Indeed, almost all proofs in this book can be omitted without loss of continuity.

The question of non-mathematical prerequisites for this book is not nearly so easy to define as the question of mathematical prerequisites. The reader of the book is not expected to be an expert in the social, biological, or environmental sciences. Indeed, the material has been used in courses populated by a great many mathematics students with little background in the applied areas discussed. Most of the basic terminology of the applied fields, when used, is explained. Indeed, it is often the purpose of the development to make precise definitions of terms from another discipline which are not defined too carefully in practice. (Examples are balance in a social structure, status in an ecological chain, etc.) Of course, any student who seriously wishes to pursue the interactions between mathematics and an applied subject had better gain some understanding of that subject as well.

This book has many exercises, usually some at the end of each section. I have long felt that the best way to learn mathematics is to do it, and so I feel that the exercises are a very important part of this book. Many of them contain additional material, not presented in the text. I have tried to arrange them in order from exercises simply asking the reader to repeat computations made in a section to more difficult theoretical ones. The hardest ones should be tractable only for the most advanced students. Some of these harder ones are marked with an asterisk *. Finally, I have added a few discussion problems and a few projects at the end of some of the sections. Some of the projects suggest a mathematical or mathematical modelling research problem which goes in a new or untried direction.

My interest in the applications of mathematics to social, biological, and environmental problems, goes back to my days as an undergraduate. This interest was nurtured along the way by many people, and in a sense they planted the seed from which this book developed. I would especially like to thank John Kemeny, Duncan Luce, Robert Norman, Dana Scott, and Patrick Suppes.

I would also like to thank the following institutions for their financial and other support of my research prior to and during the development of this book: the Institute for Advanced Study, Rutgers University, the Sloan Foundation, the RANN Program of the National Science Foundation, and the Rand Corporation, which gave me permission to use various materials I originally developed as a Rand researcher.

Prentice-Hall supplied chapter-by-chapter technical reviews which, I feel, significantly improved the quality of the book. Arthur Wester, the former Mathematics Editor at Prentice-Hall, took an early interest in the work, and his conviction that I had something different and important to say were a source of encouragement.

Many individuals supplied comments and criticisms and I cannot possibly acknowledge them all. But I would like to single out Duncan Luce and Victor Klee, who sent detailed comments on an early draft, and Kenneth Bogart, who supplied a detailed review of the next-to-last draft. William Lucas and Lloyd Shapley gave me very helpful comments about Chapter 6 (*n*-person Games). (Bill Lucas also provided me with several forums for presentation of this material to potential users.) Jeffrey Ullman, Kenneth Bogart, David Rosen, Frank Norman, Peyton Young, and Duncan Luce provided detailed comments on other individual chapters.

Judy Johnson and Rochelle Leibowitz found many errors in each draft, made countless useful suggestions, and carefully worked all of the exercises in the final draft. Their conscientious and enthusiastic help is hard to measure.

I alone, however, am responsible for all errors which may remain.

Finally, I would like to thank my wife Helen. As a college teacher of mathematics and statistics, and a student of mathematics and its relation to social, biological, and environmental problems, she was able to help me with many technical questions. As a wife, her patience, encouragement, understanding, and love helped me to finish this project, I hope successfully.

New Brunswick, N.J.   FRED S. ROBERTS

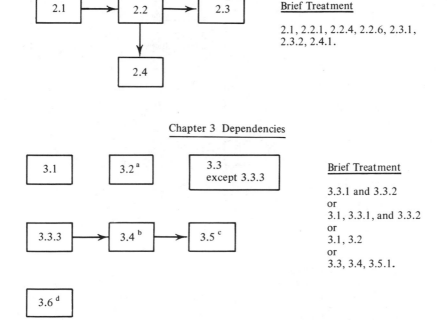

Chapter 2 Dependencies

2.1 → 2.2 → 2.3
2.2 → 2.4

Brief Treatment

2.1, 2.2.1, 2.2.4, 2.2.6, 2.3.1, 2.3.2, 2.4.1.

Chapter 3 Dependencies

3.1   3.2 [a]   3.3 except 3.3.3

3.3.3 → 3.4 [b] → 3.5 [c]

3.6 [d]

Brief Treatment

3.3.1 and 3.3.2
or
3.1, 3.3.1, and 3.3.2
or
3.1, 3.2
or
3.3, 3.4, 3.5.1.

Notes: [a] Depends on Sec. 2.2.5.

[b] Secs. 3.4.3 and 3.4.4 depend in part on Theorem 3.4 of Sec. 3.2.

[c] Sec. 3.5.2 is independent of Sec. 3.5.1 and of most of the chapter.

[d] Part of Sec. 3.6.2 is dependent on Sec. 3.4.

Chapter 4 Dependencies

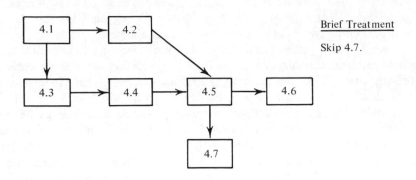

Brief Treatment

Skip 4.7.

Chapter 5 Dependencies

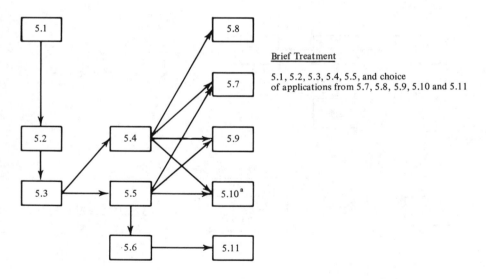

Brief Treatment

5.1, 5.2, 5.3, 5.4, 5.5, and choice
of applications from 5.7, 5.8, 5.9, 5.10 and 5.11

Note:

[a] Corollary 2 to Theorem 5.12 of Sec. 5.6 is used.

Chapter 6 Dependencies

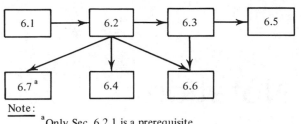

Brief Treatment

6.1, 6.2, 6.7 or
6.1, 6.2, 6.3, 6.7.

Note:

[a] Only Sec. 6.2.1 is a prerequisite

Chapter 7 Dependencies

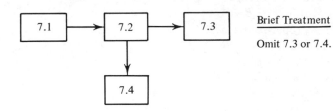

Brief Treatment

Omit 7.3 or 7.4.

Chapter 8 Dependencies

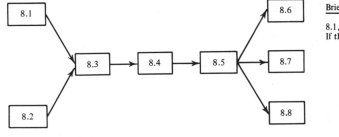

Brief Treatment

8.1, 8.2, 8.3, 8.4, 8.5.
If there is time, add 8.6, 8.7, or 8.8.

Notation

Set Theoretical Notation

A^c	the complement of A
$A \cup B$	the union of A and B
$A \cap B$	the intersection of A and B
$A - B$	the difference between A and B, namely $A \cap B^c$
$A \subseteq B$	A is a subset of B
$A \subsetneqq B$	A is a proper subset of B
$A \nsubseteq B$	A is not a subset of B
$x \in B$	x is a member of B
$x \notin B$	x is not a member of B
$\varnothing$	the empty set
$|A|$	the cardinality of A, the number of elements in A

Logical Notation

$\sim$	not
$\Rightarrow$	implies
$\Leftrightarrow$	if and only if (equivalence)
iff	if and only if
$\forall$	for all
$\exists$	there exists

Miscellaneous

$\mathfrak{R}$	the set of real numbers
$\mathfrak{R}^+$	the positive real numbers
$\mathfrak{R}^n$	Euclidean n-space, i.e., the set of all n-tuples of real numbers
$[a, b]$	the closed interval consisting of all real numbers c with $a \leq c \leq b$.
(a, b)	the open interval consisting of all real numbers c with $a < c < b$.
$\binom{n}{k}$	the binomial coefficient $n!/k!(n-k)!$
$\equiv$	congruent to

INTRODUCTION

The Scope
of the Book

The interaction between mathematics and any field of application goes two ways. One way, the obvious one, is that mathematics can be applied to the other field. It can do this in a variety of ways, from solution of specific practical problems to development of broad theories. The second way, the one which is disregarded by many people, is that the applied field can be "applied to mathematics." It can do this by stimulating the development of new mathematics or by helping solve old mathematical problems.

The relations between mathematics and physics over the years clearly demonstrate this two-way interaction. Not only have many types of mathematics been used in physics, but also physical problems have been a stimulus for the development of new mathematics, for example, the calculus. Moreover, occasionally, physical models have been used to solve mathematical problems, for example certain optimization problems. (For a discussion of this point, see Polya [1954, Ch. II].)

In this book, I explore the interrelations between mathematics and applied problems from such areas as the social, biological, and environmental sciences. These interrelations are, for the most part, much newer than those between mathematics and problems from the physical sciences, and involve fields of mathematics which until recently have not often been included in mathematical education. I hope to show first that mathematics can be useful in solving problems in these newer areas of application, and second, that these areas can be a stimulus to new and interesting forms of mathematics.

The first point will be made by presenting many examples throughout the book. Some of these come from real applications, for example the routing of traffic in the National Park System, the scheduling of garbage trucks in New York City, the analysis of pair comparison experiments in psychology, and the laboratory study of the fine structure inside the gene. Other examples presented involve the use of mathematics in systematizing concepts and in developing theories. Although such concepts and theories will, hopefully, help us come up with long-term solutions to social, biological, or environmental problems, their practical significance is of course hard to judge in the short run. In many of the applications in this book, we will have to make significant simplifying assumptions. In such cases, the conclusions will not be completely convincing, though they will be suggestive and indicate directions for further work. It is my thesis that in many of these cases, more convincing application of mathematics awaits the development of new mathematical tools, sufficiently powerful to handle the difficult mathematical questions which arise.

This brings me to the second point, namely, that social, biological, and environmental problems have been or should be a stimulus for the development of new mathematics. The mathematics developed as a result of this stimulus can be interesting in and of itself. That it is applicable should not necessarily be a criterion used in evaluating it, though of course we can hope that it will eventually be useful. We should keep in mind that many areas of mathematics were developed purely for mathematical reasons and later became very useful. A good example is Boolean algebra, which was applied to the design of computing machines more than fifty years after its invention. (It suggested the idea that complicated messages could be encoded by using on-off switches.) Other areas of mathematics were developed by stimulus from one application and were later found useful in other applications. Much of the theory of graphs has this sort of history. The subject was started by the famous Königsberg Bridge problem (see Chapter 2), was stimulated later by questions in chemical bonding and in analysis of games and puzzles, and is today being applied to a large number of problems in genetics, ecology, traffic flow, communications, and so on.

Since this book was written to illustrate both directions of interaction between mathematics and various applied areas, it could have been organized either by type of application or by type of mathematics. In general, I have chosen to organize it in the latter way. Thus, the book is written as a mathematics book, organized around mathematical topics. However, though each topic is developed as mathematics, the applications are never too far away, and indeed several of the chapters are organized around a rather specific applied theme, such as energy, group decisionmaking, or measurement and utility.

This book contains very few examples of the "traditional" applied mathematics, the mathematics of the physical sciences. I have chosen to emphasize the "discrete" types of mathematics, because I feel that they are the relevant mathematics for many social, biological, and environmental problems. To be more precise in what is meant by discrete, let me observe that the problems we are concerned with usually deal with finite sets: alternative plans for rapid transit systems, different social groups' opinions, etc. Sometimes these problems allow infinitely many possibilities, but these possibilities can only be whole numbers (for example number of houses, number of workers, etc.) To include such sets of possibilities as well as the finite sets, we use the term *discrete*. It might even make sense to allow other than whole numbers in a discrete set, but in general we do not allow gradations between elements in the set to get infinitely small. For example, if we are speaking of the amount of money to spend on a given project, we can speak in terms of millions of dollars, hundreds of dollars, even hundredths of dollars (i.e., pennies), but we don't have much finer gradations than that (though sometimes it is a reasonable assumption that we do). Sometimes the problems we are concerned with deal with situations where change comes only at discrete times, or is recorded only at discrete times, for example every hour or every year. For many purposes, unemployment statistics are reported only monthly. On the other hand, velocity of a projectile is usually thought of as changing continuously. Thus, the mathematics which is relevant to the problems we are concerned with finds its home on finite or discrete sets or sets which change only at discrete times. (Sometimes we allow infinite gradations or continuous changes in certain aspects of a problem being considered. For example, in Chapter 6, we allow payoffs in a game to be at least in principle infinitely divisible. However, we deal with games on a finite set of players. Similarly in Chapter 8, we allow things to be measured by using all the real numbers as possible scale values, but we develop scales over finite sets of alternatives.)

Problems on small finite sets can often be worked out by enumerating all possibilities. For example, suppose four people are each rated as to their speed in performing four different tasks, and we want to assign people to tasks so that the total time required is as small as possible. We can try out all possible assignments. There are $4 \cdot 3 \cdot 2 \cdot 1 = 4! = 24$ possibilities. (The first person could have any one of four different tasks, the second person any of the remaining three, etc.) But, as is commonly known, the number of possibilities (like so many other things) grows rapidly as the size of the set increases, and so it is usually impossible to solve problems by enumeration for even sets of modest size. If there are 10 people and 10 tasks, for example, there are already $10! = 3,628,800$ possibilities to try! More efficient techniques and quite different mathematical tools are needed, and many of the techniques discussed below have been developed to deal with problems where there are many possibilities. With the increasing availability of the computer, the range of applicability of these new tools is rapidly increasing.

The concepts of traditional applied mathematics, for example continuity, differentiability, rate of change, equations among rates of change (differential equations), etc., are sometimes the appropriate concepts for applications to social, biological, and environmental problems. These concepts are used in such areas of application as population growth or competition among species, where it is usually a good approximation to think of changes taking place continuously rather than at discrete times; and in studies of economic quantities, where it is often a useful idea to think that they may take all real values, even though the concept of one-half a house, say, does not make much sense. Thus, large bodies of the mathematics of population growth, of competition among species, of economic equilibrium, and so on, topics which might naturally have been included in this book, have been omitted, because the mathematics is so different from that emphasized here.

Finally, even the treatment of the discrete techniques is by no means complete. The probability theory in this book is for the most part limited to a chapter on Markov chains. The book would just have been too long if large amounts of probability theory were included. Besides, there are a large number of good treatments of the subject, including its applications to the social and biological sciences. (See, for example, Dwass [1969], Feller [1950], Goldberg [1960], or Parzen [1960].) In addition, I have omitted discussions of linear and nonlinear programming and of the theory of matrix games (though the less well-known theory of n-person games is included). Again, these were omitted because there already is a large literature on them. (See for example Dantzig [1963], Hillier and Lieberman [1967], Luce and Raiffa [1957], Owen [1968], Rapoport [1966], Spivey and Thrall [1970], or Wagner [1969].)

Social, biological, and environmental problems are often extremely complex, involving many hard-to-pinpoint variables, interacting in hard-to-pinpoint ways. Often, as we have pointed out, it is necessary to make rather severe simplifying assumptions in order to be able to handle such problems. If these assumptions are stated in mathematical terms, then much of the ambiguity of natural language can be avoided, and much of the power of mathematical reasoning can be applied. The procedure which translates assumptions about a problem, situation, or phenomenon into mathematical terms, and then analyzes the problem using mathematical tools, is called in this book *mathematical modelling*. We shall describe this procedure in more detail in Chapter 1, and we shall use it throughout the book. Mathematical modelling is a continuing process. It tries to replace simple models with more and more sophisticated ones. Most of the mathematical models constructed in this book are left after the first or second round, leaving the impression that the problem being studied has not been completed. This is often a correct impression. Frequently, the next step has not been taken because it would involve mathematical questions which are too hard. In any case, I have usually tried at least to suggest what the next step could or would be, or have asked the reader to think about it. Thus, the mathematical models suggested here are simply a beginning.

After introducing mathematical models in Chapter 1, we turn to graph theory in Chapter 2. Graphs were invented in the eighteenth century and, as we have remarked, have since been applied to study such diverse topics as social groups, communication networks, transportation, electrical wiring diagrams, chemical bonds, etc. The basic techniques of graph theory, summarized in Secs. 2.1 through 2.4, form a foundation for Chapters 3 through 5 and parts of Chapter 6. After that, the topics in the book are essentially independent, and may be studied in any order. A table of the dependencies of the various sections is appended to the Preface. Chapter 3 contains applications of graph theory to such topics as the theory of balance in Sociology, food webs in ecology, status in ecosystems, scheduling sanitation trucks in New York City, designing a traffic flow system and scheduling traffic lights, disrupting communications in an organization etc. Chapter 4 develops the theory of weighted digraphs (directed graphs) and pulse processes, and has as its primary application problems of energy use. However, the techniques of this chapter are being applied to problems of water resources, government decisionmaking about funding of science, and even analysis of history. Chapter 5 introduces Markov chains, and gives many applications. Included are genetics, learning theory in psychology, money flow, influence in a group, and flow of phosphorus through an ecosystem, to name just a few.

Chapter 6 introduces the theory of n-person games, and applies it to various bargaining situations, for example, land use, deterrence, and water pollution; and to various voting situations, for example in the Australian government and the United Nations Security Council. Chapter 7 passes from different groups bargaining with each other to groups trying to make decisions together. This chapter presents the famous Arrow Impossibility Theorem, which deals with the difficulties involved if groups want to make "democratic" decisions, and it presents the Kemeny-Snell ranking procedure for finding consensus among a panel of experts. The last chapter (Chapter 8), on Measurement and Utility, formalizes what it means to measure things and studies such problems as measurement of preference, loudness, etc. It develops a theory of scale type and of the meaningfulness of statements involving scales.

I hope that these topics will give the reader an introduction to a rather, new, but rapidly expanding field. Large numbers of references are included at the end of each chapter, to guide the reader into the literature.

References

DANTZIG, G. B., *Linear Programming and Extensions*, Princeton Univ. Press, Princeton, N.J., 1963.

DWASS, M., *Probability Theory and Applications*, W. A. Benjamin, New York, 1969.

FELLER, W., *An Introduction to Probability Theory and its Applications*, John Wiley & Sons, Inc., New York, 1950, 1957, 1968.

GOLDBERG, S., *Probability: An Introduction*, Prentice-Hall, Inc., Englewood Cliffs, N.J., 1960.

HILLIER, F. S., and LIEBERMAN, G. J., *Introduction to Operations Research*, Holden-Day, Inc. San Francisco, 1967, 1974.

LUCE, R. D., and RAIFFA, H., *Games and Decisions*, John Wiley & Sons, Inc., New York, 1957.

OWEN, G., *Game Theory*, W. B. Saunders, Philadelphia, 1968.

PARZEN, E., *Modern Probability Theory and its Applications*, John Wiley & Sons Inc., New York, 1960.

POLYA, G., *Induction and Analogy in Mathematics*, Vol. I, Princeton Univ. Press, Princeton, N.J., 1954.

RAPOPORT, A., *Two-Person Game Theory: The Essential Ideas*, Univ. of Michigan Press, Ann Arbor, Mich., 1966.

SPIVEY, W. A., and THRALL, R. M., *Linear Optimization*, Holt, Rinehart & Winston, Inc., New York, 1970.

WAGNER, H. M., *Principles of Operations Research with Applications to Managerial Decisions*, Prentice-Hall, Inc., Englewood Cliffs, N.J., 1969.

ONE

Mathematical Models

1.1. The Cyclical Nature of Mathematical Modelling

Much of this book is about mathematical models. The process of mathematical modelling involves a continuing four-stage cycle, which can be idealized as in Fig. 1.1.[1]

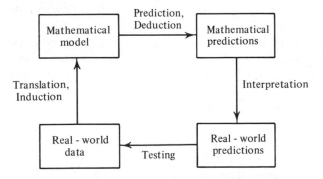

FIGURE 1.1. The cyclical nature of mathematical modelling.

[1] Our discussion of the cyclical nature of mathematical modelling is based in large part on that of Kemeny and Snell [1962, Ch. 1]. For other discussions of mathematical modelling, see Freudenthal [1961], Kemeny [1959], or Maki and Thompson [1973].

In the initial stage, one gathers data about the real-world phenomenon one is trying to understand. Then, one formulates certain assumptions about that phenomenon in a precise language—mathematics. This is the mathematical model. In the precise language used to describe the model, one can, in principle, state general assumptions, laws, or theories in such a way that arguments are about substantive matters and not about differences in usage of vague terms. Furthermore, the tools developed for centuries to reason in this precise language can be applied to reason about real-world phenomena.[2]

In any case, the mathematical assumptions formulate a mathematical model, and the next two stages of the cycle are intended to test this model, and modify it if need be. To test the model, one wants to draw certain conclusions about the real world. Such conclusions are of two types, those related to previously observed situations (these are *explanatory* in nature) and those related to new, not previously observed situations (these are *predictive* in nature). Both types are important for testing a mathematical model, though for purposes of discussion, it is reasonable to refer to both types of conclusions as *predictions*. To obtain these predictions, one first makes mathematical predictions, using mathematical tools which have been previously developed or are developed expressly to handle the particular mathematical model.

These mathematical predictions are then translated back from the language of the model to the language of the real world, and hence interpreted as real-world predictions or conclusions.

In the final stage, the predictions are checked against real data, either old (in the case of tests of the explanatory power of the model) or new (in the case of tests of its predictive power). On the basis of the new data, which includes the performance of the model's predictions, the model is modified and the cycle starts all over. Thus, no mathematical model is ever accepted except as tentative. The cyclical process continues all the time, and new pieces of data must be pitted against the explanatory power of the model, or against its predictive power. The reader should note that not every mathematical model is built in steps identical to the ones we have described, i.e., steps are skipped, repeated, etc. But as an idealization, our four-stage procedure does pretty well.

1.2. An Example: One-Way Streets

To illustrate this procedure, let us discuss a traffic flow problem. Imagine that a city has a number of locations, some of which are joined by two-way streets. The number of cars on the road has markedly increased, resulting in traffic jams and increased air pollution, and it has been suggested that the

---

[2]Further discussion of the nature of mathematics is included later in this chapter.

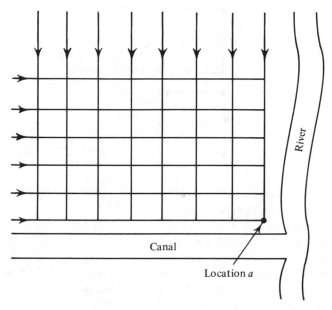

FIGURE 1.2. A one-way street assignment for the streets of a city which leaves travellers stuck at location *a*.

city should make all its streets one way. This would, presumably, cut down on traffic congestion. The question is: Can this always be done? If not, when? The answer is: Of course it can always be done. Just put a one-way sign on each street! However, it is quite possible that we will get into trouble if we make the assignment arbitrarily, for example by ending up with some places which we can get into and never leave. (See, for example, Fig. 1.2, which shows an assignment of directions to the streets of a city that is satisfactory only for someone who happens to own a parking lot at location *a*.) We would like to make every street one-way in such a manner that for every pair of locations *x* and *y*, it is possible (legally) to reach *x* from *y* and reach *y* from *x*. Can this always be done?

To solve this problem, we build a simple mathematical model for the transportation network of the city. Let the locations in question be represented as dots or points and draw a line between two locations *x* and *y* if and only if *x* and *y* are joined by a (two-way) street. The resulting picture, a collection of points and lines, is a mathematical object called a *graph* or an *undirected graph*, *G*. A simple example of such a graph is shown in Fig. 1.3. In terms of the model, our problem can now be restated as follows: Is it possible to put a direction or arrow (a one-way sign) on each line of the graph *G* in such a way that by following arrows in the resulting figure (called a

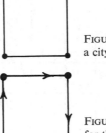

FIGURE 1.3. A two-way street graph for a city.

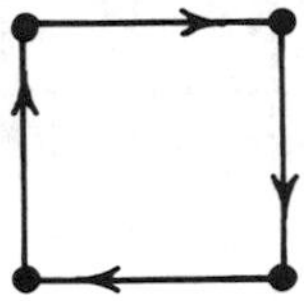

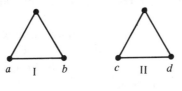

FIGURE 1.4. An assignment of directions for the graph of Fig. 1.3.

directed graph or *digraph*), one can always get from any point x to any point y? If it is not always possible, when is it possible?

In the graph of Fig. 1.3, we can certainly assign a direction to each line in the desired manner. We have done this in Fig. 1.4. To get from any point to any other point, one simply goes around.

However, it is not always possible to obtain a satisfactory assignment of arrows. For example, if our graph has two separate pieces, as in the graph of Fig. 1.5, there is no way of assigning directions to lines which will make it possible to go from points in section I of the graph to points in section II. Translated into a real-world conclusion, this mathematical conclusion says that in certain cities, for example cities whose two-way street networks look like that of Fig. 1.5, there is no satisfactory one-way street assignment. This conclusion can easily be subjected to test! Returning to our model, let us use this example to introduce some terminology about graphs. We say a graph is *connected* if it has only one piece, or equivalently, if for every pair of points x and y, there is a "chain" of lines joining x and y. (In the graph of Fig. 1.6, a and d are joined by the following chain: use the line from a to b, then the line from b to c, and finally the line from c to d. In the graph of Fig. 1.5, a and d are not joined by any such chain. We shall make the notion of "chain" more precise in Chapter 2.) To have a satisfactory assign-

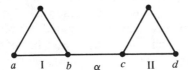

FIGURE 1.5. A graph representing a disconnected city.

FIGURE 1.6. α is a bridge.

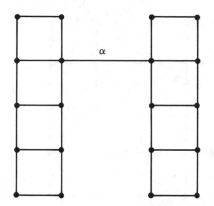

FIGURE 1.7.   α is again a bridge.

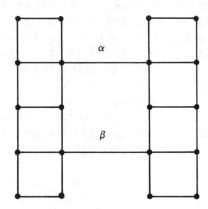

FIGURE 1.8.   Line α is no longer a bridge.

ment of directions, our graph must certainly be connected. However, even a connected graph may not have a satisfactory assignment of directions. Consider the graph of Fig. 1.6. This is connected. We must put a direction on line α. If α is directed from section I to section II, then from no point in II can we reach any point in I. If α is directed from II to I, then from no point in I can we reach any point in II. What is so special about the line α? The answer is that it is the only line joining two separate "pieces" of the graph. Put another way, the removal of α (but not the two points it joins) results in a disconnected graph. Such a line in a connected graph will be called a *bridge*. Figure 1.7 gives another example of a bridge α. It is clear that for a graph *G* to have a satisfactory assignment of directions, *G* must not only be connected, but it also must have no bridges.

In Fig. 1.7, suppose we add another bridge β joining the two separate pieces which are joined by the bridge α, obtaining the graph of Fig. 1.8. Can the lines of the new graph be assigned directions in a satisfactory way? The answer is yes. A satisfactory assignment is shown in Fig. 1.9. Doesn't this

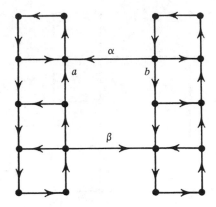

FIGURE 1.9. A satisfactory one-way assignment for the graph of Fig. 1.8.

violate the observation we have just made, namely that if a graph *G* has a satisfactory assignment of directions, it can have no bridges? The answer here is no. For we were too quick to call *β* a bridge. In the sense of our model, neither *β* nor *α* is a bridge in the graph of Fig. 1.8. For removal of *β* or *α* does not disconnect this graph. In building mathematical models, if we use suggestive terminology such as the term "bridge," we have to be careful to be consistent in our usage.

Suppose now that *G* has the following properties: It is a connected graph without bridges. Are these properties sufficient to guarantee that the lines of *G* can be assigned directions in the desired manner? The answer turns out to be yes. To verify that will take a bit of arguing, and a bit of graph theory, and so we defer it to Sec. 3.3.[3] Moreover, in that section, we shall give an algorithm, a step-by-step procedure, for deciding how to direct the lines of a connected, bridgeless graph in order to satisfy the reachability criteria we have set up.

Let us now translate our mathematical conclusions into real-world conclusions. Suppose a city has only one piece, i.e., every pair of locations is joined by a chain of two-way streets. Moreover, suppose there is no two-way street whose removal would make it impossible to get from some location to some other location by following two-way streets. Then the city has a satisfactory one-way street assignment. (Moreover, techniques for finding this assignment can be described.) This conclusion can be tested, and of course it works out.

Not every one-way street assignment for a city's streets is very efficient. Consider for example the one-way street assignment shown in the graph of Fig. 1.9. If a person at location *a* wanted to reach location *b*, he used to be able to do it very quickly, but now he has to travel a long way around. In short, this assignment meets the criterion we have set up, of being able to go

---

[3]This result was first obtained by Robbins [1939], and was a forerunner of much mathematical work in the field of transportation.

from any point to any other point, but it does not give a very efficient solution to the traffic flow problem. Here, another round of mathematical modelling is called for. The concept of "efficient" one-way street assignment needs to be defined, and algorithms or procedures for finding efficient one-way street assignments need to be discovered. The first round of mathematical modelling is only a beginning.

It should be remarked that efficiency is not always the goal of a one-way street assignment. In the National Park System throughout the United States, traffic congestion has become a serious problem. A solution being implemented by the U.S. National Park Service is to try to discourage people from driving during their visits to national parks. This can be done by designing very inefficient one-way street assignments, which make it hard to get from one place to another by car, and by encouraging people to use

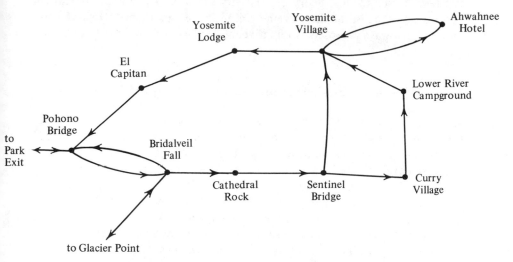

FIGURE 1.10. One-way street assignment for Yosemite Valley automobile traffic, Summer of 1973. (*Note:* Buses may go both ways between Yosemite Lodge and Yosemite Village and between Yosemite Village and Sentinel Bridge.)

alternatives, for example bicycles or buses. Figure 1.10 shows approximately the one-way street assignment used in the Yosemite Valley section of Yosemite National Park during the summer of 1973.[4] (Note that there is a two-way street; the idea of making *every* street one-way is obviously too simple.[5])

[4]In addition to making most roads one-way, the Park Service has closed off others to cars entirely.

[5]In this situation, there is no satisfactory assignment in which every street is one-way. Why?

This is a highly inefficient system. For example, to get from Yosemite Lodge to Yosemite Village, a distance of less than one mile by road, it is necessary to drive all the way around via Pohono Bridge, a distance of over eight miles! However, the Park's new buses are allowed to go directly from Yosemite Valley to Yosemite Lodge. Many people are riding them!

Another round of mathematical modelling might call for a definition of inefficient assignment, and the development of algorithms or procedures for finding inefficient assignments.

From this example, it is easy to see that the process of mathematical modelling is constantly ongoing, with successively more and more complex models. This is true even in such a developed field as physics. (Eventually, models can get so complex that they are scrapped in favor of alternative simpler models which have equal or greater explanatory power. One reason Einstein pressed for his theory of relativity in place of classical Newtonian physics was because of the extreme simplicity of this theory.) In any case, models are always regarded as tentative, and subject to continued evaluation and validation.

1.3. The Steps in the Mathematical Modelling Cycle

As we have described it, the mathematical modelling process consists of four steps. The steps of translation (model-building) and prediction should be distinguished in the following way. The first is *induction*: A general law is guessed at on the basis of a number of observations. The second is *deduction*: Certain conclusions are derived on the basis of specified assumptions and well-known rules of inference. The distinction between induction and deduction is one which has embroiled philosophers of science for many years, and it is not our intention to enter the debate here.[6] Suffice it to say that induction in general is not based on precise laws. Guessing the general pattern from having observed a number of instances is at best something which is learned from experience and at worst something of which we can never be sure.[7] Certainly it is impossible to speak of one "right" model. Thus, the art of mathematical model-building is, in a sense, something which cannot be taught. On the other hand, at least ideally, deduction or deductive reasoning is based on very precise rules of inference, which in principle at least can be taught to anyone and which, once understood, provide an ironclad test against which to check an argument. This feature of deduction,

[6]The reader interested in this issue could consult Kemeny [1959, Ch. 6], Hume [1946], Reichenbach [1949], or Madden [1960, Part 6].

[7]The story is told of the poor chicken, who, every day when she saw the sun rise, received her corn meal. One day, the sun rose and she went to the feeding pen, only to meet the farmer with a hatchet in his hand.

indeed, is one of the major advantages of mathematical model-building. We translate an imprecise situation into a precise one, and then we no longer can argue about our claims; these claims are either true or false, granted the assumptions of the model. The only thing which can be argued about is the translation of the imprecise situation into a model, or in particular, the accuracy of the assumptions. It is the cyclical procedure of mathematical modelling which is intended to test these assumptions. One other advantage of mathematical modelling arises from deduction: In a sense all of mathematics is a collection of tools for making deductions, and so the entire power of mathematics is available to deduce possible conclusions about our model.[8]

Sometimes when the mathematical problems arising in analysis of a model and in making of predictions are too difficult to handle, the analysis of the model can be based on simulation on a computer. In one approach to computer simulation, things happening with certain probabilities have their occurrence determined by consulting a table of "random numbers." Based on the assumptions of the model, the computer makes certain predictions. In such a situation, computer runs do not necessarily give deductive predictions. Computer simulation is becoming very important in analyzing mathematical models which cannot be analyzed deductively, usually because the mathematics is too hard. However, computer simulation is sufficiently different from the other tools we consider here so as to warrant its exclusion from the text.

The third step in modelling, the transition from mathematical conclusions to real-world conclusions, is not as hard as the first step, the translation procedure from the real-world phenomenon to the model. If this first step is spelled out carefully enough, then the mathematical conclusions or predictions should be readily interpretable as real-world conclusions or predictions.

To test these conclusions against data involves a decision about how accurate the predictions (and hence the model) must be. The answer to this depends on the use to which these will be put. Thus, suppose a model describes the carbon monoxide concentration in the air at a certain place. If this is being used to make decisions about possible curtailment of automobile traffic, it need not be too accurate, but only within a certain range. If, however, the model applies to a particular hospital room where delicate heart-lung surgery will take place, it had better be more accurate.

Among the techniques used to test the accuracy of a model, statistics

---

[8]The reader will note that within the space of one short chapter, we have now referred to mathematics as both a language and as a collection of tools for making deductions. Both are reasonable descriptions of different roles mathematics plays, and of different aspects of its nature. For a discussion of the philosophy of mathematics, the reader might consult Cohen and Nagel [1934], Hempel [1953], Wilder [1952], Whitehead [1953], Carnap [1939], or Kemeny [1959, Ch. 2].

is a primary tool. This book does not deal with such tools, or pay much attention to the testing step. Rather, it is mostly concerned with the development of mathematical language in which to state mathematical models and with the development of mathematical tools sufficiently powerful to analyze and make predictions about such mathematical models.

In trying to derive mathematical predictions from a model, one is often led to the development of new types of mathematical tools. In the process of studying physics, Newton, Leibnitz and others were led to the development of the calculus. Many of the mathematical topics treated in this book were developed by scientists in the process of studying social, biological, or environmental problems, and building mathematical models in these areas. In turn, new mathematical tools developed in modelling can be applied to other problems, often far removed from the problems which motivated them. We shall see a number of examples of this later on.

1.4. Types of Models

There are many types of mathematical models. One type simply makes precise an imprecise concept. The process of building such a model is called *explication* by Carnap [1950], Hempel [1952], Kemeny [1959] and others. The theory of balance in small groups discussed in Sec. 3.1 is an example of explication. So is the theory of measurement, developed in Chapter 8.

Some mathematical models are *deterministic* and others are *probabilistic*. The deterministic models make an exact prediction; the probabilistic ones make a prediction that something will happen a certain percentage of the time or with certain probability. Most of the models discussed in this book are deterministic. This is especially true of the graph theory models, such as those about traffic flow and food webs. On the other hand, the Markov chain models in genetics and learning presented in Chapter 5 are probabilistic in nature.

Some mathematical models are prescriptive and some descriptive. A *prescriptive* model describes how a person, a group, a society, a government agency, should behave in some idealized situation. A *descriptive* model describes how they actually behave. Thus, if preference or choice is being considered in a prescriptive way, the axioms of Chapter 8 which state conditions necessary and sufficient for measurement or scaling of preference to occur must be interpreted as conditions of "rationality," and a "rational" person's, group's, society's, government agency's preferences are expected to satisfy the rules given in the axioms. On the other hand, if preference or choice is being considered in a descriptive way, then these conditions are considered as testable, and should be checked against data (how the person, group, society, or agency makes choices). Then, if the axioms are not satisfied, one can try to modify the model of measurement or scaling. The original model has failed to describe the preferences. Of course, the same model can be studied as either a prescriptive or a descriptive model. Both viewpoints are frequently useful, and we take both viewpoints in this book.

As a final comment, let us note that there are many types of models other than mathematical ones. For example, we can often model a process by making a physical model and subjecting it to test. Scale models of new airplanes are examples. Treadmill tests model potential problems with newly designed automobiles and wind tunnel tests model potential problems with airplanes, automobiles, and other structures. Often we can model something by drawing pictures. Road maps are examples. In all such cases, a model is

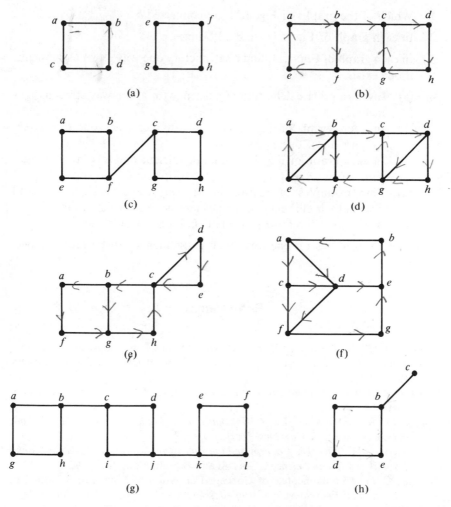

FIGURE 1.11. Graphs for exercises of Chapter 1.

a device for learning about a process, a phenomenon, a situation, a behavior, etc. What distinguishes a mathematical model from these other models is that we use the deductive power of mathematics to learn about the model, and so about the process we are studying. In the following chapters we shall present many newer tools for mathematical deductions and apply them to analyze mathematical models of social, biological, and environmental problems.

EXERCISES

1. Which of the graphs of Fig. 1.11 are connected?

2. In each graph of Fig. 1.11, find all bridges.

3. In each graph of Fig. 1.11, find a satisfactory one-way street assignment, if one exists.

4. (a) Invent a precise definition of efficiency for a one-way street assignment.

   (b) In each graph of Fig. 1.11, find an inefficient one-way street assignment, if one exists.

   (c) In each graph of Fig. 1.11, find an efficient one-way street assignment, if one exists.

   (d) Find examples of real-world one-way street assignments and discuss their efficiency or inefficiency using your definition.

   (e) Are there alternative possible definitions of efficiency?

(*Note:* More exercises on one-way street assignments can be found in Sec. 3.3.)

References

CARNAP, R., "Foundations of Logic and Mathematics," in *International Encyclopedia of Unified Science*, Vol. 1, No. 3, Univ. of Chicago Press, Chicago, 1939, 1947.

CARNAP, R., *Logical Foundations of Probability*, Univ. of Chicago Press, Chicago, 1950.

COHEN, M. R. and NAGEL, E., *An Introduction to Logic and Scientific Method*, Harcourt, Brace & Co., New York, 1934.

FREUDENTHAL, H. (ed.), *The Concept and the Role of the Model in Mathematics and Natural and Social Sciences*, Gordon and Breach, New York, 1961.

HEMPEL, C. G., "Fundamentals of Concept Formation in Empirical Science," in *International Encyclopedia of Unified Science*, Vol. 2, No. 7, Univ. of Chicago Press, Chicago, 1952.

HEMPEL, C. G., "On the Nature of Mathematical Truth," in *Readings in the Philosophy of Science*, H. Feigl and M. Brodbeck (eds.), Appleton-Century-Crofts, New York, 1953.

HUME, D., *An Enquiry Concerning Human Understanding*, Open Court, La Salle, Ill., 1946.

KEMENY, J. G., *A Philosopher Looks at Science*, D. Van Nostrand Co., Inc. Princeton, N. J., 1959.

KEMENY, J. G. and SNELL, J. L., *Mathematical Models in the Social Sciences*, Blaisdell Publishing Co., Waltham, Mass., 1962; reprinted by M.I.T. Press, Cambridge, Mass., 1972.

MADDEN, E. H., *The Structure of Scientific Thought*, Houghton Mifflin Company, Boston, 1960.

MAKI, D. P. and THOMPSON, M., *Mathematical Models and Applications*, Prentice-Hall, Inc., Englewood Cliffs, N.J., 1973.

REICHENBACH, H., "On the Justification of Induction," in *Readings in Philosophical Analysis*, H. Feigl and W. Sellars (eds.), Appleton-Century-Crofts, New York, 1949.

ROBBINS, H. E., "A Theorem on Graphs, with an Application to a Problem of Traffic Control," *Amer. Math. Monthly*, **46** (1939), 281–283.

WHITEHEAD, A. N., "The Abstract Nature of Mathematics," in *Readings in the Philosophy of Science*, P. P. Wiener (ed.), Charles Scribner's Sons, New York, 1953.

WILDER, R. L., *Introduction to the Foundations of Mathematics*, John Wiley & Sons, Inc., New York, 1952.

TWO

Graphs

2.1. Some Examples

In Chapter 1 we introduced the notion of a mathematical object called a graph in connection with our discussion of the street network of a city. The theory of graphs is an old subject which has been undergoing a tremendous growth in interest in recent years. From the beginning, the subject has been closely tied to applications. It was invented by Euler [1736] in the process of settling the famous Königsberg Bridge problem.[1] Graph theory was later applied by Kirchhoff [1847] to the study of electrical networks, by Cayley [1857, 1874] to the study of organic chemistry, by Hamilton to the study of puzzles, and by many mathematicians and nonmathematicians in the study of maps and map coloring. In the twentieth century, graph theory has been increasingly used in the social, biological, and environmental sciences. For example, Euler's solution to the Königsberg Bridge problem and related mathematical theorems on eulerian "chains" and "circuits" have recently been applied to schedule street sweepers in large cities, at significant savings. (See Tucker and Bodin [1975].) These theorems also have applications in

[1]The city of Königsberg had seven bridges linking islands in the River Pregel to the banks and to each other, as shown in Fig. 2.1. The residents wanted to know if it was possible to take a walk which starts at some point, crosses each bridge exactly once, and returns to the starting point. The famous mathematician Euler translated this into a graph theory problem. For a further discussion, see Euler [1776, 1953] or Harary [1969, Ch. 7].

telecommunications and cryptography (Berge [1962, p. 168]) and in genetics (Hutchinson [1969], Hutchinson and Wilf [1975]). To illustrate applications of graph theory, and to motivate the formal definitions of graph and directed graph which we shall introduce, let us consider several examples.

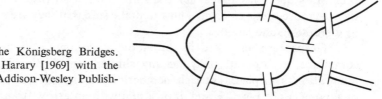

FIGURE 2.1. The Königsberg Bridges. Adapted from Harary [1969] with the permission of Addison-Wesley Publishing Company.

   Graphs and directed graphs arise in many transportation problems other than the one-way street problem discussed in Chapter 1. For example, consider any set of locations in a given area, between which it is desired to transport goods, people, cars, etc. The locations could be cities, warehouses, street corners, airfields, and so on. Represent the locations as points, as shown in the example of Fig. 2.2, and draw an arrow or directed line (or curve)

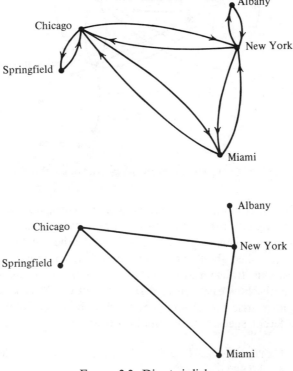

FIGURE 2.2. Direct air links.

from location x to location y if it is possible to move the goods, people, etc. directly from x to y. The situation where all the links are two-way, which we have already encountered, can be more simply represented by drawing a single undirected line between two locations which are directly linked, rather than drawing two arrows for each pair of locations (again see Fig. 2.2). Interesting questions about such transportation networks are how to design them to move traffic efficiently, how to make sure that they are not vulnerable to disruption, and so on.

Graphs are also used in the study of communications. Consider a committee, a corporate body, or any similar organization in which communication takes place. Let each member of the organization be represented by a point, as in Fig. 2.3, and draw a line with an arrow from member x to

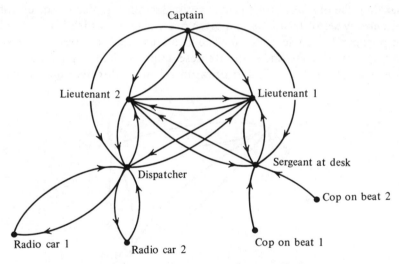

FIGURE 2.3. Communication network of part of a police force.

member y if x can communicate directly with y. Thus, for example, in the police force of Fig. 2.3, the captain can communicate directly with the dispatcher, who in turn can reach the captain via either of his lieutenants.[2] Typical questions asked about such a "communication network" are similar to questions about transportation networks: How can the network be designed efficiently, how easy is it to disrupt communications, and so on.

Graphs also arise in the study of food webs in ecology. Consider a number of different species[3] which make up an ecosystem. Represent the

[2]See Kemeny and Snell [1962, Ch. 8] for a more detailed discussion of a similar communication network of a police force.

[3]The word "species" is being used rather loosely here.

species as points, as in Fig. 2.4, and draw an arrow from species x to species y if x is a predator of y. Again with regard to such food webs, it is natural to ask how vulnerable they are to disruption by removal or addition of new species. It is also interesting to try to measure their complexity, for an old ecological principle is that increased complexity of an ecosystem corresponds to decreased vulnerability. And it is interesting to pursue questions of "status" of various species in the food web, and to try to designate "important" or critical species.

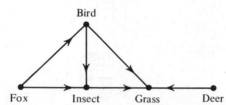

Bird

FIGURE 2.4.   Food web.   Fox   Insect   Grass   Deer

To give yet another application of graphs, consider a round-robin tournament[4] in tennis, where each player must play every other player exactly once, and no ties are allowed. One can represent the players as points and draw an arrow from player x to player y if x "beats" y, as in Fig. 2.5. Similar tournaments arise in a surprisingly large number of places in the social, biological, and environmental sciences. Psychologists perform a pair com-

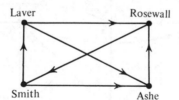

Laver   Rosewall

FIGURE 2.5.   A tournament.   Smith   Ashe

parison preference experiment on a set of alternatives by asking a subject, for each pair of alternatives, to state which he prefers. This also defines a tournament, if we think of the alternatives as corresponding to the players and "prefers" as corresponding to "beats." Tournaments also arise in biology. In the farmyard, for every pair of chickens, it has been found that exactly one "dominates" the other. This "pecking order" among chickens again defines a tournament. In studying tournaments, a basic problem is to decide on the "winner" and to rank the "players." Graph theory will help with this problem too.

More generally, consider any game, such as chess, bridge, etc. Let the various positions of the game be represented as points, as in Fig. 2.6(a),

---

[4]This is not the more common elimination tournament.

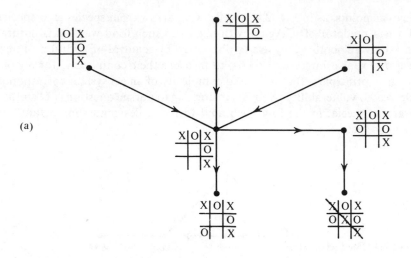

(a)

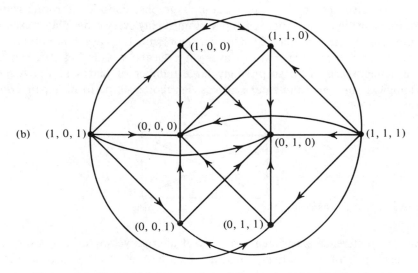

(b)

FIGURE 2.6. Games. (a) is a portion of tic-tac-toe, with X going first. (b) shows winnings as a vector, with the first component showing the winnings of player 1, the second component showing the winnings of player 2, and the third component showing the winnings of player 3. Shown are preferences of the group consisting of players 1 and 3 together.

and draw an arrow from position x to position y if some legal move of the game takes it from position x to position y. The game-ending positions are those with no arrows leaving them and the rules of the game define the winner at each game-ending position. Alternatively, consider a game with many players, say n. Let the possible outcomes of the game (distributions of winnings) be represented as points. For each group S of players, draw a different diagram by putting an arrow from potential outcome x to potential outcome y if group S prefers x to y as in Fig. 2.6(b). (More specifically, game theory talks of "effective preference," i.e., preference which the group has the power to enforce.) The theory of two-person games deals with questions of optimal strategies of particular players. The theory of n-person games, which we shall discuss in Chapter 6, deals with questions of bargaining to determine the distribution of winnings among the players. Such n-person games are very general. For in some sense, all of the economic marketplace is a game, all of legislative decisionmaking is a game, all of bargaining in international affairs, for example over disarmament, is a game, etc.

These examples all give rise to directed or undirected graphs. To be precise, let us define a *directed graph* or *digraph* D as a pair (V, A), where V is a set and A is a set of ordered pairs of elements of V. V will be called the set of *vertices* and A the set of *arcs*. (Some authors use the terms node, point, etc. in place of vertex, and the terms arrow, directed line, directed edge, or directed link in place of arc.) If more than one digraph is being considered, we will use the notation $V(D)$ and $A(D)$ for the vertex set and the arc set of D, respectively. Usually, digraphs are represented by simple diagrams such as those of Fig. 2.7, which should give an indication of the great variety of digraphs. Here, the vertices are represented by points and there is a directed line (or curve or arrow) heading from u to v if and only if (u, v) is in A. For example in the digraph D_5 of Fig. 2.7, V is $\{u, v, w, x\}$ and A is the set

$$\{(u, v), (u, w), (v, w), (w, x), (x, u)\}.$$

If there is an arc from vertex u to vertex v, we shall say that u is *adjacent* to v. Thus in Fig. 2.7, in digraph D_5, v is adjacent to w, w is adjacent to x, and so on.

The reader should notice that the particular placement of the vertices in a diagram of a digraph is unimportant. The distances between the vertices have no significance, the nature of the lines joining them is unimportant, etc. Moreover, whether or not two arcs cross is also unimportant; the crossing point is not necessarily a vertex of the digraph. (One of the interesting problems we shall discuss in Chapter 3 is to give criteria for when a digraph might be drawn with no arcs crossing except at vertices.) All the information in a diagram of a digraph is included in the observation of whether or not a given pair of vertices is joined by a directed line or arc and in which direction the arc goes. Thus, for example, digraphs D_5 and D_6 of Fig. 2.7 are the same digraph, only drawn differently.

In a digraph, it is perfectly possible to have arcs in both directions, from u to v and from v to u, as shown in digraph D_8 of Fig. 2.7, for example. It is possible to have an arc from a vertex to itself, as is also shown in digraph

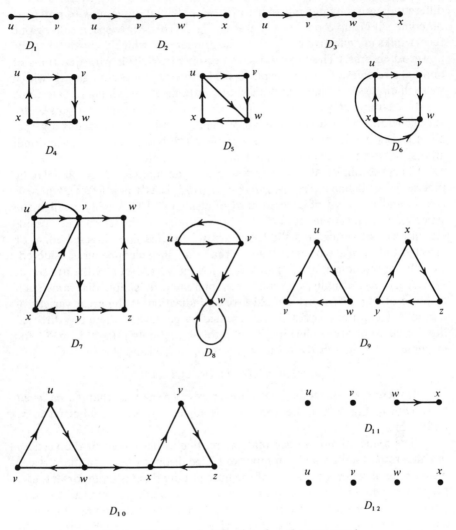

FIGURE 2.7. Digraphs. (Continued on the next page.)

D_8. Such an arc is called a *loop*. It is not possible, however, to have more than one arc from u to v. Often in the theory and applications of digraphs, such multiple arcs are useful—this is true in the study of chemical bonding, for example—and then one studies *multigraphs* or, better, *multidigraphs*, rather than digraphs.

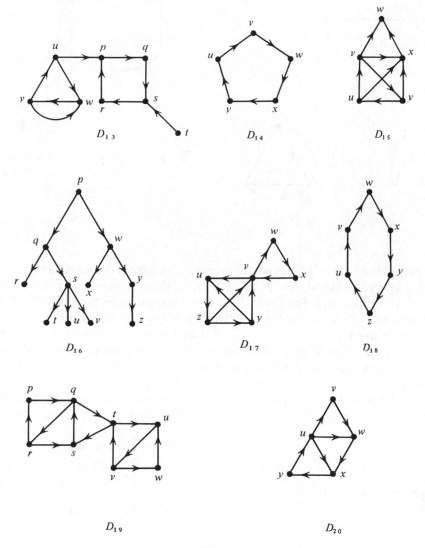

FIGURE 2.7 continued.

Very often, there is an arc from u to v whenever there is an arc from v to u. In this case we shall say that the digraph (V, A) is a *graph*. Figure 2.8 shows several graphs. In the diagram of a graph, it is convenient to disregard the arrows, and to replace a pair of arcs between vertices u and v by a single nondirected line joining u and v. (In the case of a directed loop, it is replaced

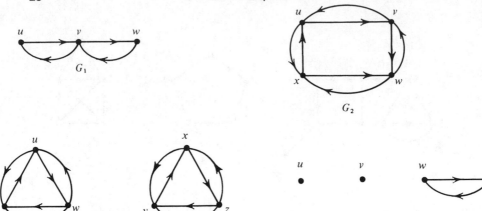

FIGURE 2.8. Graphs.

by an undirected one.) We shall call such a line an *edge* of the graph, and think of it as an unordered pair of vertices $\{u, v\}$. (The vertices u and v do not have to be distinct.) The graph diagrams obtained from those of Fig. 2.8 in this way are shown in Fig. 2.9. Thus, a graph may be defined as a pair (V, E), where V is a set of vertices and E is a set of unordered pairs of elements from V, the edges. If more than one graph is being considered, we will use the notation $V(G)$ and $E(G)$ for the vertex set and the edge set of G, respectively.

It is convenient to make several assumptions about our digraphs and graphs. Many graph theorists make explicit the following assumption: There

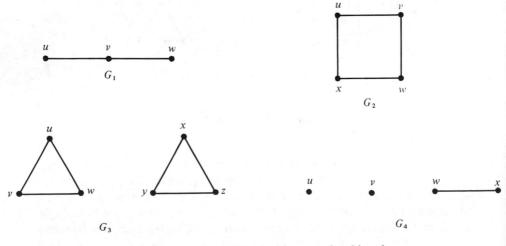

FIGURE 2.9. Graphs of Fig. 2.8 with arcs replaced by edges.

are no multiple arcs or edges, i.e., no more than one arc or edge from vertex u to vertex v. For us, this assumption is contained in our definition of a digraph or graph. We shall assume, at least at first, that digraphs and graphs have no loops. (Almost everything we say for loopless digraphs and graphs will be true for digraphs and graphs with loops.) We shall also limit ourselves to digraphs or graphs with finite vertex sets. Let us summarize these assumptions as follows:

Assumption: *Unless otherwise specified, all digraphs and graphs referred to in this book have finite vertex sets. All digraphs and graphs referred to in Chapters 2 and 3 have no loops. Digraphs and graphs are not allowed to have multiple arcs or edges.*

In Chapter 1, we showed how graph theory could solve a problem in traffic flow. In this chapter and the next, we shall apply graph theory to problems of transportation, communications, food webs, tournaments, genetics, archaeology, and numerous other problems. In later chapters, we shall also make use of graph theory, though our graphs and digraphs will have more structure. In Sec. 3.1 we add a sign ($+$ or $-$) to each edge of a graph, and use the resulting signed graph to model the sociological concept of balance in small groups of people. In Chapter 4, we add real numbers (weights) to each arc of a digraph and use the resulting weighted digraphs to study such societal problems as increasing energy demand. In Chapter 5 we interpret the weights of a weighted digraph as probabilities and develop models to study ecosystems, genetics, money flow, the process of learning, the assertion of influence, etc. In Chapters 6 and 8, digraphs will represent preference relations among alternatives in numerous situations involving games, bargaining, social choice, and decisionmaking.

Before doing applications or building mathematical models, it is necessary to learn the language of the mathematical tool which is relevant. That language is what we concentrate on in this chapter.

EXERCISES

1. In the digraph of Fig. 2.4:
    (a)   Identify the set of vertices.
    (b)   Identify the set of arcs.

2. Repeat Exer. 1 for the digraph of Fig. 2.5.

3. Repeat Exer. 1 for the digraph D_{10} of Fig. 2.7.

4. In each of the graphs of Fig. 2.9:
    (a) Identify the set of vertices.
    (b) Identify the set of edges.

5. In digraph D_7 of Fig. 2.7, find a vertex adjacent to vertex y.

6. Draw a transportation network with the cities New York, Paris, Vienna, Washington, D.C., and Algiers as vertices, and an edge joining two cities if it is possible to travel between them by road.

7. Draw a communication network for a team fighting a forest fire.

8. Draw a food web for an ecosystem of your choice.

9. Draw a digraph representing the following football tournament. The teams are Princeton, Harvard, and Yale. Princeton beats Harvard, Harvard beats Yale, and Yale beats Princeton.

10. Draw a digraph to represent the following game. A coin is flipped until either two heads or two tails have come up. If two heads come up first, player 1 wins; otherwise player 2 wins.

11. Translate the Königsberg Bridge problem into a graph problem. (*Hint:* Use a multigraph.) Decide if it is possible to make the desired type of walk. Do this by first considering the simpler situations shown in Fig. 2.10.

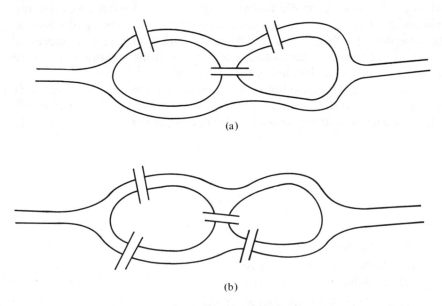

(a)

(b)

FIGURE 2.10. Islands and bridges for Exer. 11 of Sec. 2.1.

2.2.   Connectedness[5]

2.2.1.   REACHING

In a communication network, a natural question to ask is: Can one person initiate a message to another person? In a transportation network, an analogous question is: Can a car move from location u to location v? In a game, we might ask whether a given position can ultimately lead to a winning one. In ecology, we are interested in tracing from one species to another, through a chain of predators.

All of these examples have in common the following idea of reachability in a digraph $D = (V, A)$: Can we reach vertex v by starting at vertex u and following arrows?

To make this concept precise, let us introduce some definitions. A *path* in D is a sequence

$$u_1, a_1, u_2, a_2, \ldots, u_t, a_t, u_{t+1}, \qquad (1)$$

where $t \geq 0$, each u_i is in V, i.e., is a vertex, and each a_i is in A, i.e., is an arc, and a_i is the arc (u_i, u_{i+1}). That is, arc a_i goes from u_i to u_{i+1}. In the path (1) we follow from u_1 to u_2 to u_3, and on to u_{t+1}. Since t might be 0, u_1 alone is a path, a path from u_1 to u_1. The path (1) is called a *simple path* if we never use the same vertex more than once.[6] For example, in digraph D_7 of Fig. 2.7, $u, (u, v), v, (v, w), w$ is a simple path and $u, (u, v), v (v, y), y,$ $(y, x), x, (x, v), v, (v, w), w$ is a path which is not a simple path since it uses vertex v twice. Naming the arcs is superfluous when referring to a path, and so we simply speak of (1) as the path $u_1, u_2, \ldots, u_t, u_{t+1}$.

A path (1) is called *closed* if $u_{t+1} = u_1$. In a closed path, we end at the starting point. If the path (1) is closed and the vertices $u_1, u_2, \ldots, u_t$ are distinct, then (1) is called a *cycle* (a simple closed path). (The reader should note that if the vertices u_i are distinct, the arcs a_i must also be distinct.)

To give some examples, the path u, v, w, x, u in digraph D_5 of Fig. 2.7 is a cycle, as is the path u, v, w, x, y, u in digraph D_{20}. But the closed path u, v, y, x, v, y, x, u of D_7 is not a cycle, since there are repeated vertices. In general, in counting or listing cycles of a digraph, we shall not distinguish two cycles which use the same vertices and arcs in the same order, but start at a different vertex. Thus, in digraph D_{20} of Fig. 2.7, the cycle w, x, y, u, v, w

---

[5]The discussion in this and the following two sections is based heavily on that in Harary, Norman, and Cartwright [1965].

[6]One of the difficulties in learning graph theory is the large number of terms which have to be mastered early. To help the reader overcome this difficulty, we have included the terms path, simple path, etc. in a Glossary at the end of this chapter. We have also listed the terms dealing with reachability in Table 2.1.

TABLE 2.1

Reaching and Joining

Digraph D $u_1, a_1, u_2, a_2, \ldots, u_t, a_t, u_{t+1}$		Graph G $u_1, e_1, u_2, e_2, \ldots, u_t, e_t, u_{t+1}$
Reaching	Joining	
Path a_i is (u_i, u_{i+1})	Semipath a_i is (u_i, u_{i+1}) or (u_{i+1}, u_i)	Chain e_i is $\{u_i, u_{i+1}\}$
Simple path path and u_i distinct	Simple semipath semipath and u_i distinct	Simple chain chain and u_i distinct
Closed path $u_{t+1} = u_1$	Closed semipath $u_{t+1} = u_1$	Closed chain $u_{t+1} = u_1$
Cycle (simple closed path) path $u_{t+1} = u_1$ u_i distinct, $i \leq t$ $(a_i$ distinct)*	Semicycle (simple closed semipath) semipath $u_{t+1} = u_1$ u_i distinct, $i \leq t$ a_i distinct	Circuit (simple closed chain) chain $u_{t+1} = u_1$ u_i distinct, $i \leq t$ e_i distinct

*This follows from u_i distinct, $i \leq t$.

is considered the same as the cycle u, v, w, x, y, u. The *length* of a path, simple path, cycle, etc. is the number of *arcs* in it. Thus, the path (1) has length t. In digraph D_7 of Fig. 2.7, u, v, y, x, v is a path of length 4, u, v, y, z is a simple path of length 3, and u, v, y, x, u is a cycle of length 4.

We say v is *reachable* from u if there is a path from u to v. By definition, the singleton u defines a path, so each vertex u is reachable from itself.

Theorem 2.1.   If v is reachable from u, then there is a simple path from u to v.

Proof.[7]   Since there is a path from u to v, pick one which is as short as possible. Denote this path $u_1, u_2, \ldots, u_t, u_{t+1}$, where $u = u_1$ and $v = u_{t+1}$. If this path is not simple, there must be some vertex which is repeated. That is, there are $i < j$ with u_i and u_j the same. Then it is easy to verify that $u_1, u_2, \ldots, u_i, u_{j+1}, u_{j+2}, \ldots, u_{t+1}$ is also a path from u to v, and it is shorter than the original path, which is a contradiction. Thus, the original path must have been a simple path.   Q.E.D.

[7]Throughout this book, proofs may be omitted without loss of continuity. However, the reader is encouraged to follow as many proofs as he can.

In communication networks, Theorem 2.1 has the following straight-forward interpretation: If a person can send a message to another person, he can do it without anyone hearing the message twice.

2.2.2. DISTANCE[8]

If u and v are two vertices in a digraph $D = (V, A)$, we say that the *distance* from u to v is the length of a shortest path from u to v. Notice that there may be several paths from u to v which are of the same shortest length. Also, there may be no path at all from u to v, in which case we say that the distance is undefined. (By the proof of Theorem 2.1, a shortest path must be simple.) We usually write $d(u, v)$ to denote the distance from u to v. The reader will note that $d(u, v)$ does not satisfy all of the usual conditions which distance satisfies. In particular, the distance is not necessarily symmetric: It is possible that $d(u, v) \neq d(v, u)$. To give an example, consider the digraph D_{18} of Fig. 2.7. We have $d(u, v) = 1$, since u, v is a path; but $d(v, u) = 5$, since the shortest path from v to u is v, w, x, y, z, u. In D_{17}, $d(u, y) = 2$, for the shortest path from u to y is u, z, y. But $d(y, u) = 1$, for y, u is a path from y to u. In D_{10}, $d(u, z) = 4$, for u, w, x, y, z is a shortest path from u to z. But $d(z, u)$ is undefined, as there is no path from z to u. Finally, since u alone is a path, we have $d(u, u) = 0$, for all u. Under certain circumstances the distance d does satisfy one of the traditional properties of a distance function, namely the triangle inequality. For we have the following result:

Theorem 2.2. If v is reachable from u and w is reachable from v, then $d(u, w) \leq d(u, v) + d(v, w)$.

Proof. If $d(u, v) = s$ and $d(v, w) = t$, let $u, u_2, u_3, \ldots, u_s, v$ be a shortest path from u to v and $v, v_2, v_3, \ldots, v_t, w$ be a shortest path from v to w. Then $u, u_2, u_3, \ldots, u_s, v, v_2, v_3, \ldots, v_t, w$ is a path of length $s + t$ from u to w, and we conclude that $d(u, w) \leq s + t = d(u, v) + d(v, w)$.   Q.E.D.

2.2.3. JOINING

Our ideas about reaching in a digraph can be illustrated by considering a communication network. To be precise, let us consider a police force, and return to that part of the communication network of a police force shown in the digraph of Fig. 2.3. The cops on the beat can communicate to the captain, by calling the sergeant at the desk. Thus, the captain is reachable from each cop on the beat. But the cops on the beat are not reachable from the captain. The captain can, however, still communicate with a cop on the beat, using an intermediary. He can call the sergeant at the desk, and then wait until the cop on the beat calls in. In this sense, the captain and each cop on the beat are "joined" in the communication network or digraph. This notion

[8]This subsection may be omitted at this point.

of joining can be made precise by saying that vertices u and v in a digraph are *joined* if you can go from u to v by following arcs, not necessarily always going with the arrows.[9] To make this notion even more precise, let us say that if $t \geqq 0$, then

$$u_1, a_1, u_2, a_2, \ldots, u_t, a_t, u_{t+1} \tag{2}$$

is a *semipath* if each u_i is a vertex, each a_i is an arc, and a_i is always either the arc (u_i, u_{i+1}) or the arc (u_{i+1}, u_i). The semipath is called a *simple semipath* if we never use the same vertex more than once. It is *closed* if $u_{t+1} = u_1$.[10] A closed semipath (2) in which the vertices $u_1, u_2, \ldots, u_t$ are distinct *and* the arcs $a_1, a_2, \ldots, a_t$ are distinct is called a *semicycle*.[11] Finally, the *length* of a semipath, simple semipath, semicycle, etc. is the number of arcs in it. For example, in digraph D_{15} of Fig. 2.7, $u, (u, x), x, (y, x), y, (v, y), v, (v, x),$ $x, (x, w), w$ is a semipath. It is not a simple semipath, since vertex x is repeated. In $D_7, u, (u, v), v, (\dot{v}, w), w, (z, w), z$ is a semipath, indeed a simple semipath. This is different from the semipath $u, (v, u), v, (v, w), w, (z, w), z$ in the same digraph: The former uses the arc (u, v), the latter the arc (v, u). Also in $D_7, u, (u, v), v, (x, v), x, (x, u), u$ is a semicycle of length 3, but the closed semipath $v, (v, w), w, (v, w), v$ is not a semicycle since arc a_1 equals arc a_t.[12]

Formally, we say that vertex u is *joined* to vertex v if there is a semipath from u to v. Notice that if u is joined to v, then v is joined to u: A semipath from u to v may be reversed to obtain one from v to u. Thus, we may as well refer to u and v as being joined, independently of order. In the police force communication network of Fig. 2.3, the vertex lieutenant 2 and the vertex cop on beat 1 are joined, for the following sequence is a semipath: lieutenant 2, (lieutenant 2, sergeant at desk), sergeant at desk, (cop on beat 1, sergeant at desk), cop on beat 1.

In a graph, i.e., a digraph where $(u, v) \in A$ iff $(v, u) \in A$, the notions of reaching and joining are the same. That is, u is reachable from v if and only if u is joined to v. Why? We return to the case of graphs below.

2.2.4. CONNECTEDNESS CATEGORIES

One reason that graph theory is so useful is that its geometric point of view allows us to define various structural concepts. One of these concepts is connectedness. The degrees of connectedness we shall introduce make precise the idea that some digraphs "hang together" more than others. For

---

[9] The remainder of this subsection may be omitted at this reading.

[10] These concepts are also summarized in Table 2.1.

[11] We shall see below why the restriction that the arcs be distinct is added.

[12] Note that the vertices $u_1, u_2, \ldots, u_t$ are distinct, which is why we added the requirement that the arcs $a_1, a_2, \ldots, a_t$ must also be distinct.

example, in Fig. 2.7, digraph D_9 is disconnected, with two separate pieces, one consisting of vertices $u, v,$ and w and the other consisting of vertices $x, y,$ and z. Each piece seems to hang together well. Digraph D_{10} hangs together better than D_9, but not as well as D_4, in which each vertex can reach each other vertex. In D_{10}, x cannot reach w, for example.

To make such distinctions precise, we define several notions of connectedness. A digraph $D = (V, A)$ is *strongly connected* or *strong* if for every pair of vertices u and v, v is reachable from u and u is reachable from v. Thus, digraph D_4 of Fig. 2.7 is strongly connected, but digraph D_{10} is not. In the traffic flow problem we considered in Chapter 1, we sought an assignment of directions which gave rise to a strongly connected digraph. If a communication network is strongly connected, then every person can *initiate* a communication to every other person. If a game is strongly connected, there are no game-ending positions, for these are positions with no arcs leaving them. Thus, such a game goes on forever. Food webs, as we shall see, are usually not strongly connected.

A digraph is *unilaterally connected* or *unilateral* if for every pair of vertices u and v, either v is reachable from u or u is reachable from v, but not necessarily both. Thus, digraph D_{10} of Fig. 2.7 is unilaterally connected. Of course, every strongly connected digraph is unilaterally connected. Digraph D_3 is not unilateral, since neither u nor x is reachable from the other. A communication network is unilaterally connected if for every pair of members u and v, at least one can initiate a message to the other.

A digraph is *weakly connected* or *weak* if every pair of vertices u and v is joined. Thus, every unilateral digraph (and hence every strong digraph) is weakly connected. The digraph D_3 of Fig. 2.7 is weakly connected. But the digraph D_9 is not, since u and x are not joined. Finally, a digraph is *disconnected* if it is not weakly connected.

In a graph, the concepts of strong, unilateral, and weak coincide. Why?

Using these concepts, we can define *degrees* or *categories* of *connectedness*: A digraph D is *connected of degree 0* if it is not weak; it is *connected of degree 1* if it is weak but not unilateral, *of degree 2* if it is unilateral but not strong, and *of degree 3* if it is strong. Thus, in Fig. 2.7, D_9 has degree 0, D_3 has degree 1, D_{10} has degree 2, and D_4 has degree 3. (The degree of connectedness of a digraph defines what is called an *ordinal scale*, to use the terminology we shall introduce in Chapter 8. Only the order of connectedness is important, and the numbers 0, 1, 2, 3 carry no significance in and of themselves.)

2.2.5. CRITERIA FOR CONNECTEDNESS[13]

Verifying strong, unilateral, or weak connectedness directly from the

---

[13]This section may be omitted without serious loss of continuity. However, Theorem 2.4 is used in several places in the following.

definition can get tedious, for in a digraph of n vertices there are

$$\binom{n}{2} = \frac{n!}{2!(n-2)!} = \frac{n(n-1)}{2}$$

pairs of vertices. In this subsection, we provide criteria for membership in
the three classes of digraphs, strong, unilateral, and weak. The following
theorems are proved in Harary, Norman, and Cartwright [1965], and our
proofs are similar. The reader should note that there are available more
efficient tests for strong, unilateral, and weak connectedness. We shall
mention these briefly in Sec. 3.3.1. We shall use the terminology *complete
path*, *complete semipath*, etc to mean one which goes through all the vertices.

To characterize strong connectedness, note first that a cycle such as
D_4 of Fig. 2.7 is strongly connected. So is any digraph obtained from a cycle
by adding arcs, for example D_5. Thus, if there is a cycle through all the
vertices, then the digraph is strongly connected. Not every strongly connected
digraph has a cycle which goes through all vertices. (Can you give an example?) However, as the next theorem shows, it does have a closed path which
goes through all the vertices.

Theorem 2.3.   A digraph D is strongly connected if and only if it has
a complete closed path.

Proof.   Let $u_1, u_2, \ldots, u_t, u_1$ be a complete closed path. Given u and
v, since the path is complete, we can say that u is u_i, for some i, and v is u_j,
for some j. We assume $i < j$. Then $u_i, u_{i+1}, \ldots, u_j$ is a path from u to v
and $u_j, u_{j+1}, \ldots, u_t, u_1, \ldots, u_i$ is a path from v to u. Thus, D is strongly
connected.

Conversely, suppose D is strongly connected. List the vertices of D
as $u_1, u_2, \ldots, u_n$. Then there are paths P_1 from u_1 to u_2, P_2 from u_2 to
$u_3, \ldots, P_{n-1}$ from u_{n-1} to u_n, and P_n from u_n to u_1. A complete closed path
for D can be constructed by putting together these paths in the order $P_1, P_2,$
$\ldots, P_{n-1}, P_n$.   Q.E.D.

To illustrate this theorem, let us take the digraph D_{17} of Fig. 2.7. It
is strongly connected, because v, u, z, y, v, w, x, v forms a complete closed
path. So is D_{19}, for $p, q, r, s, q, t, u, v, w, u, v, t, s, q, r, p$ is a complete closed
path.

To use Theorem 2.3 to test for strong connectedness, we simply list
the vertices of D as $u_1, u_2, \ldots, u_n$. Then we test whether or not there is a
path from u_1 to u_2, from u_2 to $u_3, \ldots$, from u_{n-1} to u_n, and from u_n to u_1.

Thus, instead of two tests for each of $\binom{n}{2}$ pairs of vertices, i.e., $n(n-1)$
tests, we only need to make n tests. This is a big improvement if n is even

moderate in size. For example, with $n = 8$, $n(n - 1)$ is already 56, which is 48 more tests than with the new technique!

In terms of communication networks, Theorem 2.3 says that in order for every person to be able to initiate a message to every other person, it is necessary and sufficient to have a sequence of people with the following properties: (1) Each one of them can contact the next; (2) all the people are represented; and (3) the last person can contact the first. Theorem 2.3 can be used in the design of communication networks; for example we want such a network to be strongly connected, and we want to use as few links as possible. (See Exer. 12.)

Turning from strong connectedness, let us recall that a digraph is unilaterally connected if for every pair of vertices u and v, there is a path from u to v or a path from v to u.

Theorem 2.4.   A digraph D is unilaterally connected if and only if it has a complete path.

To illustrate this theorem, we note that the digraph D_7 of Fig. 2.7 is unilateral, since it has a complete path x, v, u, v, y, z, w. (Is this digraph strong?)

To prove the theorem, suppose $u_1, u_2, \ldots, u_t$ is a complete path. Then, given $u \neq v$, since the path is complete, u is u_i for some i and v is u_j for some j. If $i < j$, then $u_i, u_{i+1}, \ldots, u_j$ is a path from u to v. If $i > j$, then $u_j, u_{j+1}, \ldots, u_i$ is a path from v to u. Thus, D is unilateral.

Conversely, suppose D is unilateral. We first establish the following preliminary result.

Lemma.   In any set of vertices of a unilateral digraph D, there is a vertex which can reach (using arcs of D) all others in the set.

The digraph D_{10} of Fig. 2.7 illustrates the lemma. It is unilateral. In the set $\{u, v, x\}$, the vertex u can reach all the others. In the set $\{x, y\}$, the vertex x can. In the set $\{u, v, w, x, y, z\}$, the vertex u can. And so on. To prove the lemma, we argue by induction on k, the number of vertices in U. The lemma clearly holds if $k = 1$. (Why?) Let us assume that it holds for all sets U with k vertices and pick a set U of $k + 1$ vertices. Let us list the elements of U as $v_1, v_2, \ldots, v_{k+1}$. By the induction hypothesis, there is a vertex v_i in $U - \{v_{k+1}\}$ which can reach all v_j for $j < k + 1$. Now since D is unilateral, either v_i reaches v_{k+1} or v_{k+1} reaches v_i. If v_i reaches v_{k+1}, then v_i reaches all vertices of U. If v_{k+1} reaches v_i, then v_{k+1} reaches all vertices of U. This proves the lemma.

Let us now prove the theorem by means of the lemma. Pick u_1 in $V = V(D)$ so that u_1 can reach all other vertices in V. Pick u_2 in $V - \{u_1\}$ so that it can reach all vertices in $V - \{u_1\}$. Pick u_3 in $V - \{u_1, u_2\}$ so that it can reach all vertices in $V - \{u_1, u_2\}$. And so on. Now u_1 can reach u_2 via a path P_1, u_2 can reach u_3 via a path P_2, etc. Putting these paths together gives a complete path for the digraph D.

Digraph D_7 of Fig. 2.7 illustrates the proof of Theorem 2.4. Pick $u_1 = y$, $u_2 = x, u_3 = u, u_4 = v, u_5 = z, u_6 = w$. Then P_1 is y, x, P_2 is x, u, P_3 is u, v, P_4 is v, y, z, and P_5 is z, w. The complete path is given by y, x, u, v, y, z, w. (In this digraph, there is even a complete simple path. Is there one in every unilateral digraph?)

Finally, let us recall that a digraph D is weakly connected if every pair of vertices of D is joined.

Theorem 2.5.   A digraph D is weakly connected if and only if it has a complete semipath.

Proof.   Left to reader (Exer. 21).

To illustrate this theorem, note that digraph D_{15} of Fig. 2.7 is weak, since u, v, w, x, y is a complete semipath (the arcs are unambiguous here, so we need not list them). (D_{15} is also unilateral. Why?)

2.2.6.   The Case of Graphs

Suppose $G = (V, E)$ is a graph. Terminology analogous to that for digraphs can be introduced. A *chain* in G is a sequence

$$u_1, e_1, u_2, e_2, \ldots, u_t, e_t, u_{t+1}, \tag{3}$$

where $t \geqq 0$, each u_i is a vertex, and each e_i is the edge $\{u_i, u_{i+1}\}$. A chain is called *simple* if all the u_i are distinct and *closed* if $u_{t+1} = u_1$. A closed chain (3) in which $u_1, u_2, \ldots, u_t$ are distinct and $e_1, e_2, \ldots, e_t$ are distinct is called a *circuit* (a simple closed chain). The *length* of a chain, circuit, etc. of form (3) is the number of edges in it.

To give some examples, in the graph of Fig. 2.11, $r, \{r, t\}, t, \{t, w\}$, $w, \{w, u\}, u, \{u, t\}, t, \{t, r\}, r, \{r, s\}, s, \{s, u\}, u, \{u, t\}, t, \{t, w\}, w$ is a chain. This chain can be written without reference to the edges as $r, t, w, u, t, r, s, u, t, w$. A simple chain is given by r, t, u, w, x. A circuit is given by r, t, u, s, r. Finally, $p, \{p, r\}, r, \{r, p\}, p$ is not considered a circuit, since $e_1 = e_t$. (The edges are unordered, so $\{p, r\} = \{r, p\}$.)

Let us say a graph is *connected* if between every pair of vertices u and v there is a chain. This notion of connectedness, which we encountered in Chapter 1, coincides with the one used in topology: The graph has one "piece." In Fig. 2.9, graphs G_1 and G_2 are connected while G_3 and G_4 are not. A result analogous to Theorem 2.5 is the following: A graph is connected if and only if it has a complete chain.

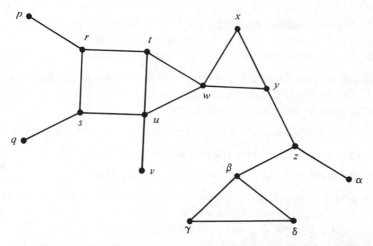

FIGURE 2.11. Graph for Exers. 2 and 23 of Sec. 2.2 and Exer. 16 of Sec. 2.4.

EXERCISES

1. For the digraph D_{19} of Fig. 2.7:
   (a) Find a path which is not a simple path.
   (b) Find a semipath which is not a simple semipath.
   (c) Find a closed path.
   (d) Find a simple path of length 4.
   (e) Determine if u, (u, v), v, (v, w), w, (w, u), u is a cycle.
   (f) Find a cycle of length 3 containing vertex t.

2. For the graph of Fig. 2.11:
   (a) Find a closed chain which is not a circuit.
   (b) Find the longest circuit.
   (c) Find a chain different from the one in the text which is not simple.
   (d) Find a closed chain of length 6.

3. Give an example of a digraph and a path in that digraph which is not a simple path but has no repeated arcs.

4. For the digraph D_{20} of Fig. 2.7, calculate $d(i, j)$ for all i and j.

5. For each digraph of Fig. 2.7:
   (a) Determine if it is strongly connected.
   (b) Determine if it is unilaterally connected.
   (c) Determine if it is weakly connected.
   (d) Determine its connectedness category (0, 1, 2, or 3).

6. For each digraph D of Fig. 2.7:

  (a) Find a complete closed path if D is strongly connected.
  (b) Find a complete path if D is unilaterally connected.
  (c) Find a complete semipath if D is weakly connected.

7. Discuss the reaching and joining properties of the police force of Fig. 2.3. In particular, determine the connectedness category, interpret major gaps in reaching or joining, etc.

8. Do the same as Exer. 7 for the transportation network of Fig. 2.2.

9. Do the same as Exer. 7 for the tic-tac-toe game of Fig. 2.6(a).

10. Do the same as Exer. 7 for the game of Fig. 2.6(b).

11. A digraph is *unipathic* if whenever v is reachable from u, then there is exactly one simple path from u to v.
  (a) Is the digraph D_{10} of Fig. 2.7 unipathic?
  (b) What about the digraph D_{14}?
  (c) Draw a unipathic digraph which has no semicycles.

12. For a digraph which is strongly connected and has n vertices, what is the least number of arcs? What is the most? (Observe that a digraph which is strongly connected with the least number of arcs is very vulnerable to disruption. How many links is it necessary to sever in order to disrupt communications? We return to the concept of vulnerability in Sec. 3.3.)

13. (a) Give an example of a strongly connected digraph which has no complete cycle.
  (b) Does every unilaterally connected digraph have a complete simple path?

14. Prove that vertices u and v are joined if and only if there is a simple semipath between u and v.

15. (Harary, Norman, and Cartwright [1965].) Refer to the definition of unipathic in Exer. 11. Can two cycles of a unipathic digraph have a common arc? (Give proof or counterexample.)

16. (Harary, Norman, and Cartwright [1965].) If D is strongly connected and has at least two vertices, does every vertex have to be on a cycle? (Give proof or counterexample.)

17. Suppose a digraph D is in connectedness category 0.
  (a) If D has 4 vertices, what is the maximum number of arcs?
  (b) What if D has n vertices?

18. Do Exer. 17 for connectedness category 2.

19. Do Exer. 17 for connectedness category 1.

20. (Harary, Norman, and Cartwright [1965].) If D is a digraph, define the

complementary digraph D^c as follows: $V(D^c) = V(D) = V$ and an ordered pair (u, v) from $V \times V$ (with $u \neq v$) is in $A(D^c)$ if and only if it is not in $A(D)$. For example, if D is the digraph of Fig. 2.12, then D^c

FIGURE 2.12. A digraph D and its complementary digraph D^c.

is the digraph shown. Give examples of digraphs D which are weakly connected, not unilaterally connected, and such that:

(a) D^c is strong.

(b) D^c is unilateral but not strong.

(c) D^c is weak but not unilateral.

21.   Prove Theorem 2.5.

22.   A graph G is a *tree* if it is connected and has no circuits.

    (a)   Draw several trees.

    (b)   Prove that if G is a tree, then there is at least one vertex u which is joined to at most one other vertex by an edge.

  *(c)   Prove that if G is a tree, then the number n of vertices of G is exactly one more than the number e of edges of G.

  *(d)   Prove that if G is a connected graph and $n = e + 1$, then G is a tree.

23.   A chain or circuit in a graph (multigraph) is called *eulerian* if it uses each edge exactly once. (Compare the Königsberg Bridge problem.)

    (a)   Show that each graph of Fig. 2.13 has an eulerian closed circuit.

    (b)   Show that if a graph G has an eulerian closed circuit, then each vertex is on an even number of edges. (The converse is also true, if G is connected.)

    (c)   Show that the graph of Fig. 2.11 does not have an eulerian closed circuit.

24.   A path or cycle in a digraph is *eulerian* if it uses each arc exactly once.

    (a)   Find an eulerian path in digraph D_5 of Fig. 2.7.

    (b)   Show that D_5 does not have an eulerian closed path.

    (c)   Guess at necessary (and sufficient) conditions for a digraph to have an eulerian closed path. (*Hint:* See Exer. 23b.)

*The symbol ∗ will indicate a more difficult exercise.

25. Consider a transportation network for a city in which all streets are
     two-way. Certain designated streets are to be swept in a given period of
     time. A schedule or order in which the street sweeper should go from
     place to place is called for. An efficient schedule is one in which the street
     sweeper never uses a street which does not need to be swept or which
     has been swept before. Under what circumstances does there exist such
     a schedule? (*Hint:* Consider the graph which uses as edges only
     designated streets. (See Tucker and Bodin [1975].))

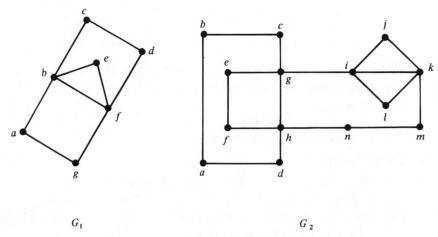

FIGURE 2.13. Graphs for Exer. 23, Sec. 2.2.

2.3. Strong Components and the Vertex Basis

2.3.1. VERTEX BASES AND COMMUNICATION NETWORKS

Suppose we wish to plant a message in a communication network in
such a way that it can reach all members. If the network is strongly connected,
it suffices to plant the message with one person. (Indeed, by Theorem 2.4, it
still suffices if the network is unilaterally connected.) However, in general,
it will not necessarily suffice to plant the message with one person. We would
like to find a set of people (vertices) from which it is possible to reach all
other people (vertices). Moreover, we would like this set to be as small as
possible. A collection B of vertices of a digraph D is called a *vertex basis*
of D if every vertex not in B is reachable from some vertex in B and B is minimal. Here, *minimal* means that no proper subset of B can reach all vertices.
Among all vertex bases, we would like to find one which has a *minimum*

number of elements. To give an example, in the digraph D_{13} of Fig. 2.7, the vertex t has no incoming arcs. Thus, to build a vertex basis, we must include t. A vertex basis may be obtained by adding either u, v, or w. Note that the set $\{t, u, q\}$ also can reach all other vertices, but it is not a vertex basis since the subset $\{t, u\}$ already has the required property. Indeed, the sets $\{t, u\}$, $\{t, v\}$, and $\{t, w\}$ form all the vertex bases. For none of u, v, and w is reachable from any vertex other than u, v, or w. Thus, one of these three vertices must be in every vertex basis.

The reader will observe that all these vertex bases have the same number of vertices. That will turn out to be no accident. Thus, the search for a vertex basis with a minimum number of elements will be over once we find any vertex basis.

In this section, we shall develop a procedure for finding all vertex bases of a given digraph. Most of the results of this section are due to König [1936]. To describe König's procedure, we first introduce some preliminary definitions. If $D = (V, A)$ is a digraph, a *subgraph* of D is a digraph (W, B) whose vertex set W is a subset of V and whose arc set B is a subset of A. (This should technically be called a *subdigraph*.) Since (W, B) is a digraph, it is implicit in the definition that B cannot be an arbitrary subset of A, but rather must consist of ordered pairs of elements of W. Figure 2.14 shows a digraph and some of its subgraphs. Of particular interest are the subgraphs *generated* by a subset W of V, namely those subgraphs which contain all arcs of $A(D)$ joining vertices of W. For example, in Fig. 2.14, the subgraph generated by $W - \{u, v, x, y\}$ is digraph D_1. To find all vertex bases of a digraph D, the trick is to lump together certain vertices and pass to a smaller, more tractable digraph. Then it will turn out that this new digraph, D^*, the *condensation* of the original one, will have an easily determined, unique vertex basis B^*, and moreover, it will be easy to identify all vertex bases in D from the vertex basis B^* of D^*.

To define the condensation D^*, we consider generated subgraphs of D which are strongly connected. Such a subgraph with a maximal number of vertices is called a *strong component*.[14] Strong components will define the sets of vertices to be lumped in obtaining the condensation. (Strong components are also useful in understanding the structure of a digraph. And they will be used in Chapter 4, when we calculate stability of an energy demand signed digraph.) Figure 2.15 shows the strong components of the digraph given. Two strong components may have different numbers of elements; that is why we used the term maximal rather than the term maximum

---

[14]In the sequel, we shall sometimes identify a strong component with its underlying vertices. Maximal here means that if we add more vertices, the resulting generated subgraph is not strongly connected.

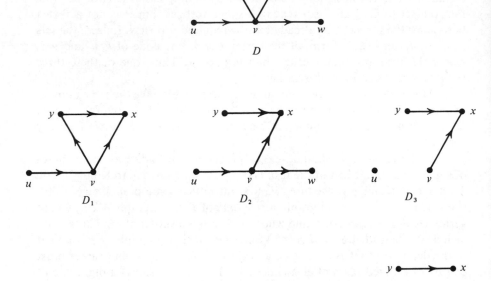

FIGURE 2.14. A digraph D and some of its subgraphs.

in the definition. Note that a single vertex may constitute a strong component.

Theorem 2.6.   In a digraph $D = (V, A)$, each vertex u is in one and only one strong component.

Proof.[15]   The vertex u is in at least one strong component. For, the digraph generated by u alone is strong. Pick the strongly connected generated subgraph containing u with the largest number of vertices. This is in a strong component containing u. Suppose next that u is in strong components K and L. Consider the subgraph generated by vertices of K and L. This subgraph is strongly connected. For if a is in K and b is in L, then a can reach b via vertices of $K \cup L$, for a can reach u via vertices of K and u can reach b via vertices of L. Similarly, b can reach a

[15]As usual, the proof may be omitted without loss of continuity.

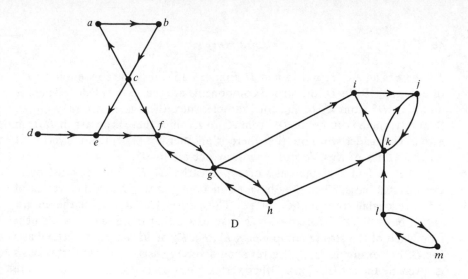

D

Strong components: $\{a, b, c\}$ K_1

$\{d\}$ K_2

$\{e\}$ K_3

$\{f, g, h\}$ K_4

$\{i, j, k\}$ K_5

$\{l, m\}$ K_6

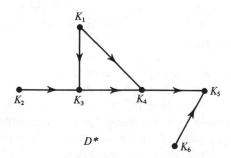

D*

Vertex Basis B^* of D^*: $\{K_1, K_2, K_6\}$

FIGURE 2.15. A digraph D, its strong components, and its condensation.

via vertices of $K \cup L$. By maximality of K and of L, we have $K \cup L = K$ and $K \cup L = L$, so $K = L$.   Q.E.D.

If D is a digraph, we can now define a new digraph D^*, the *condensation* (by strong components) of D, as follows. Let $K_1, K_2, \ldots, K_p$ be the strong components of D. Then $V(D^*) = \{K_1, K_2, \ldots, K_p\}$ and we draw an arc from K_i to K_j if and only if $i \neq j$ and for some vertices $u \in K_i$ and $v \in$

K_j, there is an arc from u to v in D. Figure 2.15 shows the condensation D^* of a digraph D. In D, the strong components are the six sets listed. There is an arc in D^* from K_3 to K_4, for example, since there is an arc (e, f) in D. Similarly, there is an arc in D^* from K_4 to K_5 since there is an arc in D from g to i. The reader will note that there is another arc (h, k) from a vertex of K_4 to a vertex of K_5. We still only draw one arc in D^*.

Let us find a vertex basis of D^*. Clearly, K_1, K_2, and K_6 must be in every vertex basis, since they have no incoming arcs. Every other vertex of D^* is reachable from K_1, K_2, or K_6. Thus, $B^* = \{K_1, K_2, K_6\}$ is the unique vertex basis of D^*. Moreover, it is easy to see that if we take one element from each of the strong components K_1, K_2, K_6 of B^* we get a vertex basis for D. For example, $\{a, d, l\}$ gives such a vertex basis. Another vertex basis is given by the set $\{a, d, m\}$. Other vertex bases of D obtained from B^* are given as follows:

$$\{b, d, l\}$$
$$\{b, d, m\}$$
$$\{c, d, l\}$$
$$\{c, d, m\}.$$

It turns out that D^* always has a unique vertex basis B^* consisting, as in this example, of all vertices with no incoming arcs. And it turns out that every vertex basis of D may be obtained from one of D^* by taking one vertex from each strong component of D which is in B^*. Thus, in our example, the vertex bases of D which we have given are all the vertex bases.

Before proving these assertions, let us make the following observation: The fact that a message planted with the members of a vertex basis can reach all members of a communication network does not mean that it will. For, the reader will recall, the appearance of an arc from u to v means that u *can* communicate directly with v, but not that u *will* communicate directly with v. If we consider a directed graph as a mathematical model of a communication network, we see that this model (given our interpretation of arc) is not adequate for more sophisticated analysis. To take account of the possibility that a message might not be passed along a given arc, we might think of trying to assign to each arc (u, v) a probability that a message reaching u will be passed along to v. Thus, we would go around the mathematical modelling cycle a second time. We shall pursue this idea in several exercises of this and other sections, and also in some of the exercises of Chapter 5.

2.3.2. THEOREMS AND PROOFS[16]

We turn now to proofs of our assertions about how to find all vertex bases. We begin by proving a theorem about the structure of the condensation D^*.

[16]The reader may omit the details of this subsection, but he should read the theorems.

Theorem 2.7. If D is a digraph, its condensation D^* is acyclic, i.e., it has no cycles.

Proof. Suppose there is a cycle $K_{i_1}, K_{i_2}, \ldots, K_{i_t}, K_{i_1}$ in D^*. Let u be a vertex in K_{i_1} and v be a vertex in K_{i_2}. Using the cycle $K_{i_1}, K_{i_2}, \ldots,$ K_{i_t}, K_{i_1}, it is easy to prove that u is reachable from v and v is reachable from u. Thus, it follows that u and v are in the same strong component and hence that $K_{i_1} = K_{i_2}$. (This argument uses Theorem 2.6). This violates the definition of a cycle. Q.E.D.

Theorem 2.8. An acyclic digraph D has a unique vertex basis, consisting of all vertices with no incoming arcs.

Proof. Let B be the set of all vertices with no incoming arcs. Clearly every u in B must be in every vertex basis. It suffices to prove that every vertex v not in B is reachable from some vertex of B. To show this, suppose $v \notin B$. Let $v = v_0$. Since $v_0 \notin B$, there is an incoming arc (v_1, v_0) with $v_1 \neq v_0$. If $v_1 \in B$, we are done. If not, there is an incoming arc (v_2, v_1) with $v_2 \neq v_1$. Continuing in this way, we build a path $v_t, v_{t-1}, \ldots, v_1, v_0$ such that no v_i is in B. All the vertices of this path are distinct. For, if $v_i = v_j$, $i > j$, and $v_i, v_{i-1}, \ldots, v_{j+1}$ are distinct, then $v_i, v_{i-1}, \ldots, v_{j+1}, v_j$ is a cycle, which is contrary to the assumption that the digraph is acyclic. Since D has only a finite number of vertices, we cannot continue building the path $v_t, v_{t-1}, \ldots,$ v_1, v_0 indefinitely. We must eventually reach a vertex v_t of B. Thus, $v = v_0$ is reachable from v_t. Q.E.D.

Corollary. In an acyclic digraph, there is a vertex with no incoming arcs.

The next theorem describes exactly how to construct all vertex bases. If D is a digraph, its condensation D^* is acyclic, by Theorem 2.7. Thus, it has a unique vertex basis B^*, by Theorem 2.8.

Theorem 2.9. If D is a digraph, let B^* be the unique vertex basis of D^*. Then the vertex bases of D are those sets B consisting of one vertex from each strong component of D which is in B^*.

Proof. Suppose B^* is the unique vertex basis of D^* and B consists of one vertex from each strong component of B^*. Clearly, every vertex of D is reachable from B. (Why?) We must show that B is a minimal set with the property that every vertex of D is reachable from B. To prove minimality, it is sufficient to show that no $v \in B$ is reachable from another $u \in B$. If this were possible, then the strong component containing v would be reachable in D^* from the strong component containing u, and this would contradict the minimality of B^*.

To complete the proof, we show that if B is any vertex basis, it consists of exactly one vertex from each strong component of D which is in B^*. Certainly B must contain at least one vertex from each such strong component, as well as possibly other vertices. By minimality, no other vertices are needed.   Q.E.D.

As a corollary of Theorem 2.9, we have the following theorem.

Theorem 2.10.   Every two vertex bases of a digraph have the same number of vertices.

2.3.3. Some Graph-Theoretical Terminology

Some of the concepts introduced in Sec. 2.3.1 have analogues for graphs which will be important in Chapter 3. If $G = (V, E)$ is a graph, a *subgraph* of G is a graph $H = (W, F)$ where W is a subset of V and F is a subset of E. H is a *generated subgraph* if H contains all edges of E which join vertices in

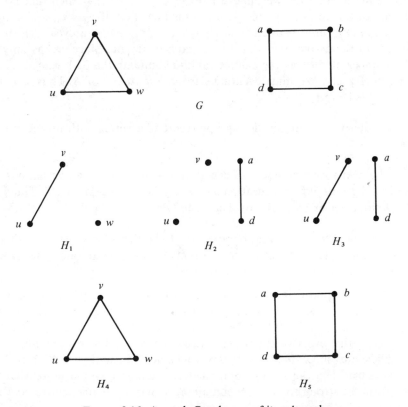

FIGURE 2.16. A graph G and some of its subgraphs.

W. A (*connected*) *component* of *G* is a maximal connected generated sub-graph. Figure 2.16 shows a graph *G* and five of its subgraphs. H_3, H_4, and H_5 are generated subgraphs. H_4 and H_5 are connected components. The reader will note that every vertex of *G* is in one and only one connected component.

EXERCISES

1. For each digraph of Fig. 2.17:
    (a)   Find a subgraph which is not a generated subgraph.
    (b)   Find the subgraph generated by vertices 5, 8, and 9.
    (c)   Find a strongly connected generated subgraph which is not a strong component.
    (d)   Find all strong components.
    (e)   Find its condensation.
    (f)   Find a vertex basis.
    (g)   Determine the number of vertex bases.

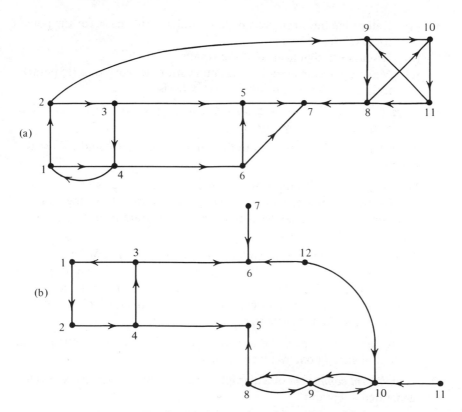

FIGURE 2.17. Digraphs for exercises of Secs. 2.3. and 2.4.

2. For the graph of Fig. 2.18:
   (a) Find a subgraph which is not a generated subgraph.
   (b) Find a generated subgraph which is connected but not a connected component.
   (c) Find all connected components.

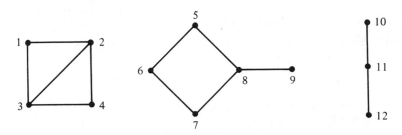

FIGURE 2.18. Graphs for Exers. 2 and 6, Sec. 2.3.

3. (a) Determine the strong components and a vertex basis for the police force of Fig. 2.3.
   (b) Discuss the significance of the results.
   (c) Make up a more detailed communication network of a (hypothetical) police force and find its vertex basis.

4. Does the idea of a vertex basis make any sense for a transportation network, such as that in Fig. 2.2?

5. Apply the idea of vertex basis to the game of Fig. 2.6(b) and discuss the results.

6. For an (undirected) graph, we can define a vertex basis as a minimal set of vertices which are joined to every other vertex. Describe a procedure for finding a vertex basis in a graph. Illustrate it on the graph of Fig. 2.18.

7. A *weak component* of a digraph is a maximal weakly connected generated subgraph, that is, a weakly connected generated subgraph with the property that if we add any vertices, the subgraph generated is not weakly connected. Find all weak components of the digraph D_9 of Fig. 2.7. Do the same for each digraph of Fig. 2.17.

8. Using the definition in the previous exercise, show that every vertex of D is in exactly one weak component.

9. A *unilateral component* of a digraph is a maximal unilaterally connected generated subgraph.

(a) Find a unilateral component with five vertices in the digraph (b) of Fig. 2.17.

(b) Is every vertex of a digraph in at least one unilateral component?

(c) Can it be in more than one?

10. Suppose D is an acyclic digraph, i.e., D has no cycles, and D^* is its condensation. Is it true that D and D^* always have the same number of vertices? Why?

11. Prove that in a unilateral digraph, every vertex basis has one element.

12. The *converse* D' of a digraph D is the digraph defined as follows: $V(D') = V(D)$ and $(u, v) \in A(D')$ iff $(v, u) \in A(D)$. That is, to form D', we reverse all arcs of D. Using the idea of converse, show that an acyclic digraph has a vertex with no outgoing arcs.

13. A set A of vertices in a digraph D is called a *vertex contrabasis* if every vertex $u \in D$ can reach some vertex of A and A is minimal in the sense that no proper subset of A has this property. Find all vertex contrabases of the digraphs (a) and (b) of Fig. 2.17. (*Hint*: Does the notion of converse (Exer. 12) help here?)

14. Does the idea of vertex contrabasis (Exer. 13) have any interpretation for communication? For transportation? For games? Comment on a vertex contrabasis for the police force of Fig. 2.3 and on a vertex contrabasis for the tic tac toe game of Fig. 2.6(a).

15. A *vertex duobasis* of a digraph D is a set of vertices which is at the same time both a vertex basis and a vertex contrabasis.

(a) Give an example of a digraph which has no vertex duobasis.

(b) Give an example of a digraph with n vertices which has a vertex duobasis.

(c) Under what circumstances does D have a vertex duobasis? (*Hint*: Use D^*.)

16. (Harary, Norman, and Cartwright [1965].) Suppose vertex u has the property that no vertex can reach (by a path) more vertices than u can. Does u necessarily have to be in some vertex basis?

17. Refer to the definition of tree given in Exer. 22, Sec. 2.2. Let us say a *spanning subgraph* of a graph G is a subgraph H such that $V(H) = V(G)$. For example, Fig. 2.19 shows two spanning subgraphs H_1 and H_2 of the graph G. Note that H_1 is a tree.

(a) Show that not every graph has a spanning tree, i.e., a spanning subgraph which is a tree.

(b) Show that every connected graph has a spanning tree.

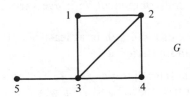

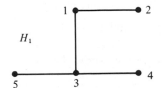

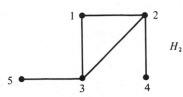

FIGURE 2.19. H_1 and H_2 are spanning subgraphs of G and H_1 is a spanning tree of G.

18. Formalize a mathematical model which takes into account the possibility that a message might not be passed along a given arc of a communication network. For example, answer the following questions:

   (a) What is the probability that a message starting at individual u will follow a particular path $u, u_2, u_3, \ldots, u_t, v$ to individual v?

   (b) If a message starts at individual u, how would you calculate (or approximate) the probability that the message will ever reach individual v?

2.4. Digraphs and Matrices

Much of the information concerning a digraph D (and hence a graph) can be conveniently summarized using certain matrices corresponding to D. These matrices are also useful for studying the digraph. One of them, the adjacency matrix, will be important in our study of signed digraphs and their applications to energy demand, in Chapter 4.

   In our study of these matrices associated with a digraph, we will use certain matrix operations of which we wish to remind the reader. Suppose $A = (a_{ij})$ and $B = (b_{ij})$ are two $n \times n$ matrices. Then we define the following new matrices:

$$A + B = (a_{ij} + b_{ij}),$$
$$A \times B = (a_{ij} \times b_{ij}),$$

where $\times$ denotes ordinary multiplication, and

$$AB = (c_{ij}),$$

where

$$c_{ij} = \sum_{k=1}^{n} a_{ik} b_{kj}.$$

The matrices $A \times B$ and AB should be distinguished; the former we shall call the *elementwise product* of A and B, the latter the *product*. The *transpose of A*, A', is the matrix (b_{ij}), where $b_{ij} = a_{ji}$.

2.4.1. THE ADJACENCY MATRIX

Suppose $D = (V, A)$ is a digraph and $u_1, u_2, \ldots, u_n$ is a list of its vertices. The adjacency matrix $A(D)$ associated with D is the matrix (a_{ij}) defined by

$$a_{ij} = \begin{cases} 1 & (\text{if } (u_i, u_j) \in A) \\ 0 & (\text{otherwise}). \end{cases}$$

Figure 2.20 shows a number of digraphs and their associated adjacency matrices.[17] The following theorem shows one of the many uses of the adjacency matrix.

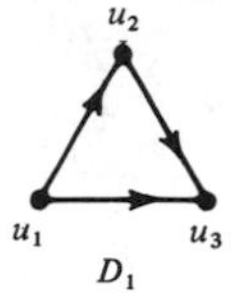

$$A(D_1) = \begin{matrix} & 1 & 2 & 3 \\ 1 \\ 2 \\ 3 \end{matrix} \begin{pmatrix} 0 & 1 & 1 \\ 0 & 0 & 1 \\ 0 & 0 & 0 \end{pmatrix}$$

$$R(D_1) = \begin{matrix} & 1 & 2 & 3 \\ 1 \\ 2 \\ 3 \end{matrix} \begin{pmatrix} 1 & 1 & 1 \\ 0 & 1 & 1 \\ 0 & 0 & 1 \end{pmatrix}$$

$$(d_{ij}) = \begin{matrix} & 1 & 2 & 3 \\ 1 \\ 2 \\ 3 \end{matrix} \begin{pmatrix} 0 & 1 & 1 \\ x & 0 & 1 \\ x & x & 0 \end{pmatrix}$$

x = undefined

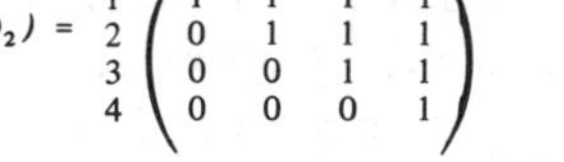

$$A(D_2) = \begin{matrix} & 1 & 2 & 3 & 4 \\ 1 \\ 2 \\ 3 \\ 4 \end{matrix} \begin{pmatrix} 0 & 1 & 0 & 0 \\ 0 & 0 & 1 & 0 \\ 0 & 0 & 0 & 1 \\ 0 & 0 & 0 & 0 \end{pmatrix}$$

$$R(D_2) = \begin{matrix} & 1 & 2 & 3 & 4 \\ 1 \\ 2 \\ 3 \\ 4 \end{matrix} \begin{pmatrix} 1 & 1 & 1 & 1 \\ 0 & 1 & 1 & 1 \\ 0 & 0 & 1 & 1 \\ 0 & 0 & 0 & 1 \end{pmatrix}$$

$$(d_{ij}) = \begin{matrix} & 1 & 2 & 3 & 4 \\ 1 \\ 2 \\ 3 \\ 4 \end{matrix} \begin{pmatrix} 0 & 1 & 2 & 3 \\ x & 0 & 1 & 2 \\ x & x & 0 & 1 \\ x & x & x & 0 \end{pmatrix}$$

x = undefined

FIGURE 2.20. Digraphs with their adjacency, reachability, and distance matrices. (Continued on the next page.)

[17]Of course, the specific matrix obtained depends on the way we list the vertices. But by an abuse of language, we shall call any of these matrices *the* adjacency matrix.

$$A(D_3) = \begin{array}{c} \\ 1 \\ 2 \\ 3 \\ 4 \\ 5 \end{array} \begin{array}{ccccc} 1 & 2 & 3 & 4 & 5 \\ \begin{pmatrix} 0 & 1 & 0 & 0 & 0 \\ 0 & 0 & 1 & 0 & 0 \\ 0 & 0 & 0 & 1 & 0 \\ 0 & 0 & 0 & 0 & 1 \\ 1 & 0 & 0 & 0 & 0 \end{pmatrix} \end{array} \quad R(D_3) = \begin{array}{c} \\ 1 \\ 2 \\ 3 \\ 4 \\ 5 \end{array} \begin{array}{ccccc} 1 & 2 & 3 & 4 & 5 \\ \begin{pmatrix} 1 & 1 & 1 & 1 & 1 \\ 1 & 1 & 1 & 1 & 1 \\ 1 & 1 & 1 & 1 & 1 \\ 1 & 1 & 1 & 1 & 1 \\ 1 & 1 & 1 & 1 & 1 \end{pmatrix} \end{array}$$

$$(d_{ij}) = \begin{array}{c} \\ 1 \\ 2 \\ 3 \\ 4 \\ 5 \end{array} \begin{array}{ccccc} 1 & 2 & 3 & 4 & 5 \\ \begin{pmatrix} 0 & 1 & 2 & 3 & 4 \\ 4 & 0 & 1 & 2 & 3 \\ 3 & 4 & 0 & 1 & 2 \\ 2 & 3 & 4 & 0 & 1 \\ 1 & 2 & 3 & 4 & 0 \end{pmatrix} \end{array}$$

$$A(D_4) = \begin{array}{c} \\ 1 \\ 2 \\ 3 \\ 4 \end{array} \begin{array}{cccc} 1 & 2 & 3 & 4 \\ \begin{pmatrix} 0 & 1 & 0 & 0 \\ 1 & 0 & 1 & 0 \\ 0 & 0 & 0 & 1 \\ 1 & 0 & 0 & 0 \end{pmatrix} \end{array} \quad R(D_4) = \begin{array}{c} \\ 1 \\ 2 \\ 3 \\ 4 \end{array} \begin{array}{cccc} 1 & 2 & 3 & 4 \\ \begin{pmatrix} 1 & 1 & 1 & 1 \\ 1 & 1 & 1 & 1 \\ 1 & 1 & 1 & 1 \\ 1 & 1 & 1 & 1 \end{pmatrix} \end{array}$$

$$(d_{ij}) = \begin{array}{c} \\ 1 \\ 2 \\ 3 \\ 4 \end{array} \begin{array}{cccc} 1 & 2 & 3 & 4 \\ \begin{pmatrix} 0 & 1 & 2 & 3 \\ 1 & 0 & 1 & 2 \\ 2 & 3 & 0 & 1 \\ 1 & 2 & 3 & 0 \end{pmatrix} \end{array}$$

FIGURE 2.20 continued.

Theorem 2.11. If D is a digraph with adjacency matrix $A = (a_{ij})$, then the i, j entry of A^t gives the number of paths of length t in D which lead from u_i to u_j.

Proof. We argue by induction on t. If $t = 1$, the result is obvious. Assuming it for t, let us prove it for $t + 1$. Let $a_{ij}^{(t+1)}$ represent the number of paths of length $t + 1$ from u_i to u_j. Similarly, let $a_{kj}^{(t)}$ represent the number of paths of length t from u_k to u_j. To go from u_i to u_j in $t + 1$ steps, one must go from u_i to some u_k directly and then from u_k to u_j in t steps. The number of ways to go from u_i to u_j in $t + 1$ steps with the first step going through u_k is $a_{ik}a_{kj}^{(t)}$. For this term is $a_{kj}^{(t)}$ if $(u_i, u_k) \in A$ and it is 0 if $(u_i, u_k) \notin A$. To obtain $a_{ij}^{(t+1)}$, we simply sum the terms $a_{ik}a_{kj}^{(t)}$ for all k. Thus,

$$a_{ij}^{(t+1)} = \sum_{k=1}^{n} a_{ik}a_{kj}^{(t)}.$$

$$
A = \begin{array}{c} \\ 1 \\ 2 \\ 3 \\ 4 \end{array}
\begin{array}{c} \begin{array}{cccc} 1 & 2 & 3 & 4 \end{array} \\
\begin{pmatrix} 0 & 1 & 0 & 0 \\ 1 & 0 & 1 & 0 \\ 0 & 0 & 0 & 1 \\ 1 & 0 & 0 & 0 \end{pmatrix}
\end{array}
$$

$$
A^2 = \begin{array}{c} \\ 1 \\ 2 \\ 3 \\ 4 \end{array}
\begin{array}{c} \begin{array}{cccc} 1 & 2 & 3 & 4 \end{array} \\
\begin{pmatrix} 1 & 0 & 1 & 0 \\ 0 & 1 & 0 & 1 \\ 1 & 0 & 0 & 0 \\ 0 & 1 & 0 & 0 \end{pmatrix}
\end{array}
$$

$$
A^3 = \begin{array}{c} \\ 1 \\ 2 \\ 3 \\ 4 \end{array}
\begin{array}{c} \begin{array}{cccc} 1 & 2 & 3 & 4 \end{array} \\
\begin{pmatrix} 0 & 1 & 0 & 1 \\ 2 & 0 & 1 & 0 \\ 0 & 1 & 0 & 0 \\ 1 & 0 & 1 & 0 \end{pmatrix}
\end{array}
$$

$$
A^4 = \begin{array}{c} \\ 1 \\ 2 \\ 3 \\ 4 \end{array}
\begin{array}{c} \begin{array}{cccc} 1 & 2 & 3 & 4 \end{array} \\
\begin{pmatrix} 2 & 0 & 1 & 0 \\ 0 & 2 & 0 & 1 \\ 1 & 0 & 1 & 0 \\ 0 & 1 & 0 & 1 \end{pmatrix}
\end{array}
$$

FIGURE 2.21. Powers of the adjacency matrix of digraph D_4 of Fig. 2.20.

By inductive assumption, $a_{kj}^{(t)}$ is the k, j entry of A^t. Hence, $a_{ij}^{(t+1)}$ is the i, j entry of $AA^t = A^{t+1}$.   Q.E.D.

Figure 2.21 shows the matrices A^2, A^3, and A^4 corresponding to the digraph D_4 of Fig. 2.20. Since the 2, 1 entry of A^3 is 2, there are 2 paths of length 3 from u_2 to u_1 in D_4. These paths are u_2, u_1, u_2, u_1 and u_2, u_3, u_4, u_1.[18]
Theorem 2.11 is an example of a result showing how matrices can be used to learn something about digraphs. Sometimes digraphs can be used to

[18]A much harder problem is to determine the number of simple paths of length t from u_i to u_j. This problem was first discussed in Luce and Perry [1949]. The solution was extended in Ross and Harary [1952] and a general procedure was found by Parthasarathy [1964].

learn something about matrices. In Sec. 4.2, we shall show how to translate a matrix into a digraph and use the strong components of the digraph to calculate eigenvalues associated with the matrix.

2.4.2. The Reachability Matrix[19]

The second matrix we associate with a digraph D is its *reachability matrix* $R(D) = (r_{ij})$. This is the matrix defined as follows:

$$r_{ij} = \begin{cases} 1 & \text{(if } u_j \text{ is reachable from } u_i) \\ 0 & \text{(otherwise).} \end{cases}$$

Note that each vertex is reachable from itself, since u_i alone is a path, so $r_{ii} = 1$, all i. Figure 2.20 gives the reachability matrices of a number of digraphs. The reachability matrix can be expressed in terms of the adjacency matrix if we introduce the notion of *Boolean addition*. In Boolean addition, $0 + 0 = 0$, $0 + 1 = 1 + 0 = 1$, but $1 + 1 = 1$. To extend this to the class N of all nonnegative integers, we define the *Boolean function* $B: N \longrightarrow \{0, 1\}$ as follows:

$$B(x) = \begin{cases} 0 & \text{(if } x = 0) \\ 1 & \text{(if } x > 0). \end{cases}$$

We define $B(A)$, where $A = (a_{ij})$ is a matrix, in the obvious way: The i, j entry of $B(A)$ is $B(a_{ij})$. Thus,

$$B \begin{pmatrix} 1 & 8 & 5 \\ 0 & 2 & 0 \\ 1 & 1 & 0 \end{pmatrix} = \begin{pmatrix} 1 & 1 & 1 \\ 0 & 1 & 0 \\ 1 & 1 & 0 \end{pmatrix}.$$

Theorem 2.12. Suppose D is a digraph of n vertices with reachability matrix R and adjacency matrix A. Then

$$R = B(I + A + A^2 + \cdots + A^{n-1}) = B[(I + A)^{n-1}],$$

where I is the identity matrix

$$\begin{pmatrix} 1 & 0 & 0 & \cdots & 0 \\ 0 & 1 & 0 & \cdots & 0 \\ 0 & 0 & 1 & \cdots & 0 \\ & & \cdot & \cdot & \cdot & \\ 0 & 0 & 0 & \cdots & 1 \end{pmatrix}$$

and B is the Boolean function.

---

[19]This subsection may be omitted without loss of continuity, though it is a good application of the ideas in earlier sections.

Proof. By Theorem 2.1, u_i can reach u_j if and only if there is a simple path from u_i to u_j. Since a simple path must use distinct vertices, its length is at most $n - 1$. This proves the first equality. Proof of the second equality follows by expanding and using the definition of B, and is left to the reader (Exer. 13).   Q.E.D.

The reachability matrix has a number of applications. The first is to the calculation of types of connectedness. The reader will find it useful to prove the following theorem as an exercise.

Theorem 2.13 (Ross and Harary [1959]). Suppose D is a digraph with reachability matrix R and adjacency matrix A. Then:

(a) D is strongly connected if and only if $R = J$, the matrix of all 1's.
(b) D is unilaterally connected if and only if $B(R + R') = J$, where B is the Boolean function and R' is the transpose of R.
(c) D is weakly connected if and only if $B[(I + A + A')^{n-1}] = J$, where n is the number of vertices in D and A' is the transpose of A.

A second application of the reachability matrix is to the calculation of strong components, as indicated in the next theorem.

Theorem 2.14. Suppose D is a digraph with reachability matrix $R = (r_{ij})$ and suppose R^2 is the matrix (s_{ij}). Then:

(a) The strong component containing a vertex u_i is given by the entries of 1 in the ith row (or column) of $R \times R'$, the elementwise product of R and R', the transpose of R.
(b) The number of vertices in the strong component containing u_i is s_{ii}.

Proof.

(a) A vertex u_j is reachable from the vertex u_i if and only if $r_{ij} = 1$. It reaches u_i if and only if $r_{ji} = 1$. Thus, u_i and u_j are mutually reachable if and only if $r_{ij} \times r_{ji} = 1$.
(b) The quantity s_{ii} is equal to $\sum_{j=1}^{n} r_{ij} r_{ji}$, where n is the number of vertices. Now $r_{ij} r_{ji}$ is 1 if and only if u_i and u_j are mutually reachable. Thus, summing these numbers over all j gives the number of u_j which are mutually reachable from u_i.   Q.E.D.

Figure 2.22 shows the matrices R, $R \times R'$, and R^2 for the digraph D shown. The row corresponding to u in the matrix $R \times R'$ indicates the strong component $\{u, v, w\}$. The u, u entry in the matrix R^2, namely, 3, gives the number of elements in this strong component.

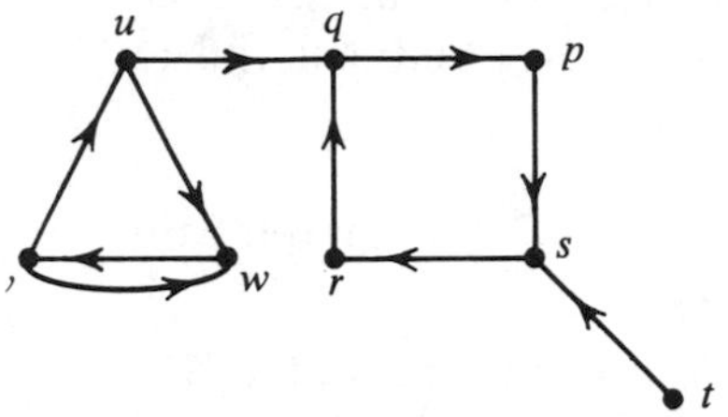

$$R = R(D) = \begin{array}{c} \\ u \\ v \\ w \\ p \\ q \\ r \\ s \\ t \end{array} \begin{pmatrix} u & v & w & p & q & r & s & t \\ 1 & 1 & 1 & 1 & 1 & 1 & 1 & 0 \\ 1 & 1 & 1 & 1 & 1 & 1 & 1 & 0 \\ 1 & 1 & 1 & 1 & 1 & 1 & 1 & 0 \\ 0 & 0 & 0 & 1 & 1 & 1 & 1 & 0 \\ 0 & 0 & 0 & 1 & 1 & 1 & 1 & 0 \\ 0 & 0 & 0 & 1 & 1 & 1 & 1 & 0 \\ 0 & 0 & 0 & 1 & 1 & 1 & 1 & 0 \\ 0 & 0 & 0 & 1 & 1 & 1 & 1 & 1 \end{pmatrix}$$

D

$$R \times R' = \begin{array}{c} \\ u \\ v \\ w \\ p \\ q \\ r \\ s \\ t \end{array} \begin{pmatrix} u & v & w & p & q & r & s & t \\ 1 & 1 & 1 & 0 & 0 & 0 & 0 & 0 \\ 1 & 1 & 1 & 0 & 0 & 0 & 0 & 0 \\ 1 & 1 & 1 & 0 & 0 & 0 & 0 & 0 \\ 0 & 0 & 0 & 1 & 1 & 1 & 1 & 0 \\ 0 & 0 & 0 & 1 & 1 & 1 & 1 & 0 \\ 0 & 0 & 0 & 1 & 1 & 1 & 1 & 0 \\ 0 & 0 & 0 & 1 & 1 & 1 & 1 & 0 \\ 0 & 0 & 0 & 0 & 0 & 0 & 0 & 1 \end{pmatrix}$$

Strong components:

$\{u, v, w\}$
$\{p, q, r, s\}$
$\{t\}$

$$R^2 = \begin{array}{c} \\ u \\ v \\ w \\ p \\ q \\ r \\ s \\ t \end{array} \begin{pmatrix} u & v & w & p & q & r & s & t \\ 3 & 3 & 3 & 7 & 7 & 7 & 7 & 0 \\ 3 & 3 & 3 & 7 & 7 & 7 & 7 & 0 \\ 3 & 3 & 3 & 7 & 7 & 7 & 7 & 0 \\ 0 & 0 & 0 & 4 & 4 & 4 & 4 & 0 \\ 0 & 0 & 0 & 4 & 4 & 4 & 4 & 0 \\ 0 & 0 & 0 & 4 & 4 & 4 & 4 & 0 \\ 0 & 0 & 0 & 4 & 4 & 4 & 4 & 0 \\ 0 & 0 & 0 & 5 & 5 & 5 & 5 & 1 \end{pmatrix}$$

FIGURE 2.22. Use of the reachability matrix to study the strong components of a digraph D.

2.4.3. THE DISTANCE MATRIX[20]

A third matrix which is usefully studied in connection with digraphs is the distance matrix (d_{ij}), where d_{ij} is the distance from u_i to u_j, which was defined in Sec. 2.2.2 as the length of the shortest path from u_i to u_j. (Recall that d_{ij} is undefined if there is no path from u_i to u_j.) Figure 2.20 shows the distance matrices asssociated with a number of digraphs.

Theorem 2.15. If D is a digraph with adjacency matrix A and distance matrix (d_{ij}), then for $i \neq j$, if d_{ij} is defined, then d_{ij} is the smallest number k such that the i, j entry of $B(A^k)$ is 1, where B is the Boolean function.

[20]This subsection may be omitted without loss of continuity.

Proof.   Left to the reader (Exer. 14).

EXERCISES

1. For each digraph of Fig. 2.17, find:
    (a) Its adjacency matrix.
    (b) Its reachability matrix.
    (c) Its distance matrix.

2. For digraphs D_3, D_5, and D_{17} of Fig. 2.7, find the number of paths of length 5 from u to v by using the adjacency matrix. Identify the paths.

3. For digraphs D_3 and D_5 of Fig. 2.7:
    (a) Calculate the reachability matrix R.
    (b) Calculate $B(I + A + A^2 + \cdots + A^{n-1})$.
    (c) Calculate $B[(I + A)^{n-1}]$.
    (d) Show from your calculations that these three matrices are the same.

4. (a) For digraph D_3 of Fig. 2.7, use Theorem 2.15 to calculate $d(u, w)$.
    (b) Repeat for digraph D_5 of Fig. 2.7 and $d(u, x)$.

5. If G is a graph, we define its *adjacency matrix* $A(G) = (a_{ij})$ in the obvious way: a_{ij} is 1 if there is an edge between u_i and u_j, and a_{ij} is 0 otherwise. Calculate $A(G)$ for each of the graphs of Fig. 2.9.

6. If G is a graph with adjacency matrix $A = A(G)$, what is the interpretation of the i, j entry of A^t?

7. For each digraph of Fig. 2.17, use the reachability matrix to find the strong components.

8. If R is the reachability matrix of a digraph and $c(i)$ is the ith column sum of R, what is the interpretation of $c(i)$?

9. If G is a graph, how would you define its reachability matrix $R(G)$?

10. If G is a graph and $R = R(G)$ is its reachability matrix (Exer. 9):
    (a) Show that $R \times R' = R$.
    (b) What is the interpretation of the 1, 1 entry of R^2?

11. Suppose R is a matrix of 0's and 1's with 1's down the diagonal (and perhaps elsewhere). Is R necessarily the reachability matrix of some digraph? (Give proof or counterexample.)

12. Explain how to use the reachability matrix to determine if a digraph has a unique vertex basis.

13. Prove the second equality in Theorem 2.12.

14. Prove Theorem 2.15.

15. Prove Theorem 2.13.

16. Let A be a set and $\mathfrak{F}$ a family of subsets of A. Define the *point-set incidence matrix* M as follows: List elements of A as $u_1, u_2, \ldots, u_n$; list sets in $\mathfrak{F}$ as $S_1, S_2, \ldots, S_m$. Then M is an $n \times m$ matrix, and its i, j entry is 1 if element u_i is in set S_j and 0 if element u_i is not in set S_j. For example, if $A = \{1, 2, 3, 4\}$, and $\mathfrak{F} = \{\{1,2\}, \{2,3\}, \{1\}\}$, then M is

$$
\begin{array}{c}
\quad \{1, 2\} \quad \{2, 3\} \quad \{1\} \\
\begin{array}{c} 1 \\ 2 \\ 3 \\ 4 \end{array}
\left(
\begin{array}{ccc}
1 & 0 & 1 \\
1 & 1 & 0 \\
0 & 1 & 0 \\
0 & 0 & 0
\end{array}
\right).
\end{array}
$$

Now if G is a graph, its *vertex-edge incidence matrix* B is the point-set incidence matrix for the set $A = V(G)$ and the set $\mathfrak{F} = E(G)$. Draw the vertex-edge incidence matrix for the graph G of Fig. 2.11.

17. (Harary [1969].) Suppose B is the vertex-edge incidence matrix of a graph G (Exer. 16) and B' is the transpose of B. What is the significance of the i, j entry of the matrix $B'B$?

18. (Harary [1969].) Let G be a graph. The *circuit matrix* C of G is the point-set incidence matrix with $A =$ the set of edges of G and $\mathfrak{F} =$ the family of circuits of G. Let B be the vertex-edge incidence matrix of G (Exer. 16). Then every entry of BC is $\equiv 0 \pmod 2$.

References

BERGE, C., *The Theory of Graphs and its Applications*, Methuen, London, 1962.

CAYLEY, A. L., "On the Theory of Analytical Forms Called Trees," *Philos. Mag.*, **13** (1857), 19–30; reprinted in *Mathematical Papers*, Cambridge, **3** (1891), 242–246.

CAYLEY, A. L., "On the Mathematical Theory of Isomers," *Philosophical Mag.*, **67** (1874), 444–446; reprinted in *Mathematical Papers*, Cambridge, **9** (1895), 202–204.

EULER, L., "Solutio Problematis ad Geometriam Situs Pertinentis," *Comment. Academiae Sci. I. Petropolitanae*, **8** (1736), 128–140; reprinted in *Opera Omnia*, Series I-7 (1766), 1–10.

EULER, L., "The Königsberg Bridges," *Sci. Amer.*, **189** (1953), 66–70. (A translation of the previous article.)

HARARY, F., *Graph Theory*, Addison-Wesley Publishing Co., Inc., Reading, Mass., 1969.

HARARY, F., NORMAN, R. Z., and CARTWRIGHT, D., *Structural Models: An Introduc-*

tion to the Theory of Directed Graphs, John Wiley & Sons, Inc., New York, 1965.

HUTCHINSON, G., "Evaluation of Polymer Sequence Fragment Data using Graph Theory," *Bull. Math. Biophysics*, **31** (1969), 541–562.

HUTCHINSON, J. P., and WILF, H. S., "On Eulerian Circuits and Words with Prescribed Adjacency Patterns," *J. Comb. Th.* (A), **18** (1975), 80–87.

KEMENY, J. G., and SNELL, J. L., *Mathematical Models in the Social Sciences*, Blaisdell Publishing Co., New York, 1962. Reprinted by M.I.T. Press, Cambridge, Mass., 1972.

KIRCHHOFF, G., "Uber die Auflösung der Gleichungen, auf Welche man bei der Untersuchung der Linearen Verteilung Galvanischer Ströme Geführt Wird," *Ann. Phys. Chem.*, **72** (1847), 497–508.

KÖNIG, D., *Theorie des Endlichen und Unendlichen Graphen*, Akademische Verlagsgesellschaft M.B.H., Leipzig, 1936; reprinted by Chelsea, New York, 1950.

LUCE, R. D., and PERRY, A. D., "A Method of Matrix Analysis of Group Structure," *Psychometrika*, **14** (1949), 95–116.

PARTHASARATHY, K. R., "Enumeration of Paths in Digraphs," *Psychometrika*, **29** (1964), 153–165.

ROSS, I. C., and HARARY, F., "On the Determination of Redundancies in Sociometric Choices," *Psychometrika*, **17** (1952), 195–208.

ROSS, I. C., and HARARY, F., "A Description of Strengthening and Weakening Members of a Group," *Sociometry*, **22** (1959), 139–147.

TUCKER, A., and BODIN, L., "A Model for Municipal Street Sweeping Operations," in CUPM Applied Mathematics Modules Project, Berkeley, California, 1975.

Glossary of Frequently Used Terms

I. *Paths, Chains, etc.*

 A. Suppose $D = (V, A)$ is a digraph.

  1. A *closed path* in D is a path $u_1, u_2, \ldots, u_t, u_{t+1}$ such that $u_{t+1} = u_1$.

  2. A *closed semipath* in D is a semipath $u_1, a_1, u_2, a_2, \ldots, u_t, a_t, u_{t+1}$ such that $u_{t+1} = u_1$.

  3. A *complete closed path* in D is a complete path which is closed.

  4. A *complete path* or *semipath* in D is a path or semipath going through all the vertices of D.

  5. A *cycle* in D is a closed path $u_1, u_2, \ldots, u_t, u_1$ for which the vertices $u_1, u_2, \ldots, u_t$ are distinct.

  6. The *distance* $d(u, v)$ from vertex u to vertex v in D is the length of a shortest path from u to v, and is undefined if there is no path from u to v.

  7. The vertices u and v are *joined* if there is a semipath from u to v.

8. The *length* of a path, simple path, cycle, semipath, simple semipath, or semicycle of D is the number of arcs in it.
9. A *path* in D is a sequence $u_1, a_1, u_2, a_2, \ldots, u_t, a_t, u_{t+1}$, where $t \geq 0$, so that each u_i is in V and each a_i is in A and a_i is always the arc (u_i, u_{i+1}). The path is usually written as $u_1, u_2, \ldots, u_t, u_{t+1}$.
10. Vertex v is *reachable* from vertex u if there is a path from u to v.
11. A *semicycle* in D is a closed semipath $u_1, a_1, u_2, a_2, \ldots, u_t, a_t, u_1$ in which vertices $u_1, u_2, \ldots, u_t$ are distinct and arcs $a_1, a_2, \ldots, a_t$ are distinct.
12. A *semipath* in D is a sequence $u_1, a_1, u_2, a_2, \ldots, u_t, a_t, u_{t+1}$, where $t \geq 0$, so that each u_i is in V and each a_i is in A and a_i is always the arc (u_i, u_{i+1}) or the arc (u_{i+1}, u_i).
13. A *simple path* in D is a path with no repeated vertices.
14. A *simple semipath* in D is a semipath with no repeated vertices.

B. Suppose $G = (V, E)$ is a graph.
1. A *chain* in G is a sequence $u_1, e_1, u_2, e_2, \ldots, u_t, e_t, u_{t+1}$, where $t \geq 0$, so that each $u_i \in V$ and each $e_i \in E$ and e_i is always the edge $\{u_i, u_{i+1}\}$. The chain is usually written $u_1, u_2, \ldots, u_t, u_{t+1}$.
2. A *circuit* in G is a closed chain $u_1, e_1, u_2, e_2, \ldots, u_t, e_t, u_1$ in which vertices $u_1, u_1, \ldots, u_t$ are distinct and edges $e_1, e_2, \ldots, e_t$ are distinct.
3. A *closed chain* in G is a chain $u_1, u_2, \ldots, u_t, u_{t+1}$ in which $u_{t+1} = u_t$.
4. The *distance* between vertices u and v is the length of the shortest chain between u and v, and is undefined if there is no such chain.
5. The *length* of a chain, simple chain, or circuit in G is the number of edges in it.
6. A *simple chain* in G is a chain with no repeated vertices.

II. Subgraphs and Connectedness

A. Suppose $D = (V, A)$ is a digraph.
1. D is in *connectedness category* 3 if D is strong.
2. D is in *connectedness category* 2 if D is unilateral but not strong.
3. D is in *connectedness category* 1 if D is weak but not unilateral.
4. D is in *connectedness category* 0 if D is not weak.
5. A *generated subgraph* of D is a subgraph (W, B) so that B consists of all arcs in A joining vertices in W.

6. D is *strongly connected* (*strong*) if for every pair of vertices u and v in D, u is reachable from v and v is reachable from u.

7. A *strong component* of D is a maximal strongly connected (generated) subgraph.

8. A *subgraph* of D is a digraph (W, B) such that $W \subseteq V$ and $B \subseteq A$.

9. D is *unilaterally connected* (unilateral) if for every pair of vertices u and v in D, u is reachable from v or v is reachable from u.

10. D is *weakly connected* (*weak*) if every pair of vertices u and v in D is joined (by a semipath).

B. Suppose $G = (V, E)$ is a graph.

   1. A *(connected) component* of G is a maximal connected (generated) subgraph.

   2. G is *connected* if every pair of vertices u and v in G is joined by a chain.

   3. A *generated subgraph* of G is a subgraph (W, F) so that F consists of all edges in E joining vertices in W.

   4. A *subgraph* of G is a graph (W, F) where $W \subseteq V$ and $F \subseteq E$.

THREE

Applications
of Graphs

3.1. Signed Graphs and the Theory of Structural Balance

3.1.1. SIGNED GRAPHS

Frequently a graph or digraph will seem to be an appropriate model for a relationship, for example, the relationship u knows v. However, we might discover that such a relationship has more possibilities than just whether or not u knows v. For example, u may like or dislike v. For this reason, we consider ways to add additional information to the structure of vertices and edges or arcs. In this section, we consider adding a sign, $+$ or $-$, to each edge or arc. The resulting *signed graph* or *signed digraph* will be applied here to study the theory of structural balance in sociology and in Chapter 4 to study such diverse problems as the energy crisis and the levelling off of funding for scientific research.

All of the terminology of Chapter 2 can be applied to signed graphs or digraphs, with the understanding that we are referring to the underlying graph or digraph. We shall use one additional piece of terminology specific to signed graphs and digraphs. The *sign* of a path, chain, closed path, closed chain, cycle, circuit, etc. is the product of the signs of its arcs or edges, if a $+$ sign is replaced by $+1$ and a $-$ sign by -1. In the signed digraph D of Fig. 3.1, the path u, v, w, y has sign $+$. The sign of the cycle u, v, w, x, u is $-$. In the

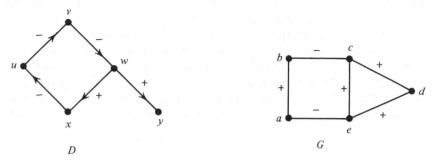

FIGURE 3.1. A signed digraph *D* and a signed graph *G*.

signed graph *G* of Fig. 3.1, the sign of the chain *a*, *b*, *c*, *d* is —. The sign is — if the path, chain, etc. has an odd number of — signs, and + otherwise.

3.1.2. BALANCE IN SMALL GROUPS

Before any discipline can develop broadly applicable theories, it must have precise concepts. We shall illustrate the process of taking an imprecise concept and replacing it by a precise one. This process, a necessary prerequisite for theoretical development, is sometimes called *explication*. Specifically, we illustrate the process with an example from small group sociology. The small groups studied could be groups of decisionmakers, panels of experts, groups of nations, different socioeconomic classes, etc. These groups are called "balanced" if, in a vague sense, they exhibit "absence of tension" and have an "ability to perform tasks well, work well together, etc."

We shall try to replace this imprecise concept of "balance" with a more precisely stated one, and in the process illustrate the four-stage cycle of mathematical modelling described in Chapter 1. At this point, it will be well for the reader to review that cycle, which is summarized in Fig. 1.1.

In forming a mathematical model of balance, one begins by forming a preliminary mathematical model of a small group. The people are represented as vertices of a graph and there is an edge {*a*, *b*} drawn between vertices *a* and *b* if there is a (strong) liking or disliking between people *a* and *b*. A similar model results if instead of using liking-disliking, we use associates with-avoids, agrees with-disagrees with, etc. As a simplification, let us assume that liking is a symmetric relation, namely that whenever *a* likes *b*, *b* likes *a*. Then on each edge {*a*, *b*}, we can put a sign + or —, telling us whether or not *a* and *b* like each other. As a result, the group is represented in this model by a signed graph. (Since we disregard loops, we don't consider whether or not a person likes himself.) Having formed such a preliminary model of a small group, we go back for more data on balance. Thus, we do not follow all around the mathematical modelling cycle of Fig. 1.1, but we skip back to the first step for more information before going on to the second step.

Relevant data can be found in the work of Heider [1946], who studied groups of three people, in particular, groups in which each member had strong likes or dislikes for each other member. All possible such groups are shown in Fig. 3.2. Heider observed that groups of types I and III were balanced, those of types II and IV were not. (In group I, all work well

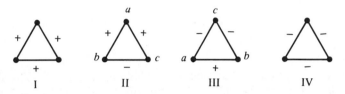

FIGURE 3.2. The possible groups of three members in which each member has strong likes or dislikes for each other member.

together. In group III, *a* and *b* like each other, and both mutually dislike *c*. Everyone is content to have *a* and *b* work together, and have *c* work alone. The group works well together. In group II, *a* likes both *b* and *c* and would like to work with both of them. They would both like to work with *a* but they dislike each other, so there is tension. Formation of working groups is difficult and the group is unbalanced. In group IV, everyone dislikes everyone else, no cooperation can take place, and the group is unbalanced.) Using Heider's data, Cartwright and Harary [1956] observed that groups of types I and III form circuits with an even number of − signs and groups of types II and IV form circuits with an odd number of −signs. Analysis of this and other examples suggested to them the following mathematical model of balance. A small group is represented by its signed graph and the group is considered *balanced* if each circuit in its signed graph is positive. Let us also call the signed graph *balanced* in this situation. To illustrate the definition, let us note that the signed graph of Fig. 3.3 is balanced. For its circuits are *u, v, w, x, u*; *u, v, x, u*; and *v, w, x, v*; and these are all positive. The step from the four

FIGURE 3.3. A balanced signed graph.

Heider examples (and others) to this definition is not mathematics; it is an example of the "art" of mathematical model-building.

The next step, the predictive step, is to make predictions about which particular groups are balanced. To do this, it is necessary to develop techniques for determining balance—testing the definition directly can get very complicated for large groups. For example, the reader should try applying the definition to determine whether or not the signed graph of Fig. 3.4 is balanced. Harary [1954] discovered the following criterion for balance, which is often easier to work with than the definition: A signed graph is balanced if and only if its vertices can be divided into two classes such that each edge within a class has sign $+$ and each edge between two classes has sign $-$.

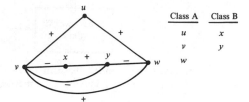

Class A	Class B
u	x
v	y
w	

FIGURE 3.4. A balanced signed graph and a two-set partition for its vertices.

Using this result, we easily see that the signed graph shown in Fig. 3.4 is balanced. For, it can be divided into two classes as shown in the figure. A proof of Harary's Theorem, sometimes called the *Structure Theorem*, is included in the next subsection. The theorem has interpretations for political science. A legislature is said to have an "idealized two-party structure" if its members can be divided into two groups in such a way that all liking-disliking relations within groups are positive and all liking-disliking relations between groups are negative. Thus, according to the theorem, a legislature is balanced if and only if it has a two-party structure. (One party can be empty.) Kemeny and Snell [1962, Ch. VIII] suggest that this result might explain why the French parliament of the 1950's, with many political parties, was unbalanced.

To fit Harary's Theorem into our schematic diagram of the process of mathematical modelling, one would have to add a fifth box, as shown in Fig. 3.5. The problem of analyzing the mathematical model gives rise to the development of new mathematical tools, which are in turn applied to make predictions concerning this (and other) models.

Using Harary's Theorem, one can make predictions about which graphs are balanced. In turn, these mathematical predictions can be translated into predictions about which small groups are balanced. For example, a small group whose liking-disliking structure can be represented in a signed graph

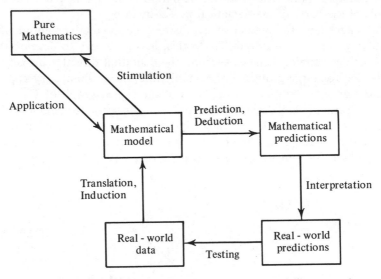

FIGURE 3.5. Expanded diagram of mathematical modelling procedure, indicating interrelation with "pure mathematics."

like that of Fig. 3.4 should be balanced, i.e., exhibit cooperation and lack of tension. The last step, the test of the model, is to study groups which have the given structure, and see if they are balanced. Often liking and disliking relationships can be induced among strangers in an artificial group in a laboratory situation if the strangers are told things about each other beforehand. For information on laboratory and other tests of the balance model, which come out fairly well, see Morissette [1958] and Taylor [1970].[1]

It should be stressed that the precise notion of balance introduced in the model need not agree one hundred percent with our imprecise notion of balance. In such a situation, *the new concept may in fact be more useful than the old one, even if different.* For it can become universal, and all sociologists can agree on its meaning. Moreover, in terms of the new concept, we can state precise theories and precise predictions, and perhaps we are better able to think about small group behavior using it. The only thing we would have to consider doing, should the new concept be markedly different from the old, is to change its name. Such differences between the model and the original situation would not be tolerated if we were trying to build an accurate model

[1]Morissette presented subjects with hypothetical situations involving four-sided relationships, and asked the subjects to rate the unpleasantness of the situation. He equated unpleasantness and tension, and discovered that balanced situations tended to be pleasant, unbalanced ones unpleasant. (Technically, he used the idea of local balance; see Exer. 8.)

of a certain kind of behavior, for example. But they are frequently useful in the case of models of imprecise concepts.

Before leaving this preliminary discussion of balance, we should observe that the same concept, once developed, can be applied to a larger number of situations than originally studied. For example, the concept of balance (in its precise form) has been applied (in a very tentative way) to the study of international relations, in particular to an analysis of the alliance-disalliance relations among the countries involved in the Middle East (including the U.S., U.S.S.R., etc.) The countries are the vertices and there is a + edge between two countries if they are allies, a − edge if they are "enemies." (See Harary [1961a].) Balance theory has also been applied to international relations by Abell and Jenkins [1967], Axelrod [1972, 1973], Bonham and Shapiro [to appear], Doreian [1969], Dowly [1970], Hart [1974 a, b], Healy and Stein [1973], and Sherwin [1972].

Harary, Norman, and Cartwright [1965] suggest that the notion of balance can even be applied to analysis of literature; if at one stage of a novel, play, short story, etc., the relations among characters form an unbalanced signed graph, then it is predictable that one of the signs will have to change. (Underlying this suggestion is the hypothesis that authors like to have "balanced" or tension-less endings. It is also suggested that authors build up tension by creating unbalanced situations.)

The mathematical model of balance, among other things, stimulated the study of signed graphs, which have become a subject of interest in their own right. In turn, the mathematics developed to handle signed graphs has become useful in other situations. In Chapter 4, we apply entirely different signed graph models to the study of the growing demand for energy, to the comparison of alternative policies for health care delivery, etc. This illustrates the point made in the Introduction to this book, namely that mathematics arising from a particular problem in the social, biological, or environmental sciences and developed for its own sake might eventually be relevant to other areas of application as well.

There are several weaknesses in the model of balance proposed so far. We mention four. First, the model assumes that relations such as "liking" are symmetric. Second, the model ignores the fact that some liking relations are stronger than others. Third, the model ignores the difference between the types of imbalance described in the second and fourth situations in Fig. 3.2. Fourth, the model treats balance as a "black and white" concept, ignoring the fact that there might be gradations or degrees of balance. In the exercises, we shall discuss how to eliminate the first weakness by dealing with signed digraphs rather than signed graphs and the fourth weakness by measuring degree of balance rather than just speaking of balance and imbalance. The second weakness could be eliminated by passing to the weighted graphs or digraphs we discuss in Chapter 4.

3.1.3. Proof of the Structure Theorem

In this section we present a proof of Harary's Theorem on balance, which we call the *Structure Theorem*.[2] First, let us state the result more completely.

Theorem 3.1 (Harary).   Suppose $G = (V, E)$ is a signed graph. The following statements are equivalent:

(a) G is balanced.
(b) Every closed chain in G is positive.
(c) Any two chains between vertices u and v have the same sign.
(d) The set V can be partitioned into two sets A and B so that every positive edge joins vertices of the same set and every negative edge joins vertices of different sets.

Proof.   We prove the sequence of implications: (a) $\Rightarrow$ (b) $\Rightarrow$ (c) $\Rightarrow$ (d) $\Rightarrow$ (a). The proof that (d) $\Rightarrow$ (a) is very simple. Any circuit can have only an even number of edges going between the sets A and B, since every time the circuit leaves the set in which it starts, it must return. The only negative edges are those between sets A and B. This completes the proof.

To prove that (a) $\Rightarrow$ (b), suppose there is a negative closed chain in G. Let $u_1, u_2, \ldots, u_t, u_1$ be one of minimum length. Since G is balanced, this cannot be a circuit, and so there are $i < j$ so that $u_i = u_j$. The reader will readily verify that $u_i, u_{i+1}, \ldots, u_j$ and $u_1, u_2, \ldots, u_i, u_{j+1}, u_{j+2}, \ldots, u_t, u_1$ are both closed chains, are both shorter than $u_1, u_2, \ldots, u_t, u_1$, and that one of them must be negative. This contradicts the minimality of $u_1, u_2, \ldots, u_t, u_1$.

To prove that (b) $\Rightarrow$ (c), suppose that $u, u_2, \ldots, u_t, v$ and $u, v_2, \ldots, v_s, v$ are two chains between u and v. Then $u, u_2, \ldots, u_t, v, v_s, v_{s-2}, \ldots, u$ is a closed chain. The sign of this closed chain is $-$ if the two starting chains have different signs.

Finally, to prove that (c) $\Rightarrow$ (d), we might as well assume that the graph G is connected. For, prove the theorem for a connected component (see Sec. 2.3.3) and then combine the sets A for different components and the sets B for different components. Assuming that G is connected, pick an arbitrary vertex u in $V(G)$ and let A be the set consisting of u plus all vertices joined to u by some positive chain. Let B consist of all remaining vertices. If vertices x in A and y in B are joined by an edge, this edge must be negative. For x is joined to u by a positive chain $u, v_2, \ldots, v_t, x$. If the edge $\{x, y\}$ is positive, then y is also joined to u by a positive chain, $u, v_2, \ldots, v_t, x, y$. Thus, y would

---

[2] As usual, the proof may be omitted without loss of continuity. However, the discussion following the proof should be included.

have to be in A, not B. Next, suppose vertices x and y are in A and are joined by an edge. This edge must be positive, for otherwise we easily obtain a negative chain from u to y from a positive chain from u to x. This violates condition (c). Finally, suppose vertices x and y are in B and are joined by an edge. By connectedness of G, there is a chain joining u to x. By definition of A and B, and condition (c), this chain must be negative. Finally, the edge joining x and y must be positive, for otherwise we obtain a positive chain from u to y from the negative chain from u to x. Q.E.D.

Harary's Theorem gives a constructive procedure for finding a two-set partition for a signed graph G, if one exists. We describe the procedure for a connected signed graph.

Lemma. If a connected signed graph G is unbalanced, then for every vertex u there is a vertex v such that there are two simple chains between u and v of opposite sign.

Proof. Since G is unbalanced, there is a negative circuit C. If u is on the circuit C, pick v to be any other vertex on C, and use chains in opposite directions along C. If u is not on C, connectedness of G implies that there is a simple chain S from u to some vertex on C. Pick a shortest such simple chain S, and let it go from u to the vertex w on C. Let v be any other vertex of C. There are simple chains S_1 and S_2 from w to v of opposite sign, following C in opposite directions. Then S followed by S_1 and S followed by S_2 are the desired simple chains. Q.E.D.

Suppose G is a connected signed graph. Let u be any vertex of G. To find a two-set partition, put u in the first set A. Starting from u, follow simple chains of length 1 to other vertices. If a vertex v is reached, put v in A if the chain joining v to u is positive, and put v in the second set B otherwise. Next, repeat the procedure with simple chains of length 2, of length 3, and so on up through simple chains of length $n - 1$, where n is the number of vertices in G. If G is balanced, the two sets A and B obtained will give the desired partition. If G is not balanced, it will by the lemma eventually be necessary to put a vertex into two different sets, and when this difficulty is encountered, we conclude that G is unbalanced. We illustrate the procedure on the signed graph of Fig. 3.4. We start by putting vertex u in A. Since u is joined to v by a positive edge, we put v in A. Similarly, we put w in A. Next, looking to simple chains of length 2, we see that x is joined to u by a negative chain, u, v, x, so x goes into B. Similarly, y goes into B. Thus, $A = \{u, v, w\}$ and $B = \{x, y\}$ is a two-set partition. Considering other simple chains of lengths 2, 3, and 4 does not lead to any difficulties with this partition. We conclude

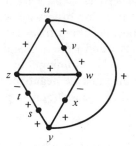

FIGURE 3.6. An unbalanced signed graph.

that G is balanced. On the other hand, suppose G is the signed graph of Fig. 3.6. Then simple chains of length 1 from u put z, v, and y in A and simple chains of length 2 put w, s, and x in A and t in B. But one simple chain of length 3 puts w in B, which cannot be. Thus, we conclude that G is not balanced. There must be a negative circuit, and it is easy to find one.

EXERCISES

1. For the signed graphs of Fig. 3.7, determine the signs of all simple chains of length 4 starting at a and the signs of all circuits.

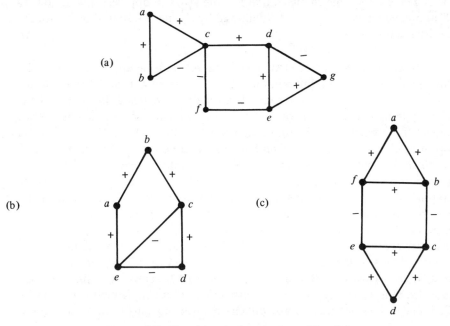

FIGURE 3.7. Signed graphs for exercises of Sec. 3.1.

2. For the signed digraph of Fig. 3.8, determine the signs of all simple paths of length 4 starting at *a* and the signs of all cycles.

3. Which of the graphs of Fig. 3.9 are balanced? For those which are, exhibit a two-set partition.

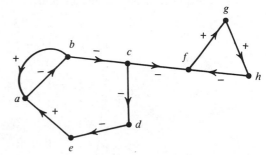

FIGURE 3.8. Signed digraph for Exer. 2, Sec. 3.1.

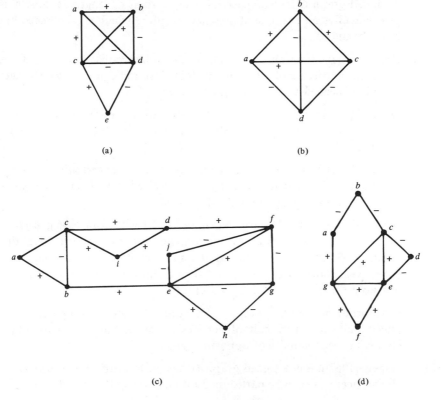

(a)

(b)

(c)

(d)

FIGURE 3.9. Signed graphs for Exer. 3, Sec. 3.1.

4. The *adjacency matrix* of a signed graph G is given by $A = (a_{ij})$ with

$$a_{ij} = \begin{cases} +1 & \text{(if edge } i, j \text{ exists and is } +) \\ -1 & \text{(if edge } i, j \text{ exists and is } -) \\ 0 & \text{(if no edge } i, j \text{ exists).} \end{cases}$$

Suppose G has adjacency matrix

$$A = \begin{array}{c} \\ 1 \\ 2 \\ 3 \\ 4 \\ 5 \\ 6 \end{array} \begin{array}{cccccc} 1 & 2 & 3 & 4 & 5 & 6 \\ \begin{pmatrix} 0 & -1 & -1 & -1 & +1 & -1 \\ -1 & 0 & +1 & +1 & 0 & 0 \\ -1 & +1 & 0 & +1 & 0 & 0 \\ -1 & +1 & +1 & 0 & 0 & 0 \\ +1 & 0 & 0 & 0 & 0 & -1 \\ -1 & 0 & 0 & 0 & -1 & 0 \end{pmatrix} \end{array}.$$

Is G balanced?

5. A signed graph G is *N-balanced* if every circuit of length at most N is positive. Give an example of a signed graph G which is 3-balanced but not 4-balanced.

6. Draw a signed graph representing the alliance-disalliance relations among countries involved in the Middle East and discuss what changes would be needed to balance the situation.

7. Draw signed graphs representing the liking-disliking relations at various stages of some novel, play, or short story. (A hypothetical one is fine.) Discuss balance.

8. A signed graph is *locally balanced* at vertex u if every circuit containing u is positive. Show that a signed graph may be locally balanced at some vertex u without being (globally) balanced.

9. It is a theorem (see Harary, Norman, and Cartwright [1965, p. 345]) that if a connected signed graph G has no vertices whose removal disconnects G, then G is balanced if and only if it is locally balanced (Exer. 8) at some vertex. Show that the conclusion of this theorem does not hold for all connected signed graphs.

10. Give proof or counterexample: If a connected signed graph is unbalanced, then for *every* pair of vertices u and v, there are two simple chains between u and v of opposite sign.

11. (Davis [1967]). Say a signed graph G has an *idealized party structure* if the vertices of G can be partitioned into classes so that all edges joining

vertices in the same class are $+$ and all edges joining vertices in different classes are $-$.

    (a)   Give an example of a signed graph which does not have an idealized party structure.

    (b)   Give an example of a signed graph which is not balanced but which has an idealized party structure.

*12.   (Davis [1967].) Show that a signed graph has an idealized party structure if and only if no circuit has exactly one $-$ sign.

*13.   (Davis [1967].) Suppose G is a *complete* signed graph, i.e., a signed graph with an edge between every pair of vertices. Show that if G has an idealized party structure, then it has a unique idealized party structure.

14.   Exercises 14 through 20 discuss the elimination in our model of the restriction that the strong liking or disliking relation in a small group is symmetric.[3] A natural generalization of our model for a small group is to represent it by a signed digraph rather than a signed graph, with arcs and signs having the obvious interpretation. A natural generalization

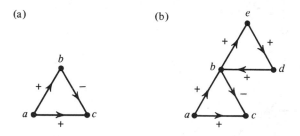

FIGURE 3.10. Two signed digraphs all of whose cycles are positive.

of the notion of balance is the following: A signed digraph is balanced if and only if every cycle is positive. There are several things wrong with this definition. For example, the signed digraph of Fig. 3.10(a) is balanced under this definition. (It has no cycles, so "all positive cycles" are balanced.) But in practice such a group is unbalanced. Tension builds because a likes b and c and so expects b to like c. (If the reader does not like this example because it has no cycles, he should consider the signed digraph of Fig. 3.10(b).) A second problem with the proposed definition is that, if it is used, the obvious analogue of the Structure Theorem is false. (The reader is asked to show that below.) It turns out that for

---

[3]The reader who has omitted the latter part of Sec. 2.2.3 should also omit these exercises.

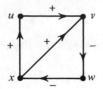

FIGURE 3.11. A balanced signed digraph.

(a)

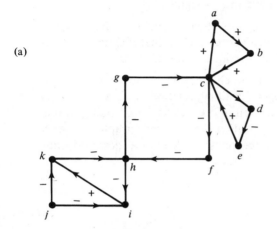

(b)

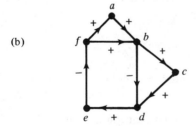

(c)

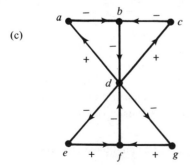

FIGURE 3.12. Signed digraphs for exercises of Sec. 3.1.

signed digraphs, balance is best defined in terms of semicycles rather than cycles. In particular, we say a signed digraph is *balanced* if every semicycle is positive. (Some writers omit semicycles of length 2, but we shall include them in this discussion.) The signed digraph of Fig. 3.11 is balanced, since its semicycles, $u, v, w, x, u; u, v, x, u;$ and $v, w, x, v,$ are all positive. (We did not include arcs in describing these semicycles, since arcs are unambiguous here.) Of the signed digraphs of Fig. 3.12, which are balanced?

15. Harary's Theorem has a natural analogue for signed digraphs: If $D = (V, A)$ is a signed digraph, then the following statements are equivalent: (a) D is balanced; (b) every closed semipath in D is positive; (c) any two semipaths between vertices u and v have the same sign; and (d) the set V can be partitioned into two sets A and B so that every positive arc joins vertices of the same set and every negative arc joins vertices of different sets. By this theorem, the signed digraph of Fig. 3.13 is balanced, for it has the two-set partition shown. For the balanced signed digraphs of Fig. 3.12, exhibit a two-set partition.

16. Show that the equivalence of conditions (a) and (d) of Exer. 15 is false if we define a signed digraph to be balanced if and only if every cycle is positive.

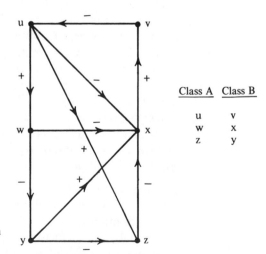

Class A	Class B
u	v
w	x
z	y

FIGURE 3.13. A balanced signed digraph and its two-set partition.

17. (Harary, Norman, and Cartwright [1965].) Suppose every strong component of a signed digraph D is balanced. Is every cycle of D positive? (Give proof or counterexample.) Is D balanced?

*18.   (Harary, Norman, and Cartwright [1965].) Suppose D is a strongly connected signed digraph each of whose cycles is positive. Is D balanced? (Give proof or counterexample.) (*Hint:* From a negative semicycle $u_1, a_1, u_2, a_2, \ldots, u_t, a_t, u_1$, build a negative closed path by reversing those arcs a_i which go from u_{i+1} to u_i.)

19.   (Harary, Norman, and Cartwright [1965].) If D is a signed digraph, the subgraph P has as vertex set all vertices lying on some positive arc and as arcs all positive arcs of D. Give proof or counterexample of the following: If D is weakly connected and P is disconnected, then D is balanced.

20.   Prove the theorem of Exer. 15.

21.   Exercises 21 through 26 deal with the observation that the models of balance described in the text and the earlier exercises make a sharp distinction between balanced and unbalanced, but do not allow for some things to be "more balanced" than others. We would like to "measure" balance by assigning to each signed graph G a number $b(G)$ which will correspond to G's *relative balance*.[4] The specific value $b(G)$ will not be important to us; we will only be interested in whether $b(G)$ is greater than, less than, or equal to another $b(G')$. A number of measures of relative balance have been described in the literature of sociology and graph theory. The book by Taylor [1970] gives an account of many of them. One idea is the following. If p is the number of positive circuits of G and t the total number of circuits, we might want to measure relative balance as $b(G) = p/t$. Or we might want to use $b'(G) = p/n$, where n is the number of negative circuits. (If $n = 0$ and $p \neq 0$, $b'(G)$ is thought of as infinite, i.e., as large as possible. If $n = 0$ and $p = 0$, then neither $b(G)$ nor $b'(G)$ is defined.)

   (a)   Calculate $b(G)$ and $b'(G)$ for each signed graph of Fig. 3.7.
   (b)   Show that the measures $b(G)$ and $b'(G)$ are *essentially the same*, i.e., that for all signed graphs G_1 and G_2 (with either $n \neq 0$ or $p \neq 0$), $b(G_1) > b(G_2)$ iff $b'(G_1) > b'(G_2)$.

22.   In measuring relative balance, we might want to take length of a circuit into account. Assuming that circuits of length k become less important in their contribution to balance as k gets larger, we could weight the effect of length by using a factor $1/k$. Thus, if p_k is the number of positive circuits of length k, t_k is the total number of circuits of length k, and c is the length of the longest circuit in G, we might measure relative

---

[4]A similar discussion applies to signed digraphs, if the term circuit is replaced by the term semicycle.

balance by

$$b(G) = \frac{\displaystyle\sum_{k=3}^{c} \frac{1}{k} p_k}{\displaystyle\sum_{k=3}^{c} \frac{1}{k} t_k}.$$

This measure is illustrated in Fig. 3.14, where it is used to calculate balance.

(a) Calculate $b(G)$ for each signed graph of Fig. 3.7.

(b) Calculate $b(G)$ for the signed graph obtained from that of Fig. 3.14 by changing the sign of the edge $\{a, b\}$ from $+$ to $-$. Relative to the original, is this new signed graph more balanced, less balanced, or equally balanced?

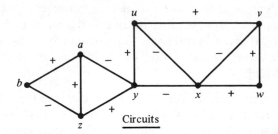

Circuits

Length 3	Sign	Length 4	Sign	Length 5	Sign
u, v, x, u	$+$	u, v, w, x, u	$-$	u, v, w, x, y, u	$-$
u, x, y, u	$+$	u, v, x, y, u	$+$		
v, w, x, v	$-$	a, b, z, y, a	$+$		
a, b, z, a	$-$				
a, y, z, a	$-$				

FIGURE 3.14. A signed graph and its circuits. The balance measure of Exer. 22, Sec. 3.1 is calculated as follows: $p_2 = 0, t_2 = 0, p_3 = 2, t_3 = 5$, $p_4 = 2, t_4 = 3, p_5 = 0, t_5 = 1, p_i = t_i = 0$ for $i > 5$. Thus,

$$b(G) = \frac{(\frac{1}{3})2 + (\frac{1}{4})2 + (\frac{1}{5})0}{(\frac{1}{3})5 + (\frac{1}{4})3 + (\frac{1}{5})1} = \frac{70}{157}.$$

23. (a) Is the balance measure $b(G)$ of Exer. 22 essentially the same (in the sense of Exer. 21) as either of the measures of Exer. 21?

(b) Show that $b(G)$ is essentially the same as the measure

$$b'(G) = \frac{\displaystyle\sum_{k=3}^{c} \frac{1}{k} p_k}{\displaystyle\sum_{k=3}^{c} \frac{1}{k} n_k},$$

where n_k is the number of negative circuits of length k.

(c) Is $b(G)$ essentially the same as the measure

$$b''(G) = \frac{\sum_{k=3}^{c} \left(\frac{1}{k}\right)^2 p_k}{\sum_{k=3}^{c} \left(\frac{1}{k}\right)^2 t_k} ?$$

(*Note:* Since there are many alternative measures of balance, we would like some systematic method for choosing among them. One method which is frequently used to produce a measure when there are many alternatives is the axiomatic method. We list certain *axioms* or properties which an acceptable measure (of balance) should satisfy. Then, a given measure is deemed acceptable if it satisfies all of the axioms we have listed. We shall describe several such axiomatic approaches in this book. An axiomatic approach to measuring balance is described in Norman and Roberts [1972a] and a balance measure obtained from the axioms is applied to the theory of distributive justice in sociology in Norman and Roberts [1972b].

24. Using the balance measure of Exer. 22, prove or disprove the following sociological theorem: It is always at least as good to evaluate as not to. Specifically, prove or disprove the following: If a signed graph G has one edge $\{a, b\}$ with undecided sign, then either an assignment of $+$ to $\{a, b\}$ or an assignment of $-$ to $\{a, b\}$ results in at least as high a measure of balance b as not assigning a sign and deleting the edge.

*25. (Harary, Norman, and Cartwright [1965].) Suppose a signed graph G is connected, G has no vertex whose removal disconnects it, and G has more than one circuit. Show that $b(G) > 0$, where b is the measure of Exer. 22.

26. This exercise develops still one other alternative measure of balance, called the *line index*. (Cf. Harary, Norman, and Cartwright [1965].)
   (a) Suppose G is a signed graph. The *negation* of a set of edges is obtained by changing the sign of each edge in the set. Show that every signed graph has a (possibly empty) set of edges whose negation results in balance.
   (b) A set of edges is called *negation-minimal* if its negation results in balance, but the negation of any proper subset of it does not. Give an example of a negation-minimal set of edges in the signed graph of Fig. 3.14.
   (c) A set of edges is called *deletion-minimal* if deletion of all edges in the set results in balance, but deletion of all edges in a proper subset

does not. Give an example of a deletion-minimal set of edges in the
signed graph of Fig. 3.14.

(d) Prove that every negation-minimal set of edges contains a deletion-minimal set.

*(e) Prove that every deletion-minimal set of edges contains a negation-minimal set. (*Hint:* Use the Structure Theorem.)

(f) Conclude that every deletion-minimal set is negation-minimal, and conversely.

(g) The *line index* of G, $\lambda(G)$, is the size of the smallest deletion-minimal (or negation-minimal) set. Calculate the line index of the signed graph of Fig. 3.14.

(h) It would be nice if $\lambda(G)$ were inversely related to $b(G)$, the measure of Exer. 22. (Why?) Unfortunately, things are not so simple. Give an example to show that $\lambda(G)$ may be less than $\lambda(G')$, while $b(G)$ is also less than $b(G')$.

27. Discuss relative balance for alternative signed graphs representing the alliance-disalliance relations among countries involved in the Middle East.

28. Discuss relative balance for signed graphs representing liking-disliking relations at various stages of a novel, play, or short story. Does relative balance increase or decrease at various stages?

3.2. Tournaments

Let (V, A) be a digraph and assume that for all $u \neq v$ in V, (u, v) is in A or (v, u) is in A, but not both. Such a digraph is called a *tournament*. Figure 3.15 shows all tournaments with two, three, or four vertices.

Tournaments arise in many different situations. There are the obvious ones, the round-robin competitions in tennis, baseball, etc.[5] In a round-robin competition, every pair of players (pair of teams) competes and one and only one member of each pair beats the other. (We assume that no ties are allowed.) Tournaments also arise from *pair comparison experiments*, where a subject is presented with each pair of alternatives from a set and he is asked to say which of the two he prefers, which is more beautiful, which is more qualified, etc. Tournaments also arise in nature, where certain individuals in a given species develop dominance over others of the same species. In such situations, for every pair of animals of the species, one and only one is domi-

---

[5]The reader should distinguish a round-robin competition from the elimination-type competition which is more common.

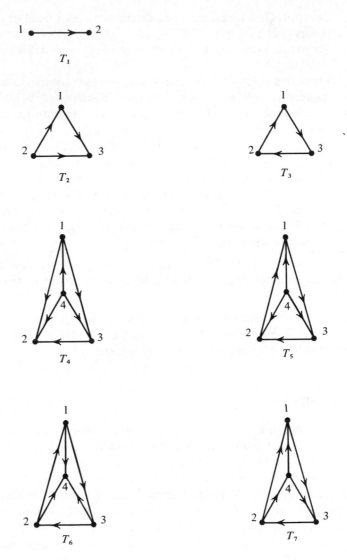

FIGURE 3.15. All tournaments with 2, 3, or 4 vertices (up to relabelling of vertices).

nant over the other. The dominance relation defines what is called a *pecking order* among the individuals concerned.[6]

[6]In the case of wolves, "no member of a pack is equal to any other, and social order is based on a 'dominance hierarchy,' or pecking order. The pack leader, usually a male, is

In this section, we prove some results about tournaments, specifically about methods for determining a winner in a tournament. Mostly, we follow Harary, Norman, and Cartwright [1965, Ch. 11] in our mathematical development. The reader who is interested in a more complete treatment of tournaments is referred to the book by Moon [1968].

If (V, A) is a tournament, we can associate with the player u his *score* $s(u)$, namely the number of players he beats. In terms of the digraph, this is the number of vertices v so that there are arcs from u to v. (In contexts other than tournaments, this number is frequently called the *outdegree* of u.) Let us assume that the players (the vertices) of the tournament have been labelled $1, 2, \ldots, n$, where n is the number of elements of V, and let $s(i)$ be the score of player i. The sequence $(s(1), s(2), \ldots, s(n))$ will be called the *score sequence* of the tournament.

Figure 3.16 shows a (mixed) tennis tournament with six players. The tournament is rather difficult to work with in this form, and it is easier to work with the adjacency matrix, which we also show in Fig. 3.16. The score

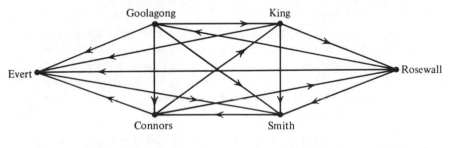

	Evert	Smith	Connors	Rosewall	King	Goolagong	Score (row sum)
Evert	0	1	0	0	0	0	1
Smith	0	0	1	0	0	0	1
Connors	1	0	0	1	1	0	3
Rosewall	1	1	0	0	0	1	3
King	1	1	0	1	0	0	3
Goolagong	1	1	1	0	1	0	4

FIGURE 3.16. A tennis tournament on 6 players and its adjacency matrix.

known as the 'alpha male.' When two wolves meet, one usually demonstrates its dominance —its tail erect, mane and ears bristling and posture straight. The other offers itself in submission, dropping its tail between its legs, lowering its hind quarters and drawing back its lips and ears. The submissive wolf may even roll over on its back, sometimes urinating and whining." (Fisher [1975].) Biologists are becoming increasingly interested in social organization among animals. A book on the developing field of sociobiology (from an evolutionary point of view) is Wilson [1975]. See also Wilson [1971].

sequence can be calculated directly from the adjacency matrix: $s(i)$ is the row sum of the ith row.

The first problem we consider concerns identification of the player or players with maximum score. Such a player or players would most likely be considered the winner or winners. The theorem we shall state says that if u is a player with maximum score, then for every other player v, either u beats v or u beats some player who beats v. The theorem was discovered by Landau [1955] during a study of pecking orders among chickens. In the case of chickens, the theorem says: If no chicken pecks more chickens than u does, then for every other chicken v, either u pecks v or u pecks a chicken which pecks v. In the case of preference, the theorem says that if no object is preferred to more objects than u, then for every other object v, either u is preferred to v or u is preferred to some object which is preferred to v. For example, in the tournament of Fig. 3.16, Goolagong has a maximum score. She beats all but Rosewall, and she beats King who beats Rosewall.

Theorem 3.2 (Landau).   In a tournament (V, A), if vertex u has maximum score, then for all other vertices v, either $(u, v) \in A$ or for some vertex w, $(u, w) \in A$ and $(w, v) \in A$.

Proof.   Let $s = s(u)$ and let $u_1, u_2, \ldots, u_s$ denote the vertices to which there are arcs from u. Given vertex v, suppose u does not beat v. Then by definition of a tournament, we have an arc from v to u. If v beats every u_i, the score of v is at least $s + 1$, for v also beats u. This contradicts the maximality of u's score. Thus, for some i, v does not beat u_i, so we have $(u, u_i) \in A$ and $(u_i, v) \in A$. We take $w = u_i$.   Q.E.D.

The condition stated in Theorem 3.2 is a necessary condition for vertex u to be a "winner." It is not sufficient, i.e., there are examples of tournaments with vertices u which beat all other v in at most two steps but which do not have maximal score. The construction of an example is left to the reader (Exer. 11).

Sometimes we want to *rank* the players of a tournament rather than just pick a winner. This might be true if we are giving out second prize, third prize, etc. It might also be true, for example, if the "players" are alternative candidates for a job and the tournament represents preferences among candidates. Then our first choice job candidate might not accept and we might want to have a second choice, third choice, and so on, chosen ahead of time. One way to rank the players is to use the score sequence. However, this can lead to ties. Other procedures for ranking can be described by studying connectedness of tournaments. A tournament can certainly be strongly connected, though it does not have to be. (Consider the two tournaments of three players shown in Fig. 3.15.) Can a tournament be in connectedness categories

0 or 1? The answer is no. For every u and v, there is an arc from u to v or from v to u, and hence a path. Thus, every tournament is unilateral, and in category 2 or 3. Actually, we can say more. The reader will recall that by Theorem 2.4, every unilateral digraph has a complete path, a path going through all the vertices. In a tournament, there is always not only a complete path, but a complete simple path.[7] This result, which we shall prove shortly, implies that we can label the players as $u_1, u_2, \ldots, u_n$ in such a way that u_1 beats u_2, u_2 beats $u_3, \ldots,$ and u_{n-1} beats u_n. Such a labelling gives us a ranking of the players: u_1 is ranked first, u_2 second, and so on. To illustrate, consider the tournament of Fig. 3.16. Here, a complete simple path and hence a ranking is given by:

Goolagong, King, Rosewall, Smith, Connors, Evert.

Before asking whether this is a "good" final ranking of the players, let us formulate and prove the theorem we have stated.

Theorem 3.3 (Rédei [1934]).   Every tournament (V, A) has a complete simple path.

Proof.   We proceed by induction on the number n of vertices. If $n = 2$, then (V, A) is the tournament T_1 of Fig. 3.15, and 1, 2 is a complete simple path. Assuming the result for tournaments of n vertices, let us consider a tournament (V, A) with $n + 1$ vertices. Let u be an arbitrary vertex and consider the subgraph generated by $V - \{u\}$. It is easy to verify that this subgraph is still a tournament. Hence, by inductive assumption, it has a complete simple path $u_1, u_2, \ldots, u_n$. If there is an arc from u to u_1, then $u, u_1, u_2, \ldots, u_n$ is a complete simple path of (V, A). If there is no arc from u to u_1, let i be the largest integer such that there is no arc from u to u_i. If i is n, then since (V, A) is a tournament, there must be an arc from u_n to u, and we conclude that $u_1, u_2, \ldots, u_n, u$ is a complete simple path of (V, A). If $i < n$, then since (V, A) is a tournament, there is an arc from u_i to u and, moreover, there is an arc from u to u_{i+1} by definition of i. Thus, $u_1, u_2, \ldots, u_i, u, u_{i+1}, u_{i+2}, \ldots, u_n$ is a complete simple path of (V, A).   Q.E.D.

If $u_1, u_2, \ldots, u_n$ is a complete simple path in a tournament (V, A), we can use it to define a ranking of the players. The reader will notice that in the tournament of Fig. 3.16, the ranking obtained from the complete simple path Goolagong, King, Rosewall, Smith, Connors, Evert does not agree with the ranking obtained from the score sequence, Goolagong, King-Rosewall-Connors, Smith-Evert. (A dash - indicates a tie.) Thus, the two ranking procedures

---

[7]A complete simple path in a digraph is often called a *hamiltonian path*, after Sir William Rowan Hamilton, a nineteenth century Irish mathematician. Hamiltonian paths have many applications. (See, for example, Exers. 19 and 20.)

lead to different rankings, which makes it problematical as to how we will decide upon a final ranking of players (or of alternatives in the case of preference). Moreover, to make matters even worse, there can be other complete simple paths in the tournament, for example King, Smith, Connors, Rosewall, Goolagong, Evert. In this situation, then, there are many possible rankings of the players. The situation can get even worse. In the tournament of Fig. 3.17, in fact, for each vertex a and each possible rank r, there is some complete simple path for which a has rank r.[8]

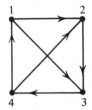

Score Sequence:

$(2, 1, 1, 2)$

FIGURE 3.17. A tournament and its score sequence. For each vertex a and rank r, there is some complete simple path for which a has rank r.

In general, given a set of possible rankings, we might want to try to choose a "consensus" ranking. The problem of finding a consensus among alternative possible rankings is a rather difficult one, which deserves a longer discussion than we can give it here. We return to it in Chapter 7.

Here, let us ask whether there are any circumstances where the problem of having many different rankings does not occur. That is, are there circumstances when a tournament has a unique complete simple path? And are there circumstances when the ranking from a complete simple path agrees with the ranking from the score sequence? The answers will both be "yes," and the circumstances will be the same for both questions. Unfortunately, many tournaments will not fit these circumstances.

To get at the answers, let us consider the tournaments of two, three, and four vertices as shown in Fig. 3.15. Tournament T_1 has a unique complete simple path. Of the two tournaments with three vertices, only T_2 has a unique complete simple path. Of the tournaments with four vertices, only T_4 does. (The path is 4, 1, 3, 2.) For T_5 has the complete simple paths 4, 2, 1, 3 and 4, 1, 3, 2 (as well as others), T_6 has the complete simple paths 1, 3, 2, 4 and 3, 2, 1, 4 (as well as others) and T_7 has the complete simple paths 4, 1, 3, 2 and 4, 3, 2, 1 (as well as others).

Note that of the tournaments in Fig. 3.15, all but T_1, T_2, and T_4 have a cycle of length 3. This suggests the theorem that a tournament has a unique complete simple path if and only if it has no cycle of length 3. We shall shortly prove this result. (It is also a theorem that a tournament has a unique com-

[8] A slightly more complicated example is given by Harary, Norman, and Cartwright [1965, p. 292].

plete simple path if and only if it has no cycles at all, i.e., it is acyclic. We shall leave the proof of this statement to the reader (Exer. 14).)

Let us observe that u, v, w, u is a cycle of length 3 if and only if $(u, v) \in A$, $(v, w) \in A$, and $(w, u) \in A$. Now if there is no cycle of length 3, then whenever $(u, v) \in A$ and $(v, w) \in A$, it is not the case that $(w, u) \in A$; thus

$$(u, v) \in A \quad \text{and} \quad (v, w) \in A \quad \text{imply} \quad (u, w) \in A. \tag{1}$$

A digraph which satisfies the condition (1) for all $u, v, w \in V$ with $u \neq w$ is called *transitive*.[9] Thus, we have shown that every tournament with no cycles of length 3 is transitive. The converse will also be true. In the real world, many tournaments are not transitive. This is certainly true in sports. Indeed, if all sports tournaments were transitive, things would be rather dull. Even in the case of preference, there are violations of transitivity, situations where you prefer u to v and prefer v to w, but prefer w to u. For example, suppose price is more important to you than quality, but you are willing to choose on the basis of quality if prices are fairly close. Suppose u and v are close in price, with u higher in quality, so that you prefer u to v. Suppose moreover that v and w close in price, with v higher in quality, so that you prefer v to w. Now u might be sufficiently higher in price than w for you to prefer w to u. (This example is due to Krantz, *et al.* [1971].)

We are now ready to state the conditions which we have been seeking.

Theorem 3.4. If (V, A) is a tournament, the following statements are equivalent.

(a)   There is a unique complete simple path.
(b)   There are no cycles of length 3.
(c) (V, A) is acyclic.
(d) (V, A) is transitive.

Moreover, if (V, A) is transitive, then the ranking obtained from the unique complete simple path agrees with that obtained from the score sequence.

Proof.   We shall show the equivalence of statements (a), (b), and (d), leaving the proof of the equivalence of these statements to (c) to the reader (Exer. 14). In our proof, we shall demonstrate the cycle of implications (a) $\Rightarrow$ (b) $\Rightarrow$ (d) $\Rightarrow$ (a).

We have already demonstrated the implication (b) $\Rightarrow$ (d). Next we show (d) $\Rightarrow$ (a). Suppose (V, A) is transitive. Then it is easy to prove by

[9]In relation theory (Sec. 8.2), transitivity is said to hold if (1) is satisfied for all $u, v, w \in V$. We require $u \neq w$ because, by convention, loops are disregarded in our digraphs, and (u, w) would be a loop if $u = w$. In tournaments, we cannot have arcs (u, v) and (v, u), so this assumption is superfluous. It will also be superfluous when we study transitive orientations in Sec. 3.3.3.

induction that whenever v is reachable from u, we have $(u, v) \in A$. Now suppose there are two different complete simple paths in (V, A). Since they are different, we must have, for some pair of distinct vertices u and v, u following v in one and v following u in the second. But then v is reachable from u in the second and u is reachable from v in the first, and it follows that $(u, v) \in A$ and $(v, u) \in A$. This violates a basic property of tournaments. We conclude (V, A) has at most one complete simple path. But by Theorem 3.3, the tournament has a complete simple path, and so we have proved that (d) $\Rightarrow$ (a).

To prove that (a) $\Rightarrow$ (b), suppose (V, A) has a unique complete simple path $u_1, u_2, \ldots, u_n$. We shall prove that $(u_i, u_j) \in A$ if and only if $i < j$. This implies that there are no cycles of length 3, for if u_i, u_j, u_k, u_i were such a cycle, then $i < j < k < i$ would follow. Since (V, A) is a tournament, it is sufficient to prove that $(u_i, u_j) \in A$ if $i < j$. If this is false for some $i < j$, there must be a smallest i so that for some j, $i < j$ and $(u_i, u_j) \notin A$. Now given this i, pick j as large as possible. Let us first assume that $i > 1$ and $j < n$. Now $(u_{i-1}, u_{i+1}) \in A$ by choice of i as small as possible and $(u_i, u_{j+1}) \in A$ by choice of j as large as possible. Furthermore, $(u_j, u_i) \in A$ since we have a tournament. Then

$$u_1, u_2, \ldots, u_{i-1}, u_{i+1}, u_{i+2}, \ldots, u_j, u_i, u_{j+1}, u_{j+2}, \ldots, u_n$$

is a second complete simple path. If $i = 1$, we simply eliminate the initial sequence $u_1, u_2, \ldots, u_{i-1}$ and if $j = n$ we simply eliminate the final sequence $u_{j+1}, u_{j+2}, \ldots, u_n$, and the same argument works. In any case, the uniqueness of the complete simple path is violated. This proves that $(u_i, u_j) \in A$ whenever $i < j$ and so the argument that (a), (b), and (d) are equivalent is finished.

Finally, from the proof of the implication (a) $\Rightarrow$ (b) we know that if $u_1, u_2, \ldots, u_n$ is the unique complete simple path, then $(u_i, u_j) \in A$ if and only if $i < j$. (This observation is important enough to be stated as a corollary below.) Thus, $s(u_i) = n - i$, and the rankings obtained from the unique complete simple path and from the score sequence are the same. This completes the proof of Theorem 3.4.   Q.E.D.

Corollary. If the tournament (V, A) has a unique complete simple path $u_1, u_2, \ldots, u_n$, then $(u_i, u_j) \in A$ if and only if $i < j$. Moreover, $s(u_i) = n - i$. The unique complete simple path can be found by choosing vertices in decreasing order of score.

It should be remarked that digraph theory can be useful in analyzing the structure of a proof like the one just given. Given a collection of n statements which we wish to prove equivalent, we draw a digraph with the statements as vertices and an arc from statement u to statement v if we demonstrate the direct implication $u \Rightarrow v$. The resulting digraph is called an *implication digraph.* The *n* statements are equivalent if and only if this digraph is strongly connected. Thus, the least number of demonstrations needed to prove that the *n* statements are equivalent is determined by finding the strongly connected digraph on *n* vertices with as few arcs as possible. This is a cycle. (Cf. Exer. 12, Sec. 2.2.)

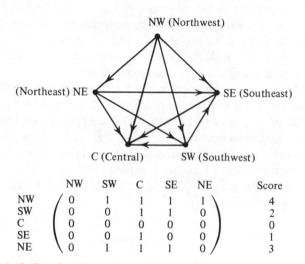

	NW	SW	C	SE	NE	Score
NW	0	1	1	1	1	4
SW	0	0	1	1	0	2
C	0	0	0	0	0	0
SE	0	0	1	0	0	1
NE	0	1	1	1	0	3

FIGURE 3.18. Results of a pair comparison experiment for preference among geographical areas. An arc from *x* to *y* indicates that *x* is preferred to *y*.

At this point, let us consider applications of Theorem 3.4 and its corollary to a decisonmaker's preferences among alternatives, or to his ratings of relative importance of alternatives, etc. The results are illustrated by pair comparison data for preference, for example, that shown in Fig. 3.18. Here, there are no cycles of length 3, and so there is a unique complete simple path. The path is

Northwest, Northeast, Southwest, Southeast, Central.

The scores correspond to the rankings in this path:

$$s(\text{Northwest}) = 4$$
$$s(\text{Northeast}) = 3$$
$$s(\text{Southwest}) = 2$$
$$s(\text{Southeast}) = 1$$
$$s(\text{Central}) = 0.$$

In studying preference, it is often reasonable to assume (or demand) that the decisionmaker's preferences define a transitive tournament, i.e., that if he prefers u to v and he prefers v to w, then he prefers u to w. This turns out to be equivalent to assuming that the decisionmaker can uniquely rank the alternatives among which he is expressing preferences. We have seen an example illustrating transitive preference in Fig. 3.18.

There are two types of theories of preference, the *prescriptive* and the *descriptive*. A prescriptive theory says that a rational man must behave in a certain way. In the case of preference, a prescriptive theory often demands that he be transitive in his preferences, and suggests that if he is confronted with a violation of transitivity in his judgments, he will immediately say: "I must have made a mistake." A descriptive theory seeks only to describe his behavior. In the case of preferences, such a theory recognizes that in practice preferences are often not transitive. We defer a detailed discussion of this point to Chapter 8, where we study preference theory and learn what to do in cases where preferences are not transitive. In Exer. 16, we introduce a potential measure of how transitive a tournament might be. Such a measure might be useful in measuring the "consistency" of preferences. For more details on this approach, see Kendall and Smith [1940] or Harary, Norman, and Cartwright [1965, p.300].

EXERCISES

1. In each tournament of Fig. 3.19, find the score of each individual.

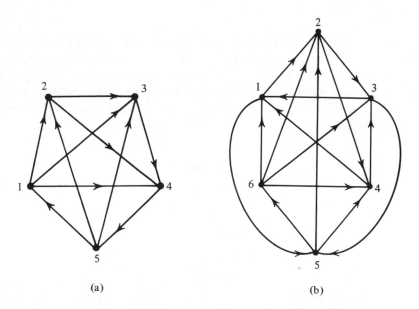

(a)

(b)

FIGURE 3.19. Tournaments for exercises of Sec. 3.2.

2. Draw the digraph of a tournament with score sequence (3, 2, 1, 0).

3. Could there be a tournament with score sequence (3, 3, 0, 0)? Why?

4. In each tournament of Fig. 3.19, use Landau's Theorem to determine all potential winners.

5. In each tournament of Fig. 3.19, find all complete simple paths.

6. For the tournament defined by the preference data of Fig. 3.20:
    (a) Do Exer. 4.
    (b) Do Exer. 5.

	New York	Boston	San Francisco	Los Angeles	Houston	Score
New York	0	0	0	0	1	1
Boston	1	0	0	1	1	3
San Francisco	1	1	0	1	1	4
Los Angeles	1	0	0	0	0	1
Houston	0	0	0	1	0	1

FIGURE 3.20. Results of a pair comparison experiment for preference among cities as a place to live. Entry i, j is 1 iff i is preferred to j.

7. Which of the tournaments with four or fewer vertices is transitive? (*Hint:* Use Theorem 3.4.)

8. Draw a digraph of a transitive tournament on five players.

9. Given the preference data of Fig. 3.20, determine whether or not it is transitive.

10. A *subtournament* of a tournament is a subgraph which defines a tournament. Of all the tournaments on four vertices, calculate for each how many transitive subtournaments there are. Which has the largest number?

11. Show that the converse of Landau's Theorem is false by building a tournament which includes a vertex u such that for all v, either u beats v or u beats some w who beats v, but such that $s(u)$ is not a maximum.

12. (a) Prove that in a tournament with vertices $u_1, u_2, \ldots, u_n$,

$$\sum_{i=1}^{n} s(u_i) = \frac{n(n-1)}{2}.$$

    (b) Could (4, 2, 2, 1, 0) be the score sequence of a tournament? Why?

13. Let T be a tournament and T^* be its condensation. Can T^* ever have more than one complete simple path? Why?

14. Show that if (V, A) is a tournament, then (V, A) has a unique complete simple path if and only if (V, A) is acyclic.

15. In a tournament T suppose $s(u) = s(v)$. Can u and v be in different strong components of T? Why?

16. If a tournament is not transitive, it is perhaps reasonable to measure how transitive the tournament is. One procedure for doing this is the following. If $\{u, v, w\}$ is a set of three vertices in the tournament $T = (V, A)$, this set defines a tournament itself. This is either a transitive tournament or not. In the former case, we call $\{u, v, w\}$ a *transitive triple* and in the latter case a *nontransitive triple*. One measure of degree of transitivity is

$$\frac{\text{Number of transitive triples}}{\text{Number of triples}}. \tag{2}$$

Recall that the denominator of (2) is

$$\binom{n}{3} = \frac{n!}{3!(n-3)!} = \frac{n(n-1)(n-2)}{6}.$$

We shall ask the reader to prove that the numerator is given by

$$\sum_{i=1}^{n} \binom{s(i)}{2} = \sum_{i=1}^{n} \frac{s(i)(s(i)-1)}{2}, \tag{3}$$

where $(s(1), s(2), \ldots, s(n))$ is the score sequence.

(a) Apply these formulas to measure the degree of transitivity of the tournaments in Fig. 3.15.

(b) Apply these formulas to measure the degree of transitivity of the preference data of Fig. 3.20.

(c) Note that there is exactly one "winner" in a transitive triple. Show that the number of transitive triples in which i is a winner is exactly $\binom{s(i)}{2}$. (Recall that $\binom{r}{s}$ is 0 if $r < s$.) Conclude that (3) holds.

17. Using the definitions of Exer. 16, show that every tournament of at least four vertices has at least one transitive triple. (*Hint:* Note that by convention, $\binom{s(i)}{2} = 0$ iff $s(i) < 2$.)

18. Suppose the vertices of a tournament can be listed as $u_1, u_2, \ldots, u_n$ so that $s(u_i) = n - i$. Does it follow that the tournament has a unique complete simple path? (Give proof or counterexample.)

19. A travelling salesman wants to take only nonstop flights, visit each city on his itinerary exactly once, and return to his starting point. He is looking for a *hamiltonian cycle*, a complete cycle in an appropriate digraph.

(a) Find an itinerary and a set of direct air routes for which there is no such cycle.

(b) Find an itinerary and a set of direct air routes for which there is no hamiltonian cycle, but for which there is a *hamiltonian path*, a complete simple path.

Note: A generalization of the problem described here is to find a hamiltonian cycle which minimizes the total distance travelled. This problem is often called the *travelling salesman problem*. In general, there is no known (efficient) procedure for finding a solution to this problem. See Lin [1965] and Held and Karp [1970] for a discussion.

20. (Berge [1962], Johnson [1954], Liu [1972].) Many scheduling problems in operations research involve finding the optimal order in which to perform a certain number of operations. Often, such problems can be solved by finding complete simple or hamiltonian paths. As an example, suppose a printer has n different books to publish. He has two machines, a printing machine and a binding machine. A book must be printed before it can be bound. Thus, time is lost only when the binding machine is idle. Let p_k be the time required to print the k-th book and b_k the time required to bind the k-th book. Assume that for all i and j, either $p_i \leq b_j$ or $p_j \leq b_i$. Show that it is possible to find an ordering of books so that if books are printed and bound in that order, the binding machine will be kept busy without idle time after the first book is printed. *Hint*: draw a digraph with an arc from i to j if and only if $b_i \geq p_j$. (More general treatment of this problem can be found in Johnson [1954].)

*21. (Camion [1959] and Foulkes [1960].) Prove that every strongly connected tournament has a complete (i.e., hamiltonian) cycle. *Hint*: Show that there is a cycle of length k, for $k = 3, 4, \ldots, n$, where n is the number of vertices.

22. Apply the ideas of this section to some sports tournament of your choice. Be careful to choose a tournament which is a tournament in the technical sense we have been using.

23. Does preference always define a tournament? Why? (Think about the digraph of Fig. 2.6(b). Is it a tournament?)

3.3. Orientability and Vulnerability

In this section we return to the problem of traffic flow discussed in Chapter 1. We supply a formal proof of the result stated there and an efficient procedure for finding a one-way street assignment. Then we discuss several related problems in graph theory which have a variety of interesting applications.

3.3.1. TRAFFIC FLOW

The traffic flow problem was the following: Given a network of two-way streets in a city, under what circumstances can we m ke each street one-way in such a way that it is possible to travel from any location to any other? Translating the problem into graph theory, we draw a graph G by taking as vertices the locations in the city and joining two locations by an edge if and only if they are joined by a two-way street. Then we ask: Can we give each edge a direction, or an orientation, in such a way that in the resulting directed graph (V, A) each vertex u can reach each other vertex v? We call A an *orientation* of G. Restated, our question asks: Under what circumstances does the graph G have a strongly connected orientation? To give the answer, we introduced the notion of a *bridge* in a connected graph. This is an edge $\{u, v\}$ with the following property: Removal of edge $\{u, v\}$ but not vertices u and v results in a disconnected graph. We now formalize the main theorem on one-way street assignments. We provide a proof below.

Theorem 3.5 (Robbins [1939]). A graph G has a strongly connected orientation if and only if G is connected and has no bridges.

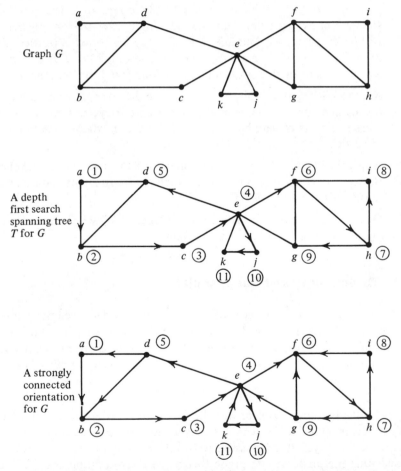

FIGURE 3.21. The one-way street algorithm. The labelling of vertices is shown by circled numbers.

If $G = (V, E)$ is connected and has no bridges, there is a constructive procedure for finding a strongly connected orientation.[10] In performing this construction, start by picking a vertex r, labelling it 1, and calling it the *root*. Since G is connected, there is a vertex joined to r by an edge. Pick any such vertex, label it 2, and orient the edge $\{1, 2\}$ from 1 to 2. If there is any vertex different from 1 and 2 and joined to 2 by an edge, pick such a vertex, label it 3, and orient the edge $\{2, 3\}$ from 2 to 3. If there is no such vertex, find any other vertex joined to vertex 1 by an edge, label this vertex 3, and orient the edge $\{1, 3\}$ from 1 to 3. In general, having labelled vertices $1, 2, \ldots, k$, and oriented some of the edges joining these vertices, continue as follows. Let j be the highest number such that there is an edge in G from the vertex labelled j to an unlabelled vertex. Since G is connected, there must be such a j. (Why?) Pick an unlabelled vertex joined to j by an edge, label it $k + 1$, and orient the edge $\{j, k + 1\}$ from j to $k + 1$. This procedure continues until all vertices have been labelled. The oriented subgraph of G obtained by this procedure is called a *depth first search spanning tree* T.[11,12] We illustrate the procedure in Fig. 3.21. Vertex a is chosen as the root and labelled 1. Then vertex b is labelled 2, and arc (a, b) is included in the depth first search spanning tree. Similarly, c is labelled 3, and arc (b, c) is included, e is labelled 4, and arc (c, e) is included, and then d is labelled 5, and arc (e, d) is included. After vertex d is labelled 5, the only vertices joined to d by edges, a, b, and e, have already been labelled. So we go back to e, which was labelled 4, and continue labelling from there. After having obtained a depth first search spanning tree of G, we complete the orientation of G by orienting all remaining edges from higher numbers to lower numbers. The orientation obtained this way is also shown in Fig. 3.21. For example, edge $\{f, i\}$ is oriented from i to f since 8 is greater than 6. The orientation is easily seen to be strongly connected for this example. We shall prove below that it always is.

Remark: The method of depth first search provides an efficient algorithm for determining whether or not a graph G is connected. For if G is not connected, the procedure described for finding a depth first search spanning tree must stop before labelling all the vertices of G. Conversely, if G is connected, the procedure labels all the vertices before stopping. A similar *directed depth first search method* can be defined for digraphs D. We simply label as

---

[10]The idea behind this procedure was suggested to the author by J. Ullman (personal communication). Tarjan [1974] has published a procedure for finding bridges, which could be used prior to this procedure.

[11]The *depth first search procedure* described here gives extremely powerful algorithms for studying graphs. See Tarjan [1972], Hopcroft and Tarjan [1973], and Aho, *et al.* [1974, Ch. 5] for a discussion.

[12]Technically, T is an oriented tree and the undirected graph obtained from T by disregarding direction of arcs is a tree. (Cf. Exer. 22, Sec. 2.2, Exer. 17, Sec. 2.3, and Exer. 20 below.)

before, but only go from a given vertex to vertices reachable from it by arcs, not edges. This method can be used to provide an efficient test for strong connectedness. For details, the reader is referred to Tarjan [1972], Hopcroft and Tarjan [1973], or Aho, *et al.* [1974, Ch. 5]. (See also Exer. 19.)

The algorithm described can, unfortunately, lead to some very inefficient one-way street assignments. We showed in Fig. 1.9 a rather inefficient one-way street assignment. This one-way assignment is attainable by our procedure (Exer. 9). As we discussed in Chapter 1, if a person at location *a* wants to reach location *b*, he used to be able to do it very quickly, but now he has to travel in a roundabout way. In short, this assignment meets the criteria we have set up, namely it gives a strongly connected orientation; but it does not give an efficient one. Here, another round of mathematical modelling is called for. The concept of "efficient" one-way street assignment needs to be defined, and algorithms or procedures for finding efficient one-way street assignments need to be discovered. The reader might like to pursue this second round for himself. (Exercise 33 provides some guidance.)

We complete this subsection by proving Theorem 3.5.[13] We first suppose that in the graph $G = (V, E)$, there is an orientation A of E such that (V, A) is strongly connected. Then it is straightforward that $G = (V, E)$ is connected. We show that G can have no bridges. We shall need the following lemma, whose proof is left to the reader (Exer. 18).

Lemma 1. If G is a connected graph, then edge $\{u, v\}$ is a bridge of G if and only if every chain from u to v in G includes edge $\{u, v\}$.

To prove that G can have no bridges, suppose $\{u, v\}$ is a bridge. Since (V, A) is an orientation, either (u, v) or (v, u) is in A. Without loss of generality, we shall assume that (u, v) is in A. Since (V, A) is strongly connected, there is a simple path from v to u in (V, A). By Lemma 1, every such path must use the arc (u, v) or the arc (v, u). Since (V, A) is an orientation, (v, u) is not in A, so every simple path from v to u must use the arc (u, v). But there cannot be a simple path from v to u which uses the arc (u, v). Thus, we have reached a contradiction. We conclude that G can have no bridges.

We next show that for a connected bridgeless graph $G = (V, E)$, the construction procedure we have described does indeed give rise to a strongly connected orientation (V, A).[14] Let $l(u)$ be the label assigned to vertex u. By the above *Remark*, since G is connected, every vertex receives a label. Let r be the vertex receiving label 1. We have called r the *root*. We shall prove that the orientation (V, A) of G is strongly connected by proving that every vertex

[13]The reader may skip the rest of this subsection.

[14]The author thanks Prof. T. Ostrand and others for helping to improve the following proof.

in V is reachable from the root and every vertex reaches the root. T will denote the (oriented) depth first search spanning tree.

Lemma 2.   Every vertex x of V is reachable in T from the root r.

Proof.   We argue by induction on $l(x)$. If $l(x) = 1$, then x is r, and we know that every vertex is reachable from itself. Suppose the lemma is true for $l(x) < t$ and suppose $l(x) = t$. Then there is an arc (y, x) in the tree T for some vertex y. Now $l(y) < l(x)$, so by the induction assumption, y is reachable from r in T. Hence, so is x.   Q.E.D.

For convenience, let us introduce the terminology *forward arc* for an arc (u, v) in T and *backward arc* for an arc (u, v) in A which is not in T.

Lemma 3.   Suppose (u, v) is a backward arc. Then u is reachable from v in T, i.e., u is reachable from v by a path of forward arcs.

Proof.   We actually prove the following stronger statement: Whenever $l(v) < l(w) \leq l(u)$, then w is reachable from v in T. The proof is by induction on $l(w)$. If $l(w) = l(v) + 1$, then since there is an edge $\{u, v\}$ in G with $l(u) > l(v)$, it follows that arc (v, w) was added to T. Thus, w is reachable from v in T. Suppose that the lemma is true for vertices with label less than t, and suppose $l(w) = t$. There is a vertex x with $l(x) < l(w)$ so that (x, w) is in T. Since there is an edge $\{u, v\}$ in G with $l(u) \geq l(w) > l(v)$, it follows that $l(x) \geq l(v)$. Now $l(x) > l(v)$, for otherwise $(v, w) \in T$ and w is reachable from v. Thus, the induction assumption applies to x, and we conclude that x is reachable from v in T. Thus, since (x, w) is in T, w is also reachable from v in T.   Q.E.D.

Lemma 4.   If x is reachable from both a and b in T, then either a is reachable from b in T or b is reachable from a in T.

Proof.   Each vertex has only one incoming arc in T. We argue by induction on the length of the path from a to x in T. If (a, x) is in T, then since (a, x) is the only arc heading into x in T, it follows that the path from b to x uses this arc. Thus, b reaches a. Arguing by induction, we know that there is an arc (w, x) in T, for some w. Now the paths a to x and b to x both must use the arc (w, x). Thus, a and b both reach w in T. By inductive assumption, either a is reachable from b or b is reachable from a.   Q.E.D.

It is a corollary of the above proof and of Lemma 2 that for every vertex x in V, there is a unique path in T from the root r to x. We can define the *depth* $d(x)$ of x as the length of this path.

Lemma 5.   If $d(x) \geq 1$, then there is a path in (V, A) from x to a vertex y with $d(y) < d(x)$.

Proof.   Since $d(x) \geqq 1$, $x \neq r$. Thus, there is an arc (w, x) in T. Let

$$R = \{u\colon u \text{ is reachable from } x \text{ in } T\}.$$

We shall show that there must be a backward arc (u, v) with $u \in R$ and $v \notin R$. To do so, we shall assume the contrary and show that $\{w, x\}$ is the only edge joining a vertex outside R to a vertex inside R. This implies that $\{w, x\}$ is a bridge, which is a contradiction. Suppose $\{v, u\}$ is an edge of G with v outside R and u inside R. Then in (V, A), either (u, v) or (v, u) is an arc. By assumption (u, v) could not be backward. But it could not be forward either, for otherwise u in R would imply v in R. Now (v, u) could not be backward. For if it were, then by Lemma 3, v is reachable from u in T, and v would be in R. Finally, suppose (v, u) is forward. Now both v and x reach u in T. Moreover, x does not reach v since v is not in R. Thus, by Lemma 4, v reaches x in T. Since the path in T from v to u is unique, and since v reaches x in T and x reaches u, it follows that v is w and u is x. Thus, $\{w, x\}$ is the only edge of G from a vertex outside R to a vertex inside R.

Suppose that (u, v) is a backward arc with u in R and v not in R. By Lemma 3, v reaches u in T. Now x and v both reach u in T. Moreover, x doesn't reach v in T since v is not in R. By Lemma 4, we conclude that v reaches x in T. Thus, $d(v) < d(x)$. Now x reaches u in T and u reaches v in A, so x reaches v in A. Pick y to be v, and the lemma is proved.   Q.E.D.

Lemma 6.   The root r is reachable in (V, A) from every vertex x.

Proof.   The proof is by induction on $d(x)$. If $d(x) = 0$, then x is r, and we know that a vertex is reachable from itself. Suppose the lemma is true for all vertices with depth less than t, and suppose $d(x) = t$. By Lemma 5, there is a path from x to a vertex y with $d(y) < t$. By inductive assumption, r is reachable from y.   Q.E.D.

Lemmas 2 and 6 complete the proof that our construction procedure gives rise to a strongly connected orientation.

3.3.2. VULNERABILITY

A bridge in a transportation network can be thought of as a vulnerable connection, one whose removal or blockage disrupts transportation. A similar concept makes sense for communication networks. Suppose a digraph $D = (V, A)$ represents a communication network and D is strongly connected. Let us say that D is *arc-vulnerable* if the removal of some arc (but not the two vertices adjoining it) results in a digraph which is no longer strongly connected. For example, the cycle D_1 of Fig. 3.22 is arc-vulnerable, for removal of any arc results in a less than strongly connected digraph.

More generally, if D is in connectedness category $i,$ we shall say that (u, v) is an i, j-*arc* if removal of (u, v) results in a digraph of connectedness category $j.$ If D has an (i, j)-arc for some $j < i,$ we say that D is *arc-vulnerable*. Exercise 12 asks the reader to give examples of i, j-arcs, whenever possible.

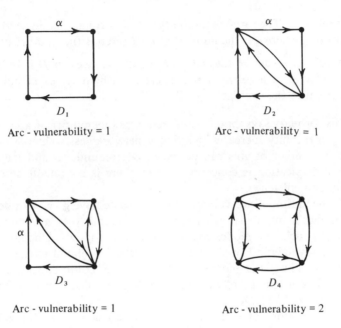

FIGURE 3.22. Some digraphs and their arc-vulnerabilities.

There are degrees of vulnerability. In a cycle, removal of any arc decreases the connectedness. Sometimes, it is necessary to remove more than one arc. Let us define the *arc-vulnerability* of a digraph in connectedness categories 1, 2, or 3 as the minimum number of arcs whose removal results in a digraph of lower connectedness. (The reader should thus observe that higher arc-vulnerability means that the network is less vulnerable. Also note that even a digraph which is not arc-vulnerable has an arc-vulnerability. The arc-vulnerability of a digraph in category 0 is undefined.)

Figure 3.22 shows four strongly connected digraphs. D_1, D_2, and D_3 have arc-vulnerability equal to 1, for removal of the arc designated α results in each case in a nonstrongly connected digraph. The fourth digraph of Fig. 3.22, D_4, has arc-vulnerability equal to 2. The arc-vulnerability of a strongly connected digraph is related to the minimum indegree of a vertex. (The *indegree* of vertex u is the number of arcs leading into $u.$)

Theorem 3.6. If D is a strongly connected digraph, then the arc-vulnerability of D is less than or equal to the minimum indegree of a vertex of D.[15]

Proof. Remove all arcs leading into the vertex of minimum indegree.

Q.E.D.

Corollary. The arc-vulnerability of a strongly connected digraph is at most a/n, where a is the number of arcs and n is the number of vertices.

Proof. The sum of the indegrees of the vertices in D is a. Thus, the average indegree of a vertex is a/n. There must be a vertex of indegree less than or equal to the average.

Q.E.D.

This corollary says that, in some sense, as the number of arcs increases, the arc-vulnerability increases (making a network less vulnerable). But an increase in number of arcs can be expensive, redundant, and inefficient as far as communication is concerned. Thus, there is a tradeoff between vulnerability and efficiency.

The definition of arc-vulnerability which we have given can be thought of as defining a mathematical model of vulnerability of a communication network. As such, the model could be tested as follows. Under reasonable assumptions, vulnerability of a network should be related to the number of breakdowns. Thus, our model predicts that a communication network designed like digraph D_1 of Fig. 3.22 would break down more often than a network designed like D_4.

Similar notions of vertex-vulnerability can be defined for graphs and digraphs. These will be pursued in the exercises.

3.3.3. TRANSITIVE ORIENTABILITY

Let us recall from the study of tournaments that a digraph (V, A) is called *transitive* if and only if it satisfies the following condition for all $u, v, w \in V$ with $u \neq w$:[16]

$$(u, v) \in A \quad \text{and} \quad (v, w) \in A \quad \text{imply} \quad (u, w) \in A. \tag{4}$$

Figure 3.23 shows six digraphs, of which the first four are transitive. Digraphs D_5 and D_6 are not transitive because arc (u, w) is missing in each.

In this subsection, we return to the notion of orientation of a graph and study graphs which have a transitive orientation. (A graph which is transitively orientable is sometimes called a *comparability graph*.) Figure 3.24 shows a transitive orientation for the graph Z_3, the circuit of length 3. (From now on, we shall use the notation Z_n for the circuit of length n.) Figure

---

[15]The same result holds for *outdegree*, the number of arcs leading out of a vertex.

[16]See footnote 9.

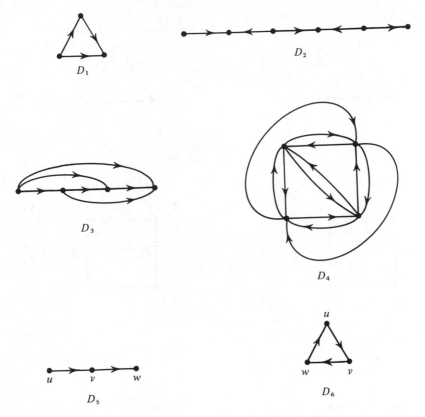

FIGURE 3.23. D_1, D_2, D_3, D_4 are transitive digraphs; D_5, D_6 are not.

3.24 also shows a transitive orientation for the graph Z_4 and a transitive orientation for the other graphs shown.

    If a digraph is transitive, then for all u and v such that u can reach v, we have $d(u, v) = 1$, where d is the distance measure defined as the length of the shortest path from u to v (if such a path exists). Hence, one-way street orientations which are transitive can be thought of as highly efficient. Unfortunately, the only graph which has an orientation which is at the same time transitive and strongly connected is the single vertex. (See Exer. 26.) The graphs which have transitive orientations are thus not very important in traffic flow. However, they will be very important in the next section, where we apply them to numerous problems.

    Graph G_1 of Fig. 3.25, the circuit Z_5 of length 5, has no transitive orientation. To prove that there is no such orientation, let us proceed by way of contradiction. The progressive steps of this argument are shown in Fig. 3.26. Supposing there is such an orientation A, we may by the symmetries of the situation assume that $(u, v) \in A$. By transitivity, $(v, w) \in A$

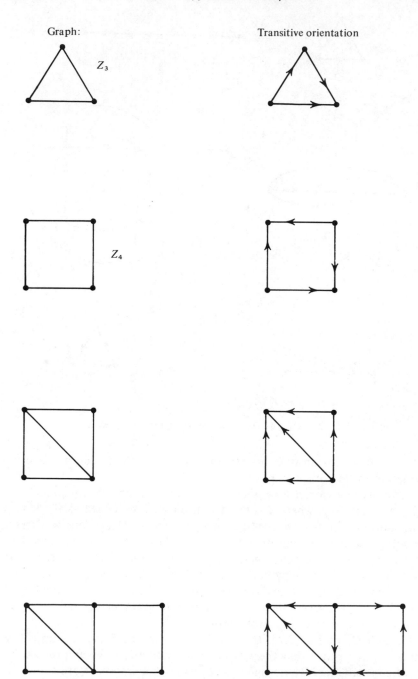

Graph: Transitive orientation

Z_3

Z_4

FIGURE 3.24. Some transitively orientable graphs and transitive orientations for them.

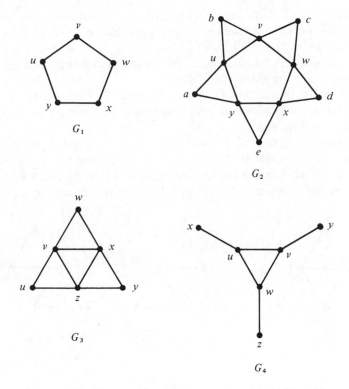

FIGURE 3.25. Some graphs which are not transitively orientable.

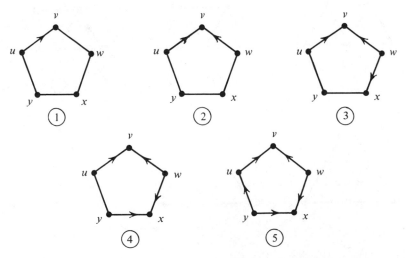

FIGURE 3.26. Proof that graph G_1 of Fig. 3.25 has no transitive orientation.

would imply $(u, w) \in A$. But $\{u, w\} \notin E(G_1)$ so (u, w) cannot be in A. We conclude that $(w, v) \in A$. Similarly, one argues that $(w, x) \in A$, that $(y, x) \in A$, and that $(y, u) \in A$. But now $(y, u) \in A$ and $(u, v) \in A$ imply $(y, v) \in A$, so $\{y, v\} \in E(G_1)$, which is false. We conclude that there is no transitive orientation A for the graph G_1.

The reader should now be able to convince himself that of the circuits Z_n of length n, those of odd length greater than 3 are not transitively orientable, and all others are.

In general, if G is transitively orientable, then every generated subgraph H of G must also be transitively orientable. For a transitive orientation of G would give a transitive orientation of H. It follows from this observation that the graph G_2 of Fig. 3.25 is not transitively orientable. For

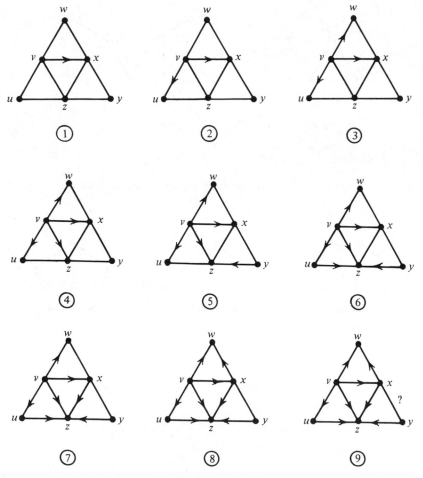

FIGURE 3.27. Proof that graph G_3 of Fig. 3.25 has no transitive orientation.

the subgraph generated by the set of vertices $\{u, v, w, x, y\}$ is Z_5, and any transitive orientation of G_2 would give a transitive orientation for Z_5. A similar argument implies that in general a transitively orientable graph cannot have Z_5 as a generated subgraph, and indeed that it cannot have any circuit of odd length greater than 3 as a generated subgraph.

We have shown that a transitively orientable graph G cannot have any Z_n with n odd and greater than 3 as a generated subgraph. However, this condition is not sufficient for G to be transitively orientable. Graph G_3 of Fig. 3.25 is not transitively orientable, and it does not contain any Z_n with n odd and greater than 3 as a generated subgraph. To show G_3 is not transitively orientable, suppose a transitive orientation A exists. This must orient the edge $\{v, x\}$. By symmetry, we can suppose the orientation goes from v to x. Figure 3.27 shows the successive steps of the following argument. If $(v, x) \in A$, then we know $(v, u) \in A$, for otherwise transitivity would imply $(u, x) \in A$, or $\{u, x\} \in E = E(G_3)$, which is not true. The rest of the arcs in Fig. 3.27 are added in successive steps, each step forced by the assumption of transitivity. In the end, we are left to decide on the orientation of the edge $\{x, y\}$. If $(y, x) \in A$, then transitivity implies $(y, w) \in A$, so $\{y, w\} \in E$. If $(x, y) \in A$, then transitivity implies $(v, y) \in A$, so $\{v, y\} \in E$. We conclude that no transitive orientation could exist.

A similar argument shows that the graph G_4 of Fig. 3.25 has no transitive orientation. We leave this argument to the reader.

To summarize, we have described two ways to prove that a graph G has no transitive orientation. First, it is sufficient to show that G has as a generated subgraph a known nontransitively orientable graph. Second, orient one arc arbitrarily and then orient others as forced by transitivity until a contradiction is reached. Similarly, it is often possible to find a transitive orientation this way. However, in some cases the proofs can get complicated and one would like to have a criterion for a graph to be transitively orientable. Such a criterion was first developed by Ghouila-Houri [1962] and Gilmore and Hoffman [1964]. We present this criterion in Exers. 27 and 28. Gilmore and Hoffman also present an algorithm for finding a transitive orientation if one exists. The papers by Gallai [1967], Pnueli, Lempel, and Even [1971], and Golumbic [1975] give various interesting results about transitive orientability.

EXERCISES

1. In each graph of Fig. 3.28, find all bridges.

2. In each graph of Fig. 3.25, find all bridges.

3. For each digraph of Fig. 3.29, determine its arc-vulnerability.

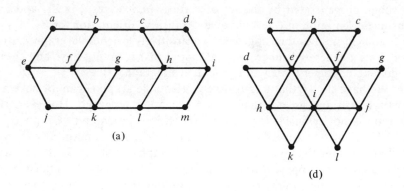

(a)

(d)

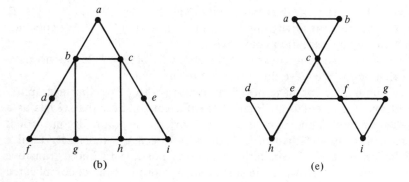

(b)

(e)

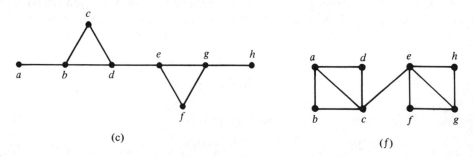

(c)

(f)

FIGURE 3.28. Graphs for exercises of Sec. 3.3.

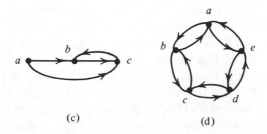

FIGURE 3.29. Digraphs for exercises of Sec. 3.3.

4. Let G be a connected graph. A *cut vertex* of G is a vertex with the following property: When you remove it and all edges to which it belongs, the result is a disconnected graph. In each graph of Fig. 3.28, find all cut vertices.

5. Which of the digraphs of Fig. 3.29 is transitive?

6. Give an example of a digraph D with arc-vulnerability equal to 4.

7. For every k, give an example of a digraph D with arc-vulnerability equal to k.

8. Apply the procedure described in the text for finding one-way street assignments to the graphs of Fig. 3.28, provided such assignments exist.

9. Show that the one-way street assignment shown in Fig. 1.9 is attainable using the algorithm described in the text.

10. Give an example of a digraph with n vertices and arc-vulnerability equal to $n - 1$. Could there be a strongly connected digraph with n vertices and arc-vulnerability equal to n?

11. Determine which of the graphs of Fig. 3.28 are transitively orientable. For those which are, supply a transitive orientation and for those which are not, explain why.

12. For each pair (i, j), $i = 0, 1, 2, 3$ and $j = 0, 1, 2, 3$, either give an example of a digraph with an i, j-arc or prove there is no such digraph.

13. A vertex u in a digraph D is called an i, j-vertex if D is in connectedness category i and $D - u$ is in connectedness category j. ($D - u$ is the di-

graph generated by all vertices of D except u.) For every pair i, j with i running from 0 to 3 and j running from 0 to 3, either give an example of a digraph with an i, j-vertex or prove that there is no such digraph.

14. Give an example of a tournament which is strongly connected and which has a 3, 3-vertex. (See Exer. 13.)

15. Is it possible to have a 3, 1-vertex in a tournament? Why?

16. (Whitney [1932].) If there is a set of vertices in a digraph D whose removal results in a digraph in a lower connectedness category than D, then we define the *vertex vulnerability* of D to be the size of the smallest such set of vertices. Otherwise, vertex vulnerability is undefined.

    (a) Show that if D is weakly connected and there is at least one pair u, v such that neither (u, v) nor (v, u) is an arc of D, then the vertex vulnerability of D is less than or equal to the minimum total degree of any vertex of D. (The *total degree* of a vertex x is the sum of the indegree (the number of incoming arcs) of x and the outdegree (the number of outgoing arcs) of x).

    (b) Show that the vertex vulnerability and the minimum total degree can be different.

    (c) Why do we need to assume that there is at least one pair u, v such that neither (u, v) nor (v, u) is an arc of D?

*17. What are the maximum and minimum number of arcs in a strongly connected digraph with n vertices which is arc-vulnerable at *every* arc?

18. Prove Lemma 1 stated in the proof of Theorem 3.5, namely that an edge $\{u, v\}$ in a connected graph G is a bridge if and only if every chain from u to v in G includes edge $\{u, v\}$.

19. Is the following test for strong connectedness of a digraph D correct? Pick any vertex of the digraph as a root and perform a *directed* depth first search (as described in the Remark on page 95) until no more vertices can be labelled. Then D is strongly connected if and only if all vertices have received a label. If this is not a correct test, how can it be modified?

20. Recall that a graph G is a *tree* if it is connected and has no circuits (Exer. 22, Sec. 2.2). (Thus, a depth first search spanning tree as we have defined it is not a tree since it is oriented; however it is a tree in the sense that if directions on edges are disregarded, it is connected and has no circuits.) Suppose G is a connected graph and H is a spanning tree (Exer. 17, Sec. 2.3). Is some orientation of H a depth first search spanning tree of G? (Give proof or counterexample.)

21. A graph G is called a *forest* if every connected component is a tree (Exer. 20).

(a) If a forest G has n vertices and k components, how many edges does it have? Why? (*Hint*: Use Exer. 22(c), Sec.2.2.)

(b) Suppose none of the connected components of a forest has a bridge. If there are n vertices, what is the maximum number of edges of such a forest? (*Hint*: What does each connected component look like?)

22. What is the maximum number of cut vertices in a connected graph with five vertices? With n vertices? (See Exer. 4 for the definition of cut vertex.)

23. Suppose D is a digraph. Its *transitive closure* is the minimal transitive digraph D^t containing D and having the same set of vertices as D. Find the transitive closures of the digraphs of Fig. 3.29. (*Hint*: Add arcs as forced by transitivity until you achieve a transitive digraph.)

24. (Harary, Norman, and Cartwright [1965].) Prove that a digraph D is strongly connected if and only if its transitive closure has all possible arcs in it.

25. Suppose D^t is the transitive closure of a digraph D and $A(D^t)$ is the adjacency matrix of D^t. Can you express $A(D^t)$ in terms of any of the common matrices related to D which were studied in Sec. 2.4?

26. Show that the only graph which has a strongly connected transitive orientation is the single vertex.

27. The next two exercises present the Ghouila-Houri and Gilmore-Hoffman characterization of transitively orientable graphs. Suppose G is a graph and $u_1, u_2, \ldots, u_{t-1}, u_t, u_1$ is a closed chain of G. A *chord* in this closed chain is an edge $\{u_i, u_j\}$ with $i \neq j$ which appears as an edge of G. A *triangular chord* is a chord $\{u_i, u_j\}$ such that i and j differ by 2. In counting how much i and j differ, we say $t - 1$ and 1 differ by 2 and t and 2 differ by 2. That is, we count modulo t. In graph (b) of Fig. 3.28, the edge $\{b, g\}$ is a chord in the closed chain $f, d, b, a, c, e, i, h, c, h, g, f$. So is the edge $\{i, h\}$, counted as an edge between the 7th vertex on this chain and the 10th. The edge $\{b, c\}$ is a triangular chord.

(a) In graph (a) of Fig. 3.28, find all chords and triangular chords in the closed chains k, f, g, h, i, m, l, k and $h, i, m, l, h, c, d, i, m, l, h$.

(b) Show that if G is transitively orientable, then every *circuit* of odd length greater than 3 must have a chord.

(c) Show that if G is transitively orientable, then every circuit of odd length greater than 3 must have a triangular chord.

(d) Show that every circuit of graph G_3 of Fig. 3.25 has a triangular chord. (However, G is not transitively orientable.)

28. A closed chain $u_1, u_2, \ldots, u_{t-1}, u_t, u_1$ in a graph G can be thought
of as a closed path in the digraph $D(G)$ obtained by replacing each edge
$\{a, b\}$ of G by two arcs (a, b) and (b, a). The closed chain is called a
generalized circuit of G if the corresponding closed path in $D(G)$ has
no repeated arcs. For example, in graph G_3 of Fig. 3.25, a generalized
circuit of length 5 is given by u, v, x, v, z, u. However, the closed chain
$u, v, x, v, z, y, x, v, z, u$ is not a generalized circuit, since arc (x, v)
is used twice. Ghouila-Houri [1962] and Gilmore and Hoffman [1964]
discovered the following theorem: A graph G is transitively orientable
if and only if every generalized circuit of G of odd length greater than 3
has a triangular chord.

(a) In graph G_3 of Fig. 3.25, show that

$$v, w, v, u, v, x, y, x, w, x, z, u, z, y, z, v$$

is a generalized circuit of length 15 with no triangular chords.

(b) In graph G_4 of Fig. 3.25, show that

$$x, u, v, y, v, w, z, w, u, x$$

is a generalized circuit of length 9 with no triangular chords.

(c) In each graph of Fig. 3.28 which is not transitively orientable, find
a generalized circuit of odd length which has no triangular chords.

29. Discuss the arc-vulnerability of the communication network of Fig.
2.3. How could this network be made less vulnerable?

30. Design a very vulnerable transportation network, i.e., one with many
cut vertices (Exer. 4).

31. *Project*: Using your results in Exer. 18, Sec. 2.3, develop a theory of
vulnerability in the more general communication networks which allow
for the possibility (probability) that a message will not be passed along
a given arc.

32. What are some oversimplifications other than omission of probabilities
in our treatment of vulnerability? How could our model be extended
to take account of these?

33. *Project*: As we pointed out in the text, some one-way street assignments
can be very inefficient. Develop the theory of efficient one-way street
assignments.

(a) One possible idea is to measure efficiency of a one-way street
assignment by calculating the *diameter* of the resulting strongly
connected digraph, that is, the maximum of $d(u, v)$ for all u, v.
Show on several examples of your choice some efficient and inefficient one-way street assignments, if this notion of efficiency is
used. ($d(u, v)$ is the length of the shortest path from u to v.)

(b) Comment on the efficiency of the orientation which alternates
one-way streets, if the original graph is a street grid of East-West
avenues and North-South streets.

(c)   Comment on whether diameter is a good measure of efficiency and offer some alternatives.

(d)   For each measure of efficiency, develop algorithms for finding efficient one-way street assignments.

34.   As was pointed out in Chapter 1, some one-way street assignments are highly inefficient on purpose.

(a)   Use diameter (Exer. 33) to measure the efficiency of the one-way street assignment for Yosemite National Park shown in Fig. 1.10.

(b)   Find algorithms for obtaining inefficient one-way street assignments.

3.4.   Intersection Graphs

3.4.1. DEFINITION OF AN INTERSECTION GRAPH

In this section we consider a problem which has a large number of applications. Suppose $\mathcal{F}$ is a family of sets. Let us define a graph G, the *intersection graph* of $\mathcal{F}$, as follows: $V(G) = \mathcal{F}$ and if $S, T \in \mathcal{F}$ with $S \neq T$, then

$$\{S, T\} \in E(G) \Longleftrightarrow S \cap T \neq \emptyset. \tag{5}$$

Figure 3.30 shows the intersection graph of the family $\mathcal{F} = \{S_1, S_2, \ldots, S_6\}$, with $S_1, S_2, \ldots, S_6$ defined as follows:

$$S_1 = \{1, 2, 3, 4\}$$
$$S_2 = \{1, 3, 6\}$$
$$S_3 = \{7, 8, 9\}$$
$$S_4 = \{3, 8, 11\}$$
$$S_5 = \{6, 7, 8, 9, 10\}$$
$$S_6 = \{1, 5\}.$$

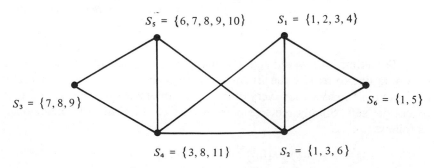

FIGURE 3.30.   An intersection graph.

The interesting question is usually the opposite: Given a graph G, does it arise as the intersection graph of a family of sets that has a certain property? The question is equivalent to the following: Given a graph $G = (V, E)$, can we find a function S from V into a family of sets with a certain property such that for all $u \neq v \in V$,

$$\{u, v\} \in E \longleftrightarrow S(u) \cap S(v) \neq \emptyset ? \tag{6}$$

If we simply ask whether a given G is the intersection graph of *some* family of sets, the answer is always yes. We have the following theorem.

Theorem 3.7 (Marczewski [1945]). Every graph $G = (V, E)$ is the intersection graph of some family of sets.

Proof. For each vertex u in G, let

$$S(u) = \{\{u, v\} : \{u, v\} \in E\}.$$

Let $\mathcal{F} = \{S(u) : u \in V\}$. Then S satisfies (6) for all $u \neq v \in V$.   Q.E.D.

For example, let G be the graph of Fig. 3.31. We have

$$S(a) = \{\{a, b\}, \{a, c\}, \{a, d\}\}$$
$$S(b) = \{\{a, b\}, \{b, c\}\}$$
$$S(c) = \{\{a, c\}, \{b, c\}, \{c, d\}\}$$
$$S(d) = \{\{a, d\}, \{c, d\}, \{d, e\}\}$$
$$S(e) = \{\{d, e\}\}.$$

Then, for example, we see that $\{d, e\} \in E$ and that $S(d) \cap S(e) \neq \emptyset$, since $\{d, e\} \in S(d) \cap S(e)$. However, $\{d, b\} \notin E$ and $S(d) \cap S(b) = \emptyset$.

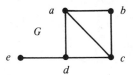

G

FIGURE 3.31. G is an intersection graph.

(Sometimes we would like the mapping S to be one-to-one, that is, we would like to require the sets assigned to each vertex to be distinct. It is easy enough to show that every graph is the intersection graph of a family of distinct sets. Simply redefine S from that of the proof of Theorem 3.7 as follows:

$$S(u) = \{\{u\}\} \cup \{\{u, v\} : \{u, v\} \in E\}.)$$

3.4.2. INTERVAL GRAPHS AND THEIR
APPLICATIONS[17]

By Marczewski's Theorem, the problem of studying what graphs are intersection graphs is uninteresting: All are. However, one gets very interesting problems, with many applications, by asking what graphs are intersection graphs of specific families of sets.

The families of sets most often considered in connection with intersection graphs have been the families of intervals on the real line. We shall concentrate on this family in the next three subsections and then study other families.

Let us say a graph G is an *interval graph* if it is the intersection graph of some family of intervals on the line.[18] That is, $G = (V, E)$ is an interval graph if and only if there is an assignment J of a real interval $J(u)$ to each $u \in V$ such that for all $u \neq v \in V$,

$$\{u, v\} \in E \Longleftrightarrow J(u) \cap J(v) \neq \emptyset. \tag{7}$$

(We use the letter J rather than the letter S here for ease of comparison with some of the work of measurement theory in Chapter 8, where J is motivated by the notion of just-noticeable-difference in psychology.)[19]

Figure 3.32 shows examples of interval graphs. To see that they are interval graphs, we also show, in Fig. 3.32, *interval assignments J* satisfying Eq. (7).

Not every graph is an interval graph. The circuit of length 4, Z_4, as shown in Fig. 3.33, is not. To prove that, suppose there is an interval assignment J satisfying Eq. (7). Then $J(u)$ and $J(v)$ must overlap. By the symmetry of the situation, we might as well take $J(v)$ a bit to the right of $J(u)$. (Why can't one of $J(u)$ and $J(v)$ be inside the other?) Now $J(w)$ must be to the right of $J(v)$, for it overlaps $J(v)$ and not $J(u)$. (See Fig. 3.33.) Similarly, $J(x)$ is to the right of $J(w)$, as shown in Fig. 3.33. We conclude $J(x)$ and $J(u)$ do not overlap, which contradicts $\{u, x\} \in E$.

Interval graphs arose in connection with a problem in genetics called Benzer's problem. Classically, geneticists have treated the chromosome as a linear arrangement of genes. Benzer [1959] was interested in whether the same was true for the fine structure inside the gene.[20] To study this fine

---

[17]The reader is referred to a forthcoming article by Victor Klee for a good summary of the applications of interval graphs.

[18]The intervals may be open or closed., i.e., of the form (a, b) or $[a, b]$, for $a, b \in \mathcal{R}$. It is not hard to prove that we may require them all to be open or all to be closed.

[19]We do not require that J be one-to-one, though it is not hard to prove that in the case of finite graphs, if any J satisfying (7) exists, a one-to-one J does also.

[20]See Benzer [1962] for more of the biological background.

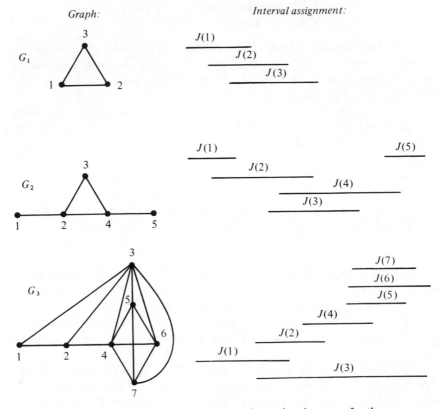

FIGURE 3.32. Some interval graphs and interval assignments for them.

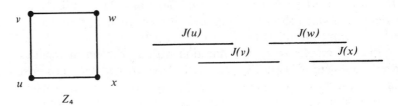

FIGURE 3.33. An interval assignment for Z_4 does not exist.

structure, which cannot be directly observed, one can study mutations, which are changes in the structure. Benzer assumed that mutations involve changes in connected substructures of the gene. By gathering mutation data, he was able to surmise whether or not two mutations (i.e., their connected substructures) overlapped. (To do this, Benzer recombined two different mutants of a given original (nonmutant) individual. In making such recombinations, one can get the original nonmutant type back provided that the mutations involve nonoverlapping substructures. Otherwise, it is inconceivable to get the original back by recombination. Thus, if he got the original back, he assumed that the mutations did not overlap. If in a large number of experiments with the two types of mutations he never got the original back, it was reasonable for him to assume that the mutations did overlap.) Given the overlap information, the problem was to decide if the mutations (substructures) studied could be part of a linear chain, i.e., was the overlap information consistent with the hypothesis that the structure in the gene is linear?

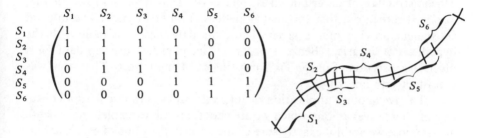

$$
\begin{array}{c@{\quad}c}
\begin{array}{c}
 \\
S_1 \\
S_2 \\
S_3 \\
S_4 \\
S_5 \\
S_6
\end{array}
&
\begin{array}{cccccc}
S_1 & S_2 & S_3 & S_4 & S_5 & S_6 \\
1 & 1 & 0 & 0 & 0 & 0 \\
1 & 1 & 1 & 1 & 0 & 0 \\
0 & 1 & 1 & 1 & 0 & 0 \\
0 & 1 & 1 & 1 & 1 & 0 \\
0 & 0 & 0 & 1 & 1 & 1 \\
0 & 0 & 0 & 0 & 1 & 1
\end{array}
\end{array}
$$

FIGURE 3.34. Overlap data and a linear chain consistent with it. (S_i and S_j overlap if and only if the i, j entry of the matrix is 1.)

For example, in Fig. 3.34 we show the overlap data from six substructures (mutations). This data is summarized in a (0, 1)-*matrix* (a matrix each of whose entries is 0 or 1) with a 1 in entry i, j if and only if observation indicates that structures S_i and S_j overlap. We give a linear arrangement compatible with the data. (That is not to say it is the only such arrangement compatible with the data; we shall mention this uniqueness problem later.) The conclusion is that the data is "consistent" with the hypothesis of linearity, rather than that it confirms linearity. For, we have made some assumptions about being able to identify overlaps. Moreover, not all substructures have been studied, so even if the given substructures can be "mapped" into substructures of a linear chain, that does not imply that the original structure is linear. In Fig. 3.35 we give a set of overlap data which is not consistent with the hypothesis of linearity; we shall show that below. In this case, assuming that we have correctly obtained overlap information, we can definitely conclude that the data refutes the linear chain hypothesis.

FIGURE 3.35. Overlap data inconsistent with the hypothesis of linearity.

$$
\begin{array}{c@{\quad}c}
\begin{array}{c}
 \\
S_1 \\
S_2 \\
S_3 \\
S_4
\end{array}
&
\begin{array}{cccc}
S_1 & S_2 & S_3 & S_4 \\
1 & 1 & 0 & 1 \\
1 & 1 & 1 & 0 \\
0 & 1 & 1 & 1 \\
1 & 0 & 1 & 1
\end{array}
\end{array}
$$

To state Benzer's problem in graph theory terms, let a graph G be
defined as follows: the vertices of G are the substructures being studied and
there is an edge between two distinct[21] substructures if and only if they are
known to overlap. Since a linear arrangement can always be straightened out
to look like a part of the real line, the problem is to decide whether or not the
vertices can be mapped into intervals in the real line in such a way that two
intervals intersect if and only if the corresponding vertices are joined by an
edge. That is, is G an interval graph? This formulation explains why the data
of Fig. 3.35 is inconsistent with the linear hypothesis: G here is Z_4.

Benzer [1959] obtained the overlap data for a small portion of the
genetic structure of a certain virus, phage $T4$. That data is shown in Fig.
3.36.[22] It turned out that this matrix defines an interval graph—an interval
assignment proving this is given. Thus, the data was consistent with the
hypothesis of linearity. (Benzer's paper has incomplete[23] overlap data for a
much larger section of phage $T4$, consisting of 145 mutations. That, too, was
consistent with linearity.)

To give another application of interval graphs, suppose V is a collection of alternatives among which we are choosing—for example, fine wines—
and suppose we are not exactly sure of the (monetary) value of each alternative. Perhaps we can assign to each u in V an interval $J(u)$ representing a
range of possible values of u. Then we would *prefer* alternative u to alternative v if the interval $J(u)$ is *strictly to the right of* the interval $J(v)$, i.e., if for
every $\alpha \in J(u)$ and $\beta \in J(v)$, $\alpha > \beta$. And we would be in doubt between
u and v, or *indifferent* between them, if their intervals $J(u)$ and $J(v)$ overlapped.
Thus, the graph which represents indifference, in this situation, is an interval
graph. The idea of preference corresponds to a directed graph. The vertices
of this digraph $D = (V, A)$ are again the elements of V, and we draw an arc
from u to v if and only if u is preferred to v. In Chapter 8, we study measurement of preference and we start with this digraph. Then we shall ask
under what circumstances there is an assignment of an interval $J(u)$ to each
u so that for all u and v in V,

$$u \text{ is preferred to } v \text{ if and only if } J(u) \text{ is strictly}$$
$$\text{to the right of } J(v),$$

or equivalently,

$$(u, v) \in A \text{ if and only if } J(u) \text{ is strictly to the} \qquad (8)$$
$$\text{right of } J(v).$$

[21]We do not take an edge from a substructure to itself, even though the substructure
overlaps with itself.

[22]Technically, Benzer used 0 where we use 1, and vice versa.

[23]Not all pairs of mutants were crossed.

Structure number	184	215	221	250	347	455	459	506	749	761	782	852	882	A103	B139	C 4	C 33	C 51	H 23
184	1	1	0	1	0	1	0	0	0	0	1	0	0	0	0	0	1	1	1
215	1	1	0	0	0	0	0	0	0	0	0	0	0	0	0	0	0	0	1
221	0	0	1	0	1	0	1	1	1	1	1	1	1	1	1	1	1	0	1
250	1	0	0	1	0	0	0	0	0	0	0	0	0	0	0	0	1	1	1
347	0	0	1	0	1	0	0	0	0	0	1	0	0	0	0	0	0	0	1
455	1	0	0	0	0	1	0	0	0	0	0	0	0	0	0	0	0	0	1
459	0	0	1	0	0	0	1	0	1	1	1	1	0	0	0	1	0	0	1
506	0	0	1	0	0	0	0	1	0	0	1	0	0	0	0	0	0	0	1
749	0	0	1	0	0	0	1	0	1	1	1	1	0	0	0	1	0	0	1
761	0	0	1	0	0	0	1	0	1	1	1	1	0	0	0	1	0	0	1
782	1	0	1	0	1	0	1	1	1	1	1	1	1	1	1	1	1	0	1
852	0	0	1	0	0	0	1	0	1	1	1	1	0	0	0	1	0	0	1
882	0	0	1	0	0	0	0	0	0	0	1	0	1	0	0	1	0	0	1
A103	0	0	1	0	0	0	0	0	0	0	1	0	0	1	1	0	0	0	1
B139	0	0	1	0	0	0	0	0	0	0	1	0	0	1	1	0	0	0	1
C 4	0	0	1	0	0	0	1	0	1	1	1	1	1	0	0	1	0	0	1
C 33	1	0	1	1	0	0	0	0	0	0	1	0	0	0	0	0	1	0	1
C 51	1	0	0	1	0	0	0	0	0	0	0	0	0	0	0	0	0	1	1
H 23	1	1	1	1	1	1	1	1	1	1	1	1	1	1	1	1	1	1	1

Interval assignment

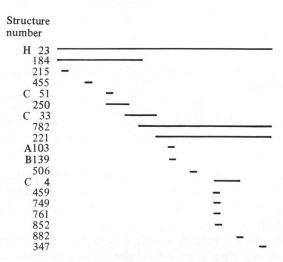

FIGURE 3.36. Overlap data and an interval assignment for 19 mutants of phage T4. From Benzer [1959]. Reprinted with permission of the author.

A digraph D for which there is an assignment of intervals satisfying (8) will be called an *interval order*. Interval orders will be studied in Sec. 8.8. In the measurement of indifference, we would ask for assignments of intervals of values which satisfy for all $u \neq v$ in V:

> You are indifferent between u and v if and
> only if $J(u)$ and $J(v)$ overlap.

We can make such an assignment if and only if "indifference" defines an interval graph.[24]

Interval graphs also arise in archaeology. Several types of pottery (or other artifacts) are found in different prehistoric graves. We would like to place the pottery in proper serial or chronological order, assigning to each type of pottery a time period. Seriation in archaeology began with the work of Flinders Petrie in the late nineteenth century [1899, 1901]. Petrie studied graves in the cemeteries of Naqada, Ballas, Abadiyeh, and Hu, located in what was prehistorical Egypt. (Recent radiocarbon dating shows the graves ranged from 4000 B.C. to 2500 B.C.) Petrie used the data from approximately 900 graves to order them and assign a time period or sequence to each type of pottery found. His techniques, further spelled out in Petrie [1920, 1921], were a mixture of subjective and objective techniques. Here, we formulate the problem of sequence dating or seriation in terms of graph theory. A related discussion can be found in Kendall [1963, 1969a, b].

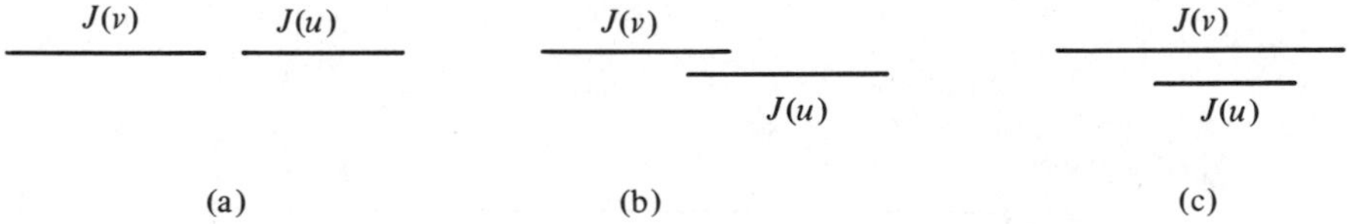

FIGURE 3.37. In (a), u strictly follows v. In (b), u weakly follows v. In (c), v over-reaches u.

To be precise, we would like to assign to each type of pottery u an interval of time $J(u)$, with the assignment of intervals satisfying the following conditions:

(1) Suppose type u pottery *strictly followed* type v pottery, i.e., type v pottery disappeared before type u pottery appeared. Then $J(u)$ should be strictly to the right of $J(v)$, as shown in part (a) of Fig. 3.37.

(2) Suppose type u pottery *weakly followed* type v pottery, i.e., type u appeared when type v was already around and did not dis-

---

[24]A similar discussion applies to the psychological concept "similarity" or "indistinguishability," as well as to the concept "indifference." See Harary [1964] or Roberts [1970].

appear until after type v did. Then $J(u)$ and $J(v)$ should overlap as in part (b) of Fig. 3.37.

(3) Suppose type v pottery *over-reached* type u, i.e., type u appeared after type v was already present and disappeared before type v did. Then $J(u)$ and $J(v)$ should overlap as shown in part (c) of Fig. 3.37.

An assignment of intervals satisfying conditions (1) through (3) is called an *admissible chronological ordering*.

If pottery type u and pottery type v appear in common in some grave, we can assume that the time periods of the two types of pottery must have overlapped. Conversely, if the selection of graves is extensive enough, we should be able to assume that if two types of pottery appeared in overlapping periods, then they will appear in some common grave. Let us now draw a graph G as follows: $V(G)$ is the set of types of pottery and there is an edge between type u and type v if and only if u and v appear in common in some grave. An admissible chronological ordering assigns to each type of pottery u a time interval $J(u)$ so that if u and v are different types, then

$$\{u, v\} \in E \longleftrightarrow J(u) \cap J(v) \neq \emptyset.$$

Thus, no admissible chronological ordering exists unless G is an interval graph. Conversely, if G is an interval graph, an interval assignment J induces

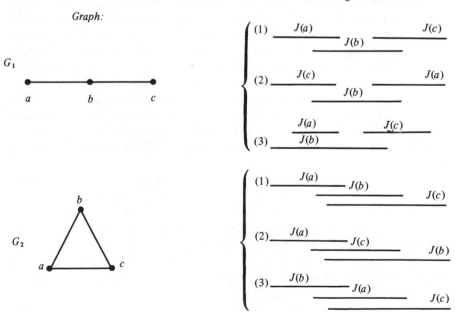

FIGURE 3.38. Possible chronological orderings.

a possible admissible chronological ordering, or a *possible chronological ordering* for short. But how do we know that it is admissible? In fact, for every possible chronological ordering assigned to an interval graph G by an interval assignment J, there is almost always a different ordering which disagrees with the first on strict following. In particular, if $E(G)$ is not all possible edges, there is always a disagreeing possible chronological ordering. To get such a disagreeing possible chronological ordering, simply reverse the order of the intervals (think of flipping over the real line). Figure 3.38 shows several possible chronological orderings for graph G_1. Ordering (2) is obtained from ordering (1) by reversing. In (1), c strictly follows a, while in (2), it is the opposite. If these two were the only possible chronological orderings, it would usually be relatively easy to choose between them: Simply decide between two "far apart" nonoverlapping types of pottery, which is older. However, there are other ways two interval assignments or possible chronological orderings can differ. In Exer. 26 we shall ask the reader to show that for every possible chronological ordering there is (almost) always a second one which disagrees on over-reaching. Figure 3.38 shows such an ordering (3) taken from the ordering (1) for graph G_1. In (1) b weakly follows a, in (3) it over-reaches a. Hence, for most interval graphs, there are many possible chronological orderings. To give a second example, we note that also on weak following there can be disagreement. Graph G_2 of Fig. 3.38 has at least the three possible chronological orderings shown, which pairwise differ on weak following. In general it is an open question of graph theory to determine for a given interval graph how many different[25] possible chronological orderings there are. Using only the overlap data, perhaps the best we can hope to do is get "strict following" right. It has only recently been determined how many possible chronological orderings there are which disagree on strict following (see footnote 27, page 121) and it is still an open question to determine which graphs have exactly two such.[26] (We have already observed that most graphs have at least two, and that if there are exactly two, choice between them should be relatively easy.) Thus, a simple problem in archaeology already taxes the ability of mathematicians to solve it.

Similar problems arise in the study of developmental psychology. (See Coombs and Smith [1973].) Here, we study various traits or characteristics. We would like to assign periods to each trait, representing the natural order in the developmental process during which the trait is present. It is assumed that each child exhibits the same natural developmental process. We study various children and observe whether or not traits u and v exist in the same child. Here, the traits correspond to the pottery types and the

---

[25]Two chronological orderings are different if they differ on strict following, weak following, or over-reaching.

[26]We shall formulate this problem more precisely later.

children correspond to the graves. If there are enough children observed, it is reasonable to assume that the periods in which traits u and v are present should overlap if and only if traits u and v are found in common in some child. Thus, the "found in common in some child" graph must be an interval graph. Again, the problem of determining the "correct" chronological ordering is not totally solved. Indeed, it is mathematically equivalent to the archae-

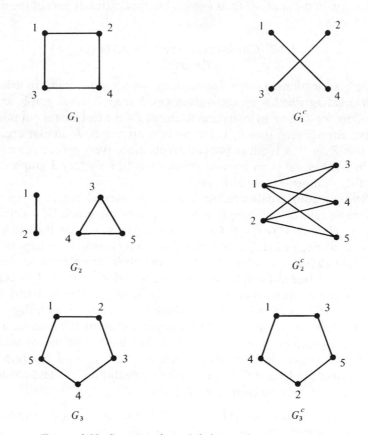

FIGURE 3.39. Some graphs and their complementary graphs.

FIGURE 3.40. The complementary graph G_2^c of the graph G_2 of Fig. 3.32, and the transitive orientation of G_2^c corresponding to the interval assignment of Fig. 3.32.

ology problem we have discussed. It is always surprising to see two such diverse disciplines come up with the same problem. It is the power of the mathematical formulation which shows that these problems are the same, and which hopefully will provide solutions.

Interval graphs also arise in connection with transportation, in particular in the phasing of traffic lights at complicated intersections. We shall return to this idea in Sec. 3.4.6. Still another application of interval graphs is in ecology, in the study of food webs. This application is part of the discussion of Sec. 3.5.

3.4.3. CHARACTERIZATION OF INTERVAL GRAPHS

Before mentioning other applications, we are now ready to ask for a test for deciding whether or not a given graph is an interval graph, and for a procedure for finding an interval assignment J if the test comes out positive. We have already seen that Z_4 is not an interval graph. A similar argument shows that Z_n, $n > 3$, is not an interval graph. Since every generated subgraph of an interval graph is an interval graph (why?), an interval graph cannot contain Z_4 as a generated subgraph.

Before stating conditions for G to be an interval graph, let us define one more notion. G^c, the *complementary graph* of the graph G, is defined to be that graph with $V(G^c) = V(G) = V$ and a pair u, v from V (with $u \neq v$) chosen as an edge in $E(G^c)$ if and only if it is not chosen as an edge in $E(G)$. Figure 3.39 shows a number of graphs and their complementary graphs. Now if G is an interval graph, we can build an orientation A of its complementary graph G^c as follows: $(u, v) \in A$ if and only if $J(u)$ is strictly to the right of $J(v)$. (Cf. Eq.(8).) This is a transitive orientation. (Why?) Figure 3.40 gives the transitive orientation of G_2^c corresponding to the interval assignment for the graph G_2 shown in Fig. 3.32. We have thus shown that if G is an interval graph, then not only does it not have Z_4 as a generated subgraph, but its complement G^c has a transitive orientation.[27] These conditions together are not only necessary but sufficient.

Theorem 3.8 (Gilmore and Hoffman [1964]). A graph G is an interval graph if and only if it satisfies the following conditions:

(a) Z_4 is not a generated subgraph of G

and

(b) G^c is transitively orientable.

[27]The problem of how many possible chronological orderings there are for a given interval graph G which disagree on strict following is now seen to be equivalent to the following more precisely stated problem: How many transitive orientations are there of G^c? This problem has recently been solved for all graphs with transitively orientable complement by Golumbic [1975]. The specific problem for interval graphs is solved in Booth and Lueker [1975].

Let us illustrate the theorem on an example, the graph G of Fig. 3.41. It is easy to see that Z_4 is not a generated subgraph of G. Moreover, condition (b) is satisfied because G^c has a transitive orientation, as shown in Fig. 3.41. We conclude that G is an interval graph. Similarly, we can show that Z_5 is not an interval graph, by showing that its complement, which is again Z_5, is not transitively orientable.

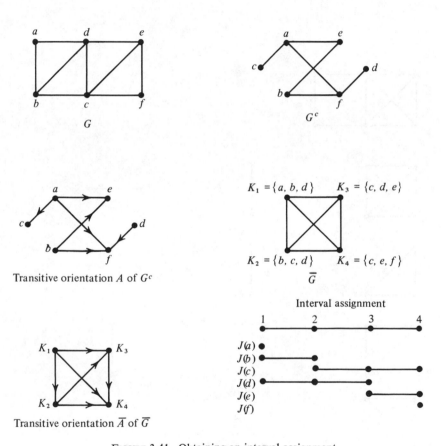

FIGURE 3.41. Obtaining an interval assignment.

Other characterizations of interval graphs have been given by Fulkerson and Gross [1965] and Lekkerkerker and Boland [1962]. We pursue these below and in Exers. 20 and 24. Sometimes the easiest way to prove that a graph is an interval graph is to exhibit an interval assignment.

The proof of the sufficiency of the conditions in Theorem 3.8 is constructive, and we shall present this constructive procedure for obtaining an interval representation for G.

As a preliminary notion, let us define a graph $G = (V, E)$ to be *complete* if for every $u \neq v \in V$, $\{u, v\} \in E$. That is, G is complete if all possible edges of G are in G. The complete graph on n vertices is denoted K_n. K_3 is of course a triangle. In an arbitrary graph G, a subgraph which is complete is called a *clique*. A clique is *maximal* if it is not contained in a larger clique. Figure 3.42 shows a graph G and its cliques. The subgraph generated by the

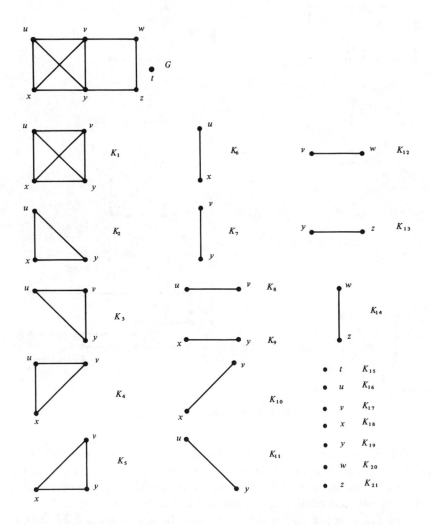

FIGURE 3.42. A graph G and its cliques. $K_1, K_{12}, K_{13}, K_{14}$, and K_{15} are maximal.

vertices u and v is a clique. It is not maximal, because it is contained in the larger clique generated by the vertices u, v, and x. This is also not maximal, as it is contained in the clique $\{u, v, x, y\}$, which is maximal. Note that while strong components are disjoint, this is not necessarily the case for maximal cliques.

Theorem 3.9 Suppose $G = (V, E)$ is a graph.

(a) If K is a maximal clique of G and a vertex $u \in V$ is not in K, then there is a vertex $v \in K$ such that $\{u, v\} \notin E$.

(b) If K and L are maximal cliques of G and $K \neq L$, then there are $u \in K$ and $v \in L$ such that $\{u, v\} \notin E$.

(c) If $\{u, v\} \in E$, there is a maximal clique K of G such that u and v are in K.

Proof.

(a) If there is no v, then the subgraph generated by u plus the vertices of K would be a clique larger than K.

(b) Since $K \neq L$, there is u in $(K - L) \cup (L - K)$. Let us suppose u is in $K - L$. Since $u \notin L$, part (a) implies that there must be v in L such that $\{u, v\} \notin E$.

(c) The subgraph generated by the vertices u and v is a clique. If it is not maximal, keep adding vertices until a maximal clique is attained. $\hspace{2cm}$ Q.E.D.

To present the procedure for finding an interval representation, we shall suppose the graph G does not have Z_4 as a generated subgraph and suppose we are given a transitive orientation A for G^c. (For a constructive procedure for finding A, the reader is referred to Gilmore and Hoffman [1964].) Let $\mathcal{C}$ be the collection of maximal cliques in G. Build a new graph $\bar{G}$ as follows: $V(\bar{G}) = \mathcal{C}$ and if K and L are distinct maximal cliques of G, let $\{K, L\} \in E(\bar{G})$. That is, $\bar{G}$ is a complete graph with $\mathcal{C}$ as the vertex set. Build an orientation $\bar{A}$ of $\bar{G}$ using the orientation A of G^c. Specifically, if $K \neq L$, then there is by Theorem 3.9, part (b), a pair $u \in K$ and $v \in L$ such that $\{u, v\} \in E(G^c)$. Give $\{K, L\}$ the orientation of $\{u, v\}$ in A, i.e., let

$$(K, L) \in \bar{A} \Longleftrightarrow (u, v) \in A.$$

It is necessary to prove that $\bar{A}$ is well-defined. For, suppose we can find another pair $u' \in K$ and $v' \in L$ such that $\{u', v'\} \in E(G^c)$ and such that $(v', u') \in A$ while $(u, v) \in A$. Then the definition of the orientation of $\{K, L\}$ in $\bar{A}$ is ambiguous. We show below (Lemma 1) that there can be no such u', v'.

To illustrate the construction of $\bar{A}$, we again consider the graph G of Fig. 3.41. Given maximal cliques K_1 and K_2, for example, we note that

$a \in K_1$ and $c \in K_2$ and $\{a, c\} \notin E$. Thus, we can use the orientation of the edge $\{a, c\}$ in G^c to define the orientation of the edge $\{K_1, K_2\}$ in $\bar{G}$. Since $(a, c) \in A$, we conclude $(K_1, K_2) \in \bar{A}$. A detailed construction of the orientation $\bar{A}$ is shown in Table 3.1 and the orientation is illustrated in Fig. 3.41.

TABLE 3.1

Orientation $\bar{A}$ for Graph $\bar{G}$ of Fig. 3.41.

Edge $\{K_i, K_j\}$	u	v	Orientation of $\{u, v\}$ in A	Orientation of $\{K_i, K_j\}$ in $\bar{A}$
$\{K_1, K_2\}$	a	c	(a, c)	(K_1, K_2)
$\{K_1, K_3\}$	a	e	(a, e)	(K_1, K_3)
$\{K_1, K_4\}$	a	c	(a, c)	(K_1, K_4)
$\{K_2, K_3\}$	b	e	(b, e)	(K_2, K_3)
$\{K_2, K_4\}$	b	e	(b, e)	(K_2, K_4)
$\{K_3, K_4\}$	d	f	(d, f)	(K_3, K_4)

In the oriented graph $(\mathcal{C}, \bar{A})$, there is a unique complete simple path. (This is because $(\mathcal{C}, \bar{A})$ is a transitive tournament; proof of transitivity is left to the reader. By Theorem 3.4, every transitive tournament has a unique complete simple path.) The unique complete simple path defines a ranking of the vertices. In our example, the unique complete simple path and ranking is given by K_1, K_2, K_3, K_4.

Now for $u \in V(G)$, let

$$J'(u) = \{K \in \mathcal{C} : u \in K\}. \tag{9}$$

Here,

$$J'(a) = \{K_1\}$$
$$J'(b) = \{K_1, K_2\}$$
$$J'(c) = \{K_2, K_3, K_4\}$$
$$J'(d) = \{K_1, K_2, K_3\}$$
$$J'(e) = \{K_3, K_4\}$$
$$J'(f) = \{K_4\}.$$

We shall observe below (Lemma 2) that for all $u \neq v \in V(G)$,

$$\{u, v\} \in E \Longleftrightarrow J'(u) \cap J'(v) \neq \emptyset. \tag{10}$$

Finally, we get an interval assignment $J(u)$ from the assignment $J'(u)$ as follows. Let $f(K)$ be the rank of K under the rank order of $\mathcal{C}$ induced by $\bar{A}$; that is, $f(K) = 1$ if K is first, $f(K) = 2$ if K is second, and so on. Here, $f(K_1) = 1$, $f(K_2) = 2$, $f(K_3) = 3$, and $f(K_4) = 4$. Let $J(u)$ be the smallest interval containing

$$\{f(K): K \in J'(u)\}.$$

Here,

$$J(a) = [1, 1]$$
$$J(b) = [1, 2]$$
$$J(c) = [2, 4]$$
$$J(d) = [1, 3]$$
$$J(e) = [3, 4]$$
$$J(f) = [4, 4].$$

It is left to prove that J is an interval assignment for G. That we shall do in general below (Lemma 3). For the specific example of Fig. 3.41, the proof is left to the reader. (The assignment is shown in the figure.)

Before closing this section, we mention a second characterization of interval graphs. We say that a ranking $K_1, K_2, \ldots, K_p$ of the maximal cliques of G is *consecutive* if whenever a vertex u is in K_i and K_j for $i < j$, then for all $i < r < j$, u is in K_r. It is easy to see that the ranking of maximal cliques obtained in our procedure is consecutive. We shall show this during the proof of Lemma 3 in the next subsection. Thus, every interval graph has a consecutive maximal clique ranking. Conversely, if a graph G has a consecutive maximal clique ranking, then Z_4 cannot be a generated subgraph of G and the ranking comes from a transitive orientation of G^c. (Proof of these statements is left to the reader, as Exer. 27.) Thus, we have the following theorem:

Theorem 3.10 (Fulkerson and Gross [1965]). A graph G is an interval graph if and only if there is a ranking of the maximal cliques of G which is consecutive.

The Fulkerson-Gross Theorem is conveniently restated in terms of matrices in Exer. 20. Another characterization of interval graphs, due to Lekkerkerker and Boland [1962], is described in Exer. 24.

3.4.4. PROOFS OF THE LEMMAS OF SEC. 3.4.3[28]

Lemma 1. $\bar{A}$ is well-defined.

Proof. Suppose that $u, u' \in K$, that $v, v' \in L$, that $\{u, v\} \notin E$, that $\{u', v'\} \notin E$, that $(u, v) \in A$ and that $(v', u') \in A$. We first observe that either $\{u, v'\} \notin E$ or $\{u', v\} \notin E$. Otherwise, G has a generated subgraph of

---

[28]This subsection may be omitted without loss of continuity.

the form shown in Fig. 3.43. For we know that $\{u, v\} \notin E$ and $\{u', v'\} \notin E$; we also know that $\{u, u'\} \in E$ and $\{v, v'\} \in E$, since u and u' are in the same clique and so are v and v'. The generated subgraph of Fig. 3.43 is Z_4, in violation of assumed condition (a) of Theorem 3.8.

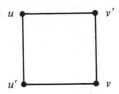

FIGURE 3.43. Graph for proof that $\bar{A}$ is well-defined.

Since either $\{u, v'\} \notin E$ or $\{u', v\} \notin E$, let us assume the former. (The proof in the case of the latter is identical.) Now either $(u, v') \in A$ or $(v', u) \in A$. If $(u, v') \in A$ then since $(v', u') \in A$, we conclude $(u, u') \in A$ by transitivity of the orientation A. If $(v', u) \in A$, we conclude by a similar argument that $(v', v) \in A$. In either case, we have a contradiction, since the edges $\{u, u'\}$ and $\{v, v'\}$ are in E.   Q.E.D.

Lemma 2. If J' is defined as in (9) then for all $u \neq v$ in $V(G)$,

$$\{u, v\} \in E \Longleftrightarrow J'(u) \cap J'(v) \neq \emptyset. \tag{10}$$

Proof. If $\{u, v\} \in E$, then by Theorem 3.9, part (c), there is $K \in \mathfrak{C}$ such that u and v are in K. Then $K \in J'(u) \cap J'(v)$. Conversely, suppose $J'(u) \cap J'(v) \neq \emptyset$ and let $K \in J'(u) \cap J'(v)$. Now u and v are in K and K is a clique, so $\{u, v\} \in E$.   Q.E.D.

Lemma 3. Suppose J' is defined as in (9), $f(K)$ is the rank of K in the unique ranking of $(\mathfrak{C}, \bar{A})$, and $J(u)$ is the smallest interval containing

$$\{f(K): K \in J'(u)\}.$$

Then for all $u \neq v \in V(G)$,

$$\{u, v\} \in E \Longleftrightarrow J(u) \cap J(v) \neq \emptyset. \tag{7}$$

Proof. If $\{u, v\} \in E$, then by Lemma 2, there is $K \in J'(u) \cap J'(v)$. Now $f(K) \in J(u) \cap J(v)$, so $J(u) \cap J(v) \neq \emptyset$. Conversely, suppose $J(u) \cap J(v) \neq \emptyset$. Now $J(u)$ and $J(v)$ are intervals each having as endpoints integers between 1 and c, the number of maximal cliques of G. Thus, there is an integer m between 1 and c in $J(u) \cap J(v)$. Now m is $f(K)$ for some maximal clique K. We shall show that u and v are in K, and then $\{u, v\} \in E$ follows. Since $f(K)$ is in $J(u)$, there are K_i and K_j in $J'(u)$ such that $f(K_i) \leq f(K) \leq f(K_j)$. We may assume that $f(K) \neq f(K_i)$ and $f(K) \neq f(K_j)$, for otherwise $K = K_i$

or $K = K_j$, and u is in K. Thus, $f(K_i) < f(K) < f(K_j)$. Suppose $u \notin K$. Then there is $a \in K$ such that $\{u, a\} \notin E$. This follows by Theorem 3.9, part (a). Now $(u, a) \in A$ since $(K_i, K) \in \bar{A}$ and $(a, u) \in A$ since $(K, K_j) \in \bar{A}$. This implies that A is not an orientation. We conclude that $u \in K$. Similarly, $v \in K$. Q.E.D.

Note that the proof that u is in K shows that the ranking of the maximal cliques is consecutive in the sense used in Fulkerson and Gross' Theorem (Theorem 3.10).

3.4.5. OTHER INTERSECTION GRAPHS

Before concluding our discussion of intersection graphs, let us note that intersection graphs of families other than intervals on the line have also been studied. Intersection graphs of unit length intervals have been studied in Roberts [1969a] and in Wegner [1967]. Tucker [1970a, 1971] has studied intersection graphs of arcs of a circle. We study these in the next subsection. Intersections of cubes and boxes in n-space have been studied in Danzer and Grünbaum [1967] and Roberts [1969b] and we return to them in Sec. 3.5. Intersection graphs of convex sets in n-space have also been studied (see Wegner [1967] and Ogden-Roberts [1970]). A summary of other results on intersection properties of various families of sets in Euclidean n-space can be found in the book by Hadwiger, Debrunner, and Klee [1964]. A number of the problems we have considered in this section, for example, Benzer's problem and the measurement problem, are interesting for other families of sets as well as intervals.

3.4.6. PHASING TRAFFIC SIGNALS

As a final application of interval graphs, let us consider the problem of phasing traffic signals at a complicated intersection. We can imagine assigning to each *traffic stream* (road, pedestrian lane, etc.) a given interval of time during which it has a green light. We shall deal only with the simple situation where the light is either green or red. The process proceeds cyclically, with the given assignment of intervals repeated over and over again. Let us imagine an s-second circular clock and let us represent the interval of green for traffic stream u as an arc on the circumference of this clock. Now certain streams are *compatible*: those which are not on a collision course or where traffic is sufficiently light to make simultaneous movement in both streams not too much of a hazard or a cause for delay. For example, in the traffic intersection shown in Fig. 3.44, traffic stream a is compatible with traffic

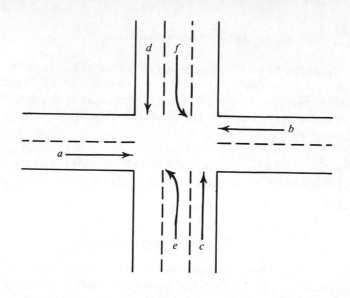

Compatibility graph

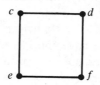

Circular arc representation

Corresponding intersection graph H

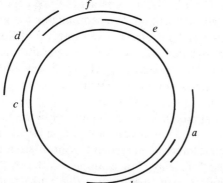

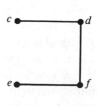

FIGURE 3.44. A traffic intersection.

130

stream b, the traffic coming in the opposite direction, but not with stream c, traffic which would collide with it. It is up to a traffic engineer to make decisions about compatibility before traffic lights are phased. In any case, given this information, the basic point is that if two streams are compatible, the corresponding green light arcs may overlap. An assignment to each traffic stream of an arc of the clock circle is called a *feasible green light assignment* if only compatible traffic streams get overlapping arcs. How does one find a feasible assignment? To answer this question, suppose we draw a graph G, the *compatibility graph*, as follows. The vertices of G are the traffic streams and two streams are joined by an edge if and only if they are compatible. We want an assignment of an arc of the clock circle to each stream such that if u and v are joined by an edge, then the corresponding arcs are allowed to intersect (but do not necessarily intersect). The intersection graph of the collection of arcs used defines a subgraph H of G with the same vertex set as G. We call such a subgraph a *spanning subgraph*. (Cf. Exer. 17, Sec. 2.3.) The intersection graph of a collection of arcs on a circle is called a *circular arc graph*. Thus, some spanning subgraph H of G is a circular arc graph. Figure 3.44 gives a possible compatibility graph corresponding to the street intersection shown. (Stoffers [1968] gives more complicated examples.) The compatibility graph G here is not a circular arc graph. (Proof of this is left to the reader.) Thus, any feasible assignment of arcs of a circle to vertices of this G must correspond to a proper spanning subgraph H of G, i.e., a spanning subgraph H different from G. Figure 3.44 shows such an assignment and the corresponding graph H.

Not every feasible green light assignment is terribly efficient. For example, there is always an assignment which gives virtually no green light time to any traffic stream. To find an "efficient" green light assignment it is first necessary to define what efficient means. One requirement is that each green interval have a certain minimal length—long enough for example to allow drivers to react. In the case of an isolated traffic intersection, we probably want to minimize total waiting time, i.e., the sum total of the red light times in a given cycle. In the case of a nonisolated traffic intersection, there is a different concept of efficiency involved, for we want to coordinate with other lights. Stoffers speaks of "ideal" starting times for a green light interval assigned to a given traffic stream; these would be determined by knowing when traffic going through other lights could be expected to arrive at the light being phased. Then, Stoffers argues, we want to minimize the (weighted) time lags between ideal and realized starting times.

In any case, to find an efficient green light assignment, an effective procedure is the following. First, given the compatibility graph G, find all circular arc spanning subgraphs of G. For each such subgraph H, find a green light assignment which is most efficient in the sense of efficiency adopted. Finally, find the most efficient of these assignments. If we are willing to assume that the last green light of a cycle ends before the next cycle begins, then each green light assignment corresponds to an interval assignment, for we can lay open the circle on a line. The intersection graph corresponding to a green light assignment is then an interval graph. In this case, we only need to find the most efficient interval assignment for each spanning subgraph which is an interval graph. Stoffers [1968] describes a specific procedure for finding a most efficient green light assignment for a given spanning subgraph H which is an interval graph. This is related to the algorithm for finding an interval assignment for an interval graph which was described in Sec. 3.4.3. First, find all maximal cliques of H. Then, find all possible rankings of these cliques which are generated by transitive orientations of H^c or which are consecutive in the sense used in the Fulkerson-Gross Theorem (Theorem 3.10). Suppose $K_1, K_2, \ldots, K_p$ is such a ranking. Now each maximal clique K_i will correspond to a *phase* during which all traffic streams making up K_i receive a green light. The phases come in the sequence $K_1, K_2, \ldots, K_p$. Phase K_i has a certain duration d_i, with the durations to be determined. We know that the traffic stream u appears in consecutive maximal cliques, $K_i, K_{i+1}, \ldots, K_j$. Thus u receives a green light just as phase K_i begins and remains green during all the phases through K_j. If K_1 starts at time 0, then K_i starts at time $d_1 + d_2 + \cdots + d_{i-1}$, and so u receives the green light interval

$$(d_1 + d_2 + \cdots + d_{i-1}, d_1 + d_2 + \cdots + d_j),$$

where $d_1 + d_2 + \cdots + d_{i-1}$ is interpreted to be 0 if $i - 1 = 0$. (We use open intervals here; a slight modification would allow closed intervals.) The most efficient assignment of durations d_i for a given ranking of maximal cliques is discovered by a linear programming procedure. For it is not hard to see that the d_i's satisfy certain inequalities (constraints) and we wish to minimize quantities expressible in terms of the d_i's.

For a given interval graph spanning subgraph H of the compatibility graph G, we can find the most efficient feasible green light assignment by choosing among those most efficient assignments constructed from different rankings of the maximal cliques of H. Finally, the most efficient assignment for G can be chosen by looking at all possible such subgraphs H.

To illustrate this procedure, suppose we consider the graph H of Fig. 3.44. The maximal cliques of H are $K = \{a, b\}$, $L = \{c, d\}$, $M = \{d, f\}$ and $N = \{e, f\}$. It is easy to see that one consecutive ranking of these maximal cliques is K, L, M, N. We wish to assign durations d_1 to K, d_2 to L, d_3 to M, and d_4 to N. Traffic stream a is only in maximal clique K, so a is green during the interval $(0, d_1)$. Stream d is in L and M, so d is green during the interval $(d_1, d_1 + d_2 + d_3)$. Other green light intervals are shown in Table 3.2.

TABLE 3.2

Green Light Intervals for Subgraph H of Fig. 3.44.

Traffic stream u	Maximal cliques containing u	Green interval for u	Total green time	Total red time
a	K	$(0, d_1)$	d_1	$d_2 + d_3 + d_4$
b	K	$(0, d_1)$	d_1	$d_2 + d_3 + d_4$
c	L	$(d_1, d_1 + d_2)$	d_2	$d_1 + d_3 + d_4$
d	L, M	$(d_1, d_1 + d_2 + d_3)$	$d_2 + d_3$	$d_1 + d_4$
e	N	$(d_1 + d_2 + d_3, d_1 + d_2 + d_3 + d_4)$	d_4	$d_1 + d_2 + d_3$
f	M, N	$(d_1 + d_2, d_1 + d_2 + d_3 + d_4)$	$d_3 + d_4$	$d_1 + d_2$
Total				$4d_1 + 4d_2 + 4d_3 + 4d_4$

If we want each traffic stream to be green for at least 10 seconds, we want the total length of the clock cycle to be $s = 60$ seconds, and we want the total red light time to be as small as possible, then we require

$d_1 \geq 10$ (green time for a and b)   (11)

$d_2 \geq 10$ (green time for c)   (12)

$d_2 + d_3 \geq 10$ (green time for d)   (13)

$d_4 \geq 10$ (green time for e)   (14)

$d_3 + d_4 \geq 10$ (green time for f)   (15)

$d_1 + d_2 + d_3 + d_4 = 60$[29]   (total green time in all phases) (16)

$R = 4d_1 + 4d_2 + 4d_3 + 4d_4$ is minimal   (total red time for all streams). (17)

Finding d_i which minimize an expression like R subject to constraints like (11) through (16) is a standard problem of linear programming. In this simple case, R is $4(d_1 + d_2 + d_3 + d_4)$, so R will be $4(60) = 240$. Thus, here the minimization problem reduces to finding any d_i which satisfy Eqs. (11) through (16). For example; taking each d_i to be 15 satisfies these conditions. In this solution, the east-west traffic a and b has green light for the first 15 seconds. In the next phase, the north-south traffic c and d gets green lights. At 30 seconds, c's light turns red and left-turning traffic f gets a green light. In the final phase, starting at 45 seconds, d's light turns red and left-turning traffic e gets a green light. Now both e and f have green lights. At 60 seconds, the cycle starts all over again. The solution is somewhat different for different choices of the d_i's.

[29]We really only require that $d_1 + d_2 + d_3 + d_4 \leq 60$. But since expanding green light durations can only decrease total red light time, we might as well leave no time during which one of the phases isn't in green light operation. The Total red time column of Table 3.2 was filled in under this assumption.

We have to repeat this same computation for each ranking of maximal cliques of H which is consecutive. We then have to repeat the computations for each H. Each such computation will give us a different set of phases. It is instructive for the reader to make several such computations and compare the sets of phases which result. Among all the possible assignments which give rise to minimal R-values for a given ranking of maximal cliques of some H, we pick that assignment which gives rise to the smallest R-value.

If we want to eliminate the assumption that the last green light of a cycle ends before the next cycle begins, we have to look for spanning subgraphs H which are circular arc graphs. We proceed as above, but now search for a "circular ranking" of cliques $K_1, K_2, \ldots, K_p, K_1$ which is consecutive in a sense analogous to that used in the Fulkerson-Gross Theorem (Theorem 3.10). The rest of the procedure is the same.

The solution we have described to the traffic light phasing problem requires a characterization of circular arc graphs (or interval graphs). The circular arc graphs, as we mentioned above, have been characterized by Tucker [1970a, 1971]. This is another example of a mathematical problem which was posed and solved with one application in mind (Benzer's problem and measurement theory) but whose solution leads to applications in another area.

EXERCISES

1. For each of the following families of sets $\mathfrak{F}$, find the intersection graph.
   (a) $\mathfrak{F} = \{S_1, S_2, S_3, S_4\}$, where

   $$S_1 = \{a, b, c, d\}$$
   $$S_2 = \{x, y, z\}$$
   $$S_3 = \{a, e, i, o, u\}$$
   $$S_4 = \{a, b, c, d, \ldots, x, y, z\}.$$

   (b) $\mathfrak{F} = \{S_1, S_2, S_3\}$, where

   $$S_1 = [1, 10]$$
   $$S_2 = [9, 15]$$
   $$S_3 = (15, 20).$$

   (c) $\mathfrak{F} = \{S_1, S_2, S_3, S_4, S_5\}$, where $S_i = \{i, i + 1\}$.

2. For each of the graphs G_i of Fig. 3.39, find a family of sets $\mathfrak{F}$ such that G_i is the intersection graph of $\mathfrak{F}$.

3. Show (without going through the construction procedure of Sec. 3.4.3) that graph G_2 of Fig. 3.39 is an interval graph.

4. Suppose the substructures S_1, S_2, and S_3 overlap as follows. Show that the overlap data is consistent with the hypothesis of linearity.

$$\begin{array}{c} \ \ S_1\ S_2\ S_3 \\ \begin{array}{c} S_1 \\ S_2 \\ S_3 \end{array}\begin{pmatrix} 1 & 1 & 0 \\ 1 & 1 & 1 \\ 0 & 1 & 1 \end{pmatrix}. \end{array}$$

5. For the overlap data of Exer. 4, find two different possible chronological orderings which differ on strict following.

6. For the overlap data of Exer. 4, are there two different possible chronological orderings which differ on over-reaching?

7. Suppose we consider four fine wines, u, v, w, and x. We feel that u is worth between $2 and $5 a bottle, v is worth between $3 and $5, w is worth between $7 and $10, and x is worth between $3 and $10. Assuming overlap defines indifference, draw a graph which represents indifference among these wines. Draw a digraph which represents preference if preference is defined by Eq. (8). Is there any wine which we definitely prefer to all others?

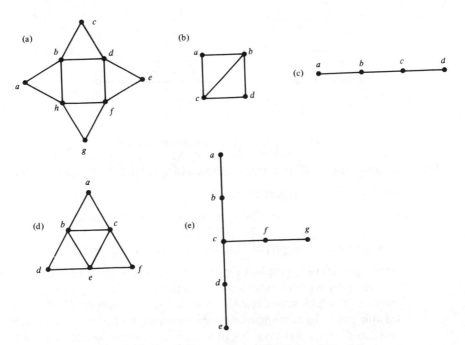

FIGURE 3.45. Graphs for exercises of Sec. 3.4 and Sec. 3.5.

8. Find a transitive orientation for the complement of each interval graph in Fig. 3.32.

9. Which of the graphs of Fig. 3.45 is an interval graph?

10. For each graph of Fig. 3.45, find all the maximal cliques and indicate one clique which is not maximal.

11. For each graph G of Fig. 3.46, find a transitive orientation for G^c and use the procedure outlined in the text to obtain an interval assignment for G.

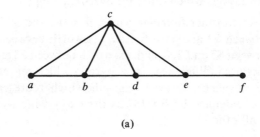

(a)

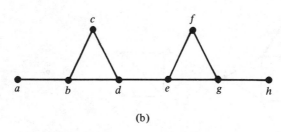

(b)

FIGURE 3.46. Graphs for exercises of Sec. 3.4.

12. Give an example of an interval graph which is transitively orientable.

13. Show that Z_4 is a circular arc graph.

14. (a) Show that the compatibility graph of Fig. 3.44 is not a circular arc graph.
    (b) Which of the graphs of Fig. 3.45 is a circular arc graph?

15. For the compatibility graph of Fig. 3.44, find another spanning subgraph H' which is an interval graph, and a (consecutive) ranking of maximal cliques of H' which arises from a transitive orientation of $(H')^c$. Find a feasible green light assignment corresponding to this ranking which minimizes total red light waiting times. Discuss the phases and compare the solution to that found in the text for the spanning subgraph H of Fig. 3.44.

16. (a) Draw a traffic intersection, its compatibility graph G, and a feasible green light assignment.
    (b) Draw the corresponding intersection graph. Is it an interval graph?
    (c) Find a spanning subgraph H of G which is an interval graph.
    (d) Find a transitive orientation of H^c and the corresponding ranking of maximal cliques of H. Set up the linear programming problem which arises from attempting to minimize total red light waiting times.
    (e) Solve the linear programming problem and discuss the traffic phases corresponding to the solution.

17. The *line graph* $L(G)$ of a graph G is the intersection graph of the family of edges of G. That is, $V(L(G)) = E(G)$ and $e \neq e'$ are joined by an edge in $L(G)$ iff they have a vertex in common in G. For example, Fig. 3.47 shows several graphs and their line graphs.

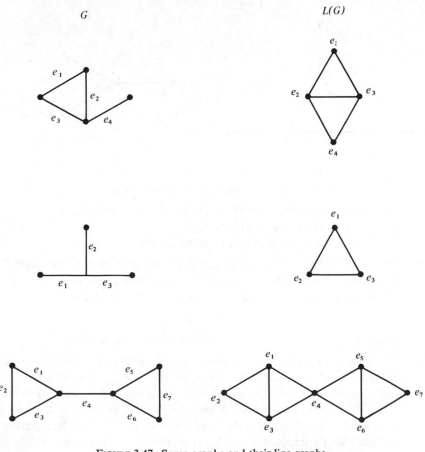

FIGURE 3.47. Some graphs and their line graphs.

(a)   What is the line graph of each of the graphs in Fig. 3.45?

(b)   Is K_4 the line graph of some graph? What about K_n, the complete graph on n vertices?

(c)   Is Z_4 the line graph of some graph? What about Z_n, the circuit of length n?

(d)   Is the graph of Fig. 3.48 the line graph of any graph?

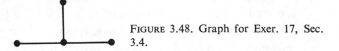

FIGURE 3.48. Graph for Exer. 17, Sec. 3.4.

18.   The *clique graph* $K(G)$ of a graph G is the intersection graph of the family of maximal cliques of G. For example, the graph of Fig. 3.49 has maximal cliques $K_1 = \{1, 2\}$, $K_2 = \{2, 3, 4\}$, $K_3 = \{3, 5\}$, and $K_4 = \{4, 6\}$ and its clique graph is as shown.

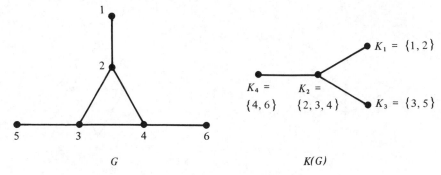

FIGURE 3.49. A graph and its clique graph.

(a)   What is the clique graph of the graph G_3 of Fig. 3.25?

(b)   What is the clique graph of K_n?

(c)   Is Z_4 ever a clique graph of some graph?

19.   (Harary [1969].) Let B be the vertex-edge incidence matrix of a graph G. (See Exer. 16, Sec. 2.4.) Suppose $L(G)$ is the line graph of G (see Exer. 17) and $A(L(G))$ is the adjacency matrix of $L(G)$. Show that

$$A(L(G)) = B'B - 2I,$$

where I is the identity matrix and B' is the transpose of B.

20.   The *vertex-maximal clique incidence matrix* $M(G)$ of a graph G is the point-set incidence matrix (see Exer. 16, Sec. 2.4) for the set $A = V(G)$ and the set $\mathcal{C} =$ the collection of maximal cliques of G. Fulkerson and Gross' Theorem (Theorem 3.10) may be restated as follows: A graph G is an interval graph if and only if the columns of $M(G)$ can be permuted so that the 1's appearing in any row appear consecutively. (Related results about this *consecutive 1's property* can be found in the papers by

Ryser [1969] and Tucker [1970b, 1972].) Using the Fulkerson and Gross Theorem, show that the graphs of Fig. 3.46 are interval graphs.

21. Guess at a characterization of circular arc graphs which is similar to the Fulkerson and Gross characterization of interval graphs of Theorem 3.10.

22. Show that if G is not an interval graph, then the orientation $\bar{A}$ on $\bar{G}$ defined in Sec. 3.4.3 is not necessarily well-defined.

23. (a) (Roberts [1969a].) A graph is a *unit interval graph* if and only if it is the intersection graph of a family of closed real intervals of unit length. Give an example of a graph which is an interval graph but not a unit interval graph.

    (b) (Roberts [1969a].) A graph is a *proper interval* graph if and only if it is the intersection graph of a family of real intervals with no interval properly contained inside another. (Two intervals may be identical.) Give an example of an interval graph which is not a proper interval graph.

24. A *triple* of vertices u, v, and w of a graph is called *asteroidal* if there are chains C_1 between u and v, C_2 between u and w, and C_3 between v and w, so that there is no edge from u to C_3, from v to C_2 or from w to C_1. Show that the graphs of Fig. 3.45 which are not interval graphs and do not contain Z_n, $n \geq 4$, as a generated subgraph all have asteroidal triples. (Lekkerkerker and Boland [1962] prove that a graph is an interval graph if and only if it does not have any Z_n, $n \geq 4$, as a generated subgraph and it has no asteroidal triples.)

*25. A graph G is called a *rigid circuit graph* if it has no circuit Z_n, $n \geq 4$, as a generated subgraph. Dirac [1961] shows that in every rigid circuit graph there is a vertex u, called a *simplicial vertex*, with the property that any two vertices joined to u by an edge are also joined to each other. (*Note:* u may have degree 0 or 1.) Show that every rigid circuit graph which is incomplete has at least two nonadjacent simplicial vertices. Show this (a) from Dirac's result and (b) without assuming Dirac's result.

26. It is clear that every interval graph is a rigid circuit graph (Exer. 25). Use Dirac's result (Exer. 25) to show that for every interval assignment for a graph G (of more than one vertex), there is another interval assignment which disagrees on over-reaching.

*27. Suppose G is a graph and $K_1, K_2, \ldots, K_p$ is a ranking of the set of maximal cliques $\mathcal{C}$ which is consecutive.

    (a) Show that Z_4 is not a generated subgraph of G.

    (b) Show that the ranking of maximal cliques arises from the ranking $\bar{A}$ on $\mathcal{C}$ for some transitive orientation A of G^c.

(c)   Show directly (without proving (a) or (b)) that G is an interval graph, by constructing an interval assignment J from the consecutive ranking.

28.   Discuss how the notion of intersection graph might apply to the problem of scheduling class meeting rooms at a university, given that certain classes cannot meet at the same time, certain classes are expected to be quite large, etc.

29.   Imagine the problem of putting the works of a famous ancient writer, for example Aristotle, in correct chronological order. Does the notion of intersection graph help with this problem?

30.   *Project.* It has been proposed that in scheduling traffic lights, it is an unnecessary restriction to assume that a given traffic stream gets only one green light interval during each cycle. Develop a mathematical theory for the situation where this restriction is eliminated.

31.   What are some other simplifications in the treatment of traffic lights presented in Sec. 3.4.6? How could our model be modified to take account of these?

3.5.   Food Webs

In this section we turn to a quite different set of applications of graph theory—the study of food webs in ecology. Ecology is the study of the relations among different creatures, including man, and their relations with their environment. In our first application, we study certain determinants of the ecological relationships.

3.5.1.   THE DIMENSION OF ECOLOGICAL PHASE SPACE[30]

We may think of a species of animal or plant as determined by certain "dimensions" which characterize its usual healthy environment. These dimensions might be such factors as temperature, moisture, pH, size of prey, nutrients needed, etc. The region in Euclidean k-space consisting of all points which meet certain specified constraints on these different dimensions then corresponds to the particular species; it is called the species' *ecological niche.* Usually the constraints on a particular dimension define an interval of possible values. Then the ecological niche is a k-dimensional rectangle with sides parallel to the coordinate axes, where k is the number of dimensions. Such a figure is called a (k-dimensional) *box.* See Fig. 3.50 for examples. The Euclidean space which has as dimensions such factors as temperature and the like is called *ecological phase space.* In general, no two species have identical

[30]The discussion in the first part of this subsection is based on some unpublished work of Joel E. Cohen.

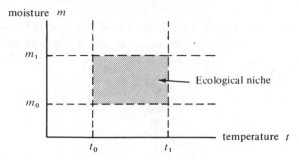

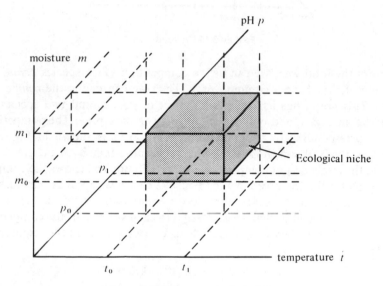

FIGURE 3.50. Some ecological niches.

niches in ecological phase space. A basic idea of ecology is that two species compete if and only if their ecological niches overlap. (If these are sufficiently similar, the species cannot coexist. This is known as the *principle of competitive exclusion* or *Gause's Principle;* cf. Wilson and Bossert [1971, p. 157ff].) A question of basic interest is the following: Starting with an independent notion of competition, what is the minimum number of dimensions which can be used to describe an ecological phase space which at least reflects competition, i.e., so that competing species have overlapping niches and noncompeting species do not?

To get at the answer to this question, we consider the notion of a *food web* of an ecological community. This is a digraph whose vertex set is the set of all species[31] in the community. An arc is drawn from species u to species v

---

[31]The word "species" will be used rather loosely here.

if u preys on v. Figure 3.51 shows a small oversimplified food web with five species (we have already seen this example in Fig. 2.4) and Fig. 3.52 shows a food web with 15 different species.[32,33] Also shown is the adjacency matrix of this food web.

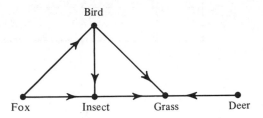

FIGURE 3.51. A food web.

From the food web, we can define competition: Two species *compete* if and only if they have a common prey. Then we can define the *competition graph*. This graph has as vertex set the set of species and two species are joined by an edge if and only if they have a common prey. The competition graph corresponding to the food web of Fig. 3.51 is shown in Fig. 3.53. Note that the bird and the insect compete, because they both eat grass. The fox and the insect do not compete. And so on. Figure 3.54 shows the competition graph for the food web of Fig. 3.52. We can now rephrase our question about ecological phase space as follows: What is the smallest number of dimensions k needed so that we can assign to each vertex u of the competition graph $G = (V, E)$ a box $B(u)$ in Euclidean k-space subject to the requirement that for all $u \neq v \in V$,

$$\{u, v\} \in E \Longleftrightarrow B(u) \cap B(v) \neq \emptyset. \tag{18}$$

If k is 1, this is the familiar condition that G be an interval graph. In general, G is the intersection graph of a family of boxes in k-space. Let us define the *boxicity* of G to be the smallest k so that G can be represented as an intersection graph of boxes in k-space. To show that boxicity is well-defined, one must verify that every graph is the intersection graph of boxes in some k-space. This we leave as an exercise (Exer. 18). Of course, an interval graph has boxicity 1.[34] It is not difficult to verify that the competition graph of Fig. 3.53 has boxicity 1, i.e., that it is an interval graph. One can do this directly, by exhibiting an interval assignment such as that shown in Fig. 3.55. More surprisingly, the competition graph of Fig. 3.54 also has boxicity 1. An interval assignment is shown in Fig. 3.56.

[32]See the previous footnote; we call "plant" a species.

[33]For further discussion of this food web, see Harary [1961b].

[34]Technically, one likes to speak of complete graphs, which are interval graphs, as having boxicity 0, since each vertex can be mapped into the same single point.

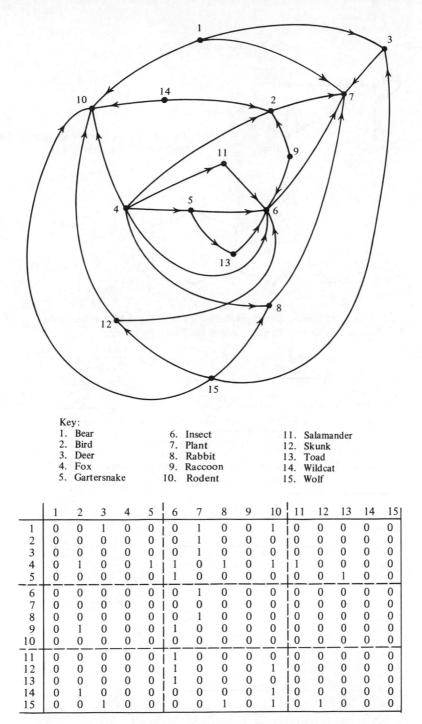

Key:
1. Bear
2. Bird
3. Deer
4. Fox
5. Gartersnake
6. Insect
7. Plant
8. Rabbit
9. Raccoon
10. Rodent
11. Salamander
12. Skunk
13. Toad
14. Wildcat
15. Wolf

	1	2	3	4	5	6	7	8	9	10	11	12	13	14	15
1	0	0	1	0	0	0	1	0	0	1	0	0	0	0	0
2	0	0	0	0	0	0	1	0	0	0	0	0	0	0	0
3	0	0	0	0	0	0	1	0	0	0	0	0	0	0	0
4	0	1	0	0	1	1	0	1	0	1	1	0	0	0	0
5	0	0	0	0	0	1	0	0	0	0	0	0	1	0	0
6	0	0	0	0	0	0	1	0	0	0	0	0	0	0	0
7	0	0	0	0	0	0	0	0	0	0	0	0	0	0	0
8	0	0	0	0	0	0	1	0	0	0	0	0	0	0	0
9	0	1	0	0	0	1	0	0	0	0	0	0	0	0	0
10	0	0	0	0	0	0	0	0	0	0	0	0	0	0	0
11	0	0	0	0	0	1	0	0	0	0	0	0	0	0	0
12	0	0	0	0	0	1	0	0	0	1	0	0	0	0	0
13	0	0	0	0	0	1	0	0	0	0	0	0	0	0	0
14	0	1	0	0	0	0	0	0	0	1	0	0	0	0	0
15	0	0	1	0	0	0	0	1	0	1	0	1	0	0	0

FIGURE 3.52. Food web and its adjacency matrix. Adapted from Burnett, Fisher, and Zim [1958], and from Harary [1961b], with permission of Western Publishing Company and the Society for General Systems Research.

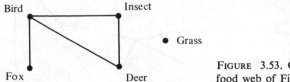

FIGURE 3.53. Competition graph for food web of Fig. 3.51.

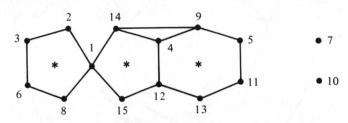

FIGURE 3.54. Competition graph for food web of Fig. 3.52. * indicates that all edges joining vertices in the *n*-gon are joined by an edge.

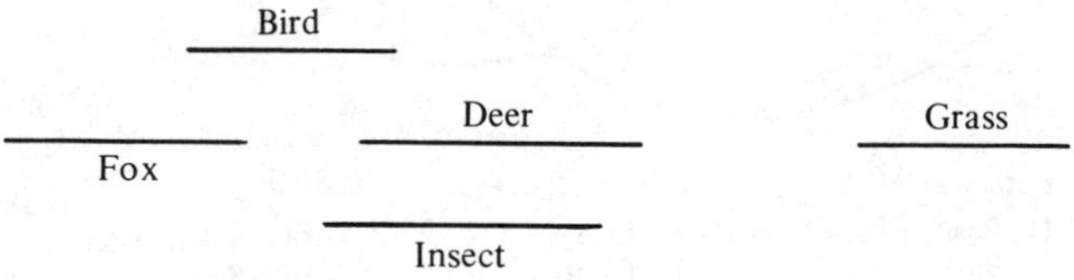

FIGURE 3.55. Interval assignment for the competition graph of Fig. 3.53.

FIGURE 3.56. Interval assignment for the competition graph of Fig. 3.54.

Strangely enough, inspection of a number of food webs has turned up the same result: Their competition graphs are interval graphs.[35] It is not known whether this is a general law, and if so, what interpretation might be put on the single dimension. Mathematically, an interesting question is to find conditions on a food web or digraph (necessary and) sufficient to guarantee that the corresponding competition graph is an interval graph. This question is still open at this time. It is even open if we assume the food web has no cycles, as we shall later in this section.

Let us calculate the boxicity of a number of other graphs. The graph Z_4, the circuit of length 4, has boxicity 2. For, by the criterion of Gilmore and Hoffman (Theorem 3.8, condition (a)), it is not an interval graph. But it can be represented as the intersection graph of boxes in 2-space, as shown in Fig. 3.57. As a second example, let us consider the graph $K(n_1, n_2, \ldots, n_p)$ defined as follows: There are $n_1 + n_2 + \cdots + n_p$ vertices, divided into p classes. The ith class has n_i vertices. Within a given class, there are no edges. Every vertex in one class is joined to every vertex in every other class. Figure 3.57 shows some graphs $K(n_1, n_2, \ldots, n_p)$ (in addition to showing the graph Z_4, which may be thought of as $K(2, 2)$. Why?). These graphs are usually called the *complete p-partite graphs*. Figure 3.57 also shows box representations for the complete p-partite graphs shown. Thus $K(1, 4)$ has boxicity 1. $K(2, 2)$ and $K(3, 3)$ are each not interval graphs. This follows easily from Theorem 3.8, since they each contain Z_4 as a generated subgraph. Thus, the box representations shown prove that these two graphs have boxicity 2. $K(2, 2, 2)$, it can be shown, has boxicity 3. In general, the boxicity of $K(n_1, n_2, \ldots, n_p)$ is the number of n_i bigger than 1.[36] This result is proved in Roberts [1969b]. (See Exer. 17.) For graphs other than $K(n_1, n_2, \ldots, n_p)$, the boxicity is in general unknown, and there is no procedure known for determining boxicity of a given graph. Once again, before we can make progess on the dimensionality of ecological phase space, we shall probably have to solve these mathematical problems. (Gabai [1974] has recently made considerable progress on them. (See Exer. 19.))

3.5.2. Trophic Status

As our second application to food webs, we mimic some results from an area of mathematical sociology called organization theory. Organization theorists have been concerned with the status of an individual in an organization. We mimic some results on measurement of status in organizations in trying to measure the "status" of a species in a food web.[37] In ecology, the

---

[35] Joel E. Cohen, personal communication.

[36] In particular, $K(1, 1, \ldots, 1) = K_p$ has boxicity 0.

[37] The idea of measuring status of species in food webs analogously to status in organizations is due to Harary [1961b], and part of our approach follows his. It is another example of an idea from the field of sociobiology which we mentioned in Sec. 3.2.

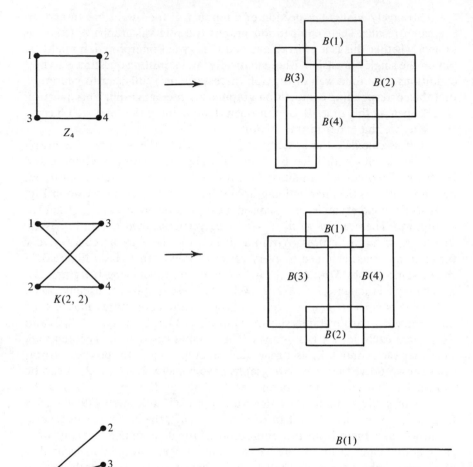

FIGURE 3.57. Box representations for some graphs. Continued on the next page.

status is usually called the *trophic level* (cf. Wilson and Bossert [1971, p. 139 ff.]). The trophic level is sometimes used to help measure the complexity or diversity of a food web. (Again, cf. Wilson and Bossert.) A web with many species at each trophic level has a high complexity. It is a widely-accepted principle of ecology, to which we return in Exer. 21, that the more complex ecosystems are more stable.

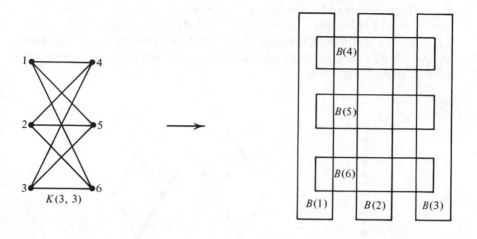

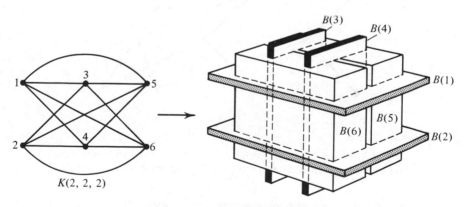

FIGURE 3.57 continued.

It is easy to define the trophic level of a species if the food web is a simple path (a food chain). It is also easy if the food web has a simple structure in which every path from a species to one with no outgoing arcs in the web has the same length. (Examples of such food webs are shown in Fig. 3.58.) Here, the trophic level can be simply defined as the length of such a path.

The question of how to define trophic level in general is difficult. The reader should think for example of how to define it in the food web of Fig. 3.52. To distinguish the idea of level in a complicated food web from that in a food chain, we shall use the term *trophic status* for the former, not trophic level.

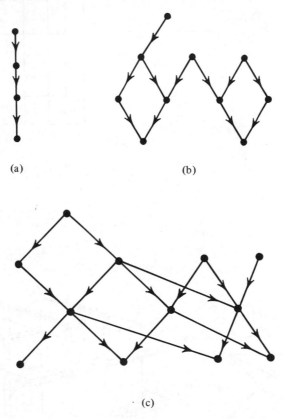

(a)   (b)

· (c)

FIGURE 3.58. Food webs in which trophic level or status can be measured as the number of levels from the "bottom." (Food web (a) is a food chain.)

To begin with, we make some general assumptions about food webs. Let us assume that our food web is acyclic (free of cycles), i.e., that there are no species $u_1, u_2, \ldots, u_t$ so that u_1 preys on u_2 which preys on $u_3 \ldots$ which preys on u_t which preys on u_1. In particular it follows that the prey relationship is *asymmetric*: If u preys on v, then v does not prey on u.

Our approach to obtaining a measure of trophic status is one we will use on a number of occasions in this book. We will write down certain conditions which a reasonable measure should satisfy. These will be stated in the form of axioms. Then a measure will be considered acceptable if and only if it satisfies the axioms. In some sense, the axioms define a mathematical model for an ideal measure of trophic status.

Now a given set of axioms can be sufficiently broad so as to allow several acceptable measures, as will be the case here. Later on in this book, we will encounter sets of axioms for certain measures which are sufficiently restrictive that there is exactly one measure satisfying them. This is an ideal situation—we no longer have a problem choosing if we accept all the axioms. Such a set of axioms is called *categorical*. We will also encounter a situation where all the axioms seem to be reasonable, but there is no measure satisfying all of them. Such a situation leads to a dilemma: Which of the axioms which we accepted are we willing to give up?

Our axiomatic approach follows that of Harary [1959] and the discussion in Kemeny and Snell [1962, Ch. 8], who deal with status in an organization. When we say that v is a (direct) prey of u, we shall think of the analogous idea that v is a direct subordinate of u, the relation which defines the organization digraph in Harary's paper. We shall say that v is an *indirect prey* of u if and only if $v \neq u$, v is not a direct prey of u, and v is reachable from u. Of two species u and v, it is quite possible that neither is a direct or indirect prey of the other. If v is a direct or indirect prey of u, let us define the *(trophic) level of v relative to u* as the length of the shortest (simple) path from u to v. With these preliminaries, we can now introduce the three axioms which a measure of trophic status will be asked to satisfy. Let $t_D(u)$ be the measure of status of species u in food web D. The axioms for t_D are as follows:

Axiom 1.   If species u in food web D has no prey, then $t_D(u) = 0$.

Axiom 2.   If, without otherwise changing the food web D, we add a new vertex which is a direct prey of u to obtain a new food web D', then $t_{D'}(u) > t_D(u)$.

Axiom 3.   Suppose the food web D is changed (by deletion of arcs or vertices) so that the level of some direct or indirect prey of u relative to u is increased and no (direct or) indirect prey of u has its level relative to u decreased. If D' is the new food web, then $t_{D'}(u) > t_D(u)$.

For motivation, the reader should think of the analogous axioms for an organization. For example, Axiom 1 says that if an individual in an organization has no direct subordinates, then his status is 0. Axiom 2 says that if we hire an additional direct subordinate of an individual, then the individual's status increases. Axiom 3 says that an individual's status is increased if he is moved further above some subordinate.

We shall prove that there is a measure of trophic status for acyclic food

webs which satisfies Axioms 1-3. In particular, one such measure is the following measure $h_D(u)$, which was suggested by Harary [1959]. If u has n_k species
at level k below it in the food web D, then

$$h_D(u) = \sum_k kn_k. \tag{19}$$

On food webs like those in Fig. 3.58, this measure differs from the measure
we originally suggested, namely the length of any path to a species with no
outgoing arcs. We are concerned with the number of species at various levels
below a species u, not just the level of u above the "bottom" species.

It should be observed that $h_D(u)$ is always a nonnegative integer. If we
add this restriction to our requirements about a measure of trophic status,
then $h_D(u)$ is in some sense a minimal measure satisfying the axioms, though
it is by no means unique. We summarize the results in the following theorem.

Theorem 3.11 (Kemeny and Snell [1962]).   The measure $h_D(u)$ defined by
Eq. (19) has the following properties:

(a)   It satisfies Axioms 1–3 for all acyclic food webs D.
(b)   If $t_D(u)$ is any other measure of trophic status of acyclic food webs
such that $t_D(u)$ is always a nonnegative integer and such that $t_D(u)$
satisfies Axioms 1–3, then for all vertices u, $t_D(u) \geqq h_D(u)$.

Proof.

(a)   Axiom 1 is immediate. If we add a direct prey of u, then $h_D(u)$
goes up 1, which verifies Axiom 2. If we move a direct or indirect
prey at level k below u to level $k' = k + p$ below u without decreasing the level relative to u of any other (direct) or indirect
prey of u, then $h_D(u)$ goes up by at least p. This verifies Axiom 3.
(b)   The proof will use the following lemma, whose proof we leave to
the reader as Exer. 11.

Lemma.   Suppose a digraph (V, A) is acyclic and $u \in V$. Then there is
a vertex $v \in V$ which is reachable from u but which has no arcs leading out
from it.

We argue by induction on $m = h_D(u)$. If $m = 0$, then u has no direct
prey and so, by Axiom 1, $t_D(u) = 0$. Assume that whenever $h_D(u) = m$ in an
acyclic food web D, then $t_D(u) \geqq h_D(u)$. Let D be an acyclic food web with
$h_D(u) = m + 1$. Thus, u has at least one direct prey in D. By the lemma, there
is a vertex v in D which is reachable from u and has no outgoing arcs. Build a
new food web D' from D as follows. If v is a direct prey of u, remove it. If
v is not a direct prey of u, consider a shortest simple path $u, u_1, u_2, \ldots, u_k, v$
from u to v in D and add an arc u_{k-1}, v to D. Then the level of v relative
to u goes down by 1 in D', and the level relative to u of all other direct or

indirect prey of u is unchanged. Hence, $h_{D'}(u) = m$. By the inductive assumption, $t_{D'}(u) \geqq m$. But by Axiom 2 in the first case or Axiom 3 in the second case, $t_D(u)$ is bigger than $t_{D'}(u)$. Thus, since t is an integer, $t_D(u) \geqq m + 1 = h_D(u)$.   Q.E.D.

The measure $h_D(u)$ has some drawbacks. It would seem reasonable to require that if t is a measure of trophic status and v is a direct or indirect prey of u, then $t_D(v) < t_D(u)$. The measure h, unfortunately, does not satisfy this property. Figure 3.59 shows an acyclic food web in which $h_D(u) = 6$ and $h_D(v) = 8$, even though v is a prey of u. (This is because all direct or indirect prey of u are direct prey of u, though there are a number of species at level two below v.)

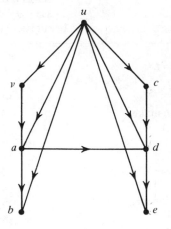

FIGURE 3.59. An acyclic food web D in which $h_D(u) = 6$ and $h_D(v) = 8$. From Harary, Norman, and Cartwright [1965, p. 273]. Reprinted with permission of John Wiley and Sons, Inc.

The problem we have encountered is avoided if we modify our definition of level of v relative to u. If v is a direct or indirect prey of u, let us define the *level of v relative to u* to be the length of the longest rather than the shortest simple path from u to v. Then we have the following theorem.

Theorem 3.12 (Kemeny and Snell [1962]).   If the level of v relative to u is defined using the length of the longest simple path, then the measure h defined by Eq. (19) satisfies Axioms 1–3 and also the following:

Axiom 4.   If v is a direct or indirect prey of u, then $t_D(v) < t_D(u)$.

Proof.   Left to the reader (Exer. 12).

Using the new definition of level, and referring again to the food web of Fig. 3.59, we find that $h_D(u) = 14$ and $h_D(v) = 8$, thus illustrating the fact that the problem with the old measure is avoided. It would be interesting to know whether there is a measure $t_D(u)$ satisfying Axioms 1–4 if the old definition of level is used.[38] (See Exer. 16.)

For illustration, we return to the food web of Fig. 3.52 and calculate the trophic status measure h, using both definitions of level, for all species. The results are shown in Table 3.3. It is interesting to note that the fox has highest trophic status in each method of calculating status, while the plant and the rodent are tied for lowest. In this example, the change from the first measure of trophic status to the second breaks some ties, as that between the bear and the salamander, but does not reverse any order of relative trophic status.

TABLE 3.3

Status of Species in the Food Web of Fig. 3.52.

Species	Status h calculated using level defined by shortest simple path*	Status h calculated using level defined by longest simple path
1 (Bear)	3	4
2 (Bird)	1	1
3 (Deer)	1	1
4 (Fox)	10	14
5 (Gartersnake)	4	6
6 (Insect)	1	1
7 (Plant)	0	0
8 (Rabbit)	1	1
9 (Raccoon)	4	4
10 (Rodent)	0	0
11 (Salamander)	3	3
12 (Skunk)	4	4
13 (Toad)	3	3
14 (Wildcat)	4	4
15 (Wolf)	8	9

*From Harary [1961b].

The problem with the measure h satisfying our first set of axioms points up one of the pitfalls of the axiomatic approach. The problem with the first axiom system was subtly hidden in the definition of level, something which could not be discovered by considering the "reasonableness" of the axioms. Ideally, a mathematical model, in particular a mathematical model making precise an imprecise term as here, will use no terms which do not have a firm axiomatic foundation. Of course, in practice, it is impossible to meet this requirement all the time. But the mathematical modeller must constantly be aware of the pitfalls involved if this requirement is not met.

[38]To the author's knowledge, this is an unsolved problem.

Indeed, this requirement is not met by the new axiom system (using the new definition of level of v relative to u) any more than it was met by the first axiom system. This throws the whole model into doubt.

Perhaps at this stage, the only reasonable requirement for a measure $t_D(u)$ of trophic status is the following:

$$\text{If } v \text{ is reachable from } u, \text{ then } t_D(v) < t_D(u). \qquad (20)$$

In Exer. 14, the reader is asked to prove that a measure of trophic status t_D satisfying (20) exists for all acyclic digraphs D. Moreover, if a measure t_D satisfying Eq. (20) exists for a given digraph D, then D must be acyclic.

EXERCISES

1. In each of the following, find the ecological niche in Euclidean 2-space determined by the following requirements.
   (a) Ozone count at least .001 and at most 2.0; temperature at least 0° C and at most 100° C.
   (b) Oxygen content in atmosphere at least 50% and at most 83%; rainfall at least 12 inches per year and at most 45 inches per year.

2. In the food web of Fig. 3.60, find all indirect prey of species 8.

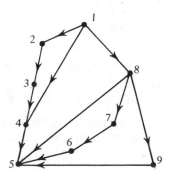

FIGURE 3.60. Food web for exercises of Sec. 3.5.

3. Find the competition graph G corresponding to the food web of Fig. 3.60.

4. Is the competition graph G of Exer. 3 an interval graph?

5. Calculate the trophic status $h_D(u)$ of Eq. (19) for every species in Fig. 3.60, using both definitions of level.

6. Give an example of a food web where every species has the same trophic status $h_D(u)$. (Do this for both definitions of level.)

7. Give an example of a food web where every species has a different trophic status $h_D(u)$. (Do this for both definitions of level.)

8. (Kemeny and Snell [1962].)

   (a) Suppose we have an organization with five members. Using the status measure $h_D(u)$ and the first definition of level, how would you organize things so that the president of the organization has as high a status as possible? How do things change if the second definition of level is used?

   (b) Repeat part (a) for as low a status as possible, under the assumption that each member is a direct or an indirect subordinate of the president.

   (c) Repeat parts (a) and (b) for an arbitrary number of members.

9. What is the boxicity of Z_n, the circuit of length n?

10. What is the boxicity of each of the graphs of Fig. 3.45?

11. Prove the lemma stated in the proof of Theorem 3.11.

12. Prove Theorem 3.12.

13. Give an example of an (acyclic) food web (hypothetical or real) whose competition graph is not an interval graph. (That is, find an acyclic digraph which gives rise to a noninterval graph when competition is defined as in food webs.)

14. (a) Show that there is a measure of trophic status t_D satisfying Eq. (20) for all acyclic digraphs D.

    (b) Show that if a measure t_D satisfying Eq. (20) exists for a digraph D, then D must be acyclic.

15. Suppose the longest simple path of an acyclic digraph D has length k. Show that there is a measure of trophic status satisfying Eq. (20) and having no more than $k + 1$ different status values.

16. If level is measured using the length of the shortest simple path, a plausible measure of trophic status is

$$t_D(u) = \sum \{h_D(v): v \text{ is reachable from } u\},$$

where h is given by Eq. (19). Show that $t_D(u)$ satisfies Axiom 4, but not all of the axioms we have adopted. Recall that u is reachable from itself.

17. (a) Show that if $b(G)$ is the boxicity of G, then

$$b[K(n_1, n_2, \ldots, n_p, 1)] = b[K(n_1, n_2, \ldots, n_p)].$$

    (b) Show that $b[K(2, 2, \ldots, 2)] = p$ if there are p classes.

    (c) Use (a) and (b) to prove that $b[K(n_1, n_2, \ldots, n_p)]$ is the number of n_i which are greater than 1. (Use the convention that $b(K_p) = 0$.)

*18. Show that boxicity is well-defined by proving that every graph G of n vertices is the intersection graph of a family of boxes in n-space. (*Hint*:

Let $J^k(u_k) = [0, 1]$ and if $i \neq k$, let

$$J^k(u_i) = \begin{cases} [1, 2] & (\text{if } \{u_i, u_k\} \in E(G)) \\ [2, 3] & (\text{if } \{u_i, u_k\} \notin E(G)). \end{cases}$$

Show that for $i \neq j$,

$$\{u_i, u_j\} \in E(G) \Longleftrightarrow \text{for all } k, J^k(u_i) \cap J^k(u_j) \neq \varnothing .)$$

19. A pair of edges in a graph is called *independent* if they have no vertices in common. An *independent set of edges* is a set of edges each pair of which is independent. Suppose G is a graph and G^c is its complementary graph. Gabai [1974] has shown that if the maximum size of a set of independent edges of G^c is m, then $b(G) \leq m$, where $b(G)$ is boxicity of G. Moreover, if G^c has a *generated subgraph* consisting of k independent edges, then $b(G) \geq k$.

(a) Find a set of independent edges in Z_4^c which is of maximum size.

(b) Find a generated subgraph of Z_4^c consisting of two independent edges.

(c) Apply one of Gabai's results to show that $b(Z_4) \leq 2$.

(d) Apply another of Gabai's results to show that $b(Z_4) \geq 2$.

(e) Apply Gabai's results to prove that if G is graph (c) of Fig. 3.45, then $b(G) \leq 2$ and $b(G) \geq 1$. Why can't you conclude from Gabai's results that $b(G) \geq 2$?

(f) Apply Gabai's results to calculate $b(G)$ exactly if G is $K(2, 2, \ldots, 2)$, with p classes.

20. One way of defining how *critical* or *important* species s is in a food web is to try to determine what happens to the food web if that species is removed. A measure of importance might be what happens to the connectedness of the food web if the vertex corresponding to species s is removed. The notion of i, j-vertex defined in Exer. 13 of Sec. 3.3 might be useful here.

(a) Assuming that food webs are acyclic, what kind of i, j-vertices can there be, for i and j running from 0 to 3?

(b) Can you think of other measures of importance?

21. It is a widely-accepted ecological principle (cf. Wilson and Bossert [1971, p. 139ff.]) that increased complexity of an ecosystem corresponds to increased stability and decreased vulnerability. If an ecosystem is represented by a food web (or a competition graph), there are various ways of measuring complexity: number of arcs or edges, number of arcs or edges divided by number of vertices, average outdegree, average indegree, etc. There are also various ways of measuring vulnerability.

(a) Decide if the notions of arc and vertex vulnerability developed in Sec. 3.3.2 and Exer. 16 of Sec. 3.3 are the appropriate ones for a food web.

(b)   For alternative measures of complexity and vulnerability of your choice, test the principle that complexity is inversely related to vulnerability. (Cf. Exer. 18, Sec. 4.6.)

(c)   How would you use the notion of trophic status developed in Sec. 3.5.2 to help measure complexity of a food web?

22.   Discuss the axioms for trophic status. Are they reasonable? Can you think of alternatives?

23.   What are some oversimplifications in our treatment of trophic status? How could these be incorporated into a more sophisticated model?

24.   How can our results about trophic status be applied to status in an organization? Invent some organizations and study the measures developed in the text.

3.6.   Garbage Trucks and Colorability

3.6.1. TOURS OF GARBAGE TRUCKS

In this section we start with a routing problem posed by the Department of Sanitation of the City of New York (cf. Beltrami and Bodin [1973] and Tucker [1973]).[39] It should be clear that techniques like those to be discussed can be applied to other routing problems, for example milk routes, air routes, etc. A garbage truck can visit a number of sites on a given day. A *tour* of such a truck is a schedule (an ordering) of sites it visits on a given day, subject to the restriction that the tour can be completed in one working day. We would like to find a set of tours with the following properties:

(1)   Each site i is visited a specified number k_i times in a week.

(2)   The tours can be partitioned among the six days of the week (Sunday is a holiday) in such a way that (a) no site is visited twice on one day[40] and (b) no day is assigned more tours than there are trucks.

(3)   The total time involved for all trucks is minimal.

In one method proposed for solving this problem, one starts with any given set of tours and successively improves the set as far as total time is concerned. (In the present state of the art, the method comes close to a minimal set, but does not always reach one.) At each step, the given improved collection of tours must be tested to see if it can be partitioned in such a way as to satisfy condition (2a), i.e., partitioned among the six days of the week in

---

[39]Other interesting applications of graph theory to sanitation can be found in Tucker and Bodin [1975].

[40]Requirement (a) is included to guarantee that garbage pickup is spread out enough to make sure that there is no accumulation.

such a way that no site is visited twice on one day. Thus, we need an efficient test for partitionability which can be applied over and over again. Formulation of such a test reduces to a problem in graph theory, and that problem will be the subject of this section. (The reader is referred to Beltrami and Bodin [1973] and to Tucker [1973] for a description of the treatment of the total problem.)

To test if a given collection of tours can be partitioned so as to satisfy condition (2a), let us define a graph G, the *tour graph*, as follows. The vertices of G are the tours in the collection, and two distinct tours are joined by an edge if and only if they service some common site. Then the given collection of tours can be partitioned into six days of the week in such a way that condition (2a) is satisfied if and only if the collection of vertices $V(G)$ can be partitioned into six classes with the property that no edge of G joins vertices in the same class. It is convenient to speak of this question in terms of colors. Each class in the partition is assigned one of six colors and we ask for an assignment of colors to vertices such that no two vertices of the same color are joined by an edge.[41]

Speaking more generally, let us say a graph G is *k-colorable* if the vertices of G can be partitioned into k classes, each corresponding to a given color, in such a way that no two vertices joined by an edge have the same color. Figure 3.61 shows a graph which is 3-colorable; the 3-coloring is shown. It is not 2-colorable (why?). The question about tours can now be rephrased as follows: Is the tour graph 6-colorable?

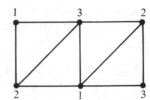

FIGURE 3.61. Graph with a 3-coloring.

If G is a graph, let $\chi(G)$ be the smallest k such that G is k-colorable. (Every graph of n vertices is obviously n-colorable: Use a different color for each vertex.) The number $\chi(G)$ is called the *chromatic number* of G. A collection of tours can be partitioned so as to satisfy condition (2a) if and only if the chromatic number of the tour graph is less than or equal to 6. In the rest of this section, we prove a number of theorems about colorability of graphs, chromatic number, and related subjects.

In general, to apply the procedure for finding a minimal set of tours, one has to have a procedure or algorithm, which can be quickly applied over and over again, for deciding whether a given graph is k-colorable. The problem of

[41]This idea is due to Tucker [1973].

whether a given graph is k-colorable reduces to a problem in the field of integer programming. (See Berge [1962, Ch. 4].) However, there is not always a "good" algorithm for solving this problem. Indeed, in general, it is not known whether there is a "good" algorithm for deciding if a given graph is k-colorable.[42] The garbage truck routing problem has thus reduced to a difficult mathematical question. However, formulation in precise mathematical terms has made it clear why this is a hard problem, and it has also given us many tools to use in solving it, at least in special cases. We present some of these tools, some results on colorability, here. Let us remark that in a real-world situation, it is not sufficient to say a problem is unsolvable or hard. Imagine the $500,000 consultant walking into the mayor's office and reporting that after careful study, he has concluded that the problem of routing garbage trucks is hard! Garbage trucks must be routed. So what can you do in such a situation? The answer is, you develop partial solutions, you develop solutions which are applicable only to certain special situations, you modify the problem, or, in some cases, you even "lie." You lie by using results which are not necessarily true, but which seem to work. We will describe these below.

3.6.2. Theorems on Colorability

Our first result on colorability is a characterization of the 2-colorable graphs. (Such graphs are sometimes called *bipartite*.)

Theorem 3.13 (König [1936, p. 170]).   A graph G is 2-colorable if and only if it has no circuits of odd length.

Proof.   This theorem follows directly from Harary's Theorem which characterizes balanced signed graphs (Theorem 3.1). We leave a proof of this to the reader (Exer. 21).

Next, we shall seek upper and lower bounds on $\chi(G)$. These bounds will be given in terms of other numbers associated with a graph. To define the first of these numbers, let us introduce the notation $\delta(u)$ for the *degree* of a vertex u, the number of vertices connected to u by an edge. Then we define $\Delta(G)$ for a graph G to be the maximum of $\delta(u)$ for vertices u in G. For example, the graph G of Fig. 3.62 has $\Delta(G) = 4$ since $\delta(y) = 4$.

Theorem 3.14. $\chi(G) \leqq 1 + \Delta(G)$.

Proof.   The proof is by induction on n, the number of elements of $V(G)$. If $n = 1$, the result is obvious. Arguing by induction, let vertex u in G have degree $\Delta(G)$. Then $\Delta(G - u) \leqq \Delta(G)$, where $G - u$ is obtained from G by

---

[42]A "good" algorithm or procedure, according to a fairly popular line of thought, is one which takes no more than $f(n)$ steps where n is the size of the input (here, the number of vertices) and $f(n)$ is a polynomial in n. This is the notion of "good" algorithm we are using. For a detailed discussion of this idea, and in particular a discussion of algorithms for determining if a graph is k-colorable, the reader is referred to Karp [1972].

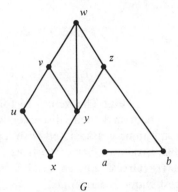

FIGURE 3.62. $\chi(G) = 3$, $\Delta(G) = 4$,
$\omega(G) = 3$.

removing u and all edges containing u. By inductive assumption, $G - u$ can
be colored in $1 + \Delta(G)$ colors. Now there is one color among these which can
be used for u, since the vertices joined to u are colored in at most $\Delta(G)$ colors.

Q.E.D.

A stronger theorem is the following:

Theorem 3.15 (Brooks [1941]). $\chi(G) \leq 1 + \Delta(G)$, with equality if and
only if G is an odd circuit or a complete graph.

It is easy enough to verify that if n is odd, then $\chi(Z_n) = 3 = 1 + \Delta(Z_n)$,
and that for all n, $\chi(K_n) = n = 1 + \Delta(K_n)$. It is not so easy to show that
$\chi(G) < 1 + \Delta(G)$ otherwise, and we shall omit the proof. Brooks' Theorem
(in its weak version, Theorem 3.14) has been generalized as follows:

Theorem 3.16 (Szekeres and Wilf [1968]). Suppose $\lambda(G)$ is a function
which assigns a real number to each graph G. Moreover, suppose that $\lambda(G)$
satisfies the following conditions:

  (a) If G' is a generated subgraph of G, then $\lambda(G') \leq \lambda(G)$.
  (b) $\lambda(G) \geq \delta(G) = \min \{\delta(u) : u \in V(G)\}$.

Then $\chi(G) \leq 1 + \lambda(G)$.

Proof. A graph H is called *critical* (with respect to colorability) if
$\chi(H - u) < \chi(H)$ for all vertices u in H. For example, Z_3 is critical. If H is
critical and $\chi(H) = k$, then $\delta(u) \geq k - 1$ for all vertices u in H. For suppose
$\delta(u) < k - 1$ for some u. $H - u$ is colorable in $k - 1$ colors, since $\chi(H - u)$
$< \chi(H)$. A $(k - 1)$-coloring of $H - u$ can now be extended to a $(k - 1)$-
coloring of H, by coloring u with a color different from that of its neighbors.

Given a graph G, it is easy to show that there is a generated subgraph
G_c of G with the same chromatic number and such that G_c is critical. Let
$k = \chi(G_c) = \chi(G)$. By our above remarks, $\delta(G_c) \geq k - 1$. Finally, using

properties (a) and (b),

$$\lambda(G) \geqq \lambda(G_c) \geqq \delta(G_c) \geqq k - 1,$$

so

$$\lambda(G) + 1 \geqq k = \chi(G). \qquad \qquad \text{Q.E.D.}$$

If is easy to see that Theorem 3.14 follows from Theorem 3.16. Other special cases of Theorem 3.16 are discussed in Exer. 28 and Exer. 9 of Sec. 4.2.

The second number associated with a graph which we shall consider is the *clique number*, the number of vertices in the largest clique of G.[43] It is denoted by $\omega(G)$ (Greek "omega"). The graph of Fig. 3.62 has $\omega(G) = 3$, since the largest clique is a triangle.

Theorem 3.17. $\chi(G) \geqq \omega(G)$.

Proof.   Each vertex of a clique must get a different color.   Q.E.D.

Figure 3.63 gives a graph G whose chromatic number $\chi(G)$ is greater than its clique number $\omega(G)$. For $\chi(G) = 4$ and $\omega(G) = 3$.

Graphs with $\chi(G) = \omega(G)$ are called *weakly γ-perfect*.[44] If $\chi(G') = \omega(G')$ for all generated subgraphs G' of G, then G is called *γ-perfect*. For weakly γ-perfect and γ-perfect graphs, the chromatic number can be computed by computing the number of vertices in the largest clique. We shall see below how this enters into the solution of the garbage truck problem. For a detailed discussion of γ-perfect graphs, see Berge [1973, Ch. 16]. We have already encountered a number of γ-perfect graphs in this chapter, as the following theorem shows.

Theorem 3.18 (Berge [1967]).   The following classes of graphs are γ-perfect:
   (a) Bipartite graphs.[45]
   (b) Transitively orientable graphs.
   (c) Interval graphs.

Proof.   The generated subgraph of a graph in one of these classes is again in the class. (Why?) Thus it is sufficient to prove that $\chi = \omega$ for G itself.

---

[43]Recall that a clique of G is a subgraph of G in which all vertices are joined by an edge.

[44]γ is used because in the literature $\gamma(G)$ often represents the minimum number of independent sets covering all the vertices of G, where a set of vertices U is called *independent* if there are no edges joining vertices of U in G. It is easy to prove that $\gamma(G) = \chi(G)$.

[45]Of course, every bipartite graph is transitively orientable. (Why?) Thus, (b) implies (a).

(a) Apply König's Theorem (Theorem 3.13). If G is bipartite, it either has $\chi = 2$ or $\chi = 1$. (One of the classes may be empty.) If $\chi = 1$, then the maximal clique has one element. If $\chi = 2$, there is an edge joining two elements of different classes and so there is a clique consisting of two elements. By König, there are no longer cliques.

(b) Suppose $G = (V, E)$ is transitively orientable and A is an orientation. For all $u \in V$, let $f(u)$ be the length of the longest simple path in (V, A) beginning at u. The number $f(u)$ is taken to be 0 if there is no such simple path. Suppose max $f(u) = k - 1$, and let $u_1, u_2, \ldots, u_k$ be a simple path of length $k - 1$ in (V, A). By transitivity of A, $(u_i, u_j) \in A$ for all $i < j$. Thus, $\{u_1, u_2, \ldots, u_k\}$ forms a clique of G. We conclude $\omega(G) \geq k$.

   Now color G in the k colors $0, 1, 2, \ldots, k - 1$, as follows: Assign color $f(u)$ to vertex u. To prove this is a coloring, suppose $\{u, v\} \in E(G)$. Then $(u, v) \in A$ implies $f(u) > f(v)$. For if $v, v_1, \ldots, v_r$ is a longest simple path beginning at v, then $u, v, v_1, \ldots, v_r$ is a longer simple path beginning at u. (It is simple since $u \neq v_i$ for all i, else by transitivity $(v, u) \in A$ and A is not an orientation.) Similarly, $(v, u) \in A$ implies $f(v) > f(u)$. Thus, $f(u) \neq f(v)$. We conclude that $\chi(G) \leq k$.

   Finally, we have $k \leq \omega(G) \leq \chi(G) \leq k$, so $\omega(G) = \chi(G)$.

(c) See Exers. 29 and 30.   Q.E.D.

FIGURE 3.63. $\chi(G) > \omega(G)$.

It has been conjectured that a graph G is γ-perfect if and only if neither G nor G^c has a graph Z_n (n odd, $n > 3$) as a generated subgraph. This conjecture, for which there is a great deal of evidence, is known as the *Strong Perfect Graph Conjecture*, and is due to Claude Berge. If the Strong Perfect Graph Conjecture is accepted as being true, then there is a much improved test for k-colorability in the sanitation scheduling situation. To decide if a given tour graph is 6-colorable fairly quickly, check first whether G or G^c has an odd circuit of length greater than 3. If so, use the old, inefficient algorithm for determining if G is 6-colorable. But if not, use the Strong Perfect Graph Conjecture, and determine if G is 6-colorable by determining if G has a clique of size greater than 6. Tucker [1973] has argued that the algorithm for routing garbage trucks would be greatly speeded up by substituting this check for 6-colorability.[46] If that is done, the user is in effect lying—he is using a result which is not known to be true. But what is the worst thing that could happen to him? The final set of tours produced by the entire procedure might not be amenable to a 6-day partition. However, the user could always check this independently at the end, for this one set of tours, at a great savings in time, since the check for 6-colorability must be repeated often. If this final set of tours is not 6-colorable, the user will have wasted some computer time, but as a byproduct he will have found a coun- terexample to the Strong Perfect Graph Conjecture, and so solved an unsolved mathematical problem!

3.6.3. The Four-Color Problem

The notion of graph coloring is historically significant in part because of its connection with the famous four-color problem. We shall discuss this

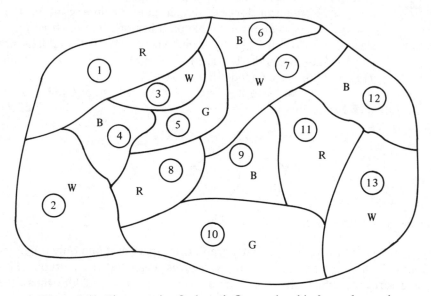

Figure 3.64. The countries ① through ⑬ are colored in four colors, red (R), white (W), blue (B), and green (G).

---

[46]Tucker's procedure is efficient only if there is a good algorithm for checking whether or not a graph has a clique of size k. In general, no good algorithm for this test is known, and it is not known whether there exists a good algorithm. However, in the context of garbage truck routing, the test for 6-colorability is applied successively to graphs which have been changed only locally, i.e., by making changes near a given vertex. After a local change is made in a graph whose chromatic number and clique number are known, determining the chromatic number requires testing the entire graph while determining the clique number does not. Thus, testing for colorability by testing for clique number is an efficient procedure here.

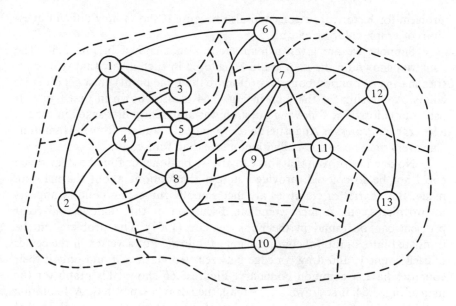

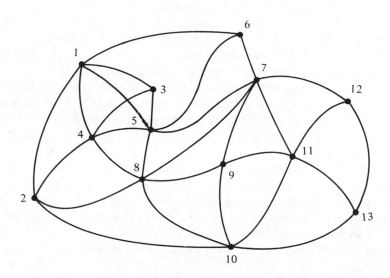

FIGURE 3.65. Graph obtained from the map of Fig. 3.64.

problem for historical reasons and also because it shows how difficult questions of graph coloring can be.

Suppose we are given a map such as that shown in Fig. 3.64. The countries shown on the map are to be colored in such a way that two countries sharing a common boundary (of more than one point) must get different colors. A coloring for the map in Fig. 3.64 is also shown in that figure. It uses four colors. The following question was apparently first asked in 1852 in a letter from Professor Augustus de Morgan to Sir William Rowan Hamilton: Can every map be so colored with no more than four colors?

No one has proven that this can always be done, or found a map which could not be so colored, provided simple assumptions about the map are made: Boundaries of countries are nice smooth curves, no country appears in two pieces, etc. A most complete discussion of this famous four-color problem can be found in the book by Ore [1967]. The problem can be translated into a graph theory problem as follows. Put a vertex in the center of each country, and draw an edge between two vertices if and only if their countries have a common boundary. Figure 3.65 shows this graph for the map of Fig. 3.64, first drawn on the map, then drawn separately. A 4-coloring of the map is equivalent to a 4-coloring of the vertices of this graph so that vertices joined by an edge get different colors. Thus, the four-color problem can be restated as follows: Is every graph which arises from a map 4-colorable?

The next question is: What graphs arise from maps? We see in Fig. 3.65 that the graph shown has been drawn with no edges crossing (except at vertices). A graph which can be so drawn is called *planar*. For example, the graph $K(2, 2)$ of Fig. 3.57 is planar, even though it is drawn in that figure with edges crossing. It is easy to redraw it so that no edges cross. Planar graphs are important in electrical network theory, in the design of printed circuits. A circuit can be printed on one board with wires crossing only at connection points if and only if the circuit defines a planar graph.

It is not too hard to show that it is exactly the planar graphs which arise from maps in the manner described. Thus, the four-color problem reduces to the following graph theory problem: Is every planar graph 4-colorable? (It is known that every planar graph is 5-colorable. For a proof of this result, see Harary [1969, p. 130].)

It is helpful, in understanding the question of 4-colorability of planar graphs, to determine which graphs are planar. One famous nonplanar graph is the complete 2-partite graph $K(3, 3)$—the *water-light-gas graph*. (See Fig. 3.57.) There are three houses and three utilities, water, light, and gas, and the object is to connect each house to each utility without crossing connecting edges. A second nonplanar graph is the complete graph[47] on five vertices,

---

[47]In the complete graph, all pairs of vertices are joined by edges.

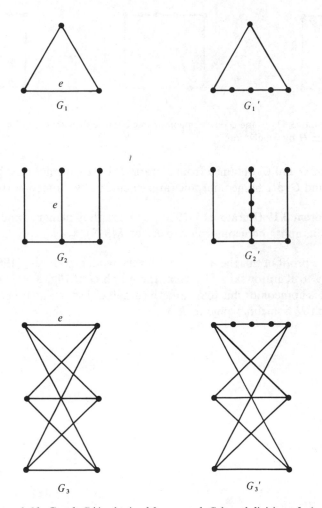

FIGURE 3.66. Graph $G_i{}'$ is obtained from graph G_i by subdivision of edge e.

K_5. (K_4 is planar.) According to a famous theorem of Kuratowski [1930], $K(3, 3)$ and K_5 are essentially the only two nonplanar graphs. To make this result precise, let us say that graph G' is obtained from graph G by *subdivision* if we obtain G' by adding vertices on some edge of G. (A vertex added can be on only one edge.) In Fig. 3.66, graph $G_i{}'$ is always obtained from graph G_i by subdivision. Two graphs G and G' are called *homeomorphic* if both can be obtained from the same graph H by a sequence of subdivisions of edges. For example, any two simple chains are homeomorphic. Figure 3.67 shows

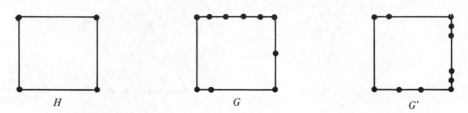

FIGURE 3.67. *G* and *G'* are homeomorphic because they are each obtained from *H* by subdivision.

two graphs *G* and *G'* obtained from a graph *H* by a sequence of subdivisions. Thus, *G* and *G'* are homeomorphic (and, incidentally, homeomorphic to *H*).

Theorem 3.19 (Kuratowski [1930]). A graph is planar if and only if it has no subgraph[48] homeomorphic to K_5 or $K(3, 3)$.

For a proof of this theorem, we refer the reader to Harary [1969, p. 109]. According to Kuratowski's Theorem, the graph *G* of Fig. 3.68 is not planar because it is homeomorphic to K_5 and the graph *G'* is not planar because it has a subgraph *H* homeomorphic to $K(3, 3)$.

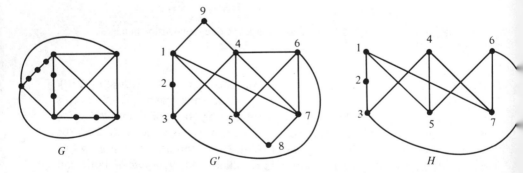

FIGURE 3.68. *G* is homeomorphic to K_5 and *G'* has the subgraph *H* which is homeomorphic to $K(3, 3)$.

[48]Not necessarily a generated subgraph.

In sum, the four-color problem reduces to the question of whether every graph with no subgraphs homeomorphic to K_5 or $K(3, 3)$ is 4-colorable. And that, in spite of a great deal of work which continues today, is approximately where it still stands.

EXERCISES

1. Consider the following four tours of garbage trucks on the West Side of New York City. Tour 1 visits sites from 21st Street to 30th, Tour 2 visits sites from 28th to 40th Streets, Tour 3 visits sites from 35th to 50th Streets, and Tour 4 visits sites from 80th to 110th Streets. Draw the corresponding tour graph.

2. In Exer. 1, can the tours each be scheduled on Monday or Tuesday in such a way that no site is visited twice on the same day?

3. For each graph of Fig. 3.69, determine if it is 3-colorable.

4. For each graph of Fig. 3.69, determine its chromatic number $\chi(G)$.

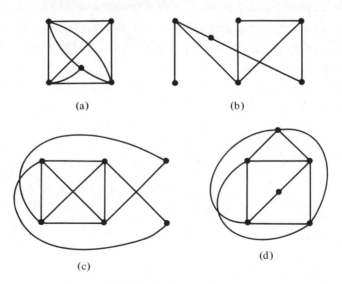

(a)   (b)

(c)   (d)

FIGURE 3.69. Graphs for exercises of Sec. 3.6.

5. A local zoo wants to take visitors on animal feeding tours, and has hit upon the following tours. Tour 1 visits the lions, elephants, and ostriches, Tour 2 the monkeys, birds, and deer, Tour 3 the elephants, zebras, and giraffes, Tour 4 the birds, reptiles, and bears, and Tour 5 the kangaroos, monkeys, and seals. If animals should not get fed more than once a day, can these tours be scheduled using only Monday, Wednesday, and Friday?

6. The following tours of garbage trucks in New York City are being considered (behind the mayor's back). Tour 1 picks up garbage at the Empire State Building, then Madison Square Garden, and then Pier 42 on the Hudson River. Tour 2 visits Greenwich Village, Pier 42, the Empire State, and the Metropolitan Opera House. Tour 3 visits Shea Stadium, the Bronx Zoo, and the Brooklyn Botanical Gardens. Tour 4 goes to the Statue of Liberty and Pier 42, Tour 5 to the Statue of Liberty, the New York Stock Exchange, and the Empire State, Tour 6 to Shea Stadium, Yankee Stadium, and the Bronx Zoo, and Tour 7 to the New York Stock Exchange, Columbia University, and the Bronx Zoo. Assuming sanitation men refuse to work more than three days a week, can these tours be partitioned so that no site is visited more than once on a given day?

7. In which of the graphs G of Fig. 3.69 is $\chi(G) = 1 + \Delta(G)$?

8. In which of the graphs G of Fig. 3.69 is $\chi(G) = \omega(G)$?

9. Show that if $\omega(G) \geqq 3$, then G is not bipartite (2-colorable).

10. Give an example of a graph G with n vertices and $\chi(G) = 2$.

11. Give an example of a graph G with n vertices and $\chi(G) = n$.

12. Which of the graphs G_1, G_2, G_3 of Fig. 3.66 is weakly γ-perfect? Which is γ-perfect?

13. Color the map of Fig. 3.70 in at most four colors. Can it be colored in fewer than four colors?

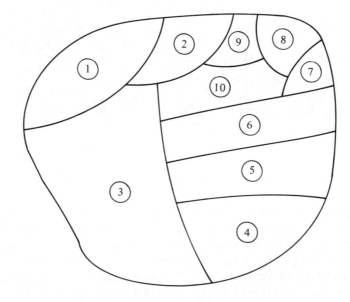

FIGURE 3.70. Map for exercises of Sec. 3.6.

14. Translate the map of Fig. 3.70 into a graph G and calculate $\chi(G)$.

15. Is $K(4, 4)$ planar?

16. Could K_6, the complete graph on six vertices, come from a map?

17. Can K_3 be obtained by subdivision from K_2?

18. Are K_2 and K_3 homeomorphic?

19. Which of the graphs of Fig. 3.69 are planar?

20. Give an example of a graph G such that $\chi(G) < 1 + \Delta(G)$.

21. Show how to prove König's Theorem about 2-colorability from Harary's Theorem about balance.

22. A graph G is *k-edge-colorable* if you can color the edges with k colors so that two edges with common vertices get a different color. Let $\chi'(G)$, the *edge chromatic number*, be the smallest k so that G is k-edge-colorable. State a relation between $\Delta(G)$ and $\chi'(G)$.

*23. A set of vertices in a graph G is *independent* if there are no edges between any two vertices in the set. The *vertex independence number* $\beta_0(G)$ is the size of the largest independent set of vertices. If G has n vertices, show that

$$\frac{n}{\beta_0(G)} \leqq \chi(G) \leqq n - \beta_0(G) + 1.$$

24. A graph G is called *k-critical* if $\chi(G) = k$ but $\chi(G - u) < k$ for each vertex $u \in V(G)$.
    (a) Find all 2-critical graphs.
    (b) Give an example of a 3-critical graph.
    (c) Can you identify *all* 3-critical graphs?

25. A graph G is *uniquely k-colorable* if any coloring of the vertices in k colors induces the same partition of vertices into color sets. For example, the graph G_1 of Fig. 3.71 is uniquely 3-colorable because any 3-coloring induces the partition $\{1\}, \{2\}, \{3\}$.
    (a) Is the graph G_2 of Fig. 3.71 uniquely 3-colorable? Why?
    (b) Let $\delta(G)$ be the minimum $\delta(u)$ over all vertices u in G. If G is uniquely k-colorable, is it true that $\delta(G) \geqq k - 1$? (Give proof or counterexample.)

26. Refer to the definition of a tree, in Exer. 22, Sec. 2.2. What is the largest possible chromatic number of a tree?

27. Give an example of a graph which is weakly γ-perfect but not γ-perfect.

28. Let $\lambda(G)$ be max $\{\delta(G'): G'$ is a generated subgraph of $G\}$, where $\delta(G') = \min \{\delta(u): u \in V(G')\}$.
    (a) Prove that $\chi(G) \leqq 1 + \lambda(G)$.
    (b) Give an example where $\lambda(G)$ is less than $\Delta(G)$, so the bound in (a) is an improvement over that in Theorem 3.14.

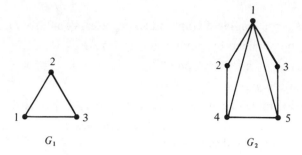

FIGURE 3.71. G_1 is uniquely 3-colorable.

29. Lovász [1972] proved the so-called *Weak Perfect Graph Conjecture*: G is γ-perfect if and only if G^c is γ-perfect. Show that Theorem 3.18(c) follows.

30. This exercise sketches a direct proof of Theorem 3.18(c).

   (a) An *articulation set* of a connected graph G is a set of vertices whose removal disconnects G. (Compare the definition of cut vertex in Exer. 4, Sec. 3.3.) Find a minimal articulation set in each of the graphs of Fig. 3.69.

   *(b) (Hajnal and Surányi [1958]). Recall the definition of rigid circuit graph from Exer. 25, Sec. 3.4. Show that every minimal articulation set in a connected rigid circuit graph is a clique.

   *(c) Let U be an articulation set in a connected graph G, let $K_1, K_2, \ldots, K_r$ be the connected components of $G - U$, the subgraph generated by $V(G) - U$, and let G_i be the subgraph generated by the vertices of U plus those of K_i. Show that if U is a clique and each G_i is weakly γ-perfect, then G is weakly γ-perfect.

   (d) Show that every rigid circuit graph is γ-perfect.

   (e) Show that every interval graph is γ-perfect.

31. Show that if G is a critical graph (cf. the proof of Theorem 3.16), then no articulation set (Exer. 30) of G is a clique.

32. How does the notion of colorability apply to the class scheduling problem of Exer. 28, Sec. 3.4?

References

ABELL, P., and JENKINS, R., "Perception of the Structural Balance of Part of the International System of Nations," *J. of Peace Res.*, **4** (1967), 76–82.

AHO, A., HOPCROFT, J., and ULLMAN, J., *The Design and Analysis of Computer Algorithms*, Addison-Wesley Publishing Co., Inc., Reading, Mass., 1974.

AXELROD, R. M., "Framework for a General Theory of Cognition and Choice," Institute of International Studies, Univ. of California, Berkeley, Research Series, No. 18, 1972.

AXELROD, R. M., "Schema Theory: An Information Processing Model of Perception and Cognition," *Amer. Pol. Sci. Rev.*, **4** (1973), 1248–1266.

BELTRAMI, E., and BODIN, L., "Networks and Vehicle Routing for Municipal Waste Collection," *Networks*, **4** (1973), 65–94.

BENZER, S., "On the Topology of the Genetic Fine Structure," *Proc. Nat. Acad. Sci. USA*, **45** (1959), 1607–1620.

BENZER, S., "The Fine Structure of the Gene," *Sci. Amer.*, **206** (1962), 70–84.

BERGE, C., *The Theory of Graphs and its Applications*, Methuen, London, 1962.

BERGE, C., "Some Classes of Perfect Graphs," in *Graph Theory and Theoretical Physics*, F. Harary (ed.), Academic Press, Inc., New York, 1967.

BERGE, C., *Graphs and Hypergraphs*, American Elsevier Publishing Co., New York, 1973 (transl. from the French edition by E. Minieka).

BONHAM, M., and SHAPIRO, M., "Explanation of the Unexpected: the Syrian Intervention in Jordan in 1970," in *The Structure of Decision*, R. M. Axelrod (ed.), Princeton University Press, Princeton, N.J., to appear.

BOOTH, K. S., and LUEKER, G. S., "Linear Algorithms to Recognize Interval Graphs and Test for the Consecutive Ones Property," *Proceedings of the Seventh Annual ACM Symposium on Theory of Computing*, May 1975, 255–265.

BROOKS, R. L., "On Coloring the Nodes of a Network," *Proc. Cambridge Philos. Soc.*, **37** (1941), 194–197.

BURNETT, R. W., FISHER, H. I., and ZIM, H. S., *Zoology: An Introduction to the Animal Kingdom*, Golden Press, New York, 1958.

CAMION, P., "Chemie et circuits hamiltoniens des graphes complets," *C. R. Acad. Sci. Paris*, **249** (1959), 2151–2152.

CARTWRIGHT, D., and HARARY, F., "Structural Balance: A Generalization of Heider's Theory," *Psych. Rev.*, **63** (1956), 277–293.

COOMBS, C. H., and SMITH, J. E. K., "On the Detection of Structure in Attitudes and Developmental Processes," *Psych. Rev.*, **80** (1973), 337–351.

DANZER, L., and GRÜNBAUM, B., "Intersection Properties of Boxes in R^d," Mimeographed, Department of Mathematics, Univ. of Washington, Seattle, August, 1967.

DAVIS, J. A., "Clustering and Structural Balance in Graphs," *Human Relations*, **20** (1967), 181–187.

DIRAC, G. A., "On Rigid Circuit Graphs," *Abhandlungen Mathematischen Seminar Universität Hamburg*, **25** (1961), 71–76.

DOREIAN, P., "Interaction under Conditions of Crisis: Applications of Graph Theory to International Relations," *Peace Res. Soc. (International) Papers*, **11** (1969), 89–107.

DOWLY, A., "Conflict in War Potential Politics," *Peace Res. Soc. (International) Papers*, **13** (1970), 85–105.

FISHER, J., "Reprieve for the Wolf," *National Wildlife*, **13** (1975), 5–10.

FOULKES, J. D., "Directed Graphs and Assembly Schedules," *Proc. Symp. Appl. Math.*, Amer. Math. Soc., **10** (1960), 281–289.

FULKERSON, D. R., and GROSS, O. A., "Incidence Matrices and Interval Graphs," *Pacific J. of Math.*, **15** (1965), 835–855.

GABAI, H., "*N*-dimensional Interval Graphs," Mimeographed, York College, C.U.N.Y., New York, 1974.

GALLAI, T., "Transitiv Orientierbare Graphen," *Acta. Math. Acad. Sci. Hungar.*, **18** (1967), 25–66.

GHOUILA-HOURI, A., "Caractérisation des graphes nonorientés dont on peut orienter les arêtes de manière a obtenir le graphe d'une relation d'ordre," *C. R. Acad. Sci. Paris*, **254** (1962), 1370–1371.

GILMORE, P. C., and HOFFMAN, A. J., "A Characterization of Comparability Graphs and of Interval Graphs," *Canadian J. Math.*, **16** (1964), 539–548.

GOLUMBIC, M. C., *Comparability Graphs and a New Matroid*, Ph.D. Thesis, Columbia University, 1975. (To appear in *J. Comb. Theory*.)

HADWIGER, H., DEBRUNNER, H., and KLEE, V., *Combinatorial Geometry in the Plane*, Holt, Rinehardt & Winston, Inc., New York, 1964.

HAJNAL, A., and SURÁNYI, J., "Über die Auflösung von Graphen in Vollständige Teilgraphen," *Ann. Univ. Sc. Budapestinensis*, **1** (1958), 113–121.

HARARY, F., "On the Notion of Balance of a Signed Graph," *Michigan Math. J.*, **2** (1954), 143–146.

HARARY, F., "Status and Contrastatus," *Sociometry*, **22** (1959), 23–43.

HARARY, F., "A Structural Analysis of the Situation in the Middle East in 1956," *J. of Conflict Resolution*, **5** (1961), 167–178, a.

HARARY, F., "Who Eats Whom," *General Systems*, **6** (1961), 41–44, b.

HARARY, F., "A Graph Theoretic Approach to Similarity Relations," *Psychometrika*, **29** (1964), 143–151.

HARARY, F., *Graph Theory*, Addison-Wesley Publishing Co., Inc., Reading, Mass., 1969.

HARARY, F., NORMAN, R. Z., and CARTWRIGHT, D., *Structural Models: An Introduction to the Theory of Directed Graphs*, John Wiley & Sons, Inc., New York, 1965.

HART, J., "Structures of Influence and Cooperation-Conflict," *International Interactions*, **1** (1974), 141–162, a.

HART, J., "Symmetry and Polarization in the European International System, 1870–79: A Methodological Study," *J. of Peace Research*, **11** (1974), 229–244, b.

HEALY, B., and STEIN, A., "The Balance of Power in International History," *J. of Conflict Resol.*, **17** (1973), 33–61.

HEIDER, F., "Attitudes and Cognitive Organization," *J. of Psych.*, **21** (1946), 107–112.

HELD, M., and KARP, R., "The Travelling Salesman Problem and Minimum Spanning Trees," *Oper. Res.*, **18** (1970), 1158–1162.

HOPCROFT, J., and TARJAN, R., "Efficient Algorithms for Graph Manipulation," *Communications of the ACM*, **16** (1973), 372–378.

JOHNSON, S. M., "Optimal Two- and Three-Stage Production Schedules with Setup Times Included," *Naval Res. Logistics Quart*, **1** (1954), 61–68.

KARP, R. M., "Reducibility Among Combinatorial Problems," in *Complexity of Computer Computations*, R. E. Miller and J. W. Thatcher (eds.), Plenum Press, New York, 1972.

KEMENY, J. G., and SNELL, J. L., *Mathematical Models in the Social Sciences*, Blaisdell Publishing Co., New York, 1962; reprinted by M.I.T. Press, Cambridge, Mass., 1972.

KENDALL, D. G., "A Statistical Approach to Flinders Petrie's Sequence Dating," *Bull. Intern. Statist. Inst.*, **40** (1963), 657–680.

KENDALL, D. G., "Incidence Matrices, Interval Graphs, and Seriation in Archaeology," *Pacific J. of Math.*, **28** (1969), 565–570, a.

KENDALL, D. G., "Some Problems and Methods in Statistical Archaeology," *World Archaeology*, **1** (1969), 61–76, b.

KENDALL, M. G., and SMITH, B. B., "On the Method of Paired Comparisons," *Biometrika*, **31** (1940), 324–345.

KLEE, V. L., "What is the Maximum Length of a *d*-dimensional Snake," *Amer. Math. Monthly*, **77** (1970), 63–65.

KÖNIG, D., *Theorie des Endlichen und Unendlichen Graphen*, Akademische Verlagsgesellschaft M.B.H., Leipzig, 1936; reprinted, Chelsea, New York, 1950.

KRANTZ, D. H., LUCE, R. D., SUPPES, P. and TVERSKY, A., *Foundations of Measurement*, Vol. I, Academic Press, Inc., New York, 1971.

KURATOWSKI, K., "Sur le Problème des Courbes Gauches en Topologie," *Fund. Math.*, **15** (1930), 271–283.

LANDAU, H. G., "On Dominance Relations and the Structure of Animal Societies: III. The Condition for a Score Structure," *Bull. Math. Biophys.*, **15** (1955), 143–148.

LEKKERKERKER, C. B. and BOLAND, J. CH., "Representation of a Finite Graph by a Set of Intervals on the Real Line," *Fund. Math.*, **51** (1962), 45–64.

LIN, S., "Computer Solutions of the Travelling Salesman Problem," *Bell Sys. Tech. J.*, **44** (1965), 2245–2269.

LIU, C. L., *Topics in Combinatorial Mathematics*, Mathematical Association of America, Washington, D.C., 1972.

LOVÁSZ, L., "Normal Hypergraphs and the Perfect Graph Conjecture," *Discrete Math.*, **2** (1972), 253–267.

MARCZEWSKI, E., "Sur deux Propriétés des Classes d'ensembles," *Fund. Math.*, **33** (1945), 303–307.

MOON, J., *Topics on Tournaments*, Holt, Rinehart & Winston, Inc., New York, 1968.

MORISSETTE, J. O., "An Experimental Study of the Theory of Structural Balance," *Human Relations*, **11** (1958), 239–254.

NORMAN, R. Z. and ROBERTS, F. S., "A Derivation of a Measure of Relative Balance for Social Structures and a Characterization of Extensive Ratio Systems," *J. Math. Psychol.*, **9** (1972), 66–91, a.

NORMAN, R. Z., and ROBERTS, F. S., "A Measure of Relative Balance for Social Structures," in *Sociological Theories in Progress*, II, J. Berger, M. Zelditch, and B. Anderson (eds.), Houghton-Mifflin Company, New York, 1972, pp. 358–391, b.

OGDEN, W. F., and ROBERTS, F. S., "Intersection Graphs of Families of Convex Sets with Distinguished Points," in *Combinatorial Structures and Their Applications*, R. Guy, H. Hanani, N. Sauer, and J. Schönheim (eds.), Gordon and Breach, New York, 1970, pp. 311–313.

ORE, O., *The Four-Color Problem*, Academic Press, Inc., New York, 1967.

PETRIE, W. M. F., "Sequences in Prehistoric Remains," *J. Anthrop. Inst.*, *N.S.*, **29** (1899), 295–301.

PETRIE, W. M. F., *Diospolis Parra*, Egypt Exploration Fund, London, 1901.

PETRIE, W. M. F., *Prehistoric Egypt*, British School of Archaeology in Egypt, London, 1920.

PETRIE, W. M. F., *Prehistoric Egypt Corpus*, British School of Archaeology in Egypt, London, 1921.

PNUELI, A., LEMPEL, A., and EVEN, S., "Transitive Orientation of Graphs and Identification of Permutation Graphs," *Can. J. Math.*, **23** (1971), 160–175.

RÉDEI, L., "Ein Kombinatorischer Satz," *Acta Litterarum ac Scientiarum (Sectio Scientarum Mathematicarum)*, Szeged, **7** (1934), 39–43.

ROBBINS, H. E., "A Theorem on Graphs, with an Application to a Problem of Traffic Control," *Amer. Math. Monthly*, **46** (1939), 281–283.

ROBERTS, F. S., "Indifference Graphs," in *Proof Techniques in Graph Theory*, F. Harary (ed.), Academic Press, Inc., New York, 1969, pp. 139–146, a.

ROBERTS, F. S., "On the Boxicity and Cubicity of a Graph," in *Recent Progress in Combinatorics*, W. T. Tutte (ed.), Academic Press, Inc., New York, 1969, pp. 301–310, b.

ROBERTS, F. S., "On Nontransitive Indifference," *J. Math. Psychol.*, **7** (1970), 243–258.

RYSER, H., "Combinatorial Configurations," *SIAM J. Appl. Math.* **17** (1969), 593–602.

SHERWIN, R. G., "The Notion of Structural Balance and the International System," unpublished manuscript, World Event Interaction Survey, Support Study #6, Univ. of Southern California, Los Angeles, Jan. 1972.

STOFFERS, K. E., "Scheduling of Traffic Lights—a New Approach," *Transportation Research*, **2** (1968), 199–234.

SZEKERES, G., and WILF, H. S., "An Inequality for the Chromatic Number of a Graph," *J. Comb. Theory*, **4** (1968), 1–3.

TARJAN, R., "Depth-first Search and Linear Graph Algorithms," *SIAM J. on Computing*, **1** (1972), 146–160.

TARJAN, R., "A Note on Finding the Bridges of a Graph," *Inf. Processing Letters*, **2** (1974), 160–161.

TAYLOR, H. F., *Balance in Small Groups*, Van Nostrand Reinhold Company, New York, 1970.

TUCKER, A. C., "Characterizing Circular-arc Graphs," *Bull. Amer. Math. Soc.*, **75** (1970), 1257–1260, a.

TUCKER, A. C., "Characterizing the Consecutive 1's Property," in *Proceedings of the Second Chapel Hill Conference on Combinatorial Mathematics and its Applications*, R. C. Bose, I. M. Chakravorti, T. A. Dowling, D. G. Kelly, and K. J. C. Smith (eds.), Univ. of North Carolina Press, Chapel Hill, North Carolina, 1970, pp. 472–477, b.

TUCKER, A. C., "Matrix Characterizations of Circular-arc Graphs," *Pacific J. Math.*, 39 (1971), 535–545.

TUCKER, A. C., "A Structure Theorem for the Consecutive 1's Property," *J. Combinatorial Theory*, 12B (1972), 153–162.

TUCKER, A. C., "Perfect Graphs and an Application to Optimizing Municipal Services," *SIAM Review*, **15** (1973), 585–590.

TUCKER, A. C., and BODIN, L., "A Model for Municipal Street Sweeping Operations," *CUPM Applied Mathematics Modules Project*, Berkeley, Calif., 1975.

WEGNER, G., *Eigenschaften der Nerven Homologisch-einfacher Familien in Rⁿ*, Doctoral Dissertation, Göttingen, 1967.

WHITNEY, H., "Congruent Graphs and the Connectivity of Graphs," *Amer. J. Math.*, **54** (1932), 150–168.

WILSON, E. O., *The Insect Societies*, Belknap Press of Harvard University Press, Cambridge, Mass., 1971.

WILSON, E. O., *Sociobiology: The New Synthesis*, Belknap Press of Harvard University Press, Cambridge, Mass., 1975.

WILSON, E. O., and BOSSERT, W. H., *A Primer of Population Biology*, Sinauer Associates, Stamford, Conn., 1971.

FOUR

Weighted Digraphs and Pulse Processes

4.1. Introduction: Energy and Other Applications

We saw in Sec. 3.1 that useful applications of graph theory to small group sociology arise from adding signs to each edge of a graph or arc of a digraph, obtaining a *signed graph* or *digraph*. There, the vertices represented people; there was an arc from person u to person v if u felt strongly about v; and there was a $+$ on the arc (u, v) if u liked v and a $-$ if u disliked v. Figure 4.1 shows an example of a signed digraph.

FIGURE 4.1. A signed digraph.

The concept of a signed digraph has numerous applications of a different kind if the vertices, arcs, and signs are given a somewhat different interpretation from that in Sec. 3.1. In this chapter, we shall make such an interpretation, and apply the results. To be concrete, we shall concentrate on the "energy crisis" and we shall see how a signed digraph can be used to begin to do such things as the following: pinpoint factors underlying the rapidly growing demand for energy; forecast future demands for energy, pollution levels from energy use, etc.; forecast the effect of new technologies or social changes on energy demand; and identify strategies or policies for modifying the growth of energy use and meeting various environmental constraints on that use. (We shall discuss these uses of signed digraphs in more detail at the end of Sec. 4.3.)

Methods similar to those of this chapter are being applied to answer a wide variety of questions, such as how should money be spent on transit in metropolitan Vancouver (Kane [1972]); how should governments decide what level of support to give scientific research (Organization for Economic Cooperation and Development [1974]); how should coastal resources be utilized to match the growing demand for water recreation (Coady, *et al.* [1973]); how should resources be allocated to health care (Kane, *et al.* [1972]); what are the possible impacts of deep water ports for oil tankers (Kruzic [1973a]); what is the effect of increased usage of coal for energy on inland waterway traffic (Antle and Johnson [1973]); how should the Navy make manpower decisions (Kruzic [1973b]); etc. We shall mention such applications from time to time, both in the text and in the exercises. We shall also sketch out in the exercises applications to simple ecosystems.

Much of the methodology sketched out in this chapter will be developed in a more general setting than a signed digraph. This is a *weighted digraph*, a digraph with a real number or *weight* $w(u, v)$ assigned to each arc (u, v). The weight may be positive, negative, or zero. In practice, in drawing a weighted digraph, we only draw arcs with nonzero weight. A weighted digraph where each $w(u, v)$ is an integer is called *integer-weighted*. Thus every signed digraph may be looked at as an integer-weighted digraph, with $w(u, v) = +1$ if (u, v) has sign $+$ and $w(u, v) = -1$ if (u, v) has sign $-$.

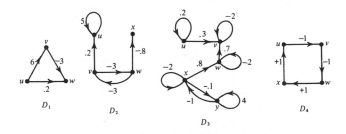

FIGURE 4.2. Some weighted digraphs.

Figure 4.2 shows several weighted digraphs. The weight is written next to the arc. D_4 is really a signed digraph, since each weight is $+1$ or -1. Most of the digraph concepts of Chapters 2 or 3 apply to signed or weighted digraphs if we apply them to the underlying digraph. Thus, for example, the concepts of reaching and joining such as path, simple path, cycle, semipath, simple semipath, and semicycle are defined as before. The concepts of connectedness, such as strong, unilateral, and weak, are also defined as before. Where we speak of weights assigned to subgraphs of a weighted digraph, we use the weights from the original weighted digraph unless specifically mentioned otherwise. In the case of a signed digraph, we shall continue to use one new piece of terminology introduced in Sec. 3.1. The *sign* of a path, closed path, cycle, etc. is the product of the signs of its arcs, if product is calculated by writing a $+$ sign as $+1$ and a $-$ sign as -1. Thus, the sign is $+$ if the path, closed path, cycle, etc. has an even number of $-$ signs, and $-$ otherwise. In the signed digraph of Fig. 4.1, the path v, w, x, u has sign $-$ and the cycle u, v, w, x, u has sign $+$. The sign of the trivial path u is not defined, and we shall disregard that path whenever we speak of signs of a path.

   In this chapter, we shall allow loops, arcs from a vertex to itself. For example, in digraph D_2 of Fig. 4.2, (u, u) is a loop. A loop is counted as an arc and so can be used in paths, cycles, semipaths, etc. In particular, a loop is counted as a cycle or semicycle of length 1. To give some examples, in digraph D_2 of Fig. 4.2, w, v, u, u is a path and u, (u, u), u, (v, u), v is a semipath. Subgraphs may include loops, and a generated subgraph D' of a digraph D has a loop (u, u) if and only if this loop is in D and u is a vertex of D'. It turns out that a digraph with some loops is strongly, unilaterally, or weakly connected if and only if the corresponding digraph without loops is. Finally, strong components of a digraph are generated subgraphs, and so include loops whenever these appear in D.

   In this chapter, we shall study in detail the application of signed and weighted digraphs to such problems of society as the energy crisis. Rather than interrupting the development later we begin in the next section by giving some mathematical preliminaries.

EXERCISES

1. Find the sign of the paths u, v, w, x and x, u, v, w in the signed digraph of Fig. 4.1.

2. Find a path of length 2 from u to v in the weighted digraph D_3 of Fig. 4.2.

3. Find all strong components in each weighted digraph of Fig. 4.2.

4.2. Excursis: Eigenvalues

In this section, we shall review some of the results about eigenvalues and eigenvectors which we shall need in our applications of signed and weighted digraphs. We shall also introduce the concept of an eigenvalue of a weighted digraph, and we shall see how weighted digraphs assist us in calculating eigenvalues of matrices. For proofs of the standard results we shall state, the reader is referred to most linear algebra texts, for example Mal'cev [1963].

Suppose M is a square $n \times n$ matrix of real numbers. An *eigenvector* (characteristic vector) is a *nonzero* $1 \times n$ row vector U so that for some scalar λ, $UM = \lambda U$. The scalar λ is called the *eigenvalue* corresponding to U. It may be a real or a complex number. For example, if

$$M = \begin{pmatrix} 2 & 0 \\ 0 & 3 \end{pmatrix}$$

and

$$U = (1, 0),$$

then

$$UM = (2, 0) = 2U.$$

Thus U is an eigenvector of M and 2 is its corresponding eigenvalue. We shall be concerned with finding eigenvalues.

Theorem 4.1. λ is an eigenvalue of an $n \times n$ matrix M if and only if det $(M - \lambda I) = 0$, where det A is determinant of A and where I is the identity matrix.

If M is an $n \times n$ matrix, then det $(M - \lambda I)$ is a sum of products of elements, one from each row and each column, and each summand with the proper sign. Thus, one gets an expression which is a polynomial in λ. This polynomial is called the *characteristic polynomial* of M and is denoted $C_M(\lambda)$ or simply $C(\lambda)$. Since $C_M(\lambda)$ is a polynomial, there are numbers $a_0, a_1, \ldots, a_n$ such that

$$C_M(\lambda) = \det (M - \lambda I) = a_n\lambda^n + a_{n-1}\lambda^{n-1} + \cdots + a_1\lambda + a_0.$$

From the definition of determinant, it is easy to see that the only place one can get a term λ^n is from the product of diagonal elements $(m_{11} - \lambda) \cdot (m_{22} - \lambda) \cdot \cdots \cdot (m_{nn} - \lambda)$. Thus, the coefficient a_n is always ± 1, specifically $(-1)^n$.

From Theorem 4.1, we see that the eigenvalues of M are exactly the roots of the characteristic polynomial, i.e., those scalars λ which satisfy

$$C_M(\lambda) = 0. \tag{1}$$

Equation (1) is called the *characteristic equation* of the matrix M. If an eigenvalue appears as a multiple root, as will occur in a number of our applications, its *multiplicity* is its multiplicity as a root of $C_M(\lambda)$.

To illustrate the ideas sketched so far, let us consider the matrix

$$M = \begin{pmatrix} 7 & -3 & 0 \\ 11 & -7 & 0 \\ 0 & 5 & 1 \end{pmatrix}. \tag{2}$$

The matrix $M - \lambda I$ is given by

$$M - \lambda I = \begin{pmatrix} 7 - \lambda & -3 & 0 \\ 11 & -7 - \lambda & 0 \\ 0 & 5 & 1 - \lambda \end{pmatrix}.$$

Then we have

$$C_M(\lambda) = \det (M - \lambda I) = -\lambda^3 + \lambda^2 + 16\lambda - 16.$$

This polynomial factors into

$$C_M(\lambda) = -(\lambda - 1)(\lambda - 4)(\lambda + 4).$$

Thus, the eigenvalues of M are $1, 4, -4$.

Let us find an eigenvector corresponding to the eigenvalue 4. Suppose $U = (r_1, r_2, r_3)$ is such an eigenvector. We have $UM = 4U$, from which we derive the three equations

$$7r_1 + 11r_2 = 4r_1, \qquad -3r_1 - 7r_2 + 5r_3 = 4r_2, \qquad r_3 = 4r_3.$$

The solutions to these equations are all the vectors of the form $(x, (-\frac{3}{11})x, 0)$. Since an eigenvector is defined to be a nonzero vector, x must be nonzero. Thus, the eigenvectors corresponding to the eigenvalue 4 are all nonzero multiples of the vector $U_1 = (1, -\frac{3}{11}, 0)$.

Similarly, one may calculate that the eigenvectors corresponding to the eigenvalue -4 are all nonzero multiples of $U_2 = (1, -1, 0)$ and the eigenvectors corresponding to the eigenvalue 1 are all nonzero multiples of the vector $(-\frac{11}{3}, 2, 1)$.

To give a second example, consider the matrix

$$M = \begin{pmatrix} 1 & 0 & 0 \\ 0 & 1 & 0 \\ 0 & 0 & 2 \end{pmatrix}.$$

The characteristic polynomial of M factors as $C_M(\lambda) = (1 - \lambda)^2(2 - \lambda)$. Thus, the eigenvalues of M are 1 and 2, with 1 appearing as an eigenvalue of multiplicity two. (In general, if M is a *diagonal matrix*, i.e., a matrix all of whose off-diagonal entries are 0, then the eigenvalues of M are the diagonal entries, as in this example.)

To give yet a third example, consider the matrix

$$M = \begin{pmatrix} 0 & 1 \\ -1 & 0 \end{pmatrix}.$$

Then $C_M(\lambda) = \lambda^2 + 1 = (\lambda + i)(\lambda - i)$, so the eigenvalues are the complex numbers i and $-i$.

Finally, it should be noted that if M is the 1×1 matrix (t), then $C_M(\lambda) = t - \lambda$, so t is the unique eigenvalue of M.

In the sequel we shall be interested in the eigenvalues of a weighted digraph D. To make this concept precise, let us list the vertices of D as $u_1, u_2, \ldots, u_n$ and define the *adjacency matrix* $A(D) = (a_{ij})$ by taking $a_{ij} = w(u_i, u_j)$. (Henceforth, we shall define $w(u, v) = 0$ if there is no arc (u, v).) We shall refer to the eigenvalues of A as the *eigenvalues of the weighted digraph* D. Thus, the weighted digraph D of Fig. 4.3 has eigenvalues $1, 4$, and -4 because its adjacency matrix is the matrix M shown in Eq. (2).

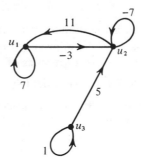

FIGURE 4.3. A weighted digraph whose adjacency matrix is the matrix M of Eq. (2).

We shall prove a very useful theorem about the eigenvalues of a weighted digraph, which helps us to reduce the calculation of eigenvalues to smaller digraphs. Conversely, this theorem may be applied to the calculation of eigenvalues of a given matrix M. This is done by constructing a weighted digraph D with M as its adjacency matrix. The vertices of D correspond to the rows of M. There is an arc from row i to row j if and only if $m_{ij} \neq 0$. The weight on the arc from row i to row j is m_{ij} if there is an arc. D is called the *Coates digraph of the matrix M* (cf. Chen [1971]). In the theorem, a strong component of a weighted digraph means the weighted digraph restricted to a strong component of the underlying digraph.

Theorem 4.2 (Harary [1959]). The set of eigenvalues of a weighted digraph is the union of the sets of eigenvalues of its strong components.[1]

Proof. We will prove that the characteristic polynomial of the adjacency matrix of a weighted digraph is the product of the characteristic polynomials of the adjacency matrices of its strong components. If the digraph in question consists of just one strong component, there is nothing to

---

[1]When multiplicity of eigenvalues is important, the theorem must be restated to say the union counting multiplicity, which means an eigenvalue is listed a number of times equal to its multiplicity.

prove. We proceed by induction on the number of strong components. Suppose the theorem has been proved for all weighted digraphs with c strong components; let A be the adjacency matrix of a weighted digraph D with $c + 1$ strong components. Choose a strong component K_1 whose corresponding vertex K_1^* in the condensation D^* of D has no incoming arcs in D^*. (Such a K_1 exists by Theorem 2.7 and the Corollary to Theorem 2.8.) Let K_2 denote the weighted digraph generated by the vertices of D which are not in K_1. If we number the vertices of D in such a way that the vertices of K_1 are $\{1, 2, \ldots, r\}$ and the vertices of K_2 are $\{r + 1, r + 2, \ldots, m\}$, then the matrix A may be written

$$A = \begin{array}{c} \\ r \\ m-r \end{array} \overset{\begin{array}{cc} r & m-r \end{array}}{\left(\begin{array}{c|c} A_1 & X \\ \hline 0 & A_2 \end{array} \right)},$$

where the $m - r$ by r submatrix in the lower left-hand corner consists entirely of zeros, A_1 is the adjacency matrix of K_1, and A_2 is the adjacency matrix of K_2. By standard properties of determinants, it follows that

$$\det (A - \lambda I) = \det (A_1 - \lambda I) \cdot \det (A_2 - \lambda I),$$

which says that the characteristic polynomial of A is the product of the characteristic polynomial of A_1 and the characteristic polynomial of A_2. But by the inductive assumption the characteristic polynomial of A_2 is simply the product of the characteristic polynomials of the adjacency matrices of its strong components, which concludes the proof.   Q.E.D.

Since the characteristic polynomial of a weighted digraph is completely determined by the characteristic polynomials of its strong components, it follows that the characteristic polynomial tells us nothing about how the individual strong components are related to one another (i.e., it tells us nothing about the structure of the condensation digraph).

Corollary 1.   The set of eigenvalues of a matrix M is the union of the sets of eigenvalues of the submatrices corresponding to the strong components of its Coates digraph.[2]

Corollary 2.   The characteristic polynomial of the adjacency matrix of a weighted digraph is the product of the characteristic polynomials of the adjacency matrices of its strong components.

Proof.   This is a corollary of the proof.

---

[2]As before, when multiplicity is important, then this should be stated as the union counting multiplicity.

To illustrate Corollary 1, let us consider the matrix

$$M = \begin{pmatrix} 0 & 1 & 0 & 0 & 0 \\ -3 & 0 & 0 & 0 & 0 \\ 0 & 5 & 8 & -2 & 0 \\ 0 & 0 & 0 & 0 & 1 \\ 0 & 0 & 0 & -1 & 0 \end{pmatrix}. \tag{3}$$

The Coates digraph of M is shown in Fig. 4.4. The strong components K_1, K_2, and K_3 of this digraph are also shown there, as are their adjacency matrices. It is a simple matter to show that K_1 has eigenvalues $+\sqrt{3}i$ and $-\sqrt{3}i$, that K_2 has eigenvalue 8, and that K_3 has eigenvalues $+i$ and $-i$. Thus, M has eigenvalues $+\sqrt{3}i$, $-\sqrt{3}i$, 8, $+i$, and $-i$.

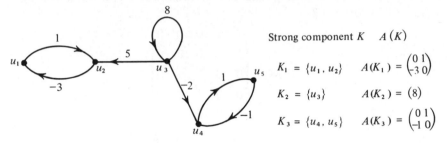

Strong component K	$A(K)$
$K_1 = \{u_1, u_2\}$	$A(K_1) = \begin{pmatrix} 0 & 1 \\ -3 & 0 \end{pmatrix}$
$K_2 = \{u_3\}$	$A(K_2) = (8)$
$K_3 = \{u_4, u_5\}$	$A(K_3) = \begin{pmatrix} 0 & 1 \\ -1 & 0 \end{pmatrix}$

FIGURE 4.4. Coates digraph of the matrix M of Eq. (3), its strong components, and their adjacency matrices.

EXERCISES

1. Calculate all eigenvalues of the matrix
$$M = \begin{pmatrix} 1 & 1 \\ 1 & 1 \end{pmatrix}.$$

2. Write the adjacency matrix of each weighted digraph of Fig. 4.2.

3. Draw a weighted digraph with adjacency matrix A given by
$$A = \begin{pmatrix} 1 & 5 & 2 \\ 0 & 0 & -1 \\ -1 & 0 & -1 \end{pmatrix}.$$

4. For each of the weighted digraphs of Fig. 4.5, calculate the eigenvalues.

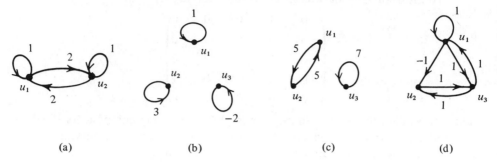

(a)   (b)   (c)   (d)

FIGURE 4.5. Weighted digraphs for Exer. 4, Sec. 4.2.

5. For each of the following matrices, calculate the eigenvalues, making use of Theorem 4.2.

(a)
$$M = \begin{pmatrix} 1 & 1 & 0 & 0 & 0 \\ 1 & 1 & 0 & 0 & 0 \\ 0 & 0 & 0 & 1 & 0 \\ 0 & 0 & 1 & 0 & 0 \\ 0 & 0 & 0 & 0 & 3 \end{pmatrix}.$$

(b)
$$M = \begin{pmatrix} 3 & 0 & 0 & 0 & 0 & 0 & 0 \\ 3 & 0 & 0 & 0 & 0 & 0 & 0 \\ 0 & 0 & 0 & 1 & 0 & 0 & 0 \\ 0 & 0 & 0 & 0 & 1 & 0 & 0 \\ 0 & 0 & 1 & 0 & 0 & 0 & 0 \\ 0 & 0 & 0 & 0 & 0 & 0 & 2 \\ 0 & 0 & 0 & 0 & 0 & 2 & 0 \end{pmatrix}.$$

(c)
$$M = \begin{pmatrix} 3 & 0 & 0 & 0 & 0 & 0 & 0 & 0 & 0 \\ 0 & 0 & 0 & 1 & 0 & 0 & 0 & 0 & 0 \\ 0 & 0 & 1 & 0 & 0 & 0 & 1 & 0 & 0 \\ 0 & 0 & 0 & 0 & 1 & 0 & 0 & 0 & 0 \\ 0 & 1 & 0 & 0 & 0 & 0 & 0 & 0 & 0 \\ 0 & 0 & 0 & 0 & 0 & 0 & 0 & 1 & 0 \\ 0 & 0 & 1 & 0 & 0 & 0 & 1 & 0 & 0 \\ 0 & 0 & 0 & 0 & 0 & 1 & 0 & 0 & 0 \\ 3 & 0 & 0 & 0 & 0 & 0 & 0 & 0 & 0 \end{pmatrix}.$$

6. For each of the following matrices M, calculate eigenvalues and an eigenvector corresponding to each eigenvalue.

(a)
$$M = \begin{pmatrix} 2 & 0 & 0 \\ 1 & 1 & 0 \\ 1 & 1 & 3 \end{pmatrix}.$$

(b)
$$M = \begin{pmatrix} -1 & 1 & 0 \\ 0 & 1 & 0 \\ 0 & 1 & 2 \end{pmatrix}.$$

7. Phillips [1967] has suggested measuring degree of balance of a small group (Exers. 21 to 26, Sec. 3.1) as follows. Draw the corresponding signed graph and calculate the largest eigenvalue — not the largest in magnitude. (All the eigenvalues of a symmetric matrix are real.)

   (a) Calculate this measure of balance for each of the signed graphs of Fig. 3.2, if an edge is replaced by two arcs in opposite directions. Comment on the results.

   (b) Investigate this measure for other signed graphs.

8. Prove that if a weighted digraph has no cycles, then all its eigenvalues are 0. (*Hint:* Consider strong components.) (*Note:* A loop counts as a cycle. The converse is considered in Exer. 28, Sec. 4.5.)

9. Every nonnegative matrix has a real eigenvalue. Let $\lambda_{max}(A)$ be the largest real eigenvalue of the nonnegative matrix A. The Perron-Frobenius theory asserts that

$$\text{min row sum of } A \leq \lambda_{max}(A) \leq \text{max row sum of } A.$$

Moreover, if $a_{ij} \geq b_{ij}$ for all i, j, and A and B are both nonnegative matrices, then $\lambda_{max}(A) \geq \lambda_{max}(B)$. Suppose G is a graph and A is the adjacency matrix, i.e., the matrix whose i, j entry is 1 if there is an edge between vertices u_i and u_j, and 0 otherwise. Wilf [1961] proved that if $\chi(G)$ is the chromatic number of G, then $\chi(G) \leq 1 + \lambda_{max}(A)$, with equality if and only if G is an odd circuit or a complete graph. This exercise explores Wilf's Theorem.

   (a) Show that $\lambda_{max}(A) \leq \Delta(G)$, so that Wilf's bound on chromatic number is at least as good as that of Brooks (Theorem 3.15).

   (b) If G is the complete 2-partite graph $K(1, 5)$, then the characteristic polynomial of the adjacency matrix A is $\lambda^4(\lambda^2 - 5)$. Show that $\lambda_{max}(A) < \Delta(G)$, so here Wilf's bound is better than Brooks'.

   (c) Show that if G is K_n, the complete graph on n vertices, then $\chi(G) = 1 + \lambda_{max}(A)$. (*Hint:* Use the Perron-Frobenius results.)

   (d) Show that if G is Z_n, the circuit of length n, for n odd, then $\chi(G) = 1 + \lambda_{max}(A)$.

(e)   Show that $\chi(G) \leqq 1 + \lambda_{\max}(A)$ by using Theorem 3.16. (*Hint*: Given G', let G'' be the subgraph of G with edges joining vertices not in G' removed. It is easy to show that $\lambda_{\max}(A) \geqq \lambda_{\max}(A'') = \lambda_{\max}(A')$.)

4.3.   The Signed or Weighted Digraph as a Tool for Modelling Complex Systems[3]

Many of the problems of society, in particular those of energy use, involve extremely complex systems. Such complex systems are made up of many variables interacting with each other, reacting to changes in each other, etc. In studying mathematical models of complex systems, one often faces a tradeoff between the precision of the model's predictions and the ability to obtain the detailed information needed to build the model. Signed and weighted digraphs can be used to build simple mathematical models of complex systems, and to analyze those conclusions attainable based on a minimal amount of information.

The basic idea we shall adopt in using a signed digraph as a model for a complex system is the following. The vertices are chosen as variables relevant to or representative of the problem at hand. There is an arc from variable u to variable v if a change in u has a significant direct effect on v. Finally, the sign of this arc is $+$ if the effect is *augmenting* (all other things being equal, an increase in u leads to an increase in v, and a decrease in u leads to a decrease in v), and the sign is $-$ if the effect is *inhibiting* (all other things being equal, an increase in u leads to a decrease in v, and a decrease in u leads to an increase in v).

With this interpretation, let us look at the signed digraph of Fig. 4.6. This signed digraph, adapted from Maruyama [1963], describes the important relationships among a number of variables related to the solid waste disposal problem in a city. The arc (P, G) is $+$ because an increase in the population of the city leads, all other things being equal, to an increase in the amount of garbage, while conversely a decrease in the population leads to a decrease in the amount of garbage. The arc (D, P) is $-$ because an increase in the number of diseases leads to a decrease in population while a decrease in the number of diseases leads to an increase in population. Other signs are determined in a similar manner. (There is no arc at all from B to M, for example. This is because the augmenting or inhibiting effect of bacteria per area on modernization is considered negligible.)

Note that the four variables P, G, B, and D form a cycle which is *deviation-counteracting*. An increase in any variable on this cycle ultimately

---

[3]The author thanks Pion Press Ltd. for permission to reproduce in this section parts of the article by Roberts [1971].

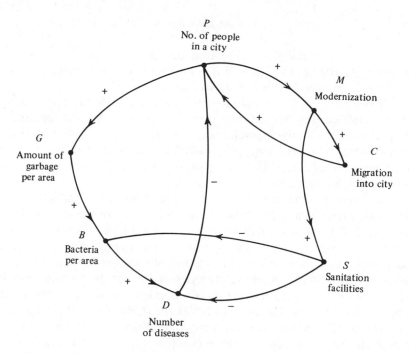

P
No. of people
in a city

M
Modernization

G
Amount of
garbage
per area

C
Migration
into city

B
Bacteria
per area

S
Sanitation
facilities

D
Number
of diseases

FIGURE 4.6. Signed digraph for solid waste disposal. Reprinted from
Maruyama [1963, p. 176] with permission of The Society of the Sigma Xi.

leads, through the other variables on the cycle, to a decrease in the variable,
and vice versa. (The more people in a city, the greater the amount of garbage;
the greater the amount of garbage, the more bacteria; the more bacteria, the
greater the number of diseases; the greater the number of diseases, the fewer
people; etc.) On the other hand, the cycle *P, M, C, P* is *deviation-amplifying.*
An increase (decrease) in any variable on this cycle ultimately leads to a
further increase (decrease) in this variable. ("The more people in a city, the
greater the pressure toward modernization; the greater the modernization,
the more appealing a city is to immigrate to; and the more immigration, the
more people in the city." (Cf. Coombs, Dawes, and Tversky [1970, p. 82]).)
Cycles correspond to *feedback loops,* deviation-amplifying cycles to *positive
feedback loops* and deviation-counteracting cycles to *negative feedback loops.*
  Deviation-amplifying cycles are easy to identify:

  Observation (Maruyama [1963]). A cycle is deviation-amplifying if
and only if it has an even number of − signs, and it is deviation-
counteracting otherwise.

This holds because if there are an even number of − signs, every time a
deviation is counteracted, the counteraction itself is counteracted. If there
is an odd number of − signs, the last deviation counteraction is not counteracted. (It is interesting to note the connection between this notion of devia-

tion amplification and that of balance discussed in Sec. 3.1: A cycle is deviation-amplifying if and only if it is balanced. We shall explore the connection with balance further in the exercises of Sec. 4.5.)

If there are mostly deviation-amplifying cycles present, initial changes can be amplified way beyond their immediate effect. Thus, the presence of many deviation-amplifying cycles suggests instability. (The presence of many deviation-counteracting cycles can also lead to instability, by leading to increasing oscillations. The theory in Sec. 4.5 will go into further detail on this point. See especially Exer. 27, Sec. 4.5.)

A signed digraph similar to that of Fig. 4.6 can be described for the (electrical) energy demand situation in a given area. For the sake of discussion, we shall use just a small number of the variables relevant to energy demand as vertices: energy capacity, energy use, energy price (per kilowatt hour), population, number of jobs, number of factories, and quality of the (physical) environment. A careful analysis of the energy demand situation either would require a much larger set of variables or would use some careful technique for choosing the basic variables for study out of this large class. Since choice of variables involves many disciplines and differing opinions, perhaps the best approach is to use a group of experts and ask each for his opinion. Then we face the problem of combining the opinions of the experts to make a choice. This is a problem we discuss at length in Chapter 7. For a discussion of the specific problem of choosing variables to study energy demand, see Roberts [1973]. (See also Sec. 8.4.3.)

Our signed digraph for energy demand is shown in Fig. 4.7. For the sake of discussion, we have included only some of the most significant direct relationships. The sign of the arc (U, Q) is $-$ because, all other things being equal, an increase in energy use leads to a decrease in the quality of the environment, and a decrease in energy use leads to an increase in the quality of the environment. The arc (J, P) is $+$ because an increase (decrease) in the number of jobs brings in people (causes people to leave). The arc (C, F) is $+$ because an increase (decrease) in energy capacity (number or size of power plants) tends to draw new factories into an area (stimulate factories to leave an area). The arc (U, R) is $-$ because according to the present electricity rate structure, the more you use, the less you pay (per kilowatt hour). This signed digraph omits loops; for example there should probably be a $+$ loop from P to P. However, it should be easy to see how loops can be added to the following analysis.

It should be noted that the cycle U, Q, P, U is deviation-counteracting. The cycles C, R, U, C and C, F, J, P, U, C are deviation-amplifying, and indeed all cycles containing the vertex C are deviation-amplifying. We shall note the significance of this observation below.

A second signed digraph for energy demand is shown in Fig. 4.8. This digraph, for energy demand in intraurban commuter transportation, was constructed from the subjective judgments of groups of experts. Details of the

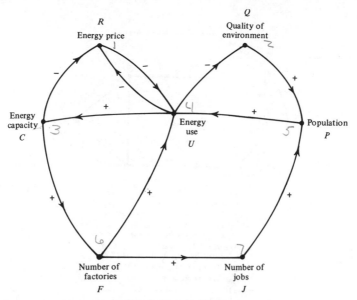

FIGURE 4.7. Signed digraph for electrical energy demand. Reprinted from Roberts [1971] with permission of Pion Press Limited.

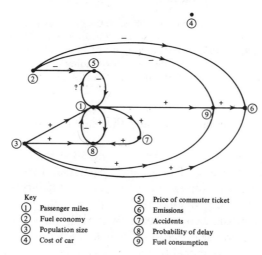

Key

①	Passenger miles	⑤	Price of commuter ticket
②	Fuel economy	⑥	Emissions
③	Population size	⑦	Accidents
④	Cost of car	⑧	Probability of delay
		⑨	Fuel consumption

FIGURE 4.8. Signed digraph for intraurban commuter transportation. The sign of arc (1, 5) is undetermined. Reprinted from Roberts [1973] with permission of Pion Press Limited.

construction can be found in Roberts [1973]. The arc (1, 5) has a question mark on it because the experts could not decide on the sign. Below we shall analyze this digraph under both choices of sign. Here, let us simply note that sometimes even the minimal information needed to build a signed digraph is hard to obtain.

Several additional signed digraphs are shown in Figs. 4.9 through 4.14.

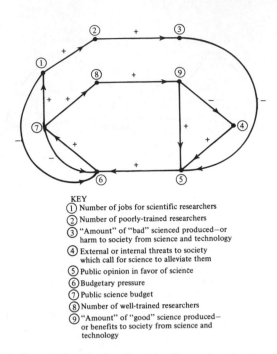

KEY
① Number of jobs for scientific researchers
② Number of poorly-trained researchers
③ "Amount" of "bad" scienced produced—or harm to society from science and technology
④ External or internal threats to society which call for science to alleviate them
⑤ Public opinion in favor of science
⑥ Budgetary pressure
⑦ Public science budget
⑧ Number of well-trained researchers
⑨ "Amount" of "good" science produced— or benefits to society from science and technology

FIGURE 4.9. Signed digraph for the scientific and technology system. Prepared by the author for an Organization for Economic Cooperation and Development study of the levelling off of support for science and of new methods for decisionmaking about the funding of scientific and technological research. The author thanks the O.E.C.D. for permission to use this material.

These have arisen in actual studies, and the reader is asked to study some of them in the exercises.

Often, a signed digraph is the most detailed mathematical model of a complex system attainable. This is true in particular if some of the variables cannot be measured, as for example the vertex "Quality of environment" in the digraph of Fig. 4.7. Many of the variables arising in societal problems are difficult or impossible to measure. Even with such a simplified model, there are still some precise conclusions which can be reached. For example, if Fig. 4.7 is an accurate model of the signs of effects in an energy demand system, then one can pinpoint certain deviation-amplifying subsystems. As we have observed earlier, the directed cycle C, R, U, C corresponds to such a subsystem. Indeed, all subsystems containing the energy capacity vertex C and corresponding to simple cycles such as C, R, U, C are deviation-amplifying. Most of the shorter closed paths containing C are deviation-amplifying as well. This observation already makes precise, from a structural point of view, why the energy capacity system is so unstable. It suggests that initial increases in energy capacity are amplified in further increases in energy capacity, and makes precise an observation made imprecisely by environmentalists.

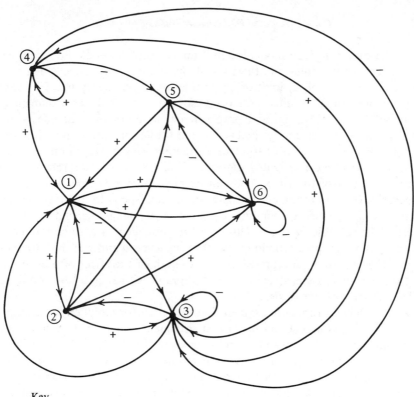

Key

①	Desired beach attendance	④	Urban population
②	Actual beach attendance	⑤	Unallocated shoreline
③	Overall urban satisfaction	⑥	Public investment in beaches

FIGURE 4.10. Signed digraph for coastal recreation. Adapted from Coady, et al. [1973], who study utilization of coastal resources in the light of growing demand for water recreation. The author thanks Dr. G. P. Johnson for permission to use this material.

It also suggests that one way to counteract the tremendous growth in energy capacity is to introduce negative "initial kicks" into some of the deviation-amplifying cycles containing C. (One way to do this is by introducing a positive initial kick into the energy price, R.) Or one could try to make some of these cycles deviation-counteracting by changing the sign of some arc. (How this is accomplished on a public policy level is of course a difficult question.) Other possibilities are analyzed below, and the question of strategies or policies is put on a more quantitative level.

These observations follow simply from the *structure* of the signed digraph. The signed digraph model is especially good for making such structural observations, for digraph theory has concerned itself over the years with just such notions of structure. (Indeed, one book about digraph theory, Harary, Norman, and Cartwright [1965], is called *Structural Models*.)

The signed digraph model has in it many simplifications. For example, some effects of variables on others are stronger than other effects. Thus, in Fig. 4.7, the effect of an increase in population P on energy use U is very strong compared to the effect of a decrease in quality of the environment Q on population P. The signed digraph model assumes that all effects are equally strong, by placing weights of equal (unit) magnitude on each arc. It might be more reasonable to place a different weight $w(u, v)$ on each arc (u, v), thus yielding a weighted digraph.[4] The weight is interpreted as the relative strength of the effect, and can be positive (for augmenting effects) or negative (for inhibiting effects). Even more realistic is to assume that the strength of an effect corresponding to the arc (u, v) changes depending on the levels of the variables u and v. This could be modelled by assigning to each arc (u, v) of a digraph a function $f_{uv}(\mathbf{u}, \mathbf{v})$ of the levels $\mathbf{u}$ and $\mathbf{v}$ of the variables u and v, with $f_{uv}(\mathbf{u}, \mathbf{v})$ interpreted as the strength of the effect of u on v if u is at the level $\mathbf{u}$ and v is at the level $\mathbf{v}$. A digraph with such a function f_{uv} is called a *functional signed digraph*.[5]

If we have a functional signed digraph, then for simplicity, we might assume that f_{uv} is a function of its first argument. This function might be increasing for small values of u and decreasing for larger values, or have other complex properties. To give an example, in the solid waste disposal situation of Fig. 4.6, the function f_{PM} has this character, with increases of population at moderate levels bringing about pressures for modernization, and increases of population at higher levels putting so much strain on a city's budget that modernization programs sit idle and considerable decay occurs. The functional signed digraphs are similar in some ways to the models used by Forrester [1961, 1969, 1971] and Meadows, *et al.* [1972] to study urban decay, environmental pollution, etc. Our vertices should be compared to their level variables and our functions $f_{uv}(\mathbf{u}, \mathbf{v})$ to their rate variables.

A second simplification in the signed (or weighted or functional signed) digraph model is that it omits the time lag involved before a change in u has an effect on v. For example, in Fig. 4.7, an increase in population P will lead almost immediately to an increase in energy use U, while there is a time lag after an increase in the number of jobs J before that attracts more population P to an area. In general, the amount of time needed before a change in u affects v will vary depending on the arc (u, v). Thus, a more accurate model would introduce a second weight on an arc, giving the time lag corresponding to the effect, or more generally a function $g_{uv}(\mathbf{u}, \mathbf{v})$ giving the time lag in the effect of u on v if u is at the level $\mathbf{u}$ and v at the level $\mathbf{v}$. Most of what we say below can probably be generalized in principle to take account

[4]In this weighted digraph, we could include arcs such as (Q, U), which are negligible compared with those included in Fig. 4.7. (This arc might have negative weight: For example, a decrease in air quality leads to increased use of air conditioning and hence to more use of electrical power.)

[5]The term functional digraph, less redundant, already is used in graph theory.

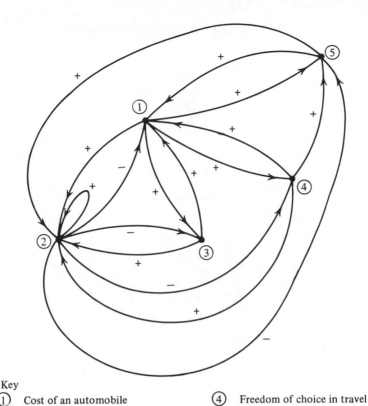

Key

①	Cost of an automobile	④	Freedom of choice in travel
②	Use of automobiles		(routes, time, etc.)
③	Comfort and convenience	⑤	Speed
	of automobile use		

FIGURE 4.11. Signed digraph for pattern of public/private transportation in Vancouver. From data in Kane [1972], who studies whether massive funding of public transit can achieve the goal of easy metropolitan access.

of time lags. A trivial solution, if the time lags are all integral, is to insert a number of additional vertices in between two vertices joined by an arc, with the number of inserted vertices corresponding to the time lag. This is not a practical solution, however, since it can give rise to a very large digraph.

It should be repeated that there is a tradeoff between the generality of the model and the possibility of estimating its parameters (weights, time lags, functions, etc.) in a realistic way. It is sometimes hard to make sense even out of weights $w(u, v)$, for example if u or v itself is hard to measure or define precisely. That is why it is important to study those conclusions which can be obtained only from knowledge of signs. An example is the conclusion that every cycle through vertex C in the signed digraph of Fig. 4.7 is deviation-amplifying.

On the other hand, various properties of the digraph can be quite sensitive to the weights or time lags involved, and small changes in these

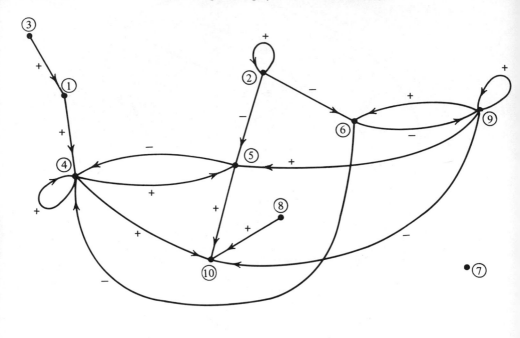

Key

①   Environmental stresses
②   Age structure
③   Population size
④   Morbidity rate
⑤   Screening efforts (searching
      out health problems)

⑥   Availability of health
      care services
⑦   Physician time available
⑧   Health care facilities
⑨   Delegation of physician
      duties to other personnel
⑩   Expenditures

FIGURE 4.12. Signed digraph for health care delivery in British Columbia. From data in Kane, *et al*. [1972], who study alternative policies for allocation of medical resources.

parameters can change some of the properties of the digraph. For example, pulse and value stability, which we shall discuss below, can be so affected. Thus, as much as possible, it is desirable to estimate at least the weights $w(u, v)$ and time lags, and perhaps even the functions $f_{uv}(\mathbf{u}, \mathbf{v})$ and $g_{uv}(\mathbf{u}, \mathbf{v})$, as accurately as can be, and to develop techniques for analyzing weighted and even functional signed digraphs and also digraphs with time lags.

In the following, we shall usually speak of a weighted digraph. It is to be understood that some of its weights are known only as signs, and others could be replaced by functions. It should be clear how to make the latter substitution for a weight in many cases. We shall disregard time lags.

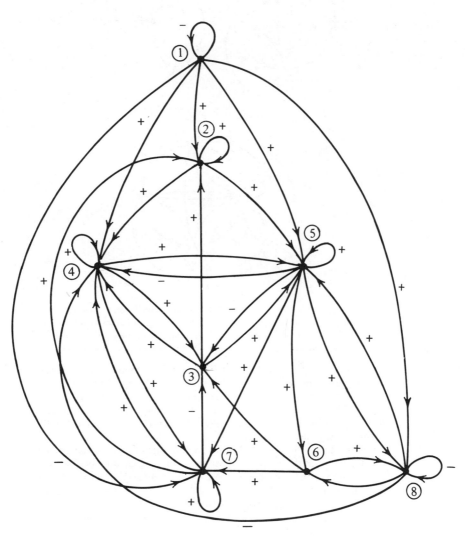

Key

① Pay, allowances, and benefits   ⑤ Retention
② Physical environment   ⑥ Military men
③ Skill opportunity   ⑦ Equipment performance
④ Job satisfaction   ⑧ Manpower budget

FIGURE 4.13. Signed digraph for Naval manpower system. From data in Kruzic [1973a], who studies effects of alternative policies on naval retention of personnel. (Adapted with permission of Office of Naval Research.)

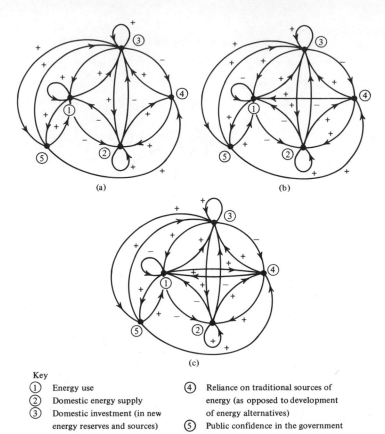

Key

① Energy use
② Domestic energy supply
③ Domestic investment (in new
   energy reserves and sources)

④ Reliance on traditional sources of
   energy (as opposed to development
   of energy alternatives)
⑤ Public confidence in the government

FIGURE 4.14. Three signed digraphs for energy use by the United States. These signed digraphs use data from Kruzic [1973b], who developed them in the process of working on an assessment of Deep Water Ports for oil tankers. The data was made available through the courtesy of the U.S. Army Corps of Engineers. Signed digraph (a) is for a system prior to 1985 in which energy use is less than 50% of an artificial maximum level estimated at 4 times 1973 use, and in which domestic energy sources supply more than 75% of the energy needs of the nation. Signed digraph (b) is for a system after 1985 in which energy use is more than 50% of the maximum level, but domestic energy sources still supply more than 75% of the needs. Signed digraph (c) is for a system after 1985 in which energy use is more than 50% of the maximum and domestic sources supply less than 50% of the energy needs.

Figures 4.15 through 4.21 give examples of weighted digraphs which have arisen in various practical applications. Figures 4.15 and 4.16 were built using (essentially) the nine variables of Fig. 4.8. We shall analyze these weighted digraphs in more detail below. Figures. 4.17 through 4.21 are weighted digraphs corresponding to the signed digraphs of Figs. 4.10 through 4.14 respectively. (Figure. 4.9 was studied purely as a signed digraph, and has no weighted counterpart in the literature.)

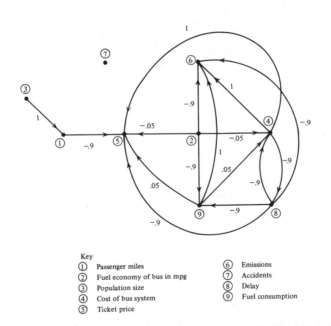

FIGURE 4.15. Weighted digraph for energy use and clean air in the transportation system of San Diego, California, under an almost-all-automobile transportation system. From Roberts [1974b]. The author thanks the Rand Corporation for permission to use this figure.

Key
① Passenger miles
② Fuel economy of car in mpg
③ Population size
④ Cost of car
⑤ Price of trip
⑥ Emissions
⑦ Accidents
⑧ Average delay
⑨ Fuel consumption

Key
① Passenger miles
② Fuel economy of bus in mpg
③ Population size
④ Cost of bus system
⑤ Ticket price
⑥ Emissions
⑦ Accidents
⑧ Delay
⑨ Fuel consumption

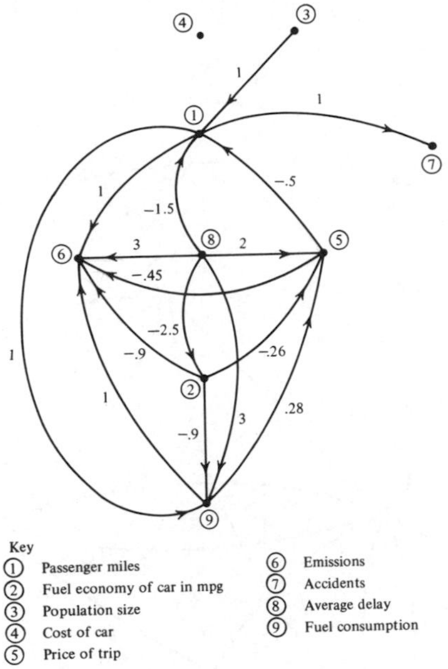

FIGURE 4.16. Weighted digraph for energy use and clean air in the transportation system of San Diego, California, under a hypothetical situation in which all cars are banned and an expensive, wide-ranging bus system has been introduced. From Roberts [1974b]. The author thanks the Rand Corporation for permission to use this figure.

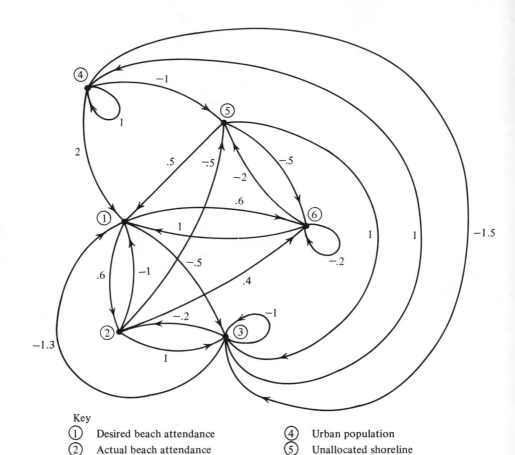

Key

① Desired beach attendance   ④ Urban population
② Actual beach attendance   ⑤ Unallocated shoreline
③ Overall urban satisfaction   ⑥ Public investment in beaches

FIGURE 4.17. Weighted digraph for coastal recreation. Weights added to signed digraph of Fig. 4.10. From data in Coady, *et al.* [1973]. The author thanks Dr. G. P. Johnson for permission to use this material.

One advantage of the weighted digraph model is that it allows us to make precise definitions of the problems we mentioned in Sec. 4.1: The problem of forecasting energy demand, the problem of forecasting the effect of a new technology on this demand, the problem of identifying and choosing possible strategy alternatives for meeting environmental constraints, etc. To make precise these problems, let us assume that each vertex u_i has a value (or level) $v_i(t)$ at each discrete time $t = 0, 1, 2, \ldots$. The times will be interpreted as hours, days, weeks, months, years, etc. The value of a variable like Population is simply its size. The value of a variable like Quality of

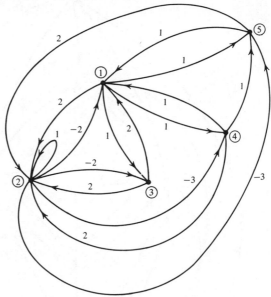

FIGURE 4.18. Weighted digraph for pattern of public/private transportation in Vancouver. Weights added to signed digraph of Fig. 4.11. From data in Kane [1972].

Key

① Cost of an automobile
② Use of automobiles
③ Comfort and convenience of automobile use

④ Freedom of choice in travel (routes, time, etc.)
⑤ Speed

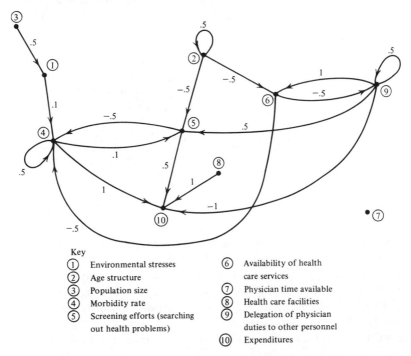

Key

① Environmental stresses
② Age structure
③ Population size
④ Morbidity rate
⑤ Screening efforts (searching out health problems)

⑥ Availability of health care services
⑦ Physician time available
⑧ Health care facilities
⑨ Delegation of physician duties to other personnel
⑩ Expenditures

FIGURE 4.19. Weighted digraph for health care delivery in British Columbia. Weights added to signed digraph of Fig. 4.12. From data in Kane, *et al.* [1972].

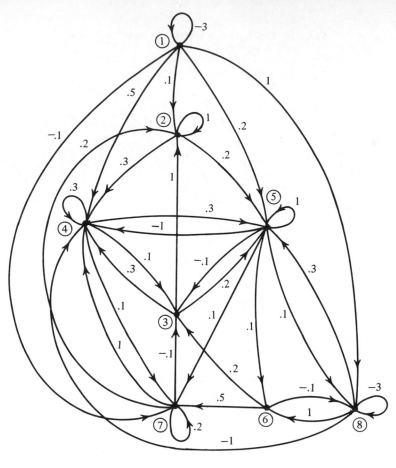

Key

①	Pay, allowances, and benefits	⑤	Retention
②	Physical environment	⑥	Military men
③	Skill opportunity	⑦	Equipment performance
④	Job satisfaction	⑧	Manpower budget

FIGURE 4.20. Weighted digraph for Naval manpower system. Weights added to signed digraph of Fig. 4.13. From data in Kruzic [1973a]. (Adapted with permission of Office of Naval Research.)

environment is some measure of environmental quality. Below, we shall introduce a specific rule, the *pulse process rule*, for how changes in these values are propagated through the system over time. We can now define the *Forecasting Problem* as follows: Predict the value $v_i(t)$ of vertex u_i at time t, or alternatively predict the change in $v_i(t)$ from its starting value $v_i(\text{start})$.

To give a sample forecast, let us consider the weighted digraph of Fig. 4.15 and assume that fuel consumption decreases by 10% at a fixed (starting) time, say by government decree. We consider a time period to be one week and forecast the level of emissions (variable 6) for the next 50

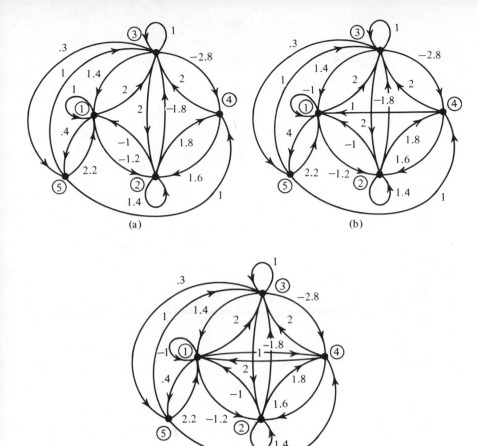

Key
① Energy use
② Domestic energy supply
③ Domestic investment (in new
   energy reserves and sources
④ Reliance on traditional sources
   of energy (as opposed to development
   of energy alternatives)
⑤ Public confidence in the government

FIGURE 4.21. Three weighted digraphs for energy use by the United States. Weights added to signed digraphs of Fig. 4.14. The data is from Kruzic [1973b] and was made available through the courtesy of the U.S. Army Corps of Engineers.

weeks. Shown in Fig. 4.22 is this forecast, with $v_6(t) - v_6(\text{start})$ expressed as a percent of $v_6(\text{start})$. The forecast was made using the pulse process rule to be introduced below. We see that the initial change in fuel consumption results in a rather rapid drop in emissions at first, but emissions start to rise again and, according to this forecast, stabilize at a level about $6\frac{1}{2}\%$ below the initial one.

There are many variations on the Forecasting Problem. One is to bring in the effect of new technologies or institutions by changing the weighted

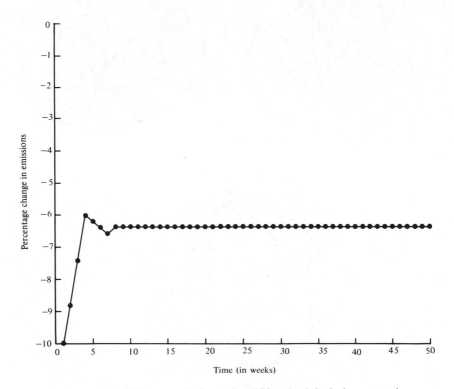

FIGURE 4.22. Effect on emissions of a 10% cutback in fuel consumption in the transportation system for San Diego, California described by the weighted digraph of Fig. 4.15. (This figure was obtained from a computer printout of percentage change at each week by joining consecutive (discrete) data points by straight lines. The same technique was used to draw all of the forecasts which appear in this chapter.)

digraph at a particular time T, the time when this technology is introduced. Another is to make probabilistic forecasts, of expected value $E(v_i(t))$, based on the expected value of T. In this chapter, we shall not be able to consider these variations. But we urge the reader to consider them.

Environmental (or other) *constraints* or *standards* can be introduced by placing limits at certain vertices above or below which the value cannot go. This is done by introducing two vectors $m = (m_1, m_2, \ldots, m_n)$ and $M = (M_1, M_2, \ldots, M_n)$, representing the lower and upper constraints m_i and M_i at each vertex u_i, with these numbers allowed to be $-\infty$ and $+\infty$ respectively. Typical constraint vectors for the signed digraph of Fig. 4.7 might have a specified m_Q and m_J with all other $m_i = -\infty$ and all $M_i = +\infty$.

Given these constraints, we try to find alternative strategies which will allow us to meet the constraints. A *strategy* is a procedure to change the system. If the system is a weighted digraph, some of the possible changes or strategies are the following:

(1) Change the value of certain vertices at specified times.
(2) Add at a given time a new vertex (institution) and new arcs to and from it (relations of interaction of the institution with existing ones).
(3) Change the sign of a given arc at a given time.
(4) Change the weight of a given arc at a given time.
(5) Add a new arc between existing vertices.
(6) Add a new cycle (deviation-amplifying or deviation-counteracting).

In this chapter, we shall concentrate on sign-change and weight-change strategies. An example of such a strategy is the strategy of changing the sign of the arc (U, R) in the signed digraph of Fig. 4.7 from $-$ to $+$. This strategy, which is increasingly being discussed, is called "inverting the rate structure" and is aimed at encouraging conversation by decreasing energy price when energy use decreases, thus it corresponds to changing from "the more is used the less is paid" to "the more is used the more is paid." Another interesting sign change strategy corresponds to changing the sign of the arc (J, P) from $+$ to $-$. This might be accomplished by opening more and more jobs to women, giving them less time to raise families. An interesting strategy which corresponds to adding an arc is to add an arc with a $+$ sign from Q to C. This amounts to tying allowable energy capacity to an index of environmental quality. Once a fixed set of strategy alternatives is defined, we can make precise the *Strategy Problem*. Briefly put, this has several versions. One version is: Find the optimal (shortest, cheapest, etc.) strategy for attaining the (environmental) standards.[6] A second version is: Find the policy which maximizes (or minimizes) the values on certain vertices (e.g., vertices measuring quality of environment), subject to constraints on values at certain other vertices.

In this book, we shall be interested in a still different version of the Strategy Problem. We shall not concentrate on how to find strategies which meet given (environmental) constraints, but rather on how to find strategies which do not allow any variable to get too large or too small. We shall call such strategies *stabilizing strategies*. To make the notion of stability precise,

---

[6]Besides finding an optimal strategy for attaining the goals (meeting the standards), it is important to see if the standards can be maintained once reached.

we introduce a specific model for the propagation of changes in the values of variables. This is the subject of the next section.

EXERCISES

1. In the signed digraph of Fig. 4.7, interpret the sign $+$ on the arc (F, U).

2. In the signed digraph of Fig. 4.7, why is the sign of the arc (Q, P) $+$?

3. In Fig. 4.7, name a loop besides (P, P) which may have been omitted.

4. In Fig. 4.8, is the cycle 1, 7, 8, 1 deviation-amplifying or deviation-counteracting? Explain and interpret your answer.

5. Repeat Exer. 4 for the cycle 1, 3, 2, 1 in Fig. 4.10. Do you agree with choice of signs?

6. In Fig. 4.7, what is the interpretation of the strategy of changing the sign of arc (C, F) from $+$ to $-$? Is this strategy reasonable as public policy?

7. Consider the signed digraphs of Figs. 4.8, 4.9, 4.10, 4.11, 4.12, and 4.13.
    (a) In each, identify a deviation-amplifying cycle (if there is one) and discuss.
    (b) In each, identify a deviation-counteracting cycle (if there is one) and discuss.

8. In each of the signed digraphs of Exer. 7, identify a not-easily-quantifiable variable or argue that there is none.

9. In each of the signed digraphs of Exer. 7, identify an arc whose sign might change depending on the level of the variables involved, or argue that there is no such arc.

10. In each of the signed digraphs of Exer. 7, identify some effects with rather long time lags and others with short ones.

11. In the signed digraph of Fig. 4.9, is there a vertex u so that every cycle though u is deviation-amplifying?

12. In the signed digraph of Fig. 4.9, is there a vertex with many negative cycles through it? If so, one might expect an oscillation in the values of this variable. Discuss this point from the point of view of your intuition or knowledge about past behavior of the values of this variable.

13. In each signed digraph of Exer. 7:
    (a) Identify a change-of-sign strategy which might be practically implemented.

(b) Identify an addition-of-arc strategy which might be practically implemented.

14. Consider the three signed digraphs of Fig. 4.14.

(a) Identify some of the new cycles which have been introduced by adding arc (4, 1) to signed digraph (a) to obtain signed digraph (b). Discuss these cycles.

(b) Do the same for the addition of arc (1, 4) to signed digraph (b) to obtain signed digraph (c).

15. In a signed digraph, one associates with a cycle its sign, which may be thought of as the product of the *signs* of its arcs. In a weighted digraph, one associates with a cycle C the product $p(C)$ of the *weights* of its arcs. Under reasonable rules for how values change, the larger the absolute value of $p(C)$, the more significant a role the cycle plays. If C is a positive or deviation-amplifying cycle, then if $|p(C)| > 1$, the deviations produced by C become increasingly amplified, while if $|p(C)| < 1$, the deviations produced by C eventually die out. If C is a negative or deviation-counteracting cycle, then if $|p(C)| > 1$, the deviations produced by C come in the form of increasing oscillations, while if $|p(C)| < 1$, the oscillations eventually die out.

(a) In each of the weighted digraphs of Figs. 4.15, 4.16, 4.17, 4.18, 4.19, and 4.20, identify a deviation-amplifying cycle which produces increasing amplifications, if there is any such cycle.

(b) In these same weighted digraphs, identify a deviation-counteracting cycle which produces increasing oscillations, if there is any such cycle.

16. Do Exer. 15 for each of the weighted digraphs of Fig. 4.21 and compare the conclusions. Have some cycles changed character?

17. The signed digraph of Fig. 4.23 is a *cognitive map*, a translation into a signed digraph of the opinions of an actual decisionmaker. Robert Axelrod [1971, and to appear] has observed that such cognitive maps tend to be balanced (see Sec. 3.1, Exer. 14) and acyclic. He takes the acyclicity to mean that the decisionmakers tend to omit important feedback loops in their decisionmaking.

(a) Verify that the cognitive map shown is balanced. (Remember to check semicycles; or use the Structure Theorem (Exer. 15, Sec. 3.1).)

(b) Comment on arcs which might have been omitted and whose inclusion would introduce feedback loops.

(*Note*: Other cognitive maps for a Middle Eastern expert and for Gouvernor Morris in the Constitutional Convention can be found in Bonham and Shapiro [to appear] and Ross [to appear], respectively.)

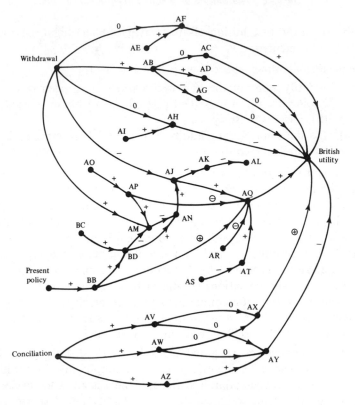

AA	Policy of complete British withdrawal from Persia	AQ	Ability of Persian government to maintain order
AB	Withdrawal from NW districts	AR	Absence of reformers in Persian parties (= no reformers)
AC	Probability of occurrence of serious disturbances in NW	AS	Amount of control of reformers by their friends
AD	Amount of chaos		
AE	Presence of Bakhtiari	AT	Strength of reformers
AF	Maintenance of status of Anglo-Persian Oil Company	AU	Policy of conciliation with Persia
AG	Maintenance of telegraph	AV	Abrogation of 1907 treaty with Russia
AH	Probability of invasion of Persia by Bolsheviks	AW	Revision of customs tariff
AI	Amount of feeling in Persia for Bolsheviks	AX	Conciliation of Persian public opinion
AJ	Amount of security in Persia	AY	Amount of Persian desire to go their own way
AK	Amount of blackmail on trade caravans		
		AZ	Amount of British interference with Persia
AL	Utility of Persian tribesman		
AM	Removal of better governors	BA	Present policy of intervention in Persia
AN	Ability of Persian governors to do anything		
		BB	Allowing Persians to have continual small subsidy
AO	Establishment of Persian constitution	BC	Amount of Persian debt to British
AP	Lack of quality of the kind of Shah at present	BD	Ability of British to put pressure on Persia

See next page for caption.

18. *Project*: For a public policy issue of your choice, construct a signed digraph and discuss it as in Exers. 7–13.

19. For a public policy issue of your choice, construct a weighted digraph, and discuss it as in Exer. 15.

4.4. Pulse Processes

To make a somewhat deeper analysis of the weighted digraph model, it is necessary to make some very specific assumptions about the effect that changes in value in one vertex have on other vertices. We shall call such assumptions *change of value rules*. The specific change of value rule assumed plays a rather subtle role in its relation to our conclusions. If we assume that the basic data (say for example initial values at each vertex and weights) are known only imprecisely, then the ultimate predictions based on a specific change of value rule will invariably be imprecise as well. Any conclusions drawn should be regarded as tentative, and subjected to a "sensitivity analysis." Such an analysis will involve redoing the modelling with changes in basic data, and perhaps using different change of value rules.[7]

To define the specific change of value rule we shall adopt, let us start with a *signed* digraph. As usual, let us list the vertices as $u_1, u_2, \ldots, u_n$. We are assuming that each vertex u_i attains a value $v_i(t)$ at discrete times $t = 0, 1, 2, \ldots$. Let us assume first that the value $v_i(t + 1)$ is determined from $v_i(t)$ and from information about whether other vertices u_j adjacent to u_i went up or down at time t. If there is an arc from u_j to u_i with a positive (negative) sign, a change in u_j at time t is reflected (reversed) in u_i at time $t + 1$. We shall assume that a unit change in u_j leads to a unit change in u_i. Thus, if (u_j, u_i) is $+$ and $p_j(t)$ is a number representing the change in u_j at time t, then the effect on u_i at time $t + 1$ of the change in u_j is an increase in u_i of $p_j(t)$. If (u_j, u_i) is $-$, then the effect on u_i at time $t + 1$ of the change in u_j is a decrease in u_i of $p_j(t)$. Of course, if $p_j(t)$ is a negative number, an

---

FIGURE 4.23. Cognitive map, based on deliberations of the British Eastern Committee of 1918 with regard to British policy toward Persia. This specific digraph represents the views of one member of the committee, Marling. The arcs represent causal assertions about how one concept variable affects another. The symbols indicate the type of relationship:

+   positive
−   negative
0   zero (no relationship)
⊕   zero or positive
⊖   zero or negative.

An arc with weight of 0 is pictured if explicit assertion of no effect ıs made.

(Adapted from Axelrod [1971] with permission of the Peace Science Society International.)

[7]For a more detailed discussion of sensitivity analysis, see Roberts [1974b] and Sec. 4.6.

increase of $p_j(t)$ units means a decrease, and a decrease of $p_j(t)$ units means an increase. The change $p_j(t)$, called a *pulse*, is given by the difference $v_j(t) - v_j(t - 1)$ if $t > 0$. It must be specified as an initial condition if $t = 0$. To summarize, suppose we denote

$$\text{sgn}(u_j, u_i) = \begin{cases} +1 & \text{(if } (u_j, u_i) \text{ is } +) \\ -1 & \text{(if } (u_j, u_i) \text{ is } -) \\ 0 & \text{(if there is no arc } (u_j, u_i)). \end{cases}$$

Then for $t \geq 0$, we define

$$v_i(t + 1) = v_i(t) + \sum_{j=1}^{n} \text{sgn}(u_j, u_i) p_j(t). \tag{4}$$

An *autonomous pulse process* on a signed digraph D is defined by the rule (4), by an initial vector of values

$$V(0) = (v_1(0), v_2(0), \ldots, v_n(0)),$$

and by a vector

$$P(0) = (p_1(0), p_2(0), \ldots, p_n(0))$$

giving the outside pulse $p_j(0)$ introduced at each vertex u_j at time 0. We shall also use the *value vector*

$$V(t) = (v_1(t), v_2(t), \ldots, v_n(t))$$

and the *pulse vector*

$$P(t) = (p_1(t), p_2(t), \ldots, p_n(t)).$$

In an autonomous pulse process, we follow the propagation of initial pulses through the system.

In an autonomous pulse process, we shall usually determine $V(0)$ as follows. Suppose we know the starting value $v_i(\text{start})$ at each vertex u_i. Then $v_i(0)$ will be defined by

$$v_i(0) = v_i(\text{start}) + p_i(0),$$

i.e., $v_i(0)$ is the starting value at vertex i plus the initial pulse introduced at vertex i. Thus, we usually define an autonomous pulse process by giving the vector

$$V(\text{start}) = (v_1(\text{start}), v_2(\text{start}), \ldots, v_n(\text{start})),$$

rather than the vector $V(0)$.

To give an example of how an autonomous pulse process works, let us consider the signed digraph of Fig. 4.24 and let us assume that $P(0) = (1, 0, 0, 0)$ and $V(\text{start}) = (0, 0, 0, 0)$. Thus, $V(0) = (1, 0, 0, 0)$. In this pulse process, a unit initial pulse is introduced at vertex 1. At time 0, vertex u_1 increased by 1 unit, so at time 1, vertices u_2 and u_3 change, vertex u_2 going up by 1, vertex u_3 going down by 1. Thus, $V(1) = (1, 1, -1, 0)$ and so $P(1) = (0, 1, -1, 0)$. Since at time 1, vertex u_2 went up 1, this leads to an

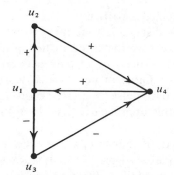

FIGURE 4.24. A signed digraph.

increase of 1 unit in vertex u_4 at time 2. But vertex u_3 went down 1 at time 1, so this leads (since arc (u_3, u_4) is $-$) to a further increase in u_4 by 1 unit at time 2. We conclude that $V(2) = (1, 1, -1, 2)$, and $P(2) = (0, 0, 0, 2)$. The increase in u_4 of two units at time 2 leads in turn to an increase of two units in u_1 at time 3. Thus, $V(3) = (3, 1, -1, 2)$. And so on.

(In this chapter, we shall limit ourselves to the autonomous pulse processes discussed so far. More generally, a pulse process might be subjected to external pulses at any time. In this case, we let $p_j^o(t)$ represent the external pulse or change in vertex u_j at time t. This amount $p_j^o(t)$ must be added to the value of vertex u_j at time t. Thus, we obtain the more general formula for a pulse process:

$$v_i(t + 1) = v_i(t) + p_i^o(t + 1) + \sum_{j=1}^{n} \operatorname{sgn}(u_j, u_i)p_j(t). \qquad (5)$$

For initial conditions we must give $p_j^o(t)$ for all j and all t, as well as the usual information. We usually have $p_j^o(0) = p_j(0)$.)

The rule (4) for an autonomous pulse process on a signed digraph generalizes in an obvious way to a change of value rule for an autonomous pulse process on a weighted digraph. We simply take

$$v_i(t + 1) = v_i(t) + \sum_{j=1}^{n} w(u_j, u_i)p_j(t), \qquad (6)$$

where as usual, $w(u, v) = 0$ if there is no arc (u, v). Here, if there is an arc from u_j to u_i with weight $w = w(u_j, u_i)$, and vertex u_j goes up by k units at time t, then as a result vertex u_i goes up by $k \times w$ units at time $t + 1$.

Since $v_i(t + 1) - v_i(t) = p_i(t + 1)$, Eq. (6) can be rewritten as follows:

$$p_i(t + 1) = \sum_{j=1}^{n} w(u_j, u_i)p_j(t). \qquad (7)$$

(This says that Eq. (6) is really a system of finite difference equations, with parameters $w(u_j, u_i)$.) The reader should also note that

$$v_i(t) = v_i(\text{start}) + \sum_{s=0}^{t} p_i(s). \qquad (8)$$

We have specified a particular change of value rule, given by Eq. (6). This rule definitely disregards time lags; it assumes that each effect takes place in unit time. As we remarked earlier, if all time lags are integral, we could take account of them by placing between two vertices joined by an arc of a digraph a number of new vertices corresponding to the time lag. Then, we could analyze pulse processes according to the rules we have given. The drawback of this approach is that the resulting digraphs can get very large.

Quite a variety of change of value rules are assumed in the literature of analysis of complex systems. We shall study two different ones in Exers. 14 and 15 of Sec. 4.7. Still a different one is used in the papers by Antle and Johnson [1973], Coady, et al [1973], Kane [1972], Kane, et al [1972], and Kruzic [1973a, b].

We close this section by proving some simple theorems about autonomous pulse processes on weighted digraphs. An autonomous pulse process for which $P(0)$ has the ith entry 1 and all other entries 0 is called *simple, starting with vertex u_i*. Here, an initial unit pulse at vertex u_i is propagated through the system over time. In a simple pulse process starting at vertex u_i of a signed digraph, the quantities $p_j(t)$ and $v_j(t)$ are related to the *signed number of paths* from u_i to u_j of length t, i.e., the difference between the number of positive paths from u_i to u_j of length t and the number of negative paths from u_i to u_j of length t. Specifically, we have the following result:

Theorem 4.3.   In a simple pulse process starting at vertex u_i of a signed digraph D, if $t > 0$, the quantities $p_j(t)$ and $v_j(t)$ are given by:

$p_j(t) =$ the signed number of paths from u_i to u_j of length $= t$
$v_j(t) = v_j(\text{start}) + p_j(0) +$ the signed number of paths from u_i to u_j of length $\leqq t$.

To illustrate this result, consider the signed digraph of Fig. 4.7 and consider the simple pulse process beginning at U. What are the pulse and value at Q at various times? We take u_i to be U and u_j to be Q. There is one path from u_i to u_j of length 1, namely U, Q. Since this path is negative, $p_j(1) = -1$ in the simple pulse process starting at vertex U. There are no paths of length 2 from U to Q, but there is a path of length 3, namely U, R, U, Q. This path is negative, so $p_j(3) = -1$. There are paths of length 4 from U to Q, namely U, Q, P, U, Q; U, C, R, U, Q; and U, C, F, U, Q. Since two of these paths are negative and one is positive, $p_j(4) = 1 - 2 = -1$. The signed number of paths of length $\leqq 4$ from U to Q is $-1 + (-1) + (-1) = -3$. Thus, $v_j(4) = v_j(\text{start}) + p_j(0) + (-3) = v_j(\text{start}) + 0 + (-3) = v_j(\text{start}) - 3$. We interpret these results as follows: If energy use is increased by 1 unit at time 0, and no other external changes are introduced into the system, then quality of the environment will go down by 1 unit at times 1, 3, and 4, and the value (level) of quality of the environment at time 4 will be 3 less than its starting value.

The second formula of Theorem 4.3 can be replaced by the equivalent formula

$$v_j(t) = v_j(0) + \text{the signed number of paths from } u_i \text{ to } u_j \text{ of length} \leq t.$$

This new formula should be used if $v_j(0)$ is known. Of course, Eq. (8) may also be used for calculating value.

Proof of Theorem 4.3. Suppose $x_0, x_1, \ldots, x_t$ is a path in D from $u_i = x_0$ to $u_j = x_t$. If a pulse of $+1$ is introduced at vertex u_i at time 0, then corresponding to this path, a pulse of $\text{sgn}(x_0, x_1)$ is introduced at vertex x_1 at time 1, a pulse of $\text{sgn}(x_0, x_1)\text{sgn}(x_1, x_2)$ is introduced at vertex x_2 at time 2, a pulse of $\text{sgn}(x_0, x_1)\text{sgn}(x_1, x_2)\text{sgn}(x_2, x_3)$ is introduced at vertex x_3 at time 3, and so on, until a pulse of $+1$ or -1 is introduced at vertex u_j at time t, with the sign $+$ or $-$ of this pulse equal to the sign of the path. Conversely, the only additions to or subtractions from u_j obtained in t steps from a unit pulse at time 0 at u_i are obtained from a path from u_i to u_j in D of length t. This gives the result for $p_j(t)$. The result for $v_j(t)$ follows from Eq. (8). Q.E.D.

A generalization of this theorem to the case of weighted digraphs is straightforward, and is left as an exercise (Exer. 12). The next theorem relates $p_j(t)$ and $v_j(t)$ to the entries of the adjacency matrix. It is stated for weighted digraphs, and so, since every signed digraph is weighted, it applies to signed digraphs as well.

Theorem 4.4. Suppose D is a weighted digraph with adjacency matrix $A = A(D)$. In a simple pulse process on D starting at vertex u_i,

$$p_j(t) = \text{the } i, j \text{ entry of } A^t$$

and

$$v_j(t) = v_j(\text{start}) + \text{the } i, j \text{ entry of } I + A + A^2 + \cdots + A^t.[8]$$

Proof. If D is signed, then the proof follows from Theorem 4.3, in an argument analogous to that in the proof of Theorem 2.11. If D is weighted, the argument follows similarly. Details are left to the reader. Q.E.D.

---

[8]The reader should note that A^t is the t-th power of A, not the transpose of A, which we are denoting in this book by A'.

We illustrate Theorem 4.4 on the signed digraph of Fig. 4.24, under the simple pulse process starting at vertex u_1, assuming that $V(\text{start}) = (0, 0, 0, 0)$. Here,

$$A = \begin{pmatrix} 0 & 1 & -1 & 0 \\ 0 & 0 & 0 & 1 \\ 0 & 0 & 0 & -1 \\ 1 & 0 & 0 & 0 \end{pmatrix}.$$

A simple calculation shows that

$$A^2 = \begin{pmatrix} 0 & 0 & 0 & 2 \\ 1 & 0 & 0 & 0 \\ -1 & 0 & 0 & 0 \\ 0 & 1 & -1 & 0 \end{pmatrix}.$$

and

$$I + A + A^2 = \begin{pmatrix} 1 & 1 & -1 & 2 \\ 1 & 1 & 0 & 1 \\ -1 & 0 & 1 & -1 \\ 1 & 1 & -1 & 1 \end{pmatrix}.$$

Since the process started at vertex u_1, $p_3(2)$ is given by the 1, 3 entry of A^2, namely 0. Similarly, $v_1(2) = v_1(\text{start}) +$ the 1, 1 entry of $I + A + A^2 = 0 + 1 = 1$. The results agree with the computation we made using the definition of pulse process.

In the simple pulse process starting at vertex u_i, we may calculate $v_j(t)$ by an alternative formula, namely

$$v_j(t) = v_j(0) + \text{the } i, j \text{ entry of } A + A^2 + \cdots + A^t.$$

(Why is this formula equivalent to that given in Theorem 4.4?)

It will be useful to restate Theorem 4.4 in vector notation. In a simple pulse process starting at vertex u_i, we have $P(0) = (0, 0, \ldots, 0, 1, 0, \ldots, 0)$, with a 1 in the ith location. Then we can restate the result as follows:

$$P(t) = P(0)A^t.$$

This equation also holds for a general autonomous pulse process. We shall state the result as a theorem.

✱ **Theorem 4.5.**   In an autonomous pulse process on a weighted digraph,

$$P(t) = P(0)A^t.$$

Theorem 4.5 follows directly from Eq. (7), which was obtained by rewriting the basic pulse process equation for weighted digraphs, Eq. (6). For Eq. (7) says that $P(t + 1) = P(t)A$. Thus, Theorem 4.5 can be proved directly. If this is done, Theorem 4.4 is an immediate corollary of Theorem 4.5. Finally, Theorem 4.3 can then be proved from Theorem 4.4.

Theorems 4.4 and 4.5 present solutions, in principle, to the Forecasting Problem. (In some sense, they also solve the first version of the strategy problem as stated in the previous section, if we know the standards are attainable.[9] We describe the solution if the allowable strategies are the introduction of a unit pulse at one vertex at some starting time. Given a desired value (set of values) of a vertex u_j (set of vertices), simply calculate sums of powers of the matrix A until some (all) i, j entry (entries) is (are) at least (at most) the desired value(s). An initial pulse at u_i gives the solution. The solution is similar if more complicated strategies are allowed or strategies with least cost are desired. The trouble with the above "solution" to the strategy problem, of course, is that once we attain a given level, we may not remain there.)

Having stated these preliminary results about pulse processes, we turn in the next section to notions of stability in pulse processes, and our main theorems will characterize stability.

EXERCISES

1. In the signed digraph of Fig. 4.25, assume that $V(\text{start}) = (0, 0, 0, 0)$ and $P(0) = (0, 0, 0, 1)$. Calculate $V(0)$, $P(1)$, $V(1)$, $P(2)$, and $V(2)$.

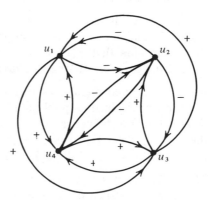

FIGURE 4.25. Signed digraph for exercises of Sec. 4.4.

2. In the signed digraph of Fig. 4.25:
   (a) Calculate the signed number of paths from u_4 to u_4 of lengths 1, 2, and 3.

[9]Of course, there is no *a priori* way of knowing this, and it would be helpful to have results characterizing "feasibility" of standards.

(b) Calculate $p_4(3)$ in the simple pulse process starting at u_4, using Theorem 4.3.

(c) Calculate $v_4(3)$ in this pulse process, using Theorem 4.3, if $V(\text{start}) = (0, 1, 1, 1)$.

3. In a pulse process on the weighted digraph of Fig. 4.26:

(a) Suppose vertex u_2 goes up by one unit at time 0. What happens to vertex u_4 at time 1?

(b) Suppose vertex u_3 and vertex u_4 both go up by one unit at time 0. What happens to vertex u_1 at time 1?

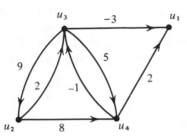

FIGURE 4.26. Weighted digraph for exercises of Sec. 4.4.

4. For the weighted digraph of Fig. 4.26, use Theorem 4.4 to calculate $p_2(3)$ and $v_2(3)$ under the simple pulse process starting at vertex u_3, under the assumption that $V(\text{start}) = (0, 0, 0, 0)$.

5. For the signed digraph of Fig. 4.25, use Theorem 4.4 to calculate $p_3(3)$ and $v_3(3)$ under the simple pulse process starting at vertex u_2, using the assumption that $V(\text{start}) = (0, 0, 0, 0)$. Check by determining all paths from u_2 to u_3 of lengths equal to 3 and at most 3, respectively.

6. For the weighted digraph of Fig. 4.21, part (a), assume unit time is one year. Assuming that public confidence is increased by one unit now, what will be the difference in the value of domestic investment between the present level and that three years from now? How does your answer change if the weighted digraph of Fig. 4.21, part (b), is used?

7. The next three exercises apply the notions of this chapter to ecosystems. For a more detailed discussion of the ecosystem in this exercise, see Kemeny and Snell [1962, Ch. 3]. For a more detailed discussion of the

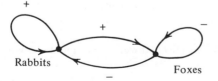

FIGURE 4.27. A rabbit and fox ecosystem.

ecosystems in Exers. 8 and 9, see Levins [1974, and to appear][10]. We return to these and other ecosystems in the exercises of Sec. 4.6. Imagine a simple ecosystem consisting of just rabbits and foxes. Assume that there is an unlimited food supply available for rabbits, and that foxes eat only rabbits. The relations among foxes and rabbits can be summarized in the signed digraph of Fig. 4.27. (As rabbits increase, this leads to more foxes, since foxes have more food. It also leads to more rabbits, through reproduction. As foxes increase, this leads to fewer rabbits, for foxes eat rabbits, and to fewer foxes, for the foxes compete for food and some die out.) If an initial increase of one unit is introduced in the rabbit population, calculate pulses and values for three time periods and see what happens. (Assume V(start) $=$ (10, 10).) Interpret the results.

8. Figure 4.28 shows an ecosystem consisting of three species, a plant, a herbivore, and a carnivore, with limited nutrients available for plants. Carnivores prey on herbivores which prey on plants. As plants increase, they compete for available nutrients and some plants die. Calculate pulses and values for four time periods following an initial unit increase in the plant population. (Assume V(start) $=$ (10, 10, 10).) Interpret the results.

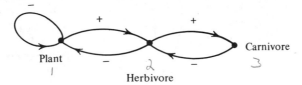

FIGURE 4.28. A plant-herbivore-carnivore ecosystem. (Adapted from Levins [to appear] with permission of Belknap Press of Harvard University Press and the author.)

FIGURE 4.29. A mice-rats-eagles ecosystem. (Adapted from Levins [to appear] with permission of Belknap Press of Harvard University Press and the author.)

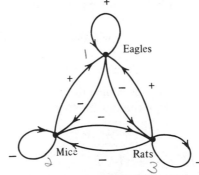

[10]Note that in Levins' work, the sign on arc (x_i, x_j) is interpreted as the effect of a change in x_i on the *rate of change* in x_j. We interpret the sign somewhat differently, and ask the reader to use our interpretation.

9.   Figure 4.29 shows a simple ecosystem with two species, mice and rats, competing for available resources. Both mice and rats are prey of the same predator, eagles. Trace pulses and values for four time periods following an initial unit increase in the mouse population. (Assume V(start) $= (10, 10, 10)$.) Interpret the results.

10. For the weighted digraph of Fig. 4.26, use Theorem 4.5 to calculate $P(3)$ under an autonomous pulse process with $P(0) = (1, 0, 1, 0)$.

11. For the signed digraph of Fig. 4.24, assume that initial unit pulses are introduced at both u_1 and u_4. Calculate $P(3)$.

12. State a generalization of Theorem 4.3 to weighted digraphs.

13. In the signed digraph of Fig. 4.30, calculate $p_2(3)$ and $v_2(6)$ under the simple pulse process starting at vertex u_1 if effect (u_1, u_2) takes two time units, effect (u_1, u_3) takes two units, effect (u_2, u_3) takes three units, and effect (u_3, u_2) takes one unit, and V (start) $= (0, 0, 0)$.

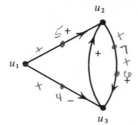

FIGURE 4.30. Signed digraph for exercises of Sec. 4.4.

*14. Consider the strategy problem for the signed digraph of Fig. 4.31, with lower constraint vector $m = (0, 5, 1)$ and upper constraint vector $M = (10, 10, 10)$, and with V(start) $= (0, 0, 0)$. (See Sec. 4.3 for definition of m and M.)

(a)   Find which simple initial pulse will realize these constraints most quickly.

(b)   Once met, will the constraints be satisfied thereafter?

(c)   Find a pair of lower and upper constraint vectors which cannot be realized by introducing a simple initial pulse.

*15. Consider the signed digraph of Fig. 4.31. Under an autonomous pulse process, write out the three difference equations of Eq. (7). Solve them

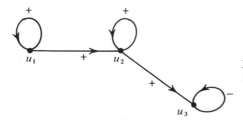

FIGURE 4.31. Signed digraph for Exers. 14 and 15 of Sec. 4.4.

explicitly in terms of $p_1(0), p_2(0)$, and $p_3(0)$. Assuming that $P(0) = (1, 0, 0)$, calculate $P(4)$ using this explicit solution.

16. For D a signed digraph, prove Theorem 4.4 from Theorem 4.3.

17. For some of the weighted digraphs of Figs. 4.15 through 4.21, identify some simple initial pulses which might correspond to feasible real-world strategies, and trace them through several time periods.

18. Invent some alternative change of value rules other than the pulse process rule.

4.5.   Stability in Pulse Processes

4.5.1.   DEFINITION OF STABILITY

There are a number of possible definitions of stability in a weighted digraph under a pulse process. We distinguish two notions. The first, *value stability*, says that the value $v_j(t)$ of a vertex u_j does not get too large in magnitude (absolute value). The second, *pulse stability*, says that the change in value, i.e., the pulse $p_j(t)$, does not get too large in magnitude. Formally, let us call a vertex u_j *pulse stable* under a pulse process if the sequence

$$\{|p_j(t)|: \ t = 0, 1, 2, \ldots\}$$

is bounded, i.e., if there is a number B so that $|p_j(t)| < B$ for all t. We say u_j is *value stable* if the sequence

$$\{|v_j(t)|: \ t = 0, 1, 2, \ldots\}$$

is bounded. The weighted digraph is *pulse (value) stable* under the pulse process if each vertex is. To give an example, if $v_j(t) = 2^t$, then vertex u_j is both pulse and value unstable, for if $t > 0$, $p_j(t) = v_j(t) - v_j(t-1) = 2^t - 2^{t-1} = 2^{t-1}$, and both sequences $\{2^t\}$ and $\{2^{t-1}\}$ are unbounded. (Similar conclusions are true in general of exponential growth $v_j(t) = a^t$, where $a > 1$. (See Fig. 4.32.) However, if $v_j(t) = a^t$, $0 < a < 1$, we have both pulse and value stability at vertex u_j. (What happens if $v_j(t) = a^t$, a negative?) If $v_j(t) = 2t + 5$, then vertex u_j is still value unstable but it is now pulse stable, since $p_j(t) = 2$, all $t > 0$. Similar conclusions are true of linear growth in general. (See Fig. 4.32.)

Under any pulse process, value stability (at u_j) implies pulse stability (at u_j), since for $t > 0$,

$$|p_j(t)| = |v_j(t) - v_j(t-1)| \leq |v_j(t)| + |v_j(t-1)|.$$

Thus, if $\{|v_j(t)|\}$ is bounded by a bound B, then $|v_j(t)| < B$ and $|v_j(t-1)| < B$, so $\{|p_j(t)|\}$ is bounded by the bound $2B$. On the other hand, pulse stability does not imply value stability: Consider for example the positive 2-cycle

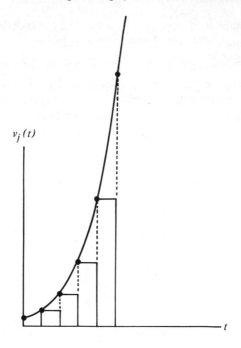

Exponential Growth

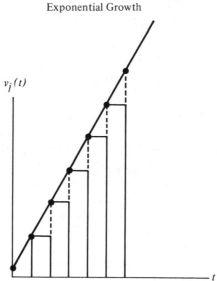

Linear Growth

FIGURE 4.32. Exponential growth $v_j(t) = a^t$ with $a > 1$ is pulse and value unstable. Linear growth $v_j(t) = at + b$ with $a \neq 0$ is pulse stable and value unstable. Pulses are shown as dotted lines. (Smooth curves are drawn through the discrete data points.)

D of Fig. 4.33. Under the simple pulse process starting at vertex u_1, $p_1(t)$ is always 0 or 1, but $v_1(t)$ grows by one unit every two time periods. Thus, vertex u_1 is pulse stable but not value stable. Instability in pulse or in value

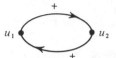

FIGURE 4.33. Positive 2-cycle.

for a given weighted digraph suggests that "something has to give." There will be forces to change the fundamental structure of the system before any values or pulses get unboundedly large.

4.5.2. EIGENVALUES AND STABILITY

In this subsection and the next, we shall present several theorems about stability of a weighted or signed digraph under a pulse process. These theorems will be applied in Sec. 4.6 to the examples we have been studying. The theorems of this subsection assert that testing for stability of a digraph D reduces to asking simple questions about the eigenvalues of D. The first theorem says that we simply have to calculate the magnitudes of the eigenvalues in order to draw some interesting conclusions. (The reader will recall that an eigenvalue is a complex number $a + bi$ and so its *magnitude* is $\sqrt{a^2 + b^2}$. A real number a has magnitude $\sqrt{a^2} = |a|$.)

Theorem 4.6.[11]   If a weighted digraph D is pulse stable under all simple pulse processes, then every eigenvalue of D has magnitude at most unity.

The proof of this theorem we shall leave for Sec. 4.7. In applying it, one usually uses the contrapositive form: if D has an eigenvalue of magnitude greater than unity, then D is pulse unstable under *some* simple pulse process. Let us illustrate this theorem with an example, the weighted digraph of Fig. 4.34. Here, an easy computation shows that $C(\lambda) = \lambda^2(\lambda^3 - 8)$. Thus, $\lambda = 2$ is an eigenvalue. By Theorem 4.6, we conclude that the weighted

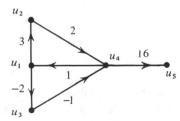

FIGURE 4.34. A weighted digraph with $C(\lambda) = \lambda^2(\lambda^3 - 8)$.

[11]This theorem was first stated in this form in Roberts and Brown [1975].

digraph is pulse unstable under some simple pulse process. This means that under an initial pulse introduced at *some* vertex, the pulse at *some* (possibly other) vertex will be unboundedly large. The weighted digraph is not pulse unstable under *every* simple pulse process. Consider for example the simple pulse process starting at u_5.

It is useful to note a corollary at this point.

Corollary. If an integer-weighted digraph D is pulse stable under all simple pulse processes, then every nonzero eigenvalue has magnitude equal to unity.

Proof. By Theorem 4.6, each nonzero eigenvalue has magnitude at most unity. Recall that the eigenvalues of D are defined to be the eigenvalues of the adjacency matrix A of D. Let $C(\lambda) = \sum_{i=0}^{n} a_i \lambda^i$ be the characteristic polynomial of A. If i is the least integer such that $a_i \neq 0$, then $C(\lambda) = \lambda^i(a_i + a_{i+1}\lambda + \cdots + a_n\lambda^{n-i})$. The nonzero eigenvalues of D are exactly the roots of the polynomial $a_i + a_{i+1}\lambda + \cdots + a_n\lambda^{n-i}$. Since the product of the roots of a polynomial is ± 1 times the constant term, it follows that the product of all the nonzero eigenvalues of D is ± 1 times a_i. Since all entries of A are integers, a_i is an integer. (Why?) Since each nonzero eigenvalue has magnitude at most unity, it follows that a_i is ± 1 and, hence, that each nonzero eigenvalue has magnitude equal to unity.   Q.E.D.

To illustrate use of the corollary, let us note that the signed digraph of Fig. 4.35 has characteristic polynomial $C(\lambda) = \lambda(\lambda^3 + \lambda + 1)$. Now the polynomial $f(\lambda) = \lambda^3 + \lambda + 1$ has $f(0) = 1$ and $f(-1) = -1$. We conclude that f has a real root strictly between 0 and -1, and so there is an eigenvalue of magnitude less than 1. (This type of reasoning is very useful in applying

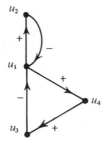

FIGURE 4.35. A signed digraph with $C(\lambda) = \lambda(\lambda^3 + \lambda + 1)$.

the eigenvalue theorems.) By the corollary, the signed digraph is pulse unstable under some simple pulse process. The fact that all weights are $+1$ or -1 allows us to draw this conclusion. Note that Theorem 4.6 can also be applied to signed digraphs. However, the Corollary gives us a stronger tool for this special case.

The converse of Theorem 4.6 is false. To give an example, consider the signed (weighted) digraph of Fig. 4.36. Its eigenvalues are 1, 1 (i.e., 1 with multiplicity 2), but it is pulse unstable under the simple pulse process starting at vertex u_1, for here $p_2(t) \longrightarrow \infty$. (Why?) The converse of Theorem 4.6 is

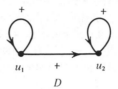

FIGURE 4.36. Signed digraph D has eigenvalues 1, 1, but it is pulse unstable under the simple pulse process starting at vertex u_1.

true if we assume that the weighted or signed digraph has its nonzero eigenvalues distinct, i.e., it has no multiple eigenvalues except possibly zero. This simplifying assumption, which is satisfied in all but the rarest practical situations, can be replaced; we do so in Section 4.7. We now state this partial converse to Theorem 4.6.

Theorem 4.7.[12]   Suppose D is a weighted digraph with all nonzero eigenvalues distinct. If every eigenvalue of D has magnitude at most unity, then D is pulse stable under all simple pulse processes.

The proof of this theorem we shall again leave for Sec. 4.7.

To illustrate the theorem, let us use the weighted digraph of Fig. 4.37. The characteristic polynomial of this weighted digraph is $C(\lambda) = \lambda^4 - .512\lambda = \lambda(\lambda^3 - .512)$. The roots are 0, .8, .4$(-1 + \sqrt{3}\,i)$ and .4$(-1 - \sqrt{3}\,i)$. The nonzero eigenvalues are distinct and all have magnitude at most unity. (For

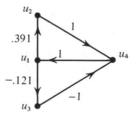

FIGURE 4.37. A weighted digraph with $C(\lambda) = \lambda^4 - .512\lambda$.

example, .4$(-1 + \sqrt{3}\,i)$ has magnitude $\sqrt{(-.4)^2 + (.4\sqrt{3}\,)^2} = \sqrt{.16 + .48} = \sqrt{.64} = .8$.) Hence, the weighted digraph is pulse stable under all simple pulse processes. The reader should not forget to check the hypothesis that nonzero eigenvalues must be distinct. If a nonzero eigenvalue is a multiple eigenvalue, the theorem does not apply. (The case of multiple eigenvalues will be discussed in Sec. 4.7.)

The next theorem characterizes value stability.

---

[12]This theorem was first stated in this form for signed digraphs in Brown, Roberts, and Spencer [1972] and more generally in Roberts and Brown [1975].

Theorem 4.8.[13]   Suppose D is a weighted digraph. Then D is value stable under all simple pulse processes if and only if D is pulse stable under all simple pulse processes and unity (1) is not an eigenvalue of D.

Once again, we defer the proof of Theorem 4.8 to Section 4.7.

According to our earlier results, the weighted digraph D of Fig. 4.37 is pulse stable under all simple pulse processes, since the eigenvalues are 0, .8, $.4(-1 + \sqrt{3}\,i)$ and $.4(-1 - \sqrt{3}\,i)$. Since 1 is not an eigenvalue, we conclude that D is also value stable under all simple pulse processes.

4.5.3. STRUCTURE AND STABILITY: ROSETTES

Before applying the results of this section in Sec. 4.6, let us note that these results have certain shortcomings. First, they can be used to *test* for stability of a signed or weighted digraph, and they can even be used to *test* whether certain changes in the digraph would be stabilizing, but they cannot help us to *discover* what kinds of changes would be stabilizing. More useful theorems, then, would relate stability to the structure of the underlying digraph, and help us to discover stabilizing strategies by showing what structural changes could lead to stability. Not much is known about how structure affects stability. However, a few results along these lines are described in this subsection. It is clear that there is much mathematical work still needed on this subject.

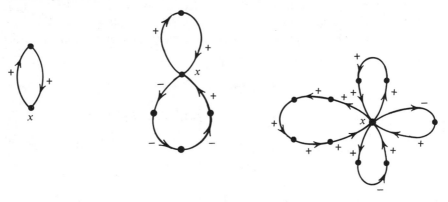

FIGURE 4.38.   Some (signed) rosettes. The central vertex is labelled x.

To give examples of results which relate stability to the structure of the underlying digraph, we shall restrict ourselves to a specific class of digraphs. Although it may appear to the reader that this class is rather special, it has turned out that many digraphs arising in the study of complex systems fall

---

[13]This theorem was first stated in this form in Roberts and Brown [1975].

into the class. To define the class, let us say that a digraph D is a *rosette* if it consists of a *central vertex* x and nonintersecting cycles leading out of x. Examples of signed digraphs which are rosettes are shown in Fig. 4.38. More generally, a digraph D is an *advanced rosette* if D is strongly connected and there is a *central vertex* x which is on all cycles of D. Of course, every rosette is an advanced rosette. Figure 4.39 shows several (signed) advanced rosettes which are not rosettes.

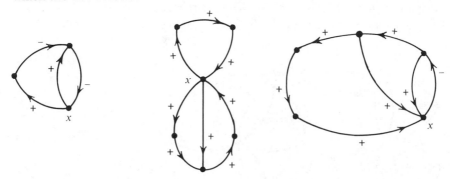

FIGURE 4.39. Some (signed) advanced rosettes which are not rosettes. The central vertex is labelled x.

Suppose D is a signed advanced rosette. Let a_i denote the sum of the signs of the cycles of length i, if $+$ is counted as $+1$ and $-$ as -1. Let s be the largest integer such that $a_s \neq 0$. If $a_i = 0$, all i, take $s = 0$. If $s = 0$, D is pulse and value stable under all simple pulse processes. (See Exer. 18). If $s > 0$, then stability properties of D are completely determined by the *rosette sequence* $(a_1, a_2, \ldots, a_s)$.

Theorem 4.9. If two advanced rosettes D_1 and D_2 have the same rosette sequence, then D_1 is pulse (value) stable under all simple pulse processes if and only if D_2 is.

Proof. Exercise 17(c).

Many results about stability can be read off from the rosette sequence. The following two theorems first appeared in Brown, Roberts, and Spencer [1972] and Roberts and Brown [1975]. The proofs are omitted; but see Exer. 14.

Theorem 4.10. Suppose D is a signed advanced rosette with $s > 0$ and rosette sequence $(a_1, a_2, \ldots, a_s)$. If D is pulse stable under all simple pulse processes, then

(a) $a_s = \pm 1$

and

(b) $a_i = (-a_s)a_{s-i}$ $i = 1, 2, \ldots, s - 1$.

Theorem 4.11. Suppose D is a signed advanced rosette with $s > 0$ and rosette sequence $(a_1, a_2, \ldots, a_s)$, and suppose D is pulse stable under all simple pulse processes. Then D is value stable under all simple pulse processes if and only if $\sum_{i=1}^{s} a_i \neq 1$.

To illustrate Theorems 4.10 and 4.11, let us consider the signed advanced rosette D_1 of Fig. 4.40. There are two cycles of length 3, one positive, and one negative, so $a_3 = 0$. There are two cycles of length 2, both

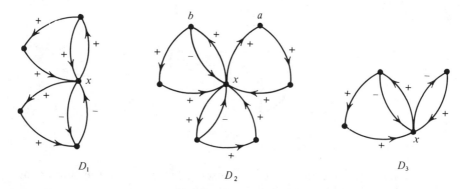

FIGURE 4.40. Some signed advanced rosettes.

positive, so $a_2 = 2$. There are no cycles of length 1. Thus, $s = 2$ and the rosette sequence is $(0, 2)$. Since $a_s \neq \pm 1$, Theorem 4.10 implies that D_1 is pulse unstable under some simple pulse process. To give another example, let us take the signed advanced rosette D_2 of Fig. 4.40. The rosette sequence is $(0, -2, 3)$, and again, since $a_s \neq \pm 1$, Theorem 4.10 implies that D_2 is pulse unstable under some simple pulse process. Let D_2' be obtained from D_2 by changing the sign of arc (x, a) from $+$ to $-$. Then D_2' has rosette sequence $(0, -2, 1)$, so the condition $a_s = \pm 1$ is satisfied. However, the second condition of Theorem 4.10 is not satisfied, since $a_1 \neq (-a_3)a_2$. Thus D_2' is still pulse unstable under some simple pulse process. (The reader should note that, by Theorem 4.9, D_2' has the same stability properties as the simpler signed advanced rosette D_3 shown in Fig. 4.40, since the latter also has rosette sequence $(0, -2, 1)$. Theorem 4.10 suggests a *potentially* stabilizing strategy: Change one of the cycles of length 2 to become positive, obtaining $a_2 = 0$. Indeed, if D_2'' is obtained from D_2' by changing the sign of arc (b, x) from $-$ to $+$, then one finds that D_2'' has rosette sequence $(0, 0, 1)$. This rosette sequence satisfies both conditions of Theorem 4.10. However, we cannot conclude pulse stability, as Theorem 4.10 gives only necessary and not sufficient conditions. It is necessary to check for pulse stability by using other methods, for example Theorem 4.7. It is not too hard to do this if we use the fact (Roberts and Brown [1975]) that the characteristic polynomial of the advanced rosette is given by $C(\lambda) = (-1)^n \lambda^{n-s}(\lambda^s - \sum_{i=1}^s a_i \lambda^{s-i})$. Here, $C(\lambda) = -\lambda^4(\lambda^3 - 1)$. Thus, we discover that D_2'' is indeed pulse stable under all simple pulse processes. For its eigenvalues are $0, 0, 0, 0, 1, -\frac{1}{2} + (\sqrt{3}/2)i, -\frac{1}{2} - (\sqrt{3}/2)i$. The nonzero eigenvalues are distinct and have magnitude unity. (This result about pulse stability also follows by another argument: D_2'' has the same stability properties as a single positive cycle of length 3.) Next, we check for value stability, using Theorem 4.11. Since $a_1 + a_2 + a_3 = 1$, we conclude that D_2'' is value unstable under some simple pulse process. (This again is clear from the observation that D_2'' is like a single positive cycle of length 3, or from the observation that 1 is an eigenvalue.) If we start with D_2, then changing the sign of arc (x, a) from $+$ to $-$ and also changing the sign of arc (x, b) from $+$ to $-$ will give a *potentially* value stable signed digraph. For here, the rosette sequence is $(0, 0, -1)$, so the conditions (a) and (b) of Theorem 4.10 are satisfied and so is the condition $\sum_{i=1}^s a_i \neq 1$ of Theorem 4.11. Theorem 4.7 now implies that the changes lead to pulse stability, since $C(\lambda) = (-1)^n(\lambda^s - \sum_{i=1}^s a_i \lambda^{s-i}) = -\lambda^4(\lambda^3 + 1)$ and so the eigenvalues are $0, 0, 0, 0, -1, \frac{1}{2} + (\sqrt{3}/2)i, \frac{1}{2} - (\sqrt{3}/2)i$. Theorem 4.11 then gives value stability, since $a_1 + a_2 + a_3 = -1$. (These results also follow from the fact that the new advanced rosette has the same stability properties as a single negative cycle of length 3.)

In sum, Theorems 4.10 and 4.11 can be used to find potentially pulse- and value-stabilizing strategies. Any change of signs which gives rise to a rosette sequence satisfying conditions (a) and (b) of Theorem 4.10 is potentially pulse-stabilizing. If in addition the change leads to a rosette sequence satisfying $\sum_{i=1}^s a_i \neq 1$, then the change is also potentially value-stabilizing. (An alternative potentially-stabilizing strategy is one which changes s to be 0.)

EXERCISES

1. Suppose in a weighted digraph under a pulse process, $v_j(t)$ is as follows. Is the vertex u_j value stable under the pulse process?
   (a) $v_j(t) = 3^t$.   (b) $v_j(t) = 5t + 6$.   (c) $v_j(t) = 5$.
   (d) $v_j(t) = (-2)^t$.   (e) $v_j(t) = (\frac{1}{2})^t$.   (f) $v_j(t) = (-\frac{1}{2})^t$.
   (g) $v_j(t) = (-1)^t$.   (h) $v_j(t) = (\frac{1}{2})t + 100$.   (i) $v_j(t) = (\frac{1}{2})^t + 100$.
   (j) $v_j(t) = \sin t$.

2. In each case of Exer. 1, calculate $p_j(t)$ and determine if vertex u_j is pulse stable under the pulse process.

3. In each of the following, the eigenvalues of a weighted digraph are listed. What can you conclude about pulse stability under simple pulse

processes? About value stability? Why?

(a) $0, 1, -1, i, -i, 5.$ (b) $0, \sqrt{3}\,i, -\sqrt{3}\,i.$ (c) $\frac{1}{2}, i, -i.$

(d) $\frac{1}{2}, 1, -1.$ (e) $\frac{1}{2}, \frac{1}{2}, i, -i.$ (f) $0, 0, \frac{1}{2}, i, -i.$

(g) $0, \frac{1}{2}i, (-\frac{1}{2})i.$ (h) $0, -1, 1, i, -i.$ (i) $0, 6, 6, 1.$

4. Suppose you begin to calculate eigenvalues of a signed digraph and obtain the numbers listed in Exer. 3. (These are not necessarily the complete set of eigenvalues.) In each case, what can you conclude about pulse stability under simple pulse processes? About value stability?

5. Calculate the rosette sequence for each of the signed rosettes in Fig. 4.38.

6. Calculate the rosette sequence for each of the signed advanced rosettes in Fig. 4.39.

7. Suppose the rosette sequence of a signed advanced rosette is given as follows. What can you say about pulse stability?
    (a) $(1, 0, 2)$.   (b) $(0, -1, 2, 1)$.   (c) $(0, 1, 0, 1)$.   (d) $(0, 5, -5, 0, 1)$.

8. Suppose the rosette sequence of a signed advanced rosette D is $(0, 1, 0, -1, 1)$. Can you conclude without further information whether or not D is value stable? What if the rosette sequence is $(0, 0, 0, 2)$? What if it is $(0, 1, 0, 1)$?

9. Give an example of a strongly connected digraph which is not an advanced rosette.

10. Restate in words the condition $\sum a_i \neq 1$ in Theorem 4.11.

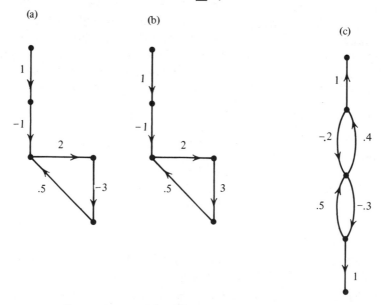

(a)   (b)   (c)

FIGURE 4.41. Weighted digraphs for Exer. 11, Sec. 4.5.

11. For each weighted digraph of Fig. 4.41:
     (a) Determine if it is pulse stable under all simple pulse processes.
     (b) Determine if it is value stable under all simple pulse processes.

12. For each signed digraph of Fig. 4.42, make use of rosette theory to:
     (a) Determine if it is pulse stable under all simple pulse processes.
     (b) Determine if it is value stable under all simple pulse processes.
     (c) If it is not pulse stable, find a strategy for changing one sign which
          is potentially pulse-stabilizing in the sense that the necessary condi-
          tions of Theorem 4.10 are satisfied, or show that no such strategy
          exists. (If you found a potential strategy, how would you determine
          if it is indeed pulse-stabilizing?)
     (d) If it is not value stable, find a strategy for changing one sign which
          is potentially value-stabilizing in the sense that the necessary con-
          ditions of both Theorems 4.10 and 4.11 are satisfied, or show that

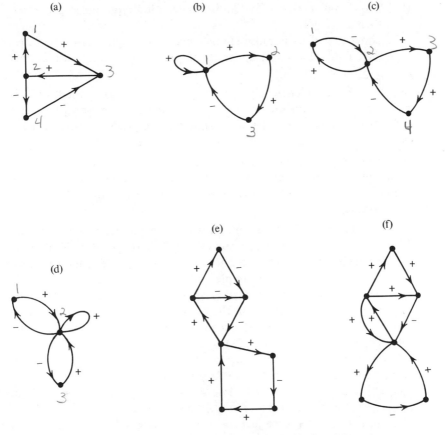

FIGURE 4.42. Signed digraphs for Exer. 12, Sec. 4.5.

no such strategy exists. (If you found a potential strategy, how would you determine if it is indeed value-stabilizing?)

13. (a) Prove that if a signed digraph has any nonzero eigenvalues, it must have at least one whose magnitude is greater than or equal to unity. (*Hint:* Read the proof of the Corollary to Theorem 4.6.)

 (b) For each set of eigenvalues (c) through (g) in Exercise 3, show that this set could not be the complete set of eigenvalues of a signed digraph. (*Hint:* Once again, make use of the proof of the Corollary to Theorem 4.6.)

14. We have observed that if a signed advanced rosette D has rosette sequence $(a_1, a_2, \ldots, a_s)$, then its characteristic polynomial $C(\lambda)$ is given by

$$C(\lambda) = (-1)^n \lambda^{n-s}\left(\lambda^s - \sum_{i=1}^{s} a_i \lambda^{s-i}\right).$$

Assuming this,

 (a) Show that if D is pulse stable under all simple pulse processes, then $a_s = \pm 1$.

 (b) Show that if D is pulse stable under all simple pulse processes, then D is value stable under all simple pulse processes if and only if $\sum_{i=1}^{s} a_i \neq 1$.

15. Suppose D is a weighted digraph and some strong component of D is pulse (value) unstable under some simple pulse process. Show that D is pulse (value) unstable under some simple pulse process. (*Hint:* Once a path leaves a strong component, it can never return.)

*16. If every strong component of a weighted digraph D is pulse stable under all simple pulse processes, does it follow that D is pulse stable under all simple pulse processes? (Give proof or counterexample.) Repeat for value stable.

*17. In Exers. 17 to 26, the reader might want to make use of Theorem 4.3. Most of these exercises state results from Roberts [1974a]. A vertex u_i of a weighted digraph D is said to be *self-value stable* if the sequence $\{|v_i(t)|\}$ is bounded in the simple pulse process starting at u_i, with $v_i(\text{start}) = 0$, all i. Similarly, u_i is said to be *self-pulse stable* if the sequence $\{|p_i(t)|\}$ is bounded in the simple pulse process starting at u_i. For example, in the signed advanced rosette D_1 of Fig. 4.40, if $x = u_i$, then Theorem 4.3 says that $p_i(2s) = 2^s$. For given any path P of length $2s$ from u_i to u_i which uses one of the cycles of length 3, there is a corresponding path Q of length $2s$ and opposite sign which uses the second cycle of length 3 each time P uses the first cycle and the first cycle each time P uses the second. And there are 2^s paths of length $2s$ from u_i to u_i which use just the cycles of length 2. Thus, vertex x is not self-pulse stable. Hence, it is not self-value stable.

(a)   Show that a signed advanced rosette is value stable under all simple pulse processes if and only if it is self-value stable at the central vertex.

(b)   Do the same for pulse stable.

(c)   Conclude that if two advanced rosettes have the same rosette sequence, they have the same stability properties (Theorem 4.9).

18.   Suppose D is a signed advanced rosette with $s = 0$. Show that D is pulse and value stable under all simple pulse processes. (*Hint:* Use Exer. 17.)

19.   Show that a vertex of a weighted digraph might be self-value stable but not value stable under all simple pulse processes.

20.   Use the signed digraph of Fig. 4.43 to show that self-value stability at every vertex does not imply value stability under all simple pulse processes.

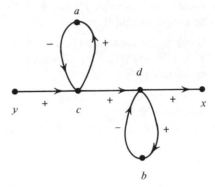

FIGURE 4.43. Signed digraph for Exer. 20, Sec. 4.5. (Reprinted from Roberts [1974a] with permission of Springer Verlag Inc.)

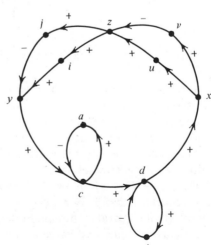

FIGURE 4.44. Signed digraph for Exer. 21, Sec. 4.5. (Reprinted from Roberts [1974a] with permission of Springer Verlag Inc.)

21. Use the signed digraph of Fig. 4.44 to show that even if a signed digraph is strongly connected, self-value stability at every vertex does not imply value stability under all simple pulse processes. (This example is due to Joel Spencer.)

22. Suppose a signed digraph D is balanced in the sense of Sec. 3.1, Exer. 14. Is D necessarily value unstable under some simple pulse process? (Give proof or counterexample.) Is this true if D has a cycle?

23. Show that if a signed digraph D is balanced and has a strong component with more than one vertex, then D is value unstable under some simple pulse process.

24. Prove that if a signed digraph D is balanced, then a vertex x of D is value unstable under some simple pulse process if and only if x is reachable from a vertex y which is on a cycle.

25. Is the result of Exer. 24 true if we substitute the hypothesis that each cycle of D is positive for the hypothesis that D is balanced? (Give proof or counterexample.)

26. If every semicycle of a signed digraph D is negative, does this imply that D is value stable under all simple pulse processes? (Give proof or counterexample.)

27. Using the theory of advanced rosettes, explain why there is usually instability if most of the cycles are of the same sign. (*Hint:* Consider closed paths from the central vertex to itself.)

*28. If all eigenvalues of a weighted digraph D are 0, can D have any cycles? Why?

(*Note:* Exercises applying the results of this section can be found in Sec. 4.6.)

4.6.   Applications of the Theory of Stability

In this section, we shall apply the theorems characterizing stability to a number of signed and weighted digraphs, those of Figs. 4.7, 4.8, 4.15, and 4.16. The conclusions should be regarded merely as tentative and suggestive—they depend on the restrictive assumptions of the model.

Let us begin by considering the signed digraph of Fig. 4.7. By computation, one determines that its characteristic polynomial $C(\lambda)$ is $-\lambda^2(\lambda^5 - \lambda^3 - \lambda^2 - 1)$. Now $f(\lambda) = \lambda^5 - \lambda^3 - \lambda^2 - 1$ has a real root strictly between 1 and 2. For $f(1) = -2$ and $f(2) = 19$. Thus, the signed digraph has an eigenvalue of magnitude greater than 1. By Theorem 4.6, the signed digraph is pulse and value unstable under some simple pulse process.

An external pulse at some vertex will lead to ever larger and larger pulses and values at some (possibly other) vertex. This result can be interpreted as follows: If we assume that no variable can get arbitrarily large, and the system as it is presently structured leads to arbitrarily large pulses and values somewhere, something in the system's structure must change. It would be wise to try to minimize the impact of that change by foreseeing it, or consciously choosing "the best" of a number of possible changes.

One such change which we have mentioned is "inverting the rate structure," charging less for energy when use decreases, rather than more as at present. Let us now consider the effect of this change. That is, let us change the sign of the arc (U, R) from $-$ to $+$ and ask whether the resulting signed digraph is stable. Again by computation, one determines that for this new signed digraph, the characteristic polynomial $C(\lambda)$ is $-\lambda^2(\lambda^5 + \lambda^3 - \lambda^2 - 1)$. Now $C(\lambda)$ factors as $-\lambda^2(\lambda - 1)(\lambda^2 + 1)(\lambda^2 + \lambda + 1)$. Hence, the eigen-values are $0, 0, 1, +i, -i, -\frac{1}{2} + (\sqrt{3}/2)i, -\frac{1}{2} - (\sqrt{3}/2)i$. Since all the nonzero eigenvalues are distinct and the largest eigenvalue has magnitude 1, Theorem 4.7 implies that the new signed digraph is pulse stable under all simple pulse processes. Hence the change (pulse) at each vertex at each time period remains bounded, whenever an initial pulse is introduced anywhere in the system. This does not, however, mean that the values remain bounded. Indeed, since 1 is an eigenvalue, Theorem 4.8 implies that the signed digraph is value unstable under some simple pulse process. Thus, our model implies that the strategy of inverting the rate structure alone is not value-stabilizing; values at some vertices will continue to be unbounded. It is only stabilizing in the sense that each change is small, which might be sufficient temporarily until more far-reaching changes in the energy demand system are made. In particular, inverting the rate structure cuts down rapidly expanding exponential growth. Our conclusions, the reader is reminded, should be evaluated in terms of the restrictive assumptions we have made. They should be tested using more complicated weighted digraphs, as well as entirely different models of complex systems.[14]

The strategy of inverting the rate structure is value-stabilizing if we also make a second change, changing the sign of (C, F) from $+$ to $-$. This would correspond to establishing a policy that creation of a new power plant should automatically force some factories to shut down. This policy, of course, might not be so easy to implement! However, it, together with inverting the rate structure, would be value-stabilizing. To show this, consider the signed digraph obtained from that of Fig. 4.7 by changing the sign of (U, R) from

---

[14]For example, one could use the world dynamics methods of Forrester [1971], one could use the KSIM methods of Kane [1972], one could use a more traditional economic analysis, or one could use a pulse process analysis with different change of value rule, for example one of those spelled out in Exers. 14 and 15 of Sec. 4.7.

— to + *and* the sign of (C, F) from + to —. Then, the characteristic polynomial turns out to be $C(\lambda) = -\lambda^2(\lambda^5 + \lambda^3 + \lambda^2 + 1)$. This factors as $-\lambda^2(\lambda + 1)(\lambda^2 + 1)(\lambda^2 - \lambda + 1)$, and we have as roots $0, 0, -1, +i, -i,$ $\frac{1}{2} + (\sqrt{3}/2)i, \frac{1}{2} - (\sqrt{3}/2)i$. By Theorem 4.7, the signed digraph is pulse stable under all simple pulse processes. Then Theorem 4.8 implies that it is also value stable.

We have seen that the eigenvalue analysis allows us to evaluate potentially good strategies. However, how can we identify such strategies ahead of time? The rosette theorems of Sec. 4.5.3 provide useful tools for doing this. Note that the signed digraph D of Fig. 4.7 is an advanced rosette, with central vertex U. (This observation made it easy for us to obtain $C(\lambda)$ above.) There is in D one cycle of length 2, three of length 3, and one of length 5. Hence, under any change of the signs of D, $a_1 = 0, a_2 = \pm 1, a_3 = \pm 1$ or ± 3, $a_4 = 0$, and $a_5 = \pm 1$. We have $s = 5$ and $a_s = \pm 1$. In order to have pulse and value stability, according to Theorems 4.10 and 4.11, a_i must equal $(-a_s)a_{s-i}$ for $i = 1, 2, 3, 4$, and $a_1 + a_2 + a_3 + a_4 + a_5$ cannot equal 1. These are necessary conditions, not sufficient conditions, but they will help us to identify potential strategies. The first condition is always satisfied for $i = 1$ and 4, since $a_1 = a_4 = 0$. For $i = 2$ and 3, it says that $a_2 = -a_s a_3$. Since $a_2 = \pm 1$ and $a_5 = \pm 1$, this implies that $a_3 = \pm 1$. Thus, we are looking for numbers a_2, a_3, a_5 for which

$$a_2 = \pm 1, a_3 = \pm 1, a_5 = \pm 1$$

$$a_2 = -a_5 a_3$$

and

$$a_2 + a_3 + a_5 \neq 1.$$

The only numbers satisfying these conditions are $a_2 = a_3 = a_5 = -1$. Thus, the only potential strategies for obtaining pulse and value stability are those which give rise to these three cycle counts. To make $a_2 = -1$, we must change the sign of either (U, R) or (R, U), but not both. If the sign of (U, R) has been changed, then we can make $a_3 = -1$ and $a_5 = -1$ by also changing the sign of (C, F). No single additional sign change will work. If the sign of (R, U) has been changed, then we can make $a_3 = -1$ and $a_5 = -1$ by changing the sign of either (F, J) or (J, P). No single additional sign change will work. Thus, we have discovered that the only possible strategies for obtaining pulse and value stability in the signed digraph of Fig. 4.7 by changing at most the signs of two arcs are the following:

> Change the sign of arcs (U, R) and (C, F).
> Change the sign of arcs (R, U) and (F, J).
> Change the sign of arcs (R, U) and (J, P).

We leave it to the reader to interpret these and to find other possible strategies which might lead to pulse and value stability. Incidentally, a check of eigenvalues shows that all three of these strategies do lead to pulse and value stability.

Let us also apply the above methodology to the signed digraph of Fig. 4.8. The reader will recall that this signed digraph, for energy demand in intraurban commuter transportation, was constructed using the judgments of a group of experts. The experts could not agree on the sign of the arc $(1, 5)$. We consider both cases, $(1, 5) = +$ and $(1, 5) = -$. The characteristic polynomial in both cases is easier to calculate than in the previous example. The digraph has six strong components, $K_1 = \{2\}$, $K_2 = \{3\}$, $K_3 = \{4\}$, $K_4 = \{6\}$, $K_5 = \{9\}$, and $K_6 = \{1, 5, 7, 8\}$. Thus, if $C_i(\lambda)$ is the characteristic polynomial of the signed digraph generated by vertices in K_i, Corollary 2 of Theorem 4.2 implies that the characteristic polynomial $C(\lambda)$ is $\prod_{i=1}^{6} C_i(\lambda)$. Now $C_i(\lambda)$ is $-\lambda$ if $i \neq 6$. For K_6, it is easy to prove that $C_6(\lambda)$ is $\lambda(\lambda^3 + 2\lambda + 1)$ if $(1, 5)$ is $+$ and $\lambda(\lambda^3 + 1)$ if $(1, 5)$ is $-$. Thus, if $(1, 5)$ is $+$, $C(\lambda)$ is $-\lambda^6(\lambda^3 + 2\lambda + 1)$. Now the polynomial $f(\lambda) = \lambda^3 + 2\lambda + 1$ has a real root strictly between 0 and -1, since $f(0) = 1$ and $f(-1) = -2$. Thus, by the Corollary to Theorem 4.6, the signed digraph is pulse unstable and hence value unstable under some simple pulse process if $(1, 5)$ is $+$. If $(1, 5)$ is $-$, then $C(\lambda)$ is $-\lambda^6(\lambda^3 + 1)$. The nonzero eigenvalues are -1, $\frac{1}{2} + (\sqrt{3}/2)i$, $\frac{1}{2} - (\sqrt{3}/2)i$. By Theorems 4.7 and 4.8, the signed digraph is now pulse and value stable under all simple pulse processes.

A straightforward computation can be made to check, in the case that $(1, 5)$ is $+$, what simple strategies are (pulse or value) stabilizing. The simple strategies we consider are those corresponding to changing the sign of one arc. The only value-stabilizing strategies besides changing the sign of arc $(1, 5)$ turn out to be changing the signs of the arcs $(1, 8)$ and $(5, 1)$. (See Exers. 1 and 4.)

Let us interpret these conclusions, again in the light of our restrictive assumptions. It is suggested that if (average) prices of commuter tickets are set as a function of total passenger miles, with prices decreasing as total passenger miles increase, then the energy demand system in intraurban commuter transportation will be stable. If the price of commuter tickets is set to increase as passenger miles increase, this leads to an unstable situation. Here, there are two simple stabilizing strategies: arranging that passenger miles decrease as price of commuter tickets increases or arranging that probability of delay goes down as passenger miles increase. A main question is whether these strategies are feasible, i.e., whether they can be implemented. The second almost certainly is not feasible. However, the first might be, using a system of "time-pricing" or more generally "cost-of-service" pricing, as discussed in Vickrey [1968]. The idea is that ticket prices do not remain uniform, but are raised during peak demand periods (to reflect their contribution to fixed costs) and are lowered during off hours to attract riders. For a further discussion of feasibility, see Roberts [1973].

These results should again be regarded as tentative and suggestive, and checked by other means. Moreover, the reader should understand that a conclusion of stability or instability under simple pulse processes may be changed if more external pulses are introduced, to either amplify or counteract various trends.

One possibility has been left out in this discussion, and that is that the arc (1, 5) is given no sign, i.e., is omitted. This corresponds to not letting passenger miles influence the price of a commuter ticket. (It could happen if tickets are free, for example. This policy has been adopted in such cities as Rome, at least for a short period of time.) The analysis of the stability of the energy demand system under this situation is left to the reader as an exercise (Exers. 1 and 5).

Let us turn next to the weighted digraph of Fig. 4.15. This digraph represents the situation in the County of San Diego, California, under an almost-all-automobile transportation system similar to that existing in 1975. The weights for this digraph were estimated by a panel of experts studying energy use and clean air in San Diego under support of the Environmental Protection Agency. (For more details, see Roberts [1974b].)

It is easy to show that the only strong component of this weighted digraph which consists of more than one vertex is the strong component $K = \{1, 5, 9\}$. Its eigenvalues are $-.52$, $.26 + .45i$, and $.26 - .45i$. All other strong components have no loops and so all the other eigenvalues are 0. Since each eigenvalue has magnitude less than or equal to 1, and since the nonzero eigenvalues are distinct, and since 1 is not an eigenvalue, we conclude that the digraph is pulse and value stable under all simple pulse processes. However, both fuel consumption and air pollution have been growing rapidly in San Diego. How does this conclusion about pulse stability make sense in the light of these observations? Is the conclusion wrong? Is our model wrong? The answer is that the pulse stability which we have concluded holds only for simple pulse processes. If we observe rapidly increasing levels, our model says that these cannot be accounted for simply by the operation of feedback in the system. However, they can be accounted for by looking for repeated external impulses introduced into the system. The pulse process analysis has turned our attention outside the system. (What could some of the repeated pulses be?)

Let us turn next to the weighted digraph of Fig. 4.16. This digraph represents the situation in San Diego under a hypothetical situation where all automobiles have been banned from the county and an expensive, wide-ranging bus system has been introduced. The weights for this digraph were estimated by the same panel of experts which estimated weights for the digraph of Fig. 4.15. For description of the bus system, and more detailed discussion of the results, see Roberts [1974b].

The only strong component of the weighted digraph of Fig. 4.16 which consists of more than one vertex is the strong component $K = \{4, 8, 9\}$. Its eigenvalues are .92, $-.87$, and $-.05$. All other strong components have no loops and so all the other eigenvalues are 0. Thus, the system is pulse and value stable under all simple pulse processes. However, the eigenvalue of largest magnitude, .92, is quite large. This suggests that a relatively small change in weights might create eigenvalues of magnitude greater than 1. Since weights as a general rule are quite imprecisely estimated, one should do a sensitivity analysis to see if the results about stability change under a small (but reasonable) change in the weights. Indeed, they do: Changing the weight of arc (4, 8) from $-.9$ to -1.1 results in a weighted digraph with the following nonzero eigenvalues: 1.02, $-.97$, $-.05$. Thus, the new weighted digraph is pulse unstable under some simple pulse process. We cannot be sure whether or not the stability prediction or the instability prediction will hold true for the bus system designed by the experts. However, in any case, whether dealing with the original weights or the revised weights, one has a system which is very sensitive to external inputs. For even in the original case, though initial pulses do not lead to arbitrarily large pulses or values, they can lead to quite large pulses and values before things "stabilize." We shall show (in Exer. 12 of Sec. 4.7) that as the magnitude of the eigenvalue of largest magnitude increases, the sensitivity of the system to external pulses increases, in the sense that the same initial pulse will tend to result in larger changes in value. To demonstrate how sensitive the particular weighted digraph of Fig. 4.16 is to external pulses, let us make some forecasts. Suppose the cost of the bus system goes up by 10% at time 0. (10% of the starting value was considered one unit in the study being discussed.) The effect on fuel consumption, calculated by computer using the pulse process model, is shown in Fig. 4.45. The prediction is that after 50 weeks (time periods), there will be a 50% increase in the consumption of fuel! A similar forecast for emissions is shown in Fig. 4.46. Thus, although in both cases the values (and hence the pulses) remain bounded, and indeed eventually level off, they do so at such greatly changed levels that the system should be considered unacceptably sensitive to outside influences. (Just by way of comparison, if we change the weight of the arc (4, 8) from $-.9$ to -1.1, we see even more sensitivity. If the new weighted digraph is used, Fig. 4.47 shows the effect on fuel consumption of an initial 10% increase in the cost of the bus system, and Fig. 4.48 shows the effect on emissions. Rather than eventually stabilize as before, both of these variables reach skyrocketing levels.)

The experts, in building the weighted digraph of Fig. 4.16, wanted to stabilize rapidly growing fuel consumption and emission levels. In this light,

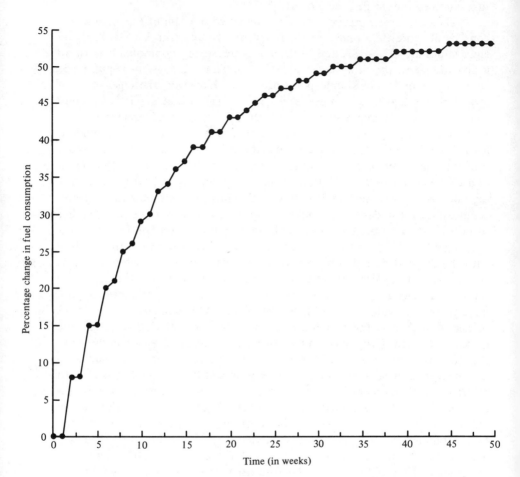

FIGURE 4.45. Effect on fuel consumption of a 10% increase in the cost of the bus system in the transportation system for San Diego, California described by the weighted digraph of Fig. 4.16. From Roberts [1974b]. The author thanks the Rand Corporation for permission to use this figure and Figs. 4.46 through 4.50.

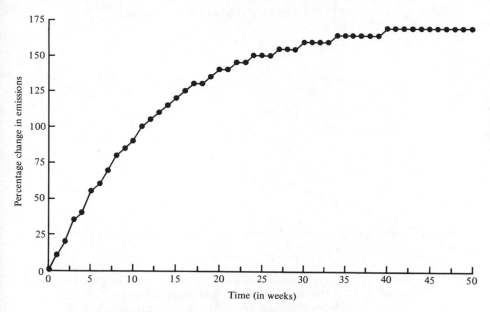

FIGURE 4.46. Effect on emissions of a 10% increase in the cost of the bus system in the transportation system for San Diego, California described by the weighted digraph of Fig. 4.16. From Roberts [1974b].

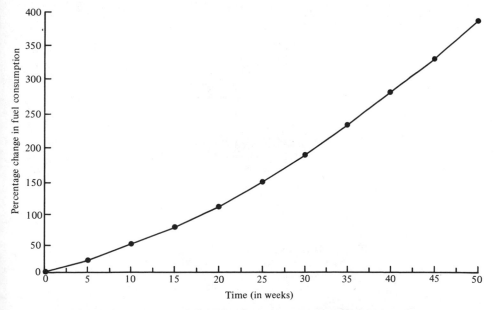

FIGURE 4.47. Effect on fuel consumption of a 10% increase in the cost of the bus system in the transportation system for San Diego, California described by modifying the weighted digraph of Fig. 4.16 so that arc (4, 8) has weight −1.1. From Roberts [1974b].

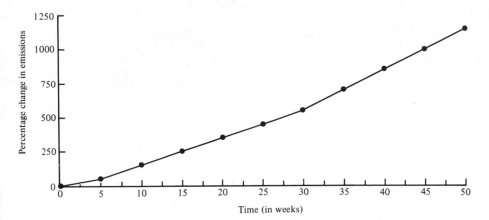

FIGURE 4.48. Effect on emissions of a 10% increase in the cost of the bus system in the transportation system for San Diego, California described by modifying the weighted digraph of Fig. 4.16 so that arc (4, 8) has weight −1.1. From Roberts [1974b].

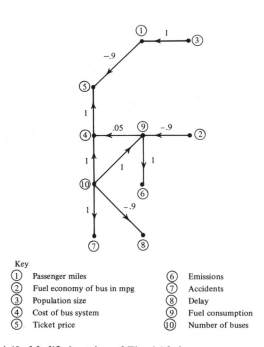

Key

①	Passenger miles	⑥	Emissions
②	Fuel economy of bus in mpg	⑦	Accidents
③	Population size	⑧	Delay
④	Cost of bus system	⑨	Fuel consumption
⑤	Ticket price	⑩	Number of buses

FIGURE 4.49. Modified version of Fig. 4.16, for energy use and clean air in the transportation system of San Diego, California, under a hypothetical situation in which all cars are banned and an expensive, wide-ranging bus system has been introduced. From Roberts [1974b].

the stability results and forecasts can be looked at in several ways. They suggest either that the model for analyzing the digraph is faulty, or that the bus system was designed very poorly, or that the experts, in building the weighted digraph of Fig. 4.16, incorrectly described the system, perhaps by introducing arcs where none should exist or incorrectly estimating the weights. Indeed, when asked to reconsider the weighted digraph they had built, the experts saw errors in their selection of arcs and weights and made many changes. (At the same time they saw errors and made changes in the weighted Fig. 4.15.) Their revised digraph is shown in Fig. 4.49. (The modified version of Fig. 4.15 is shown in Fig. 4.50.) The digraph of Fig. 4.49 uses slightly dif-

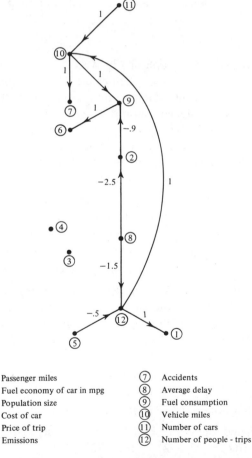

Key

①	Passenger miles	⑦	Accidents
②	Fuel economy of car in mpg	⑧	Average delay
③	Population size	⑨	Fuel consumption
④	Cost of car	⑩	Vehicle miles
⑤	Price of trip	⑪	Number of cars
⑥	Emissions	⑫	Number of people - trips

FIGURE 4.50. Modified version of Fig. 4.15, for energy use and clean air in the transportation system of San Diego, California under an almost-all-automobile transportation system. From Roberts [1974b].

ferent variables from that of Fig. 4.16 and is interesting because it has no cycles at all. The bus system as finally described by the experts has no feedback in it.[15] All eigenvalues are 0 and the system is pulse and value stable under all simple pulse processes. It is quite insensitive to external changes. This significant change in the nature of the described system points up one of the most important uses of the signed and weighted digraph as a policy planning tool. Construction of a digraph can help policymakers learn about the systems they are studying. Preliminary predictions based on the digraph model can either suggest serious problems with the system or suggest that the policymakers have not accurately described the system they would like to understand.

Note: The reader should consider when stability conclusions about *signed* digraphs could pass a sensitivity analysis like that applied to Fig. 4.16.

EXERCISES

1. Consider the signed digraph of Fig. 4.8.
   (a) Show that if (1, 5) is +, then changing (1, 8) from + to − results in a signed digraph which is pulse stable under all simple pulse processes. Is this signed digraph value stable?
   (b) Show that if (1, 5) is +, then changing (5, 1) from − to + is pulse-stabilizing. Is it also value-stabilizing?
   (c) If (1, 5) is removed, show that the resulting signed digraph is not pulse stable under all simple pulse processes.

2. For the signed digraph of Fig. 4.7, interpret the three strategies which we found in the text for obtaining pulse and value stability by changing two signs. Which of these might be implementable?

3. For the signed digraph of Fig. 4.7, find a potential strategy for obtaining pulse and value stability which changes more than two signs.

4. Use the result of Exer. 15 of Sec. 4.5 to show the following: In the signed digraph of Fig. 4.8, if the sign of (1, 5) is fixed as +, then the only potentially value-stabilizing changes involving changing the sign of an arc of the strong component {1, 5, 7, 8} are the changes of arcs (1, 8) or (5, 1). (*Hint:* Use rosette theory.)

5. If arc (1, 5) is removed from the signed digraph of Fig. 4.8, find a strategy involving change of sign of one arc which is pulse-stabilizing, or show that none exists. (*Hint:* Use the result of Exer. 15 of Sec. 4.5.)

6. For the signed digraph of Fig. 4.11, the characteristic polynomial turns out to be $C(\lambda) = \lambda^5 - \lambda^4 + \lambda^3 + 9\lambda^2 + 3\lambda$.

---

[15] Actually, this digraph includes only short-term effects. If effects with long time lags are included, there is feedback.

(a)   Determine if the digraph is pulse stable under all simple pulse processes. (*Suggestion*: Calculate $C(0)$, $C(1)$, $C(2)$, $C(-1)$, $C(-2)$, etc.)

(b)   Determine if the digraph is value stable under all simple pulse processes.

7.   Do Exer. 6 for each of the signed digraphs of Fig. 4.14 and interpret the
conclusions. $C(\lambda) = \lambda^5 - 3\lambda^4 - \lambda^2 + 6\lambda + 5$ is the characteristic polynomial for digraph (a), $C(\lambda) = \lambda^5 - \lambda^4 - 4\lambda^3 + 2\lambda^2 + 2\lambda + 6$ is for (b),
and $C(\lambda) = \lambda^5 - \lambda^4 - 5\lambda^3 + 4\lambda^2 + 2\lambda + 6$ is for (c).

8.   Consider the weighted digraph of Fig. 4.15.

(a)   The forecast shown in Fig. 4.22 was made for this digraph under
the assumption that fuel consumption drops by 10%. Explain the
qualitative nature of the forecast on the basis of the pulse and value
stability results discussed in the text and on the basis of the structure of the digraph.

(b)   When the experts reconsidered this digraph, they constructed
instead the weighted digraph of Fig. 4.50. Discuss the pulse and
value stability of this new weighted digraph.

9.   Consider the simple rabbits-foxes ecosystem introduced in Exer. 7 of
Sec. 4.4.

(a)   Show that the system is pulse and value stable under all simple
pulse processes.

(b)   Interpret pulse and value stability in ecological terms.

(c)   Show that under some choice of weights for the arcs, the system is
no longer pulse and value stable. Interpret the results.

10.   Consider the plant-herbivore-carnivore ecosystem of Exer. 8, Sec. 4.4.

(a)   Show that the system is pulse and value unstable under some simple
pulse process. (*Hint:* Show that $C(\lambda)$ has a real root of magnitude
less than 1.)

(b)   Interpret the results.

11.   Consider the mice-rats-eagles ecosystem of Exer. 9, Sec. 4.4.

(a)   Show that the system is pulse and value stable as shown.

(b)   Show that the system is pulse and value unstable if the
(eagles, eagles) loop is assumed to be — or 0, i.e., not to exist.

(c)   Discuss which sign of the (eagles, eagles) loop is most realistic.

12.   For the weighted digraph of Fig. 4.18, the characteristic polynomial
is $\lambda^5 - \lambda^4 + 16\lambda^3 + 41\lambda^2 + 7\lambda - 2$. Do parts (a) and (b) of Exer. 6.

13.   Do parts (a) and (b) of Exer. 6 for each of the weighted digraphs of Fig.
4.21 and interpret the conclusions. (This exercise is meant for the reader
with access to a computer.)

14.   For the signed digraphs of Figs. 4.6 and 4.9, use rosette theory to do the

following:
- (a) Determine for each if it is pulse stable under all simple pulse processes.
- (b) Determine for each if it is value stable under all simple pulse processes.
- (c) If it is not pulse stable, find a strategy for changing the sign of one arc which is potentially pulse-stabilizing, or show that no such strategy exists.
- (d) If it is not value stable, find a strategy for changing the sign of one arc which is potentially value-stabilizing, or show that no such strategy exists.

15. Consider the weighted digraph of Fig. 4.16.
- (a) Use the structure of the digraph to explain the forecasts of Figs. 4.45 and 4.46.
- (b) Can you discover arcs which were included in this digraph incorrectly?

16. Do any of the theorems of Sec. 4.5 apply to the signed digraph of Fig. 4.12? If not, why not?

17. The reader with access to a computer for calculation of eigenvalues might want to analyze the ecosystems of Figs. 4.51 and 4.52. See Levins [1974] for a more detailed discussion of these ecosystems.[16]
- (a) Figure 4.51 is a signed digraph illustrating control of insect pests in a cultivated field by use of insecticides. Is the system pulse and value stable? What is the effect of an initial increase in insecticide use? What about continuing, externally generated increases in insecticide use?

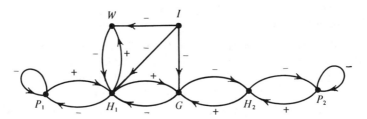

FIGURE 4.51. Insecticide use in a cultivated field. P_1 is a crop plant which is limited because of crowding. H_1 is the pest herbivore which eats the plant P_1. W is a specialized insect (for example a wasp) which kills only H_1. G is a generalized predatory insect which eats H_1 and also H_2, which is another herbivore. P_2 is another plant and I is insecticide. (Adapted from Levins [1974] with permission of the New York Academy of Sciences, and the author.)

[16]See footnote 10, page 215.

(b)   Figure 4.52 studies contamination of a lake by two nutrients, nitrate and phosphate. Is the system pulse and value stable? If there is an initial increase in phosphates, does either of the algae populations increase dramatically? (This would account for the algae "blooms" seen in many lakes.) What about continuing externally generated increases in phosphate level?

FIGURE 4.52. Two nutrients, nitrate N and phosphate P, are washed into a lake. Green algae G uses N and P, blue green algae BG uses P but releases more N. Green algae is sensitive to a toxin released by blue green algae. (Adapted from Levins [1974] with permission of the New York Academy of Sciences, and the author.)

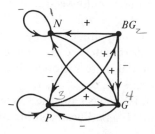

18.   In Exer. 21 of Sec. 3.5, the reader was asked to study the inverse relation between complexity of an ecosystem and its vulnerability. This exercise studies the ecological principle that complexity is inversely related to stability.

(a)   Are the notions of stability developed in this chapter the proper notions for formalizing this principle? If not, what are some proper notions?

(b)   Investigate the relation between stability and complexity for the simple ecosystems of Figs. 4.27, 4.28, 4.29, 4.51, and 4.52. (Some results on the relation between complexity and stability are contained in Levins [1974], Gardner and Ashby [1970], and May [1971, 1972, 1973]. But there is much work yet to be done to formalize this relation.)

4.7.   Proofs of Theorems 4.6 through 4.8[17]

Before proving Theorems 4.6 through 4.8, let us introduce one piece of terminology. If $X = (x_1, x_2, \ldots, x_n)$ is a vector of (real or) complex numbers, then its *norm*, $\| X \|$, is the number $\sqrt{\sum_i |x_i|^2}$, where $|x_i|$ is the magnitude of x_i. We assume that the reader who is unfamiliar with this concept can readily convince himself of the standard properties of the norm:

$$\| X + Y \| \leq \| X \| + \| Y \|$$

$$\| \alpha X \| = |\alpha| \; \| X \| \qquad \text{(if } \alpha \text{ is a (real or) complex number).}$$

[17]This is a highly technical section and may be omitted without loss of continuity. The reader who is not familiar with the Jordan Canonical Form of a matrix or has had little experience with proofs about convergence and boundedness should skip the details of the section. However, the reader may still want to try to understand Theorems 4.12 and 4.13.

If $M = (m_{ij})$ is a matrix, its norm is defined analogously, as $\| M \| = \sqrt{\sum_{i,j} |m_{ij}|^2}$, and this norm has analogous properties.

Let us begin by proving Theorem 4.6, namely that if a weighted digraph D has an eigenvalue greater than unity in magnitude, then D is pulse unstable under some simple pulse process. Let λ be an eigenvalue with $|\lambda| > 1$ and let U be an eigenvector corresponding to λ. We can find such a U with the property that $\| U \| = 1$. (Why?) The proof would be trivial if we were allowed to use U as an initial pulse vector. For if $P(0) = U$, then by Theorem 4.5, $P(t) = P(0)A^t = UA^t = \lambda^t U$. Thus, $\| P(t) \| = |\lambda^t| \| U \| = |\lambda^t| = |\lambda|^t \to \infty$. It follows that the magnitude of some component of the vector $P(t)$ gets arbitrarily large. Unfortunately, U may have noninteger components and cannot be used as an initial pulse vector. Indeed, the allowable initial pulse vectors are the vectors $E_i = (0, 0, \ldots, 0, 1, 0, \ldots, 0)$, with a 1 in the ith component. These vectors form a basis for the space of n-dimensional row vectors, and so we may write U as a linear combination $\sum_{i=1}^n \alpha_i E_i$. Then for any integer $t > 0$,

$$|\lambda|^t = \| UA^t \| = \left\| \sum_{i=1}^n \alpha_i E_i A^t \right\| \leq \sum_{i=1}^n |\alpha_i| \, \| E_i A^t \|.$$

Now each $|\alpha_i| \leq 1$, since $\| U \| = 1$. Thus,

$$|\lambda|^t \leq \sum_{i=1}^n \| E_i A^t \|.$$

If follows that for every $t > 0$, there is some i such that

$$\| E_i A^t \| \geq (1/n) |\lambda|^t.$$

Since there are only a finite number of E_i, we conclude that for at least one of them, $\| E_i A^t \| \geq (1/n) |\lambda|^t$ for arbitrarily large t. Pick $P(0) = E_i$. Then for arbitrarily large t, $\| P(t) \| = \| P(0)A^t \|$ is at least $(1/n) |\lambda|^t$. Since $|\lambda| > 1$ and n is fixed, $(1/n) |\lambda|^t \to \infty$. We conclude that with $P(0) = E_i$, $\| P(t) \|$ gets arbitrarily large as $t \to \infty$, and hence the weighted digraph is pulse unstable under the simple pulse process starting at vertex u_i. This completes the proof of Theorem 4.6.

Rather than proving Theorem 4.7, we prove a more general result. To state it, let us recall that every (real or complex) $n \times n$ matrix A is *similar*[18] to an $n \times n$ matrix J, its *Jordan Canonical Form*, which has the special form

$$J = \begin{pmatrix} \boxed{B_1} & & & \\ & \boxed{B_2} & & \text{\Large 0} \\ & & \ddots & \\ \text{\Large 0} & & & \boxed{B_r} \end{pmatrix}, \tag{9}$$

---

[18]A is *similar* to J if there is a nonsingular (i.e., invertible) matrix P so that $J = PAP^{-1}$.

where each B_j is an $(e_j + 1) \times (e_j + 1)$ diagonal block and B_j has the form

$$
\begin{pmatrix}
\lambda & \delta_j & & & & \\
 & \lambda & \delta_j & & \Large 0 & \\
 & & \lambda & \delta_j & & \\
 & & & \lambda & \delta_j & \\
 & \Large 0 & & & \lambda & \delta_j \\
 & & & & & \lambda
\end{pmatrix},
$$

where λ is an eigenvalue of A and δ_j is 0 or 1, 0 if $e_j + 1 = 1$ and 1 otherwise. (Note that λ may appear in several of the B_j's.) Specifically, if all the eigenvalues of D (including 0) are distinct, then each B_j is 1×1 and J is a diagonal matrix

$$
\begin{pmatrix}
\lambda_1 & & & & \\
 & \lambda_2 & & \Large 0 & \\
 & & \lambda_3 & & \\
 & & & \ddots & \\
 & \Large 0 & & & \ddots \\
 & & & & \lambda_n
\end{pmatrix}
$$

with the eigenvalues as diagonal entries. To give an example, suppose A is given by

$$
A = \begin{pmatrix}
3 & 1 & 0 & 0 & 0 & 0 & 0 & 0 \\
0 & 3 & 0 & 0 & 0 & 0 & 0 & 0 \\
0 & 0 & 3 & 0 & 0 & 0 & 0 & 0 \\
0 & 0 & 0 & .1 & 1 & 0 & 0 & 0 \\
0 & 0 & 0 & 0 & .1 & 1 & 0 & 0 \\
0 & 0 & 0 & 0 & 0 & .1 & 0 & 0 \\
0 & 0 & 0 & 0 & 0 & 0 & i & 0 \\
0 & 0 & 0 & 0 & 0 & 0 & 0 & -i
\end{pmatrix}. \tag{10}
$$

Then $J = A$. The blocks B_j are shown below:

$$
J = \begin{pmatrix}
\boxed{\begin{matrix} 3 & 1 \\ 0 & 3 \end{matrix}} & 0 & 0 & 0 & 0 & 0 & 0 \\
0 & 0 & \boxed{3} & 0 & 0 & 0 & 0 & 0 \\
0 & 0 & 0 & \boxed{\begin{matrix} .1 & 1 & 0 \\ 0 & .1 & 1 \\ 0 & 0 & .1 \end{matrix}} & 0 & 0 \\
0 & 0 & 0 & 0 & 0 & 0 & \boxed{i} & 0 \\
0 & 0 & 0 & 0 & 0 & 0 & 0 & \boxed{-i}
\end{pmatrix}.
$$

If J has the form (9), then J^t has the form

$$\begin{pmatrix} \boxed{B_1^t} & & & \\ & \boxed{B_2^t} & & \mathbf{0} \\ & & \ddots & \\ \mathbf{0} & & & \boxed{B_r^t} \end{pmatrix}.$$

Now if $\delta_j = 0$, then

$$B_j = (\lambda)$$

and

$$B_j^t = (\lambda^t). \tag{11}$$

If $\delta_j = 1$, then it is easy to prove by induction on t that $b_{k,l}^{j,t}$, the k, l entry of B_j^t, is given by

$$b_{k,l}^{j,t} = \begin{cases} 0 & \text{(if } k > l) \\ \binom{t}{l-k}\lambda^{t-l+k} & \text{(if } k \leq l). \end{cases} \tag{12}$$

(*Note:* By convention, $\binom{r}{s} = 0$ if $s > r$.)

Let us say that an eigenvalue λ of D is *linked in J* if there is an off-diagonal entry of 1 in some row of J in which λ appears as the diagonal element. Equivalently, λ is linked in J if it appears on the diagonal of some B_j in which $\delta_j = 1$. For example, if $A = J$ is the matrix of Eq. (10), then the eigenvalues 3 and .1 are linked while the eigenvalues i and $-i$ are not. Only multiple eigenvalues can be linked.

To prove Theorem 4.7, we shall prove the following more general theorem, from which Theorem 4.7 follows since under its hypotheses, there are no linked eigenvalues in J.

Theorem 4.12.[19]   Suppose D is a weighted digraph and J is its Jordan Canonical Form. Then the following are equivalent:

(a) D is pulse stable under all autonomous pulse processes.
(b) D is pulse stable under all simple pulse processes.
(c) Every eigenvalue of D has magnitude less than or equal to unity and every eigenvalue of D which is linked in J has magnitude less than unity.

Before proving Theorem 4.12, let us illustrate it. The following matrices A are already in Jordan form. Thus, in each case, $J = A$.

[19]This theorem was first stated in this form in Roberts and Brown [1975].

$$A_1 = \begin{pmatrix} \frac{1}{4} & 1 & 0 & 0 \\ 0 & \frac{1}{4} & 0 & 0 \\ 0 & 0 & -1 & 0 \\ 0 & 0 & 0 & -1 \end{pmatrix}, \qquad A_2 = \begin{pmatrix} \frac{1}{4} & 0 & 0 & 0 \\ 0 & \frac{1}{4} & 0 & 0 \\ 0 & 0 & -1 & 1 \\ 0 & 0 & 0 & -1 \end{pmatrix}$$

$$A_3 = \begin{pmatrix} \frac{1}{4} & 0 & 0 & 0 & 0 \\ 0 & \frac{1}{4} & 0 & 0 & 0 \\ 0 & 0 & -1 & 0 & 0 \\ 0 & 0 & 0 & -1 & 1 \\ 0 & 0 & 0 & 0 & -1 \end{pmatrix}, \qquad A_4 = \begin{pmatrix} \frac{1}{4} & 0 & 0 & 0 & 0 \\ 0 & \frac{1}{4} & 0 & 0 & 0 \\ 0 & 0 & 0 & 1 & 0 \\ 0 & 0 & 0 & 0 & 1 \\ 0 & 0 & 0 & 0 & 0 \end{pmatrix}.$$

In all four examples, the largest eigenvalue has magnitude unity. However, A_2 and A_3 have a linked eigenvalue of magnitude unity, and so correspond to weighted digraphs which are pulse unstable under some simple pulse process. In A_1 and A_4, all linked eigenvalues have magnitude less than unity, and hence the corresponding weighted digraphs are pulse stable (under all autonomous pulse processes).

The proof of Theorem 4.12 begins with two lemmas.

Lemma 1.   If A is an $n \times n$ matrix and J is its Jordan Canonical Form, then

$$\{\|A^t\|: t = 0, 1, 2, \ldots\} \tag{13}$$

is bounded if and only if

$$\{\|J^t\|: t = 0, 1, 2, \ldots\} \tag{14}$$

is bounded.

Proof.[20]   The proof uses a topological argument. The matrices A and J are related by a similarity transformation. Now such a transformation is bicontinuous under the topology induced by the norm $\|\cdot\|$, so it follows that $\{A^t\}$ is contained in a sphere if and only if $\{J^t\}$ is contained in a sphere.[21]

Q.E.D.

Lemma 2.   If a weighted digraph D is pulse stable under all simple pulse processes, then

$$\{\|J^t\|: t = 0, 1, 2, \ldots\} \tag{14}$$

is bounded. If the sequence (14) is bounded, then D is pulse stable under all autonomous pulse processes.

[20]This proof should be omitted unless the reader is familiar with point set topology.

[21]The author thanks Garrett Birkhoff for suggesting this argument.

Proof.   By Theorem 4.5,

$$P(t) = P(0)A^t.$$

Thus, pulse stability under *all* simple pulse processes implies that the sequence

$$\{\|A^t\|: t = 0, 1, 2, \ldots\} \tag{13}$$

is bounded (why?) and hence by Lemma 1 we conclude that the sequence (14) is bounded as well. Conversely, if the sequence (14) is bounded, then by Lemma 1 the sequence (13) must be bounded. Hence by Theorem 4.5, *D* must be pulse stable under all autonomous pulse processes. (Why?)

Q.E.D.

To prove Theorem 4.12, let us observe that clearly (*a*) ⇒ (*b*). We shall prove (*b*) ⇒ (*c*) and (*c*) ⇒ (*a*). We have already shown by proving Theorem 4.6 that if *D* is pulse stable under all simple pulse processes, then every eigenvalue of *D* has magnitude less than or equal to unity. To complete the proof of (*b*) ⇒ (*c*), let us show that every linked eigenvalue in *J* has magnitude less than unity. By Lemma 2, we know from pulse stability that $\{\|J^t\|\}$ is bounded. But now suppose λ is an eigenvalue of *D* which is linked in *J* and has magnitude greater than or equal to unity. If λ appears on the diagonal of block B_j, then $b^{j,t}_{1,2} = t\lambda^{t-1}$, by Eq. (12). Since $|\lambda| \geq 1$, $|b^{j,t}_{1,2}|$ gets arbitrarily large as *t* approaches ∞, and so $\|J^t\|$ becomes arbitrarily large as well. This completes the proof that (*b*) ⇒ (*c*).

To prove that (*c*) ⇒ (*a*), it is sufficient by Lemma 2 to show that under assumption (*c*), $\{\|J^t\|\}$ is bounded. To prove that $\{\|J^t\|\}$ is bounded, it is sufficient to prove that for each *j*, $\{\|B^t_j\|\}$ is bounded. Let λ be the diagonal entry of B^t_j. By hypothesis, $|\lambda| \leq 1$. If $\delta_j = 0$, then B^t_j has the form (11) and so $\|B^t_j\|$ is bounded since $|\lambda|^t$ is bounded. If $\delta_j = 1$, then by hypothesis, $|\lambda| < 1$. We show that for $k \leq l$, $\{|b^{j,t}_{k,l}|: t = 0, 1, 2, \ldots\}$ is bounded, where $b^{j,t}_{k,l}$ is given by Eq. (12). Let $e = e_j$, i.e., *e* is one less than the dimension of B_j. Observe that $l - k \leq e$. Thus, for $t > 2e$, we have

$$\binom{t}{l-k} \leq \binom{t}{e} = \frac{t(t-1)\cdots(t-e+1)(t-e)!}{e!\,(t-e)!} \leq \frac{t^e}{e!}$$

and

$$|\lambda^{t-l+k}| = |\lambda|^{t-l+k} \leq |\lambda|^{t-e},$$

since $|\lambda| < 1$. It follows that

$$|b^{j,t}_{k,l}| \leq \frac{t^e}{e!}|\lambda|^{t-e}.$$

Since $t^e|\lambda|^{t-e}$ approaches 0 as *t* approaches ∞, we conclude that $\{|b^{j,t}_{k,l}|\}$ is bounded. Q.E.D.

Corollary: Suppose *D* is an integer-weighted digraph and *J* is its Jordan Canonical Form. Then the following are equivalent:

(a) D is pulse stable under all autonomous pulse processes.

(b) D is pulse stable under all simple pulse processes.

(c) Every eigenvalue of D has magnitude less than or equal to unity and no nonzero eigenvalue of D is linked in J.

(d) Every nonzero eigenvalue of D has magnitude equal to unity and no nonzero eigenvalue of D is linked in J.

Proof. Obviously, (d) implies (c). By Theorem 4.12, (c) implies (a). Again trivially, (a) implies (b). To prove that (b) implies (d), note that (b) implies part (c) of Theorem 4.12 and (b) implies, by the Corollary to Theorem 4.6, that each nonzero eigenvalue of D has magnitude equal to unity. Thus, (b) implies (d).   Q.E.D.

To prove Theorem 4.8, let us prove the following stronger theorem.

Theorem 4.13. Suppose D is a weighted digraph. Then the following are equivalent:

(a) D is value stable under all autonomous pulse processes.

(b) D is value stable under all simple pulse processes.

(c) D is pulse stable under all simple pulse processes and unity is not an eigenvalue of D.

Proof. The proof uses lemmas analogous to Lemmas 1 and 2 for

$$\left\{ \left\| \sum_{t=0}^{N} A^t \right\| : N = 0, 1, 2, \ldots \right\} \tag{15}$$

and

$$\left\{ \left\| \sum_{t=0}^{N} J^t \right\| : N = 0, 1, 2, \ldots \right\}. \tag{16}$$

Clearly, (a) $\Rightarrow$(b). We shall prove (b) $\Rightarrow$ (c) and (c) $\Rightarrow$ (a).

To prove the latter, it is sufficient to show that under assumption (c), $\{\| \sum_{t=0}^{N} J^t \|\}$ is bounded. To show that $\{\| \sum_{t=0}^{N} J^t \|\}$ is bounded, it is sufficient to show that for each j, $\{\| \sum_{t=0}^{N} B_j^t \|\}$ is bounded. Let λ be the eigenvalue appearing on the diagonal of B_j. If $\delta_j = 0$, then $\sum_{t=0}^{N} B_j^t$ is $\sum_{t=0}^{N} \lambda^t$. Since $|\lambda| \leq 1$ by pulse stability and $\lambda \neq 1$ by hypothesis, $\sum_{t=0}^{N} \lambda^t$ is bounded. We conclude that $\{\| \sum_{t=0}^{N} \lambda^t \|\} = \{\| \sum_{t=0}^{N} B_j^t \|\}$ is bounded. Suppose next that $\delta_j = 1$. Then by pulse stability, $|\lambda| < 1$. If $k > l$, then $\sum_{t=0}^{N} b_{k,l}^t = 0$. If $k \leq l$, then by the proof of Theorem 4.12, if $t > 2e$, we have

$$|b_{k,l}^t| \leq \frac{t^e}{e!} |\lambda|^{t-e},$$

where e is one less than the dimension of B_j. Thus, for $t > 2e$,

$$\left| \sum_{t=0}^{N} b_{k,l}^t \right| \leq \frac{1}{e!} \sum_{t=0}^{N} t^e |\lambda|^{t-e}. \tag{17}$$

Applying the ratio test to the sum on the right-hand side of (17), we find that

$$\lim_{t\to\infty} \frac{(t+1)^e |\lambda|^{t+1-e}}{t^e |\lambda|^{t-e}} = \lim_{t\to\infty} \left(\frac{t+1}{t}\right)^e |\lambda| = |\lambda|.$$

Since $|\lambda| < 1$, the series converges. Thus, $\{|\sum_{t=0}^{N} b_{k,l}^{j,t}|\}$ is bounded, and hence so is $\{\|\sum_{t=0}^{N} B_j^t\|\}$.

To complete the proof, we assume (b) and prove (c). First, value stability implies pulse stability. Finally, we shall assume that unity is an eigenvalue and reach a contradiction. It is sufficient to prove that $\{\|\sum_{t=0}^{N} J^t\|\}$ is unbounded. In particular, suppose $\lambda = 1$ and B_j is a diagonal block in which λ is the diagonal entry. By pulse stability λ is unlinked, and so $\sum_{t=0}^{N} B_j^t$ is $\sum_{t=0}^{N} \lambda^t = N + 1$. Thus, $\{\|\sum_{t=0}^{N} B_j^t\|\}$ is unbounded, and hence so is $\{\|\sum_{t=0}^{N} J^t\|\}$. This contradicts value stability.   Q.E.D.

EXERCISES

1. For each of the following matrices A, observe that A is already in the special form of the Jordan Canonical Form, so $J = A$. If A is the adjacency matrix of a weighted digraph, comment on the pulse and value stability in each case.

(a)
$$A = \begin{pmatrix} 1 & 0 & 0 & 0 & 0 \\ 0 & 1 & 0 & 0 & 0 \\ 0 & 0 & 2 & 0 & 0 \\ 0 & 0 & 0 & 3i & 0 \\ 0 & 0 & 0 & 0 & -3i \end{pmatrix}$$

(b)
$$A = \begin{pmatrix} .6 & 1 & 0 & 0 & 0 & 0 \\ 0 & .6 & 0 & 0 & 0 & 0 \\ 0 & 0 & .5i & 0 & 0 & 0 \\ 0 & 0 & 0 & -.5i & 0 & 0 \\ 0 & 0 & 0 & 0 & 0 & 1 \\ 0 & 0 & 0 & 0 & 0 & 0 \end{pmatrix}$$

(c)
$$A = \begin{pmatrix} i & 1 & 0 & 0 & 0 & 0 \\ 0 & i & 0 & 0 & 0 & 0 \\ 0 & 0 & -i & 1 & 0 & 0 \\ 0 & 0 & 0 & -i & 0 & 0 \\ 0 & 0 & 0 & 0 & .8+.2i & 0 \\ 0 & 0 & 0 & 0 & 0 & .8-.2i \end{pmatrix}$$

(d)

$$A = \begin{pmatrix} 1 & 0 & 0 & 0 & 0 & 0 & 0 \\ 0 & 1 & 0 & 0 & 0 & 0 & 0 \\ 0 & 0 & .1 + .2i & 0 & 0 & 0 & 0 \\ 0 & 0 & 0 & .1 - .2i & 0 & 0 & 0 \\ 0 & 0 & 0 & 0 & .8 & 1 & 0 \\ 0 & 0 & 0 & 0 & 0 & .8 & 1 \\ 0 & 0 & 0 & 0 & 0 & 0 & .8 \end{pmatrix}$$

2. The following matrices, in Jordan Canonical Form, are the adjacency matrices of integer-weighted digraphs. Comment on the pulse and value stability in each case.

(a) $\begin{pmatrix} 1 & 0 & 0 & 0 \\ 0 & 1 & 0 & 0 \\ 0 & 0 & 0 & 1 \\ 0 & 0 & 0 & 0 \end{pmatrix}$ (b) $\begin{pmatrix} -1 & 1 & 0 & 0 \\ 0 & -1 & 0 & 0 \\ 0 & 0 & 0 & 0 \\ 0 & 0 & 0 & 0 \end{pmatrix}$ (c) $\begin{pmatrix} 1 & 0 & 0 & 0 \\ 0 & 1 & 1 & 0 \\ 0 & 0 & 1 & 0 \\ 0 & 0 & 0 & 0 \end{pmatrix}$.

3. Give an example of a weighted digraph D which has nonzero eigenvalues of magnitude less than unity but which is pulse stable under all simple pulse processes.

4. (a) Give an example of a signed digraph D which has multiple nonzero eigenvalues but which is pulse stable under all simple pulse processes.
   (b) Give another example with the same properties but which is not pulse stable under all simple pulse processes.

5. Suppose the eigenvalue of A with largest magnitude has magnitude 2. Show that if A is an $n \times n$ matrix, then there is some simple pulse process under which $\| P(t) \|$ is at least $2^t/n$ for arbitrarily large t.

6. Suppose the adjacency matrix A of a weighted digraph D is a diagonal matrix,

$$A = \begin{pmatrix} \lambda_1 & & & \\ & \lambda_2 & & \text{\Large 0} \\ & & \ddots & \\ \text{\Large 0} & & & \ddots \\ & & & & \lambda_n \end{pmatrix}.$$

Suppose that the smallest eigenvalue of A has magnitude larger than 1. Show that D is pulse unstable under all simple pulse processes.

7. Suppose a weighted digraph D is pulse stable under all simple pulse processes. Using

$$P(t) = P(0)A^t,$$

show that the sequence

$$\{\|A^t\|: t = 0, 1, 2, \ldots\}$$

is bounded.

8. Show that the conclusion of Exer. 7 does not follow if we simply assume pulse stability under *some* simple pulse process.

9. Suppose that the sequence

$$\{\|A^t\|: t = 0, 1, 2, \ldots\}$$

is bounded. Show, using Theorem 4.5, that D must be pulse stable under all autonomous pulse processes.

10. Prove that if $\delta_j = 1$, then $b_{k,l}^{j,t}$, the k, l entry of B_j^t, is given by

$$b_{k,l}^{j,t} = \begin{cases} 0 & \text{(if } k > l) \\ \binom{t}{l-k}\lambda^{t-l+k} & \text{(if } k \leq l). \end{cases}$$

(*Note:* By convention, $\binom{r}{s} = 0$ if $s > r$.)

11. Theorem 4.7 has a particularly simple proof if *all* the eigenvalues $\lambda_1, \lambda_2, \ldots, \lambda_n$ are distinct. For, by a standard theorem of linear algebra, it follows that the corresponding eigenvectors $U_1, U_2, \ldots, U_n$ are linearly independent. Hence, $P(0)$ may be written as a linear combination $\sum_{i=1}^n \alpha_i U_i$. Conclude from this and Theorem 4.5 that if every λ_i has magnitude at most unity, then $\|P(t)\|$ is bounded under all simple pulse processes.

*12. The eigenvalue of largest magnitude can be used to give bounds on pulse and value. Suppose that D is a weighted digraph and λ is the eigenvalue of D of largest magnitude. Show that there is a simple pulse process and a positive constant c so that for some i,

$$|p_i(t)| \geq c|\lambda|^t$$

in this pulse process for arbitrarily large t. (Thus, pulses grow at least exponentially, with "growth rate" the magnitude $|\lambda|$ of the largest eigenvalue.) (*Hint:* Use the proof of Theorem 4.6.)

*13. Suppose D is a weighted digraph and every eigenvalue of D has magnitude less than unity.
(a) Show that $A^t \to 0$. (*Hint:* Use the proof of Theorem 4.12 to show that $|b_{k,l}^{j,t}| \to 0$. Then use $J^t = PA^tP^{-1}$.)
(b) Show that $p_i(t) \to 0$ for all i under all autonomous pulse processes. (*Note:* The result of part (a) will be used in Exer. 16, Sec. 5.4, to show that if every eigenvalue of D has magnitude less than unity, then $V(t) \to V(\text{start}) + P(0)(I - A)^{-1}$ in every autonomous pulse process.)

*14. Most of the results about weighted digraphs described in this section depend on the specific pulse process change of value rule embodied in Eq. (6) and restated in Theorem 4.5 as

$$P(t) = P(0)A^t.$$

Note that the proof of Theorem 4.12 depends essentially only on this equation. Suppose we assume that weights have a different interpretattion than that which we have been using, and that

$$v_i(t + 1) = \sum_{j=1}^{n} w(u_j, u_i)v_j(t), \tag{18}$$

i.e., value is determined by values at previous times. We assume that $V(0) = V(\text{start})$. In matrix form, we can restate Eq. (18) as:

$$V(t) = V(0)A^t.$$

The initial condition which goes with this rule is simply to give $V(\text{start})$, which equals $V(0)$. (All pulse processes defined by the new rule are automatically autonomous; it does not make sense to introduce external pulses at different times and so to speak of nonautonomous pulse processes. We could bring in external pulses by replacing $v_j(t)$ on the right-hand side of Eq. (18) by $v_j(t) + p_j^o(t)$, where $p_j^o(t)$ is some external pulse introduced at vertex u_j just before time $t + 1$. We shall not introduce this complication.) Suppose as before that $P(t) = V(t) - V(t - 1)$ for $t > 0$, and suppose that pulse and value stability are defined as before.

(a) For the weighted digraph of Fig. 4.34, calculate $V(3)$ given that $V(\text{start}) = V(0) = (0, 1, 1, 0, 0)$, using Eq. (18). Check your answer using $V(t) = V(0)A^t$.

(b) For the weighted digraph D_1 of Fig. 4.53, give one choice of $V(\text{start}) = V(0)$ which leads to value instability and one choice of

FIGURE 4.53. Weighted digraphs for Exers. 14 and 15 of Sec. 4.7.

V(start) $= V(0)$ which leads to value stability.
(c) Repeat part (b) for pulse instability and stability.
(d) Guess at a necessary and sufficient condition for a weighted digraph
     D to be value stable under all (autonomous) pulse processes.
(e) Do part (d) for a *signed* digraph.

*15. Modifying the change of value rule of Exer. 14, let us use the following
      change of value rule:

$$v_i(t+1) = v_i(t) + \sum_{j=1}^{n} w(u_j, u_i)v_j(t). \tag{19}$$

Again assume that $V(0) = V$(start). (Once again under this rule, all
pulse processes are autonomous.)
(a) Using Eq. (19), derive a matrix equation for $V(t)$ in terms of
     V(start) $= V(0)$.
(b) For the weighted digraph of Fig. 4.34, calculate $V(3)$ given V(start)$=$
     $V(0) = (0, 1, 1, 0, 0)$, using Eq. (19). Check your answer using your
     answer to part (a).
(c) For the weighted digraph D_2 of Fig. 4.53, give one choice of
     V(start) $= V(0)$ which leads to value stability and one choice which
     leads to value instability.
(d) Repeat part (c) for pulse instability and stability.
(e) Guess at a necessary and sufficient condition for a weighted digraph
     D to be value stable under all (autonomous) pulse processes.
(f) Do part (e) for a *signed* digraph.
(g) Show that λ is an eigenvalue of A if and only if $\lambda + 1$ is an eigen-
     value of $A + I$.
(h) Suppose that D is a *signed* digraph and A is its adjacency matrix.
     Show that if A has an eigenvalue different from 0, -1, or
     $-\frac{1}{2} \pm (\sqrt{3}/2)i$, then D cannot be value stable under both rule
     (19) of this exercise and rule (18) of Exer. 14.

References

ANTLE, L. G. and JOHNSON, G. P., "Integration of Policy Simulation, Decision
    Analysis and Information Systems: Implications of Energy Conservation
    and Fuel Substitution Measures on Inland Waterway Traffic," in *Proceedings*

of Computer Science and Statistics: Seventh Annual Symposium on the Interface, Iowa State Univ., Ames, Iowa, 1973.

AXELROD, R. M., "Psycho-Algebra: A Mathematical Theory of Cognition and Choice with an Application to the British Eastern Committee in 1918," *Peace Research Society, Papers XVIII, The London Conference,* 1971, pp. 113–131.

AXELROD, R. M. (ed.), *The Structure of Decision,* Princeton University Press, Princeton, N.J., to appear.

BONHAM, M. and SHAPIRO, M., "Explanation of the Unexpected: The Syrian Intervention in Jordan in 1970," in *The Structure of Decision,* R. M. Axelrod (ed.), Princeton University Press, Princeton, N.J., to appear.

BROWN, T. A., ROBERTS, F. S., and SPENCER, J., "Pulse Processes on Signed Digraphs: A Tool for Analyzing Energy Demand," Rand Corporation Report R-926-NSF, March 1972.

CHEN, W., *Applied Graph Theory,* American Elsevier, New York, 1971.

COADY, S. K., JOHNSON, G. P., and JOHNSON, J. M., "Effectively Conveying Results: A Key to the Usefulness of Technology Assessment," Mimeographed, Institute for Water Resources, Corps of Engineers, paper delivered at the First International Congress on Technology Assessment, The Hague, May 31, 1973.

COOMBS, C. H., DAWES, R. M., and TVERSKY, A., *Mathematical Psychology: an Elementary Introduction,* Prentice-Hall, Inc., Englewood Cliffs, N.J., 1970.

FORRESTER, J. W., *Industrial Dynamics,* M.I.T. Press, Cambridge, Mass., 1961.

FORRESTER, J. W., *Urban Dynamics,* M.I.T. Press, Cambridge, Mass., 1969.

FORRESTER, J. W., *World Dynamics,* Wright-Allen Press, Cambridge, Mass., 1971.

GARDNER, M. R. and ASHBY, W. R., "Connectance of Large Dynamic (Cybernetic) Systems: Critical Values for Stability," *Nature,* **228** (1970), 784.

HARARY, F., "A Graph-Theoretic Method for the Complete Reduction of a Matrix with a View Toward Finding its Eigenvalues," *J. Math. and Physics,* **38** (1959), 104–111.

HARARY, F., NORMAN, R. Z., and CARTWRIGHT, D., *Structural Models: An Introduction to the Theory of Directed Graphs,* John Wiley & Sons, Inc., New York, 1965.

KANE, J., "A Primer for a New Cross-Impact Language—KSIM," *Technological Forecasting and Social Change,* **4** (1972), 129–142.

KANE, J., THOMPSON, W., and VERTINSKY, I., "Health Care Delivery: A Policy Simulation," *Socio-Econ. Plan. Sci.,* **6** (1972), 283–293.

KEMENY, J. G., and SNELL, J. L., *Mathematical Models in the Social Sciences,* Blaisdell Publishing Co., New York, 1962; reprinted by M.I.T. Press, Cambridge, Mass., 1972.

KRUZIC, P. G., "A Suggested Paradigm for Policy Planning," Stanford Research Institute Technical Note TN-OED-016, June 1973, a.

KRUZIC, P. G., "Cross-Impact Analysis Workshop," Stanford Research Institute Letter Report, June 23, 1973, b.

LEVINS, R., "The Qualitative Analysis of Partially Specified Systems," *Annals of the N.Y. Acad. Sci.*, **231** (1974), 123–138.

LEVINS, R., "Evolution in Communities near Equilibrium," in *Ecology of Species and Communities*, J. Diamond and M. Cody (eds.), Belknap Press of Harvard University Press, Cambridge, Mass., to appear.

MAL'CEV, A. I., *Foundations of Linear Algebra,* translated from the Russian by T. C. Brown, W. H. Freeman and Co., San Francisco, 1963.

MARUYAMA, M., "The Second Cybernetics: Deviation-Amplifying Mutual Causal Processes," *Amer. Scientist,* **51** (1963), 164–179.

MAY, R. M., "Stability in Multispecies Community Models," *Mathematical Biosciences,* **12** (1971), 59–79.

MAY, R. M., "Will a Large Complex System be Stable?", *Nature,* **238** (1972), 413–414.

MAY, R. M., *Stability and Complexity in Ecosystems*, Princeton University Press, Princeton, N.J., 1973.

MEADOWS, D. H., MEADOWS, D. L., RANDERS, J., and BEHRENS, W. W. III, *The Limits to Growth,* Universe Books, New York, 1972.

Organization for Economic Cooperation and Development, "The Slowdown in R & D Expenditure and the Scientific and Technical System," Report SPT (74) 1, Paris, France, February 1974.

PHILLIPS, J. L., "A Model for Cognitive Balance," *Psychol. Rev.*, **74** (1967), 481–495.

ROBERTS, F. S., "Signed Digraphs and the Growing Demand for Energy," *Environment and Planning,* **3** (1971), 395–410.

ROBERTS, F. S., "Building an Energy Demand Signed Digraph I: Choosing the Nodes," Rand Corporation Report R-927/1-NSF, April 1972, a.

ROBERTS, F. S., "Building an Energy Demand Signed Digraph II: Choosing Edges and Signs and Calculating Stability," Rand Corporation Report R-927/2-NSF, May 1972, b.

ROBERTS, F. S., "Building and Analyzing an Energy Demand Signed Digraph," *Environment and Planning,* **5** (1973), 199–221.

ROBERTS, F. S., "Structural Characterizations of Stability of Signed Digraphs under Pulse Processes," in *Graphs and Combinatorics*, R. Bari and F. Harary (eds.), Springer Verlag Lecture Notes #406, New York, 1974, pp. 330–338, a.

ROBERTS, F. S., "Weighted Digraph Models for Energy Use and Air Pollution in Transportation Systems," Rand Corporation Report R-1578-NSF, Dec. 1974, b. (Abridged version to appear in *Environment and Planning.*)

ROBERTS, F. S., and BROWN, T. A., "Signed Digraphs and the Energy Crisis," *Amer. Math. Monthly,* **82** (1975), 577–594.

ROSS, S., "Complexity of the Presidency: Gouvernor Morris in the Constitutional Convention," in *The Structure of Decision*, R. M. Axelrod (ed.), Princeton University Press, Princeton, N.J., to appear.

VICKREY, W. S., "Pricing in Urban and Suburban Transport," in *Readings in Urban Transportation*, G. M. Smerk (ed.), Indiana University Press, Bloomington, Ind., 1968.

WILF, H. S., "The Eigenvalues of a Graph and its Chromatic Number," *J. London Math. Soc.*, **42** (1961), 330–332.

FIVE

Markov Chains

5.1. Stochastic Processes and Markov Chains

The mathematical models we discussed in the previous chapter can be called *deterministic*. They postulate processes which are completely determined at any stage by knowing their status at previous stages. In this chapter we shall introduce elements of indeterminism into our models and deal with *probabilistic models*.[1] A probabilistic model is in many ways weaker than a deterministic model, for it does not make predictions of specific outcomes. On the other hand, frequently we can make powerful predictions of expected range of outcomes. Moreover, it is quite possible that a large number of phenomena are inherently probabilistic and it might be impossible to model them accurately using deterministic models.

We shall see that many of the mathematical tools developed in the earlier chapters for dealing with deterministic models can be modified to deal with probabilistic models as well. (This is another example of a mathematical tool having been developed for one purpose also being useful for another.) Specifically, we shall study a class of processes called Markov chains, which may be described using weighted digraphs like those of the previous chapter.[2]

[1] Probabilistic models are sometimes called *stochastic*.

[2] The reader who has not read that chapter should refer to Sec. 4.1 for the concept of weighted digraph and some of its basic properties. In particular, as in the previous chapter, **weighted digraphs in this chapter are allowed to have loops.**

Markov chains have been applied in a wide variety of areas, and we shall apply them in this chapter to such diverse topics as genetics, flow of pollutants, money flow between cities, and learning theory in psychology.

To define a Markov chain, we first define a more general process called a stochastic process. Let us imagine a sequence of experiments in which the outcome on the tth experiment depends on some chance element. We assume that if the outcomes of all the first t experiments are known, then the possible outcomes on the $(t + 1)$st experiment are also known and so are their respective probabilities. (We may not use all of the knowledge about previous outcomes, or indeed we may not use any of this knowledge, in determining the outcomes and probabilities.) If the set of possible outcomes at each trial is finite, the sequence of outcomes is called a *finite stochastic process*. All stochastic processes studied in this book are finite, so we shall drop this adjective. We give several examples of finite stochastic processes below.

Example 1: A manufacturer of a certain product markets it in boxes of 10. It is known that in a certain box, two out of 10 items are defective. The first item you use may be either defective or nondefective, with probabilities $\frac{2}{10}$ and $\frac{8}{10}$ respectively. The character of the second item depends on the first one you used. If the first was nondefective, the second may be either defective or nondefective, with probabilities $\frac{2}{9}$ and $\frac{7}{9}$ respectively. If the first was defective, the possible outcomes are the same, but the probabilities change to $\frac{1}{9}$ and $\frac{8}{9}$ respectively. If the first two were defective, the third must be nondefective. The character, defective or nondefective, of the item drawn on the tth trial defines a stochastic process. In order to know the possible outcomes on the tth trial, and their respective probabilities, it is necessary to know all the outcomes up until that time.

Example 2: In Russian roulette, there is a gun with six cylinders, one of which has a bullet in it. The chamber is spun and the gun fired at an individual's head. (Usually, there is a second individual, who is involved in betting, but we shall disregard this.) On the first trial, the possible outcomes are "alive" and "dead," and their respective probabilities are $\frac{5}{6}$ and $\frac{1}{6}$. In one version of Russian roulette, if the first individual is alive after a given trial, the gun is fired at him again without spinning the chamber. Then, on the second trial, the possible outcomes are again "alive" and "dead," but their respective probabilities are now $\frac{4}{5}$ and $\frac{1}{5}$. If the outcome of any trial is "dead," then the outcome of all trials thereafter is considered "dead." The sequence of outcomes of Russian roulette thus defines a stochastic process.

There is another version of Russian roulette, that in which the chamber is spun after each trial in which the person survived. Thus, on trial $t + 1$, the possible outcomes and their probabilities depend only on the outcome of trial t. If that outcome was "dead," then the only possible outcome on trial $t + 1$ is "dead" as well. If the outcome was "alive," then on trial $t + 1$ the outcomes are "alive" and "dead," and their probabilities are $\frac{5}{6}$ and $\frac{1}{6}$. This version of Russian roulette also defines a stochastic process. It is one where the possible outcomes and their respective probabilities on trial $t + 1$ can be determined knowing only the outcome of the previous trial, so certainly knowing the outcome of all previous trials. However, only a "memory" of one past time period is used.

Example 3: In a sequence of tosses of a fair coin, we know that on the $(t + 1)$st trial, the possible outcomes are always the same, heads (H) and tails (T), and moreover their respective probabilities are always $\frac{1}{2}$, $\frac{1}{2}$. We know this even without knowing the outcomes of the first t trials, but we certainly know it with this information. Thus, repeated tosses of a fair coin define a stochastic process. No memory is used.

Example 4: Let us next consider a more complicated gambling game in which you start by tossing a coin. If the outcome in coin tossing is H, you toss a coin again on the next trial. If it is T, you throw a die on the next trial. If the outcome in die tossing is even, you throw a die again on the next trial. If it is odd, you toss a coin next. The sequence of outcomes of this game again defines a stochastic process. Knowledge of the outcome on the tth trial is sufficient to determine the possible outcomes on the $(t + 1)$st trial, and their probabilities. Thus, again we have a process with memory only going back one trial.

Example 5: The weather today influences the weather several days or even weeks hence. However, at least on a world-wide basis, it might be reasonable to assume that we can quite accurately predict the weather tomorrow on the basis of, say, the previous week's weather. Thus, as a first approximation, the weather on the tth day defines a stochastic process, with a limited memory of, say, five to seven days.

Example 6: To take Example 5 one step further, a possible strategy in making weather forecasts is to simply use the previous day's weather. In a study of rainfall in Tel Aviv, Gabriel and Neumann [1962] showed how the sequence of wet and dry days during the rainy season over a 27-year period could be looked at as a stochastic process with a memory of only one trial. In particular, according to their data, the probability of having a wet day was .662 if the previous day was wet, and only .250 if the previous day was dry. The probability of having a dry day was .750 if the previous day was dry, and only .338 if the previous day was wet.

Stochastic processes with memory going back no trials, i.e., with no memory, are the easiest stochastic processes to deal with, but they are not very common in practice. Coin tossing is an example of such a process. The next simplest stochastic process is one where memory goes back *at most* one trial. Such a stochastic process is still easy enough to handle mathematically, but has a surprisingly large number of applications. We have already seen a number of examples of such processes, and we will study more of them in the remainder of this chapter.

Indeed, we shall study those stochastic processes with the following properties. First, there is a finite set of outcomes which includes all possible outcomes on all possible trials. Second, the probability $p_{t+1}(O)$ that outcome O will occur on trial $t + 1$ is known if we know what outcome occurred on trial t, though the knowledge of this outcome need not be used. Finally, the dependence of $p_{t+1}(O)$ on the previous outcome is independent of t. That is, it is the same for trial 2 as for trial 1000. Such a stochastic process is called a (*finite*) *Markov chain*, after the Russian mathematician A. A. Markov.[3] In a Markov chain, when the present outcome is known, information about earlier trials (and about the trial number) doesn't affect probabilities of future events. Of the examples we have discussed, coin tossing (Example 3) is certainly a Markov chain, as is the more complicated gambling game (Example 4). The version of Russian roulette with repeated spinning defines a Markov chain, though the version without repeated spinning does not, using the set of outcomes we have defined. Finally, we have seen in Example 6 that sometimes weather can be considered a Markov chain.

In this chapter, we shall study these and other examples of Markov chains and develop a general theory for studying their properties. We will use Markov chains as mathematical models for numerous phenomena, and we will derive various predictions which can be used to check whether the Markov model is an appropriate one.

EXERCISES

1. A biased coin is tossed over and over again. Suppose the probability of heads is $\frac{2}{3}$. Does the sequence of outcomes define a Markov chain?

2. Consider a modified gambling game in which we toss a coin until two heads in a row occur, then toss a die until two 3's in a row occur, then toss a coin, etc. Does the sequence of outcomes define a Markov chain?

3. A box of 10,000 items has a certain unknown number of defectives. An item is chosen at random, its character (defective or nondefective) noted, and then the item is replaced. The process stops once 20 defectives have been obtained, or once 100 nondefectives have been obtained. The outcome of the tth trial is defective, nondefective, or "stopped." Once the process has stopped, it remains stopped. (This process is

---

[3] For a discussion of Markov chains with an infinite number of states, the reader is referred to Bharucha-Reid [1960] or Howard [1971]. All Markov chains in this chapter will be finite.

important in sampling for quality control.) Does the sequence of outcomes define a Markov chain?

4. (Howard [1963].) A small town has two supermarkets, Thrifty and Saveway. The customers of each market are quite loyal. However, each week, 10% of the customers of Thrifty change to Saveway, and 20% of the customers of Saveway change to Thrifty. A marketing survey company chooses one local resident at random and asks each week what market he shopped at. Does the sequence of outcomes define a Markov chain?

5. It is an old maxim of politics that the party in power has a higher chance of winning the presidency than the party out of power. Indeed, if a party has won the presidency several times in a row, it is even more likely, according to this maxim, to win the presidency again. If this maxim is true, does the sequence of parties in power define a Markov chain?

6. (Goodman and Ratti [1971].) We often observe that the successful people seem to get "all the breaks." That is, if a person, say a businessman, has been successful in one business deal, his probability of being successful in his next deal seems to increase. Moreover, his probability of being successful after several successful deals is even higher still. Does the sequence of outcomes of business deals define a Markov chain?

7. (Bhat [1972].) A city's water supply comes from a reservoir. Careful study of this reservoir over the past twenty years has shown that if the reservoir was full at the beginning of one summer, then the probability it would be full at the beginning of the next summer is .8, independent of its condition in previous years. Similarly, if the reservoir was not full at the beginning of one summer, the probability it would be full at the beginning of the next summer is only .4. Does the sequence of observations of the reservoir ("full" or "not full") at the beginning of each summer define a Markov chain?

8. The manager of a small factory claims that of the wastes from his plant which are emptied into a nearby river, the majority are soon carried out to sea. Indeed, he claims that of a given molecule of mercury found in his wastes, its probability of being swept out to sea within a day is .999. If that molecule is still around after a certain number of days, its probability of being swept out to sea on the next day is still .999. Once it is swept out to sea, we assume it cannot return. Suppose a particular molecule of mercury could be "tagged" and on the *t*th day, we could record its presence or absence in the vicinity of the factory's disposal system. Would the sequence of observations define a Markov chain?

5.2. Transition Probabilities and Transition Digraphs

In a Markov chain, there is a set $\{u_1, u_2, \ldots, u_n\}$ of outcomes which covers all possible outcomes on all possible trials. It is traditional to refer to these outcomes as *states*, and to speak of the chain as being in state u_i at time t if the outcome at time t was u_i. We can describe the entire chain by giving, for each i and j, a number p_{ij} representing the probability that if the outcome on any trial is u_i, the outcome on the next trial will be u_j. Thus, p_{ij} is the probability that the chain will go from state u_i on one trial to state u_j on the next. For example, in the gambling game of Example 4, the states can be taken as $u_1 = H$, $u_2 = T$, $u_3 =$ even, and $u_4 =$ odd. Then $p_{12} = \frac{1}{2}$, since if you get a head on any trial, the probability of getting a tail on the next trial is $\frac{1}{2}$. On the other hand, $p_{13} = 0$, for you cannot get an even outcome (from tossing a die) on the trial following an outcome of heads (from tossing a coin).

The number p_{ij} is called a *transition probability* and it is convenient to exhibit the transition probabilities as a matrix

$$P = \begin{pmatrix} p_{11} & p_{12} & \cdots & p_{1n} \\ p_{21} & p_{22} & \cdots & p_{2n} \\ & & \cdots & \\ p_{n1} & p_{n2} & \cdots & p_{nn} \end{pmatrix}.$$

For example, in our gambling game, P is given by

$$P = \begin{pmatrix} \frac{1}{2} & \frac{1}{2} & 0 & 0 \\ 0 & 0 & \frac{1}{2} & \frac{1}{2} \\ 0 & 0 & \frac{1}{2} & \frac{1}{2} \\ \frac{1}{2} & \frac{1}{2} & 0 & 0 \end{pmatrix}.$$

P is called the *transition matrix* of the Markov chain. The entries of P satisfy

$$\sum_{j=1}^{n} p_{ij} = 1, \qquad i = 1, 2, \ldots, n, \tag{1}$$

i.e., the rows of P sum to 1. For if a Markov chain is in state u_i on a given trial, it must go to some state u_j at the next trial, and to only one such state. Every row vector of the matrix P thus has nonnegative entries and the entries sum to 1. Such a vector is called a *probability vector* and a matrix each of whose rows is a probability vector is called a *stochastic matrix*. Clearly, every Markov chain gives rise to a stochastic matrix. Conversely, every stochastic matrix gives rise to a Markov chain: Simply take the rows as states and the entries as transition probabilities.

Every stochastic matrix P can be translated into a weighted digraph in an obvious way. We let rows $1, 2, \ldots, n$ be the vertices $u_1, u_2, \ldots, u_n$ of this digraph, put an arc from u_i to u_j if $p_{ij} > 0$, and let p_{ij} be the weight on the arc (u_i, u_j). (In Sec. 4.2, this digraph was called the Coates digraph of the matrix.) Figure 5.1 shows the transition digraph for the gambling game of Example 4. This weighted digraph again has nonnegative weights and the weights satisfy Eq. (1). Such a weighted digraph will be called *stochastic*, and

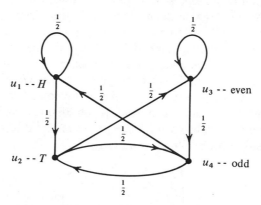

FIGURE 5.1. Transition digraph for coin tossing and die throwing.

it is obvious that there is a 1-1 correspondence between stochastic matrices and stochastic digraphs. If P is the transition matrix of a Markov chain, then the corresponding weighted digraph is called the *transition digraph*. Throughout this chapter we shall see that sometimes it is more useful to look at Markov chains using their transition matrices and sometimes it is more useful to look at them using their transition digraphs.

Let us consider the examples of Markov chains which were introduced in Sec. 5.1 as well as several more examples which we shall carry forward in the rest of this chapter. (The variety of examples should give the reader a feel for the wide applicability of the Markov chain concept.) In each case we calculate the transition matrix and digraph. We continue the numbering begun in the previous section.

Example 2: In the version of Russian roulette where the chamber is respun, the transition matrix and digraph are shown in Fig. 5.2.

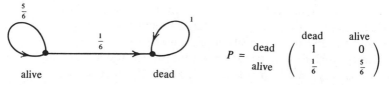

FIGURE 5.2. Transition digraph and matrix for Russian roulette.

Example 3:   In coin tossing, the transition matrix and digraph are shown in Fig. 5.3.

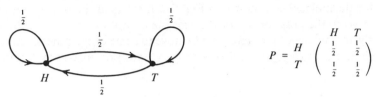

$$P = \begin{array}{c} \\ H \\ T \end{array} \begin{array}{cc} H & T \\ \left(\begin{array}{cc} \frac{1}{2} & \frac{1}{2} \\ \frac{1}{2} & \frac{1}{2} \end{array} \right) \end{array}$$

FIGURE 5.3. Transition digraph and matrix for coin tossing.

Example 6:   In the case of weather in Tel Aviv, the transition matrix and digraph are shown in Fig. 5.4.

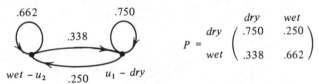

$$P = \begin{array}{c} \\ dry \\ wet \end{array} \begin{array}{cc} dry & wet \\ \left(\begin{array}{cc} .750 & .250 \\ .338 & .662 \end{array} \right) \end{array}$$

FIGURE 5.4. Transition digraph and matrix for the weather in Tel Aviv. (Data estimated by Gabriel and Neumann [1962].)

Example 7:   Let us suppose a politician tells a key aide his position (favorable or unfavorable) on a certain bill. This aide reveals the decision to another person, who reveals the decision to another person, and so on. Suppose that at each step, there is a probability p of getting the message reversed. (We assume p is positive, but not 1.) We may represent this process by a Markov chain with two states, $u_1 =$ favorable and $u_2 =$ unfavorable. Then the transition matrix and digraph are shown in Fig. 5.5. The reason this example defines a Markov chain is that at any given time, the probability that a particular person will pass on the message "he is in favor of the bill" depends only on the message that person received, and *not* on the series of steps the message has taken to get to him. We shall further discuss this Markov chain, which is modified from one of Kemeny, Snell, and Thompson [1966, p. 276 ff], in Sec. 5.5.

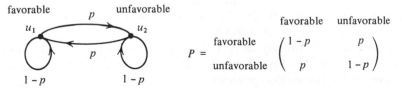

$$P = \begin{array}{c} \\ \text{favorable} \\ \text{unfavorable} \end{array} \begin{array}{cc} \text{favorable} & \text{unfavorable} \\ \left(\begin{array}{cc} 1-p & p \\ p & 1-p \end{array} \right) \end{array}$$

FIGURE 5.5. Transition digraph and matrix for message passage, with $0 < p < 1$.

Example 8: Consider a simple model for passage of a phosphorus molecule through a pasture ecosystem. For simplicity, we consider only four possible states for the molecule, as shown in Fig. 5.6. It starts in the soil. From there it can move to the grass or out of the ecosystem, or it can remain in the soil. Similarly, at any one of the states, it can move to some of the others. Once the molecule is out of the system, it stays out. A transition matrix and digraph,

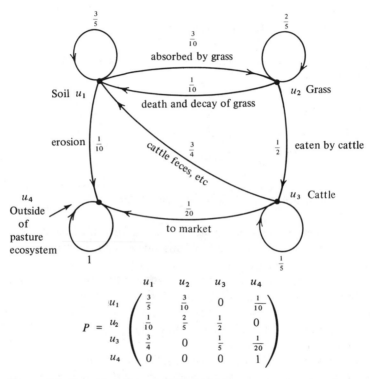

$$P = \begin{array}{c} \\ u_1 \\ u_2 \\ u_3 \\ u_4 \end{array} \begin{pmatrix} \frac{3}{5} & \frac{3}{10} & 0 & \frac{1}{10} \\ \frac{1}{10} & \frac{2}{5} & \frac{1}{2} & 0 \\ \frac{3}{4} & 0 & \frac{1}{5} & \frac{1}{20} \\ 0 & 0 & 0 & 1 \end{pmatrix}$$

FIGURE 5.6. Transition digraph and matrix for passage of a phosphorus molecule through a pasture ecosystem. Adapted from a figure of Mosimann [1968, p. 161] with permission of Prentice-Hall, Inc.

giving the probability of various transitions, are given in Fig. 5.6. (The digraph is based on a figure of Mosimann [1968].) This example defines a Markov chain because at any time, the next position of the phosphorus molecule depends entirely on its present position, and not its past history. We shall consider this system in more detail in Sec. 5.4. A similar system can be constructed for various pesticides.

Example 9: Electricity usage during the summer is closely tied to tempera-ture. Thus, in planning for day-to-day repairs and maintenance, utility companies should in principal be aware of the probabilities of hot, moderate,

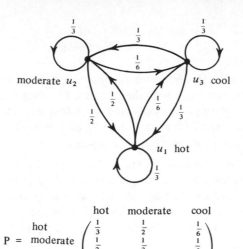

$$P = \begin{array}{c} \\ \text{hot} \\ \text{moderate} \\ \text{cool} \end{array} \begin{array}{ccc} \text{hot} & \text{moderate} & \text{cool} \\ \begin{pmatrix} \frac{1}{3} & \frac{1}{2} & \frac{1}{6} \\ \frac{1}{2} & \frac{1}{3} & \frac{1}{6} \\ \frac{1}{3} & \frac{1}{3} & \frac{1}{3} \end{pmatrix} \end{array}$$

FIGURE 5.7. Transition digraph and matrix for summer weather.

or cool weather. Let us assume as a simplified model that during the summer, whether tomorrow will be hot, moderate, or cool depends only on whether today was hot, moderate, or cool. In particular, let us assume that the transition probabilities are given by the stochastic matrix and digraph of Fig. 5.7. This defines a Markov chain which we shall study in Sec. 5.5.

Example 10: Sociologists are concerned with movement between different occupational classes from one generation to another. Following Prais [1955] and Kemeny and Snell [1960], let us treat *intergenerational occupational mobility* as a Markov chain. The states of the chain are classes of occupations,

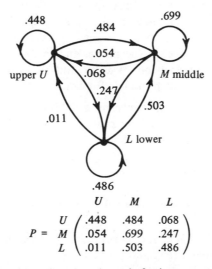

$$P = \begin{array}{c} \\ U \\ M \\ L \end{array} \begin{array}{ccc} U & M & L \\ \begin{pmatrix} .448 & .484 & .068 \\ .054 & .699 & .247 \\ .011 & .503 & .486 \end{pmatrix} \end{array}$$

FIGURE 5.8. Transition digraph and matrix for intergenerational occupational mobility (Data estimated by Glass and Hall [1954].)

say upper U, middle M, and lower L. Figure 5.8 shows the transition probabilities for a Markov chain with these states. The data is based on England and Wales for 1949 and was collected by Glass and Hall [1954]. We can interpret p_{ij} as the probability that a son of a person working in an occupation of class i gets a job in an occupation of class j.

Example 11: The stochastic matrix and digraph of Fig. 5.9 give the transition probabilities for a Markov chain corresponding to the famous *gambler's ruin problem*. A gambler, starting with $2 say, bets $1 at a time. He wins with probability p and loses with probability $1 - p$. (We assume $0 < p < 1$.) If he reaches $4 he quits. He also quits if he is ruined, i.e., if he loses all his money.

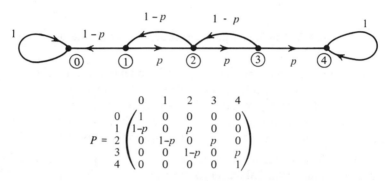

$$P = \begin{array}{c} \\ 0 \\ 1 \\ 2 \\ 3 \\ 4 \end{array} \begin{array}{ccccc} 0 & 1 & 2 & 3 & 4 \\ \begin{pmatrix} 1 & 0 & 0 & 0 & 0 \\ 1-p & 0 & p & 0 & 0 \\ 0 & 1-p & 0 & p & 0 \\ 0 & 0 & 1-p & 0 & p \\ 0 & 0 & 0 & 0 & 1 \end{pmatrix} \end{array}$$

FIGURE 5.9. Transition digraph and matrix for gambler's ruin, with $0 < p < 1$.

The gambler's ruin problem is an example of a *random walk* problem. Think of a particle (a molecule of some gas, say), moving on the line according to certain probabilities. There are two *absorbing barriers*, at 0 and 4. Random walk problems, on the line, in the plane, and in 3-space, are important in models of diffusion and Brownian motion, and in particular in air pollution models. We shall discuss such models further in Secs. 5.4 and 5.11.

A fundamental problem concerning a Markov chain is to determine the so-called *higher-order transition probability* $p_{ij}^{(t)}$, the probability that the chain is in state u_j at time t given that it starts in state u_i at time 0. Our first theorem tells us that calculation of $p_{ij}^{(t)}$ reduces to calculation of the tth power of the transition matrix P.

Theorem 5.1. If P is the transition matrix of a Markov chain, then $p_{ij}^{(t)}$ is the i, j entry of P^t.

Proof. The proof is by induction on t. The theorem is trivially true for $t = 1$. Thus we assume it for t and prove it for $t + 1$. To go from state u_i to state u_j in $t + 1$ steps, one first goes to some state u_k and then from u_k to u_j

in t steps. Hence,

$$p_{ij}^{(t+1)} = \sum_{k=1}^{n} p_{ik}p_{kj}^{(t)}.$$

By inductive assumption $p_{kj}^{(t)}$ is the k, j entry of P^t. Hence, it follows that $p_{ij}^{(t+1)}$ is the i, j entry of $PP^t = P^{t+1}$. Q.E.D.

The reader will note the similarity of this theorem to Theorem 4.4.

In the pasture ecosystem of Example 8, P is the matrix

$$P = \begin{pmatrix} \frac{3}{5} & \frac{3}{10} & 0 & \frac{1}{10} \\ \frac{1}{10} & \frac{2}{5} & \frac{1}{2} & 0 \\ \frac{3}{4} & 0 & \frac{1}{5} & \frac{1}{20} \\ 0 & 0 & 0 & 1 \end{pmatrix} = \begin{pmatrix} .6 & .3 & 0 & .1 \\ .1 & .4 & .5 & 0 \\ .75 & 0 & .2 & .05 \\ 0 & 0 & 0 & 1 \end{pmatrix}.$$

P^2, P^3, and P^4 are the following matrices:

$$P^2 = \begin{pmatrix} .390 & .300 & .150 & .160 \\ .475 & .190 & .300 & .035 \\ .600 & .225 & .040 & .135 \\ 0 & 0 & 0 & 1 \end{pmatrix}$$

$$P^3 = \begin{pmatrix} .3765 & .2370 & .1800 & .2065 \\ .5290 & .2185 & .1550 & .0975 \\ .4125 & .2700 & .1205 & .1970 \\ 0 & 0 & 0 & 1 \end{pmatrix}$$

$$P^4 = \begin{pmatrix} .384600 & .207750 & .154500 & .253150 \\ .455500 & .246100 & .140250 & .158150 \\ .364875 & .231750 & .159100 & .244275 \\ 0 & 0 & 0 & 1 \end{pmatrix}.$$

Let us assume that each step takes one day. Thus, the probability that the given molecule of phosphorus will be out of the system (in u_4) after four days given that it starts in the soil (in u_1) is given by the 1, 4 entry of P^4, namely, .253150.

Suppose we start a Markov chain off by some random device. Then we can speak of an *initial probability vector* $p^{(0)} = (p_1^{(0)}, p_2^{(0)}, \dots, p_n^{(0)})$, where $p_i^{(0)}$ gives the probability that the chain starts in state u_i. If we know a chain starts in state u_i, its initial probability vector is $p^{(0)} = (0, \dots, 0, 1, 0, \dots, 0)$, with the ith component equal to 1. Let $p_i^{(t)}$ be the probability of being in state u_i at time t, and let $p^{(t)} = (p_1^{(t)}, p_2^{(t)}, \dots, p_n^{(t)})$. We have as a corollary to Theorem 5.1 the following result, whose proof is left as an exercise (Exer. 19).

Theorem 5.2. $p^{(t)} = p^{(0)}P^t.$

Returning to the pasture ecosystem, suppose we do not know where the molecule starts. We might assume that it is equally likely to start in any state. Hence, it starts in a given state with probability $\frac{1}{4}$, and $p^{(0)} = (\frac{1}{4}, \frac{1}{4}, \frac{1}{4}, \frac{1}{4})$. Then the probability $p_4^{(3)}$ of being out of the pasture ecosystem by time 3 is given by the fourth entry of $p^{(0)}P^3 = (.330, .181, .114, .375)$. That is, $p_4^{(3)} = .375$.

EXERCISES

1. (a)   Give the transition matrix corresponding to the transition digraph of Fig. 5.10.

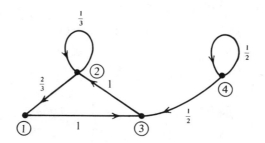

FIGURE 5.10. Transition digraph for Exer. 1, Sec. 5.2.

   (b)   Draw a transition digraph corresponding to the transition matrix

$$P = \begin{pmatrix} 1 & 0 & 0 & 0 \\ \frac{1}{2} & \frac{1}{2} & 0 & 0 \\ \frac{1}{4} & \frac{1}{4} & \frac{1}{4} & \frac{1}{4} \\ 0 & \frac{1}{8} & \frac{7}{8} & 0 \end{pmatrix}.$$

2. (a)   Give an example of a matrix which is not stochastic.
   (b)   In a probability vector, can two entries be $\frac{1}{2}$? Can more than two?

3.   Suppose a Markov chain has transition matrix

$$P = \begin{pmatrix} 1 & 0 & 0 \\ 0 & 1 & 0 \\ \frac{1}{2} & 0 & \frac{1}{2} \end{pmatrix}.$$

   (a)   If the chain starts in state 3, what is the probability it will be in state 3 after two time periods?

(b)   If the starting state of the chain is chosen at random, what is the probability of being in state 3 after two time periods?

(c)   If $p^{(0)} = (\frac{1}{2}, 0, \frac{1}{2})$, what is $p^{(2)}$?

4.   Draw the transition matrix and digraph for the supermarket Markov chain of Exer. 4, Sec. 5.1. Calculate the probability that if the customer is seen at Thrifty this week, he will still be shopping at Thrifty two weeks from now.

5.   Draw the transition matrix and digraph for the water supply Markov chain of Exer. 7, Sec. 5.1. Determine the probability that the reservoir will be full at the beginning of 1985 given that it is observed to be full at the beginning of 1982.

6.   Draw the transition matrix and digraph for the water waste Markov chain of Exer. 8., Sec. 5.1. What is the probability that a molecule of mercury released today will still be around in a week?

7.   (Bhat [1972].) The charge accounts at a given department store are given three months to pay up. Thus, an account can be in its first month with a positive balance due, in its second month with a positive balance due, or in its third month with a positive balance due. After the third month, if there is a positive balance due, the account is considered a bad debt. We build a Markov chain to describe the passage of charge accounts from one state to another. The states are 1, 2, 3, B, and P, where 1, 2, 3 represent the situation of being in the first, second, or third month with a positive balance, B represents the situation of being a bad debt, and P the situation of being paid up. (We assume accounts do not enter our accounting system until they have a positive balance due, so we do not add a state for having no balance due; and we treat paying up as having settled the transaction.) The charge accounts at a given store gave rise to the following transition probabilities:

$$
\begin{array}{c c c c c c}
 & 1 & 2 & 3 & P & B \\
1 & 0 & .7 & 0 & .3 & 0 \\
2 & 0 & 0 & .6 & .4 & 0 \\
3 & 0 & 0 & 0 & .5 & .5 \\
P & 0 & 0 & 0 & 1 & 0 \\
B & 0 & 0 & 0 & 0 & 1
\end{array}
$$

Draw the corresponding transition digraph. What is the probability that a positive balance which appears for the first time this month will still appear as a positive balance two months from now?

8.   Suppose that of the sons of Elks, 80% become Elks, and the rest become Moose. Of the sons of Moose, 60% become Moose, 20% Elks, and the rest Lions. Of the sons of Lions, 50% become Lions, 40% Moose, and

10% Elks. Assume that the son of a given individual joins one and only one of the given organizations. Suppose that each man has exactly one son.

(a)   Describe a transition matrix and digraph which corresponds to the memberships of sons.

(b)   What is the probability that the grandson of an Elk is an Elk?

(c)   Suppose that of the current population, 50% are Elks, 40% Moose, and the rest Lions (with no overlap). If a man is chosen at random from the members of these organizations, what is the probability that his grandson will be an Elk?

(d)   Why did we add the assumption that each man has exactly one son?

9.   Suppose as in the previous problem that each man has exactly one son. Suppose that a tall father will have a tall son with probability .6, a medium-height son with probability .2, and a short son with probability .2. A medium-height father will have a tall, medium-height, or short son with probability .1, .7, and .2 respectively. A short father will have a tall, medium-height, or short son with probability .4, .2, .4 respectively.

(a)   Describe the transition matrix and digraph which corresponds to this situation.

(b)   What is the probability that the great-grandson of a tall man will be short?

(c)   Suppose we don't know the height of a man's great-grandfather, but think he was either tall or of medium-height, with these two possibilities equally likely. What probability should we assign to the man's being tall?

(d)   Suppose we know that 20% of the current male population is tall, 20% short, and the rest medium. What will be the distribution of heights three generations from now?

10.   (Goodman and Ratti [1971].) Company *A* and Company *B* both manufacture and sell very good beer, and they are the only companies which sell beer in a given small town. Each year the market in this town will absorb exactly 4000 barrels of beer. Company *A* currently gets 25% of the sales, while Company *B* gets the rest. Company *A* has an advertising gimmick which it thinks in any given year will increase sales by 1000 barrels of beer with probability $\frac{3}{5}$ and decrease sales by 1000 barrels with probability $\frac{2}{5}$, except that once a company's sales are down to 0, it gives up its business in the town. Define the states for a Markov chain which will describe the situation, and give the transition matrix and digraph. What is the probability that Company *A* will be selling 3000 barrels two years after introducing the advertising gimmick?

11.   (Bhat [1972].) A market survey has described the switching of consumers among three brands of coffee, *A*, *B*, and *C*. The results show that of

those who bought Brand *A* in one month, during the next month 60%
bought *A* again, 30% switched to brand *B*, and 10% switched to brand
C. For brands *B* and *C*, the switching percentages are: 50% from *B* to
A, 30% from *B* to *B*, 20% from *B* to *C*, 40% from *C* to *A*, 40% from *C*
to *B*, and 20% from *C* to *C*. Describe a corresponding transition matrix
and digraph. What is the probability that a person will be drinking the
same brand of coffee two months from now that he drinks now?
(Assume that a consumer is equally likely to be drinking any given
brand now.)

12. In the weather Markov chain of Example 6 of the text of Sec. 5.1, what
    is the probability that it will be dry three days from now given that it is
    wet now?

13. In the electricity Markov chain of Example 9 of the text, what is the
    probability that the weather will be the same two days from now as
    today? (Assume that it is equally likely that today will be hot, moderate,
    or cool.)

14. In the gambling game of Example 4 of the text of Sec. 5.1, suppose the
    game is randomly started in one of the four states, *H*, *T*, even, odd.
    What is the probability that after two trials it is in the state *H*?

15. In the pasture ecosystem of Example 8 of the text, suppose the phos-
    phorus molecule is randomly started in either the soil or the grass.
    What is the probability that it is still in the soil two time periods later?

*16. Show that if *P* is a stochastic matrix, then so is P^t, all $t > 0$.

17. Suppose *p* is a probability vector and *P* is a stochastic matrix. Is *pP* again
    a probability vector? (Give proof or counterexample.)

18. In Markov chains with two states, *P* is given by

$$P = \begin{pmatrix} 1-a & a \\ b & 1-b \end{pmatrix}.$$

If $a \neq 0$ or $b \neq 0$, show that P^t is given by

$$P^t = \begin{pmatrix} \dfrac{b}{a+b} + \dfrac{a(1-a-b)^t}{a+b} & \dfrac{a}{a+b} - \dfrac{a(1-a-b)^t}{a+b} \\ \dfrac{b}{a+b} - \dfrac{b(1-a-b)^t}{a+b} & \dfrac{a}{a+b} + \dfrac{b(1-a-b)^t}{a+b} \end{pmatrix}.$$

19. Prove Theorem 5.2.

20. For Examples 7, 8, 9, 10, or 11 of the text, discuss whether or not you
    think the Markov model is an appropriate model. As tools for making
    predictions for Markov chains are developed in later sections, you
    will want to return to this exercise, and discuss specific ways of testing
    the appropriateness of the Markov models of these examples.

5.3.   Classification of States and Chains

It is useful to divide Markov chains into several classes. Each of these classes
of chains has very special and quite different properties. Yet, as we shall see,
understanding the behavior of specific types of Markov chains will enable us
to understand the behavior of more general chains.

To begin our classification of Markov chains, let us say that a set C of
states is (*stochastically*) *closed* if for all u_i in C and u_j not in C, $p_{ij} = 0$. Thus,
C is closed if once in C, the chain can never leave C. A set C is called *ergodic*
if it is closed but no proper subset is closed. For example, in Fig. 5.11, the set
$\{u_7, u_8, u_{11}\}$ is closed. It is not ergodic, since $\{u_{11}\}$ is also closed. The latter set,
$\{u_{11}\}$, is ergodic.

Ergodic sets can be easily characterized in terms of digraphs. The
reader will recall that a *vertex basis* of a digraph D is a minimal collection B
of vertices such that any vertex u in D is reachable from some vertex of B.
Similarly, a *vertex contrabasis* of D is a minimal set of vertices B' such that
any vertex u of D can reach some vertex of B'. (A vertex contrabasis of D is a
vertex basis in the digraph obtained from D by reversing the directions of all
arcs.)

Suppose T is a transition digraph of a Markov chain and D is the
underlying digraph, i.e., the digraph of T less the weights p_{ij}. In Sec. 2.3 we
discussed the digraph D^*, the condensation (by strong components) of D.
We showed (Theorems 2.7, 2.8, and 2.9 and the Corollary to Theorem 2.8)
that D has a strong component with no incoming arcs and that the collection
of all such strong components forms the unique vertex basis of D^*. Similarly.
it is easy to show that D has a strong component with no outgoing arcs and
the collection of all such strong components forms the unique vertex contrabasis for D^*.

Using these ideas, it is easy to see that the ergodic sets of a Markov
chain are exactly those sets of states which form a strong component of the
digraph D with no outgoing arcs in D, i.e., the strong components which form
a vertex contrabasis in D^*. For example, in the digraph of Fig. 5.11, $\{u_{11}\}$ is
such a strong component. The others are $\{u_1, u_2, u_3\}$ and $\{u_9, u_{10}\}$. Thus, these
are the ergodic sets. All strong components of D other than the ergodic ones
will be called *transient sets*. Since the strong components of a digraph uniquely
partition the vertices (Theorem 2.6), we may define a *state* of a Markov chain
as *ergodic* or *transient* according as its strong component is ergodic or
transient. Thus a state is ergodic if and only if it belongs to an ergodic set.
There is always at least one ergodic set and state. (Why?) Once a Markov
chain is in a state of some ergodic set, it can never leave that ergodic set.
Similarly, once a Markov chain has left a transient set, it can never return,
because the condensation D^* is acyclic (Theorem 2.7). Moreover, since the
ergodic sets form a vertex contrabasis for D^*, each state in a transient set can
reach some ergodic set.

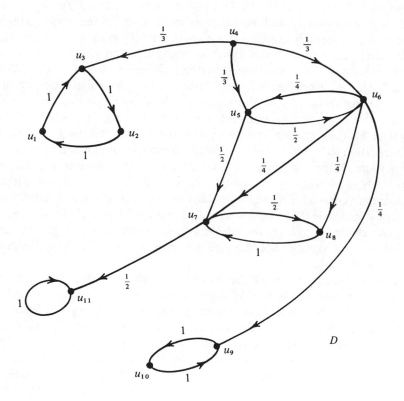

Strong components of D: $K_1 = \{u_1, u_2, u_3\}$, $K_2 = \{u_4\}$,

$K_3 = \{u_5, u_6\}$, $K_4 = \{u_7, u_8\}$, $K_5 = \{u_9, u_{10}\}$, $K_6 = \{u_{11}\}$

Condensation D^*:

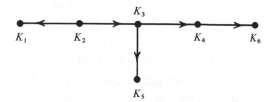

FIGURE 5.11. A transition digraph D, its strong components, and its condensation D^*.

The transition digraphs of Figs. 5.1, 5.3, 5.4, 5.5, 5.7, and 5.8 are all strongly connected and thus themselves form ergodic sets. The transition digraphs of Figs. 5.2, 5.6, and 5.9 are not strongly connected. In Fig. 5.6, the only strong component with no outgoing arcs consists of the single vertex u_4. Hence, $\{u_4\}$ is the only ergodic set and u_4 is the only ergodic state. We shall call such a state which itself forms an ergodic set an *absorbing state*. An absorbing state u_i is characterized by the fact that $p_{ii} = 1$. In Fig. 5.2, the state "dead" is absorbing. In Fig. 5.9, there are two absorbing states, 0 and 4.

Having classified sets and states of a Markov chain, we now classify chains. If a chain has no transient sets, then all its strong components are ergodic. Moreover, they have neither incoming nor outgoing arcs in the condensation D^* (why?) and a process starting in one of the ergodic sets stays there. Hence, we may consider chains on the different ergodic sets separately, and without loss of information, the study of chains without transient sets can be reduced to the study of chains whose set of states forms a single ergodic set. Such chains are called *ergodic*. Thus, a chain is ergodic if and only if it is strongly connected. We study a special class of ergodic chains, called regular chains, in Sec. 5.5, and we generalize the results to all ergodic chains in Sec. 5.6.

It is sufficient to break the study of an arbitrary Markov chain with transient sets into two parts: the stage before it enters an ergodic set and the stage after. For once the chain enters any ergodic set, it can never leave. We can study behavior once in this ergodic set as if we had an ergodic chain. Before the chain enters an ergodic set, we can disregard the structure of the ergodic sets. We can lump all of the states in a given ergodic set into one state, and treat this as an absorbing state. Once the chain enters this state, we can then treat the corresponding ergodic chain. Thus, we can usually reduce study of this first stage in the history of a Markov chain to study of a chain each of whose ergodic sets consists of a single element. Such a chain is called *absorbing*. We study absorbing chains in Sec. 5.4 and apply the theory of absorbing chains to arbitrary Markov chains in Sec. 5.10. Russian roulette of Fig. 5.2, the pasture ecosystem of Fig. 5.6, and the gambler's ruin of Fig. 5.9 are examples of absorbing chains.

EXERCISES

1. For each stochastic digraph of Fig. 5.12, identify all closed sets, all ergodic sets, and all transient sets.

2. For each stochastic digraph of Fig. 5.12, identify all ergodic states, all transient states, and all absorbing states.

3. Of the stochastic digraphs in Fig. 5.13, identify which correspond to ergodic chains, which to absorbing chains, and which to neither.

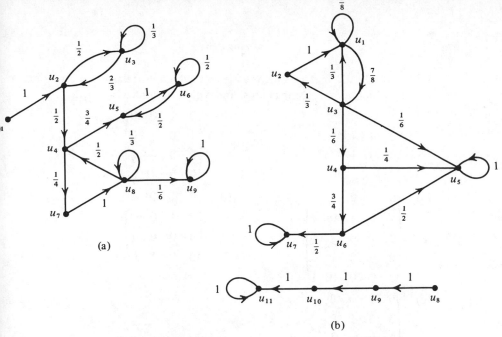

FIGURE 5.12. Stochastic digraphs for Exers. 1 and 2 of Sec. 5.3.

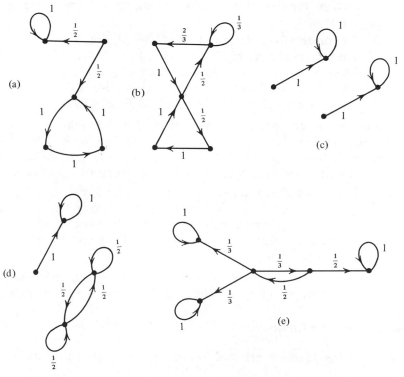

FIGURE 5.13. Stochastic digraphs for Exer. 3, Sec. 5.3, and exercises of Sec. 5.4.

4. Of the stochastic matrices below, identify which correspond to ergodic chains, which to absorbing chains, and which to neither.

(a) $\begin{pmatrix} \frac{1}{2} & \frac{1}{2} & 0 \\ 0 & 0 & 1 \\ 0 & 0 & 1 \end{pmatrix}.$

(b) $\begin{pmatrix} 0 & 1 & 0 & 0 \\ \frac{1}{2} & 0 & \frac{1}{2} & 0 \\ 0 & 0 & \frac{2}{3} & \frac{1}{3} \\ 0 & 0 & 1 & 0 \end{pmatrix}.$

(c) $\begin{pmatrix} 0 & 1 & 0 & 0 & 0 \\ \frac{2}{3} & 0 & \frac{1}{3} & 0 & 0 \\ 0 & \frac{1}{2} & 0 & \frac{1}{4} & \frac{1}{4} \\ 0 & 0 & 0 & 1 & 0 \\ 0 & 0 & 0 & 0 & 1 \end{pmatrix}.$

(d) $\begin{pmatrix} 0 & \frac{1}{2} & 0 & \frac{1}{2} \\ 0 & 1 & 0 & 0 \\ 0 & 0 & 0 & 1 \\ 0 & 0 & 1 & 0 \end{pmatrix}.$

(e) $\begin{pmatrix} 0 & 1 & 0 & 0 \\ 0 & 0 & 1 & 0 \\ \frac{1}{3} & 0 & 0 & \frac{2}{3} \\ 1 & 0 & 0 & 0 \end{pmatrix}.$

5. For each of the following Markov chains, answer the questions of Exers. 1 and 2.
    (a) The supermarket Markov chain of Exer. 4, Sec. 5.1.
    (b) The charge account Markov chain of Exer. 7, Sec. 5.2.
    (c) The Elks, Moose, Lions Markov chain of Exer. 8, Sec. 5.2.
    (d) The inheritance Markov chain of Exer. 9, Sec. 5.2.
    (e) The beer competition Markov chain of Exer. 10, Sec. 5.2.

6. Consider the communication network of Fig. 2.3. Make this into a Markov chain by assuming that if an individual has a message, he will pass it on to exactly one of the individuals (including himself) he can reach directly, and he is equally likely to pass it on to any of them.
    (a) Identify all closed, ergodic, and transient sets.
    (b) Discuss the two stages in which one would study this chain, i.e., identify the absorbing and ergodic chains discussed in the text.
    We shall return to related examples in Exer. 22 of Sec. 5.5 and Exers. 11 and 12 of Sec. 5.10.

7. How can you use the condensation D^* of the digraph D underlying the transition digraph to help in identifying closed sets in a Markov chain?

5.4. Absorbing Chains[4]

In an absorbing Markov chain, we have a number of absorbing states and

---

[4]Our approach in this section and the next follows that of Kemeny and Snell [1960, Ch. 3].

other transient or nonabsorbing states. The reader should prove the following useful characterization of absorbing chains:

Theorem 5.3.   A Markov chain is absorbing if and only if it has at least one absorbing state and from every nonabsorbing state it is possible to reach *some* absorbing state.

Proof.   Exercise 9.

For example, in the gambler's ruin of Fig. 5.9, an alternative proof that this is an absorbing Markov chain is the following. There is an absorbing state, specifically, 4. Every nonabsorbing state can reach an absorbing state. Indeed, here every nonabsorbing state can reach the absorbing state 4. In the Markov chain of Fig. 5.11, there is one absorbing state u_{11}. But u_9 cannot reach u_{11}, so this is not an absorbing Markov chain. In the Markov chain of Fig. 5.12b, state u_6 cannot reach absorbing state u_{11}. But the chain is absorbing. (Why?)

In studying absorbing chains, we shall be interested in the following information:

(1)   The probability of entering a given absorbing state u_j given that the process starts in the nonabsorbing state u_i.

(2)   The expected number of times that the process will be in non-absorbing state u_j before entering some absorbing state, given that the starting state is nonabsorbing state u_i.

(3)   The expected number of steps before the process enters some absorbing state, given that the starting state is nonabsorbing state u_i.

We shall be able to answer all of these questions in terms of the transition matrix P.

Theorem 5.4.   In any (finite) Markov chain, independent of starting state, the probability after t steps that the process is in an ergodic state approaches 1 as t approaches ∞.

Before proving this theorem, we note the following corollary.

Corollary.   In an absorbing Markov chain, the probability of absorption is 1.

Proof of Theorem 5.4.[5]   Let T be the transition digraph and D its underlying digraph. Since the ergodic sets form a vertex contrabasis for the condensation D^*, every transient vertex u_i can reach some ergodic vertex u_j.

[5]As in other chapters, proofs in this chapter may be omitted.

Since there are only finitely many vertices, there is a number r such that for each transient vertex u_i, there is some ergodic vertex u_j with $d(u_i, u_j) \leqq r$, where $d(u_i, u_j)$ is the length of the shortest path in D from u_i to u_j. Thus, independent of starting state, the probability of entering an ergodic vertex in at most r steps is a positive number p. The probability of not reaching an ergodic vertex in at most r steps is $1 - p$. Since the process is Markovian, the probability of not reaching an ergodic vertex in at most the next r steps is again $1 - p$. By such reasoning, we conclude that the probability of not reaching an ergodic vertex in at most kr steps is $(1 - p)^k$. Since p is positive and at most 1, this probability approaches 0. This completes the proof.   Q.E.D.

In an absorbing Markov chain, it is convenient to renumber the states so that the absorbing states come first. Then the transition matrix P takes on a particularly simple form, which we shall call its *canonical form*:

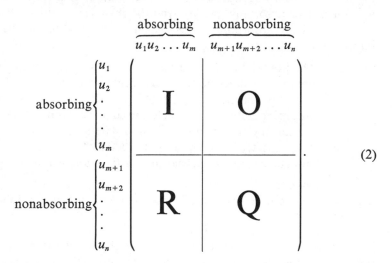

$$\tag{2}$$

Here, I is the $m \times m$ identity matrix and O is the $m \times (n - m)$ matrix all of whose entries are 0. The $(n - m) \times (n - m)$ matrix Q gives the transition probabilities from nonabsorbing states to nonabsorbing states and the $(n - m) \times m$ matrix R gives the transition probabilities from nonabsorbing states to absorbing states. Under this relabelling of states, Russian roulette of Fig. 5.2, the pasture ecosystem of Fig. 5.6, and the gambler's ruin problem of Fig. 5.9 have the following transition matrices P_1, P_2, and P_3:

$$
P_1 = \begin{array}{c} \\ \text{Dead} \\ \text{Alive} \end{array}
\begin{array}{cc} \text{Dead} & \text{Alive} \\ \left(\begin{array}{c|c} 1 & 0 \\ \hline \frac{1}{6} & \frac{5}{6} \end{array} \right) \end{array}
$$

$$
P_2 = \begin{array}{c} \\ u_4 \\ u_1 \\ u_2 \\ u_3 \end{array}
\begin{array}{cccc}
u_4 & u_1 & u_2 & u_3 \\
\left(\begin{array}{c|ccc}
1 & 0 & 0 & 0 \\ \hline
\frac{1}{10} & \frac{3}{5} & \frac{3}{10} & 0 \\
0 & \frac{1}{10} & \frac{2}{5} & \frac{1}{2} \\
\frac{1}{20} & \frac{3}{4} & 0 & \frac{1}{5}
\end{array} \right)
\end{array},
$$

$$
P_3 = \begin{array}{c} \\ 0 \\ 4 \\ 1 \\ 2 \\ 3 \end{array}
\begin{array}{ccccc}
0 & 4 & 1 & 2 & 3 \\
\left(\begin{array}{cc|ccc}
1 & 0 & 0 & 0 & 0 \\
0 & 1 & 0 & 0 & 0 \\ \hline
1-p & 0 & 0 & p & 0 \\
0 & 0 & 1-p & 0 & p \\
0 & p & 0 & 1-p & 0
\end{array} \right)
\end{array}.
$$

At this point, let us make some conventions. In particular, it is convenient to define $M^0 = I$ for M any square matrix and I the identity matrix. Moreover, we say that a sequence of matrices M_t *approaches* or *converges to* a matrix A, and write $M_t \rightarrow A$, if each component of M_t approaches the corresponding component of A as t approaches ∞. For example, suppose

$$
M = \begin{pmatrix} \frac{1}{2} & 0 \\ 0 & 1 \end{pmatrix}
$$

and

$$
M_t = M^t = \begin{pmatrix} (\frac{1}{2})^t & 0 \\ 0 & 1^t \end{pmatrix}.
$$

Since $(\frac{1}{2})^t \rightarrow 0$ and $1^t \rightarrow 1$, it follows that

$$
M_t \longrightarrow \begin{pmatrix} 0 & 0 \\ 0 & 1 \end{pmatrix}.
$$

Similarly, if

$$
M = \begin{pmatrix} 0 & 1 \\ 0 & 1/2 \end{pmatrix},
$$

and $M_t = M^t$, then it is easy to prove that

$$
M_t = \begin{pmatrix} 0 & (1/2)^{t-1} \\ 0 & (1/2)^t \end{pmatrix},
$$

and so

$$
M_t \longrightarrow \begin{pmatrix} 0 & 0 \\ 0 & 0 \end{pmatrix}.
$$

As usual, if the sequence $R_t = \sum_{s=0}^{t} M_s$ converges to a matrix B as $t \to \infty$, we say B is $\sum_{s=0}^{\infty} M_s$. For example, if

$$M = \begin{pmatrix} \frac{1}{2} & 0 \\ 0 & \frac{1}{4} \end{pmatrix}$$

and

$$M_s = M^s = \begin{pmatrix} (\frac{1}{2})^s & 0 \\ 0 & (\frac{1}{4})^s \end{pmatrix},$$

then

$$\sum_{s=0}^{t} M_s = \begin{pmatrix} \sum\limits_{s=0}^{t} (\frac{1}{2})^s & 0 \\ 0 & \sum\limits_{s=0}^{t} (\frac{1}{4})^s \end{pmatrix}$$

and

$$\sum_{s=0}^{\infty} M_s = \begin{pmatrix} 2 & 0 \\ 0 & \frac{4}{3} \end{pmatrix}.$$

Theorem 5.5. In an absorbing Markov chain whose transition matrix has canonical form (2), the following hold:

(a) $Q^t \to 0$.

(b)   The matrix $(I - Q)$ has an inverse.

(c) $(I - Q)^{-1} = I + Q + Q^2 + \cdots = \sum\limits_{s=0}^{\infty} Q^s$.

Proof. Part (a) follows by Theorems 5.1 and 5.4 since Q is the transition matrix for nonabsorbing states. To prove parts (b) and (c), we recall these properties of determinants: $\det(AB) = \det A \cdot \det B$; $\det I = 1$; M has an inverse if and only if $\det M \neq 0$. Note that

$$(I - Q)(I + Q + Q^2 + \cdots + Q^{t-1})$$
$$= (I - Q) + (Q - Q^2) + \cdots + (Q^{t-1} - Q^t).$$

Thus

$$(I - Q)(I + Q + Q^2 + \cdots + Q^{t-1}) = I - Q^t. \tag{3}$$

Now $I - Q^t \to I$, since $Q^t \to 0$. We know that $\det I = 1$. Thus, for sufficiently large t, $\det (I - Q^t) \neq 0$.[6] Since $\det (AB) = \det A \cdot \det B$, it follows from Eq. (3) that $\det (I - Q) \neq 0$, which proves part (b).

Since $(I - Q)^{-1}$ exists, we may multiply both sides of (3) by it, obtaining:

$$I + Q + Q^2 + \cdots + Q^{t-1} = (I - Q)^{-1}(I - Q^t). \tag{4}$$

As $t \to \infty$, the right hand side of (4) approaches $(I - Q)^{-1}$, which proves part (c).   Q.E.D.

[6] We are using the well-known fact that determinant defines a continuous function, and hence that $\det (I - Q^t)$ is very close to $\det I$ for t sufficiently large.

Corollary. Suppose Q is any matrix of real numbers and $Q^t \to 0$. Then the matrix $(I - Q)$ has an inverse and

$$(I - Q)^{-1} = \sum_{s=0}^{\infty} Q^s.$$

Proof. The proof of parts (b) and (c) uses only the fact that $Q^t \to 0$.
Q.E.D.

It should be emphasized that the formula of this corollary is not to be considered a practical tool for calculating $(I - Q)^{-1}$. It is a theoretical tool which we shall make use of later on. The inverse of $I - Q$ should be calculated by traditional methods.

The matrix $N = (I - Q)^{-1}$ will be called the *fundamental matrix* for the absorbing chain. We shall see that it essentially contains all of the information sought at the beginning of this section.

Theorem 5.6. The expected number of times before absorption that an absorbing chain is in nonabsorbing state u_j given that the chain starts in nonabsorbing state u_i is given by the i, j entry of the fundamental matrix N.[7]

Proof. Let e_{ij} be the expectation in question. Let $c_j^{(s)}$ be 1 if the process is in state u_j at time s and 0 otherwise. For any x, let $E_i(x)$ represent the expected value of x given that the process starts in state u_i. Then

$$e_{ij} = E_i\left(\sum_{s=0}^{\infty} c_j^{(s)}\right)$$

$$= \sum_{s=0}^{\infty} E_i(c_j^{(s)})$$

$$= \sum_{s=0}^{\infty} [(1 - p_{ij}^{(s)}) \cdot 0 + p_{ij}^{(s)} \cdot 1]$$

$$= \sum_{s=0}^{\infty} p_{ij}^{(s)}.\text{[8]}$$

Now $p_{ij}^{(s)}$ is the i, j entry of Q^s. Thus, e_{ij} is the i, j entry of $\sum_{s=0}^{\infty} Q^s$, which by Theorem 5.5(c) is N.
Q.E.D.

Corollary. The expected number of steps before absorption, given that the process starts in nonabsorbing state u_i, is given by the sum of the entries in the ith row of N.[9]

---

[7] In this section we make the convention that the i, j entry of a matrix is the entry in the row corresponding to u_i and the column corresponding to u_j. The ith row of a matrix is the row corresponding to u_i.

[8] A technical point: taking $E_i(\sum) = \sum (E_i)$ is justified because we know that $\sum (E_i) = \sum p_{ij}^{(s)}$, which is known to converge.

[9] See footnote 7.

To illustrate this theorem and corollary, we consider Russian roulette of Fig. 5.2, the pasture ecosystem of Fig. 5.6, and the gambler's ruin of Fig. 5.9. For the gambler's ruin, the matrix Q is given by

$$Q = \begin{array}{c} \\ 1 \\ 2 \\ 3 \end{array} \begin{array}{ccc} 1 & 2 & 3 \\ \left(\begin{array}{ccc} 0 & p & 0 \\ 1-p & 0 & p \\ 0 & 1-p & 0 \end{array}\right). \end{array}$$

For the case $p = \frac{1}{3}$, we find

$$I - Q = \left(\begin{array}{ccc} 1 & -\frac{1}{3} & 0 \\ -\frac{2}{3} & 1 & -\frac{1}{3} \\ 0 & -\frac{2}{3} & 1 \end{array}\right)$$

and, by calculation,

$$N = (I - Q)^{-1} = \begin{array}{c} \\ 1 \\ 2 \\ 3 \end{array} \begin{array}{ccc} 1 \quad 2 \quad 3 & & \text{row sum} \\ \left(\begin{array}{ccc} \frac{7}{5} & \frac{3}{5} & \frac{1}{5} \\ \frac{6}{5} & \frac{9}{5} & \frac{3}{5} \\ \frac{4}{5} & \frac{6}{5} & \frac{7}{5} \end{array}\right) & & \begin{array}{c} \frac{11}{5} \\ \frac{18}{5} \\ \frac{17}{5} \end{array} \end{array} . \qquad (5)$$

Thus, the expected number of steps before absorption, given that you start at \$2, is $\frac{18}{5}$. For $\frac{18}{5}$ is the row sum of the second row of the matrix N. The expected number of times you have \$3, given that you start at \$2, is given by $n_{23} \doteq \frac{3}{5}$. We interpret these results as follows. If you were to gamble until absorption a large number of times, starting with \$2 each time, and you were to keep a count of the number of rounds before absorption, it would average out to be between 3 and 4, i.e., $\frac{18}{5}$. Moreover, the number of times you have \$3 before absorption would average out to be less than one; often you will be absorbed without ever having \$3.

In the pasture ecosystem, the matrices Q, $I - Q$, and N are as follows:

$$Q = \left(\begin{array}{ccc} \frac{3}{5} & \frac{3}{10} & 0 \\ \frac{1}{10} & \frac{2}{5} & \frac{1}{2} \\ \frac{3}{4} & 0 & \frac{1}{5} \end{array}\right)$$

$$I - Q = \left(\begin{array}{ccc} \frac{2}{5} & -\frac{3}{10} & 0 \\ -\frac{1}{10} & \frac{3}{5} & -\frac{1}{2} \\ -\frac{3}{4} & 0 & \frac{4}{5} \end{array}\right)$$

$$N = \left(\begin{array}{ccc} 8.6 & 4.3 & 2.7 \\ 8.2 & 5.8 & 3.6 \\ 8.1 & 4.1 & 3.8 \end{array}\right) \begin{array}{c} \text{row sum} \\ \begin{array}{c} 15.6 \\ 17.6 \\ 16.0 \end{array} \end{array} \qquad (6)$$

The expected number of days before the phosphorus leaves the system, given that it starts in the soil, is 15.6. For 15.6 is the row sum of the first row (the row corresponding to u_1 = soil) of N. If similar calculations are done for two alternative pesticides, the information might be used to determine which has shorter expected duration in the system. This information in turn might be used to decide between the two alternatives. (One of the problems with pesticides such as DDT is the extremely long time they stay in the ecosystem after being introduced.)

For Russian roulette, the matrix Q is given by

$$\text{Alive}$$
$$Q = \text{Alive } (\tfrac{5}{6}).$$

Then, $I - Q = (\tfrac{1}{6})$ and $N = (I - Q)^{-1} = (6)$. We conclude that the expected number of trials before death given that you start alive is 6.

In a problem like the gambler's ruin, where there are two absorbing states, we would like to determine the respective probabilities of absorption in the different absorbing states. These probabilities depend on the starting state.

Theorem 5.7.   In an absorbing Markov chain with canonical transition matrix (2), let b_{ij} represent the probability of absorption in absorbing state u_j given that the process starts in nonabsorbing state u_i. If B is the $(n - m) \times m$ matrix (b_{ij}), then

$$B = NR,$$

where R is as in the canonical matrix (2) and N is the fundamental matrix $(I - Q)^{-1}$ of the chain.[10]

Proof.   We begin by obtaining a recursion for b_{ij}. Starting in state u_i, the chain can either be absorbed in absorbing state u_j, absorbed in absorbing state u_k, $k \neq j$, or go to some other nonabsorbing state u_r. These occur with respective probabilities p_{ij}, p_{ik}, and p_{ir}. From u_j, u_k, and u_r, the respective probabilities of absorption in state u_j are 1, 0, and b_{rj}. Thus,

$$b_{ij} = p_{ij} + \sum_{r=m+1}^{n} p_{ir} b_{rj}.$$

Since u_i is nonabsorbing and u_j is absorbing, p_{ij} is the i, j entry of the matrix R.[11] Since u_i is nonabsorbing and u_r is nonabsorbing, p_{ir} is the i, r entry of the matrix Q. It follows that

$$B = R + QB.$$

[10]See footnote 7.

[11]See footnote 7.

Subtracting QB and multiplying by $(I - Q)^{-1}$, we find

$$B = (I - Q)^{-1}R = NR. \qquad\qquad \text{Q.E.D.}$$

In the pasture ecosystem, the matrix R is given by

$$R = \begin{matrix} & u_4 \\ u_1 \\ u_2 \\ u_3 \end{matrix} \begin{pmatrix} \frac{1}{10} \\ 0 \\ \frac{1}{20} \end{pmatrix}.$$

Hence, using the matrix N of Eq. (6), we obtain

$$B = NR = \begin{matrix} & u_4 \\ u_1 \\ u_2 \\ u_3 \end{matrix} \begin{pmatrix} 1 \\ 1 \\ 1 \end{pmatrix}.$$

This is not surprising. Independent of starting state, the probability of being absorbed in the single absorbing state is 1. This is, in fact, the result of the Corollary to Theorem 5.4.

In the gambler's ruin problem with $p = \frac{1}{3}$, R is the matrix

$$R = \begin{matrix} & 0 & \ \ 4 \\ 1 \\ 2 \\ 3 \end{matrix} \begin{pmatrix} \frac{2}{3} & 0 \\ 0 & 0 \\ 0 & \frac{1}{3} \end{pmatrix}.$$

Hence, using N of Eq. (5), we obtain

$$B = \begin{matrix} & 0 & \ \ 4 \\ 1 \\ 2 \\ 3 \end{matrix} \begin{pmatrix} \frac{14}{15} & \frac{1}{15} \\ \frac{4}{5} & \frac{1}{5} \\ \frac{8}{15} & \frac{7}{15} \end{pmatrix}.$$

The probability of being absorbed in state 0, i.e., being ruined, varies depending on the starting state. It is $\frac{4}{5}$ if you start at state 2, and even if you start at state 3, it is greater than $\frac{1}{2}$, namely, $\frac{8}{15}$. Naturally, the closer the gambler starts to 0, the higher probability he will have of being ruined. The reader will observe that $b_{i0} + b_{i4} = 1$, all i. Again, this illustrates the Corollary to Theorem 5.4.

Finally, in Russian roulette, $R = (\frac{1}{6})$ and $B = NR = (1)$. The probability of ending up dead is, sadly, 1.

EXERCISES

1. Use Theorem 5.3 to identify which of the Markov chains of Fig. 5.13 are absorbing.

2. For each of the following transition matrices, use Theorem 5.3 to determine if it comes from an absorbing chain. Use the definition of absorbing chain as a check.

(a)
$$P = \begin{pmatrix} 1 & 0 & 0 & 0 \\ 0 & 1 & 0 & 0 \\ 0 & 0 & \frac{1}{2} & \frac{1}{2} \\ 0 & 0 & \frac{1}{2} & \frac{1}{2} \end{pmatrix}.$$

(b)
$$P = \begin{pmatrix} \frac{1}{3} & \frac{1}{3} & \frac{1}{3} \\ 1 & 0 & 0 \\ 0 & \frac{1}{2} & \frac{1}{2} \end{pmatrix}.$$

(c)
$$P = \begin{pmatrix} 1 & 0 & 0 & 0 & 0 \\ 0 & 1 & 0 & 0 & 0 \\ 0 & 0 & \frac{1}{2} & 0 & \frac{1}{2} \\ \frac{1}{3} & 0 & 0 & 0 & \frac{2}{3} \\ \frac{2}{3} & 0 & 0 & 0 & \frac{1}{3} \end{pmatrix}.$$

3. Put the following transition matrices for absorbing Markov chains into canonical form and find Q and R.

(a)
$$P = \begin{pmatrix} 0 & 0 & \frac{1}{3} & \frac{2}{3} \\ 0 & 1 & 0 & 0 \\ 0 & 0 & 1 & 0 \\ \frac{1}{2} & \frac{1}{4} & \frac{1}{4} & 0 \end{pmatrix}.$$

(b)
$$P = \begin{pmatrix} 0 & 0 & 1 \\ \frac{1}{3} & \frac{1}{3} & \frac{1}{3} \\ 0 & 0 & 1 \end{pmatrix}.$$

(c)
$$P = \begin{pmatrix} 1 & 0 \\ \frac{1}{5} & \frac{4}{5} \end{pmatrix}.$$

(d)
$$P = \begin{pmatrix} 1 & 0 & 0 & 0 & 0 \\ \frac{1}{5} & \frac{1}{5} & \frac{1}{5} & \frac{1}{5} & \frac{1}{5} \\ 1 & 0 & 0 & 0 & 0 \\ \frac{1}{5} & \frac{1}{5} & \frac{1}{5} & \frac{1}{5} & \frac{1}{5} \\ 0 & 0 & 0 & 0 & 1 \end{pmatrix}.$$

4. For the water waste Markov chain of Exer. 8, Sec. 5.1, determine the probability that a molecule of mercury will eventually be swept out to sea and the expected number of days before this happens.

5. In the advertising situation described in Exer. 10 of Sec. 5.2, what is the probability that Company A will go out of business? What is the expected number of years before the advertising campaign will either drive Company A or Company B out of business?

6. In the charge account situation described in Exer. 7, Sec. 5.2, suppose a

charge account shows a new positive balance. What is the probability that it will eventually become a bad debt?

7. In the Land of Wet, there are three types of weather, rain R, nice N, and snow S. The transition probabilities are given by the stochastic digraph of Fig. 5.14.

(a) Given that it is now snowing, find the expected number of nice days before it starts to rain perpetually.

(b) Given that it is now snowing, find the expected number of days before it starts to rain perpetually.

(c) Given that it is now nice, what is the probability that it will eventually rain?

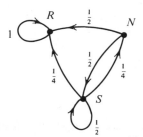

FIGURE 5.14. Transition digraph for the Land of Wet (Exer. 7, Sec. 5.4)

8. (Kemeny, Snell, and Thompson [1966].) An analysis of a recent hockey game between Dartmouth and Princeton gave rise to the following data: If the puck was in the center (C), the probabilities that it next entered Princeton territory (P) or Dartmouth territory (D) were .4 and .6 respectively. From Dartmouth territory, the puck was sent back to the center with probability .95 or into the Dartmouth goal (DG) with probability .05—this means Princeton scored a goal. From Princeton territory, the puck went back to the center with probability .9 and to Princeton's goal (PG) with probability .1—Dartmouth scored a goal. Assume that the puck begins in the center. Find the transition matrix for the 5-state Markov chain which describes action until a goal is scored. Calculate the probability that Dartmouth will score first.

9. Prove Theorem 5.3.

10. Give an example of a Markov chain which has an absorbing state but is not an absorbing chain.

11. In each of the following, find a matrix A such that $M^t \rightarrow A$.

(a)
$$M = \begin{pmatrix} 1 & 0 & 0 \\ 0 & 1 & 0 \\ 0 & 0 & 1 \end{pmatrix}.$$

(b)
$$M = \begin{pmatrix} \frac{1}{2} & 0 \\ 0 & \frac{1}{3} \end{pmatrix}.$$

(c)
$$M = \begin{pmatrix} \frac{1}{2} & 0 & 0 \\ 0 & \frac{1}{2} & 0 \\ 0 & 0 & 1 \end{pmatrix}.$$

(d)
$$M = \begin{pmatrix} -\frac{1}{2} & 0 \\ 0 & 0 \end{pmatrix}.$$

(e)
$$M = \begin{pmatrix} -\frac{1}{2} & 0 & 0 \\ 0 & \frac{1}{2} & 0 \\ 0 & 0 & 1 \end{pmatrix}.$$

(f)
$$M = \begin{pmatrix} \frac{1}{2} & \frac{1}{2} \\ 0 & 0 \end{pmatrix}.$$

(g)
$$M = \begin{pmatrix} \frac{1}{2} & \frac{1}{2} \\ 0 & \frac{1}{2} \end{pmatrix}.$$

12. In each of the following find a matrix A such that $\sum_{s=0}^{t} M^s \longrightarrow A$.

   (a)
   $$M = \begin{pmatrix} \frac{1}{2} & 0 \\ 0 & 0 \end{pmatrix}.$$
   (b)
   $$M = \begin{pmatrix} \frac{1}{2} & 0 \\ 0 & \frac{1}{2} \end{pmatrix}.$$
   (c)
   $$M = \begin{pmatrix} \frac{1}{2} & 0 \\ 0 & \frac{1}{3} \end{pmatrix}.$$

   (d)
   $$M = \begin{pmatrix} \frac{1}{2} & \frac{1}{2} \\ 0 & 0 \end{pmatrix}.$$
   (e)
   $$M = \begin{pmatrix} \frac{1}{2} & \frac{1}{2} \\ 0 & \frac{1}{2} \end{pmatrix}.$$

13. In the pasture ecosystem, calculate $\sum_{s=0}^{t} Q^s$ for $t = 1, 2,$ and 3. Does the result seem to be converging to N? Why?

14. Let

$$Q = \begin{pmatrix} \frac{1}{2} & \frac{1}{2} \\ \frac{1}{2} & \frac{1}{2} \end{pmatrix}.$$

Calculate $\sum_{s=0}^{t} Q^s$ for $t = 1, 2,$ and 3. Does the sequence of sums appear to be approaching a limit? Give an explanation.

15. Prove that $QN = NQ$.

16. By Exer. 13, Sec. 4.7, if A is the adjacency matrix of a weighted digraph and all the eigenvalues of A have magnitude less than 1, then $A^t \longrightarrow 0$. For such a digraph, let $V(t)$ be the vector of values at time t under an autonomous pulse process, to use the terminology and notation of Chapter 4. Show that $V(t) \longrightarrow V(\text{start}) + P(0)(I - A)^{-1}$, where $P(0)$ is the vector of initial pulses.

17. (Kemeny and Snell [1960].) Suppose P is the transition matrix of an absorbing Markov chain. Suppose B^* is the $n \times n$ matrix whose i, j entry (b_{ij}^*) is the probability of being absorbed in state u_j given that the process starts in state u_i. Thus, if u_i is an absorbing state, then b_{ij}^* is 1 if $i = j$ and 0 otherwise. Prove that $PB^* = B^*$.

5.5. Regular Chains

5.5.1. DEFINITION OF REGULAR CHAIN

   The reader will recall that a Markov chain is called ergodic if the underlying digraph of its transition digraph is strongly connected, that is, every state can reach every other state. If there is a number k such that every state can reach every other state in exactly k steps, then the Markov chain will be called *regular*. Not every ergodic Markov chain is regular. For example, consider the ergodic Markov chain of Fig. 5.15. State u_1 can reach state u_2

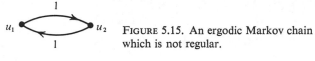

FIGURE 5.15. An ergodic Markov chain which is not regular.

only in an odd number of steps. It can reach u_1 only in an even number of steps. Thus, the chain is not regular. In a regular chain, some power k of the transition matrix P has all its entries positive. Moreover, once P^k has all entries positive, so does P^t for all $t > k$. (Why?) Thus, independent of starting state, there is a positive probability of being in any given state on any trial t after trial k. For, by Theorem 5.2, $p^{(t)} = p^{(0)}P^t$, and $p^{(t)}$ has all positive entries since P^t does and the entries of $p^{(0)}$ add to 1.[12] The ergodic chain of Fig. 5.15 does not have this property. For if we start in state u_1, then the probability of being in state u_1 on any odd trial is 0.

We have already encountered many examples of regular chains. The stochastic digraphs of Figs. 5.1, 5.3, 5.4, 5.5, 5.7, and 5.8 define regular Markov chains. In Figs. 5.3, 5.4, 5.5, 5.7 and 5.8, every vertex can reach every other vertex (including itself) in one step. (In Fig. 5.5, this uses the assumption $0 < p < 1$.) In Fig. 5.1, H cannot reach even in one step. But every vertex can reach every other vertex in two steps. For example, H can reach even in two steps via T, H can reach T in two steps by going to itself first and then to T, and so on. As a check, we note that P^2 has all entries positive, for it turns out that

$$P^2 = \begin{pmatrix} \frac{1}{4} & \frac{1}{4} & \frac{1}{4} & \frac{1}{4} \\ \frac{1}{4} & \frac{1}{4} & \frac{1}{4} & \frac{1}{4} \\ \frac{1}{4} & \frac{1}{4} & \frac{1}{4} & \frac{1}{4} \\ \frac{1}{4} & \frac{1}{4} & \frac{1}{4} & \frac{1}{4} \end{pmatrix}.$$

There are several ways of proving a Markov chain is regular. One way is to use the transition digraph and return to the initial definition. Another is to show that some power of P has all entries positive. Finally, if one knows the chain is ergodic, we shall show in the next section that it must be regular so long as any one diagonal entry p_{ii} of P is positive.

This latter result suggests that many ergodic chains will be regular. Indeed, a good many of the ergodic chains one encounters in practice are

[12]The reader will recall that $p^{(t)} = (p_1^{(t)}, p_2^{(t)}, \ldots, p_n^{(t)})$, where $p_i^{(t)}$ is the probability of being in state i at time t.

regular. Moreover, most of the theorems about regular chains generalize in some form to ergodic chains. Thus, since regular chains are easier to work with, we concentrate on them first.

For regular Markov chains, there are no absorbing states and every state can reach every other state. The questions we ask are quite different from those about absorbing chains. Some of the questions of most interest are the following:

(1) If the chain starts in state u_i, what is the probability after t steps that the chain will be in state u_j, and how does this probability depend on the starting state as t gets large?
(2) What is the expected number of steps before returning to a particular state?

5.5.2. FIXED POINT PROBABILITY VECTORS

Let us consider the weather Markov chain of Fig. 5.7. By Theorem 5.1 the successive powers P^t of the transition matrix P give the t-step transition probabilities. We calculate the powers P, P^2, P^3, P^4, obtaining:[13]

$$P = \begin{pmatrix} \frac{1}{3} & \frac{1}{2} & \frac{1}{6} \\ \frac{1}{2} & \frac{1}{3} & \frac{1}{6} \\ \frac{1}{3} & \frac{1}{3} & \frac{1}{3} \end{pmatrix} = \begin{pmatrix} .3333 & .5000 & .1667 \\ .5000 & .3333 & .1667 \\ .3333 & .3333 & .3333 \end{pmatrix}$$

$$P^2 = \begin{pmatrix} .4167 & .3889 & .1944 \\ .3889 & .4167 & .1944 \\ .3889 & .3889 & .2222 \end{pmatrix}$$

$$P^3 = \begin{pmatrix} .3982 & .4028 & .1991 \\ .4028 & .3982 & .1991 \\ .3982 & .3982 & .2037 \end{pmatrix}$$

$$P^4 = \begin{pmatrix} .4005 & .3997 & .1999 \\ .3997 & .4005 & .1999 \\ .3997 & .3997 & .2006 \end{pmatrix}.$$

The rows each seem to be approaching the same vector $w = (.4, .4, .2)$. This observation is supported by calculating higher powers of P. For example,

$$P^8 = \begin{pmatrix} .4001 & .4001 & .2001 \\ .4001 & .4001 & .2001 \\ .4000 & .4000 & .2000 \end{pmatrix}.$$

The following theorem says that this is not an isolated example.

[13]The rows don't add up to 1 because of round-off error.

Theorem 5.8.   Suppose P is the transition matrix of a regular Markov chain. Then

   (a) P^t approaches a stochastic matrix W as t approaches ∞.
   (b)   Each row of W is the same probability vector $w = (w_1, w_2, \ldots, w_n)$.
   (c)   The components of w are all positive.

We defer the proof of Theorem 5.8 to Sec. 5.5.4.

It turns out that it is easy to calculate the vector w in practice. The next theorem tells how, and also tells us several other important facts about regular Markov chains.

Theorem 5.9.   Suppose P is the transition matrix of a regular Markov chain and W and w are as in the previous theorem. Then

   (a)   For every probability vector p, pP^t approaches w as t approaches ∞.
   (b) w is the unique probability vector such that $wP = w$.
   (c) $PW = WP = W$.

Before proving Theorem 5.9, let us discuss it. Suppose we apply the first conclusion to the probability vector $p^{(0)}$, the vector of initial probabilities. Then $p^{(0)}P^t$ is $p^{(t)}$, the vector giving the probability of being in each state at time t. Theorem 5.9(a) says that if t is sufficiently large, the probability of being in state u_i at time t is close to w_i, *independent of starting state*. Part (b) of the theorem tells us how to calculate the vector w in practice. The equation $wP = w$ gives rise to n equations in n unknowns, the w_i. The equations are

$$wP_i = w_i, \tag{7}$$

where P_i is the ith column of P. These equations in general have many solutions. However, if we also add the condition that w is a probability vector, in particular that

$$\sum w_i = 1, \tag{8}$$

then the equations (7) and (8) have a unique solution $w = (w_1, w_2, \ldots, w_n)$. (Indeed, the solution is unique using Eq. (8) and any $n - 1$ of the equations (7).) The vector w is usually called the *stationary vector* or the *fixed point probability vector* of the matrix P. (The latter terminology arises because w is thought of as a fixed point of the transformation P.)[14]

---

[14]Theorem 5.9 shows that every regular Markov chain has a unique fixed point probability vector. It turns out that all Markov chains have fixed point probability vectors and, moreover, that some nonregular chains even have unique fixed point probability vectors. Thus, *just because a Markov chain has a unique fixed point probability vector does not imply that the chain is regular.*

(To use the terminology of Sec. 4.2, w is an eigenvector of P with 1 as an eigenvalue. Thus, 1 is an eigenvalue of P and there is only one eigenvector w with corresponding eigenvalue 1 and sum of its entries 1. In fact, one can prove that for regular chains, 1 is an eigenvalue of multiplicity 1, from which uniqueness of the probability vector w follows. For more details on the eigenvalue approach to Markov chains, some of which is a special case of the results in Chapter 4, we refer the reader to Romanovsky [1970], Cox and Miller [1965] or Karlin [1969a]. We also discuss this approach in more detail in Exers. 16 through 19 below and in Exer. 8 of Sec. 5.6.)

Let us illustrate the application of Theorem 5.9 by referring to the examples we have been discussing. In the case of Fig. 5.5, we have

$$P = \begin{pmatrix} 1-p & p \\ p & 1-p \end{pmatrix}, \qquad (0 < p < 1).$$

Then $(w_1, w_2)P = (w_1, w_2)$ gives rise to the equations

$$w_1(1 - p) + w_2 p = w_1 \tag{9}$$

$$w_1 p + w_2(1 - p) = w_2. \tag{10}$$

These two equations are redundant: Since $p > 0$, they are both equivalent to the equation $w_1 = w_2$. However, we also use the equation

$$w_1 + w_2 = 1. \tag{11}$$

Using Eqs. (9) and (11), and using $p > 0$, we find a unique solution: $w_1 = \frac{1}{2}$, $w_2 = \frac{1}{2}$. The interpretation of this result is as follows: Independent of the politician's initial opinion, and no matter how small p is, in the long run there is probability $\frac{1}{2}$ of hearing he is favorable to the given bill and probability $\frac{1}{2}$ of hearing he is unfavorable! (Does this explain the conflicting reports we often hear about public figures' plans?)

Let us perform a similar analysis for the summer weather of Fig. 5.7. Here,

$$P = \begin{pmatrix} \frac{1}{3} & \frac{1}{2} & \frac{1}{6} \\ \frac{1}{2} & \frac{1}{3} & \frac{1}{6} \\ \frac{1}{3} & \frac{1}{3} & \frac{1}{3} \end{pmatrix}.$$

The equation $(w_1, w_2, w_3)P = (w_1, w_2, w_3)$ gives rise to the equations

$$\tfrac{1}{3}w_1 + \tfrac{1}{2}w_2 + \tfrac{1}{3}w_3 = w_1$$

$$\tfrac{1}{2}w_1 + \tfrac{1}{3}w_2 + \tfrac{1}{3}w_3 = w_2$$

$$\tfrac{1}{6}w_1 + \tfrac{1}{6}w_2 + \tfrac{1}{3}w_3 = w_3.$$

Dropping the third equation and using instead the equation

$$w_1 + w_2 + w_3 = 1,$$

we obtain $w_1 = \frac{2}{5}$, $w_2 = \frac{2}{5}$, $w_3 = \frac{1}{5}$. These results agree with our observation earlier that P^t appears to be approaching a matrix each of whose rows is the vector (.4, .4, .2). Thus, independent of the weather today, in the long run we would predict that it will be hot with probability $\frac{2}{5}$, moderate with probability $\frac{2}{5}$, and cool with probability $\frac{1}{5}$. We can expect $\frac{2}{5}$ of the days to be hot, $\frac{2}{5}$ moderate, and $\frac{1}{5}$ cool.

Similar calculations for the examples in Figs. 5.1, 5.3, 5.4, and 5.8 give the respective fixed point probability vectors

$$(\tfrac{1}{4}, \tfrac{1}{4}, \tfrac{1}{4}, \tfrac{1}{4})$$

$$(\tfrac{1}{2}, \tfrac{1}{2})$$

$$(.575, .425)$$

and

$$(.067, .624, .309).$$

Interpretation of these results is left to the reader.

We now present a proof of Theorem 5.9. To prove part (a), note that $pW = w$, since p is a probability vector. By Theorem 5.8(a), $pP^t \longrightarrow pW = w$. To prove part (c), note that $P^{t+1} \longrightarrow W$. However, $P^{t+1} = PP^t \longrightarrow PW$ and $P^{t+1} = P^tP \longrightarrow WP$. Thus, $PW = WP = W$. Finally, to prove part (b), note that w is a probability vector, by Theorem 5.8(b). To show that $wP = w$, use part (c). To show that w is unique, suppose w and w' are both probability vectors and $wP = w$, $w'P = w'$. By part (a), $w'P^t \longrightarrow w$. But $w'P = w'$, so $w'P^t = w'$. Hence, $w'P^t \longrightarrow w'$, and $w = w'$. This completes the proof.

5.5.3. Mean First Passage[15]

We have now been able to answer questions about the long-run probability of being in a particular state of a regular Markov chain. We have seen that, unlike in the case of absorbing chains, long-term behavior in a regular chain becomes independent of starting state. We next answer questions about expected number of steps before returning to a particular state. This expected number is called the *mean first passage time*. We will calculate it by introducing a fundamental matrix Z for regular chains.

Theorem 5.10. Suppose P is the transition matrix of a regular Markov chain and W is its limiting matrix. Then

$$Z = [I - (P - W)]^{-1} \tag{12}$$

exists and

$$Z = I + \sum_{s=1}^{\infty} (P^s - W). \tag{13}$$

---

[15]This subsection may be omitted without loss of continuity.

Proof. Consider the matrix $Q = (P - W)$. We shall prove that for $t > 0$, $(P - W)^t = P^t - W$. From this $Q^t \rightarrow 0$ follows. The Corollary to Theorem 5.5 then implies the desired result.

By Theorem 5.9(c), $P^t W = W P^t = W$, all $t > 0$. Moreover, it is easy to show that $W^2 = W$. Thus, arguing by induction, we have

$$
\begin{aligned}
(P - W)^{t+1} &= (P - W)(P - W)^t \\
&= (P - W)(P^t - W) \\
&= P^{t+1} - W P^t - P W + W^2 \\
&= P^{t+1} - W - W + W \\
&= P^{t+1} - W.
\end{aligned}
$$

Q.E.D.

If P is the transition matrix of a regular chain, the matrix $Z = [I - (P - W)]^{-1}$ is called the *fundamental matrix* of the chain.

Let $E = (e_{ij})$ be a matrix where for $i \neq j$, e_{ij} is the expected number of steps before the process enters state u_j for the first time given that the process starts in state u_i and for $i = j$, e_{ij} is the expected number of steps before the chain re-enters u_i. E is called the *mean first passage matrix*.

Theorem 5.11. The mean first passage matrix E for a regular Markov chain with transition matrix P is given by

$$ E = (I - Z + J Z_{dg}) D, \tag{14} $$

where Z is the fundamental matrix, Z_{dg} is the diagonal matrix whose diagonal entries are the same as those of Z, J is a matrix of all 1's, and D is a diagonal matrix with $d_{ii} = 1/w_i$, i.e., 1 divided by the ith component of the fixed point probability vector of P.

The proof of this theorem is omitted. The reader is referred to Kemeny and Snell [1960, Sec. 4.4] for details. We shall state a corollary, whose proof is straightforward from Eq. (14).

Corollary. In a regular Markov chain, the mean first passage time from a state u_i to itself is given by

$$ e_{ii} = \frac{1}{w_i}. $$

In the case of the summer weather (Fig. 5.7), this Corollary says that on the average, the time between two hot days is $1/\frac{2}{5} = \frac{5}{2}$, i.e., $2\frac{1}{2}$ days. The time between two cool days is $1/\frac{1}{5} = 5$ days. More generally, the matrices Z and E are shown below:

$$Z = \begin{pmatrix} .95 & .09 & -.04 \\ .09 & .95 & -.04 \\ -.08 & -.08 & 1.16 \end{pmatrix}$$

$$E = \begin{pmatrix} 2.50 & 2.15 & 6.00 \\ 2.15 & 2.50 & 6.00 \\ 2.58 & 2.58 & 5.00 \end{pmatrix}.$$

5.5.4. Proof of Theorem 5.8[16]

We close this section by proving Theorem 5.8. We need the following preliminary lemma.

Lemma. Suppose P is an $n \times n$ stochastic matrix whose smallest entry is ϵ and let

$$x = \begin{pmatrix} x_1 \\ x_2 \\ \cdot \\ \cdot \\ \cdot \\ x_n \end{pmatrix}$$

be any n-dimensional column vector.[17] Suppose m_0 is the smallest x_i and M_0 the largest, and similarly m_1 the smallest component of the vector Px and M_1 the largest. Then $M_1 \leq M_0$, $m_1 \geq m_0$, and

$$M_1 - m_1 \leq (1 - 2\epsilon)(M_0 - m_0).$$

Proof. Define x' from x by replacing all components except one m_0 by M_0. Then $x \leq x'$, i.e., each component of x is $\leq$ the corresponding component of x'. Now P has row sums equal to 1, so each component of Px' is of the form $am_0 + (1 - a)M_0 = M_0 - a(M_0 - m_0)$, where $a \geq \epsilon$. Thus, each component of Px' is $\leq M_0 - \epsilon(M_0 - m_0)$. Since $x \leq x'$, we have each component of $Px \leq$ each component of Px', and we conclude that

$$M_1 \leq M_0 - \epsilon(M_0 - m_0). \tag{15}$$

Since $M_0 - m_0 \geq 0$ and $\epsilon \geq 0$, we conclude that $M_1 \leq M_0$. Applying the same to the vector $-x$, we get

$$-m_1 \leq -m_0 - \epsilon(-m_0 + M_0), \tag{16}$$

---

[16]The reader who has had little experience with proofs using limits and convergence should skip this subsection.

[17]Note that ϵ is not necessarily positive.

so $m_1 \geqq m_0$. Adding (15) and (16), we find that

$$M_1 - m_1 \leqq M_0 - m_0 - 2\epsilon(M_0 - m_0) = (1 - 2\epsilon)(M_0 - m_0).$$

Q.E.D.

To prove Theorem 5.8, note first that if P is 1×1, the results are trivial. Thus, suppose P is $n \times n$, $n \geqq 2$. Let us begin by assuming that each entry of P is positive and let ϵ be the smallest entry. Since $n \geqq 2$, it follows that $0 < \epsilon \leqq \frac{1}{2}$. Let e_j be the column vector

$$\begin{pmatrix} 0 \\ 0 \\ \cdot \\ \cdot \\ \cdot \\ 1 \\ 0 \\ \cdot \\ \cdot \\ \cdot \\ 0 \end{pmatrix}$$

with a 1 in the jth row. Let $P^t e_j$ have maximum component M_t and minimum component m_t. Since $P^t e_j = PP^{t-1}e_j$, the lemma implies that

$$M_1 \geqq M_2 \geqq M_3 \geqq \cdots$$
$$m_1 \leqq m_2 \leqq m_3 \leqq \cdots$$

and

$$M_t - m_t \leqq (1 - 2\epsilon)(M_{t-1} - m_{t-1}).$$

Letting $d_t = M_t - m_t$, we have

$$d_t \leqq (1 - 2\epsilon)^{t-1}d_1.$$

Since $0 < \epsilon \leqq \frac{1}{2}$, it follows that $d_t \rightarrow 0$ as $t \rightarrow \infty$. Thus, $M_t - m_t \rightarrow 0$ and so $P^t e_j$ tends to a vector all of whose components are the same, say w_j. Since $P^t e_j$ is the jth column of P^t, we conclude that P^t approaches a matrix W of the form

$$W = \begin{pmatrix} w_1 & w_2 & \cdots & w_n \\ w_1 & w_2 & \cdots & w_n \\ & & \cdots & \\ w_1 & w_2 & \cdots & w_n \end{pmatrix}.$$

The row sums of P^t are 1, and it follows that $\sum_{i=1}^{n} w_i = 1$. Since the m_t's are nondecreasing and the M_t's are nonincreasing, we conclude that for every t, $0 < m_1 \leqq m_t \leqq w_j \leqq M_t \leqq M_1 < 1$. Thus, $0 < w_j < 1$. This completes the proof of Theorem 5.8 in the case that entries of P are positive.

Suppose now that P is $n \times n$, $n \geq 2$, but its smallest entry is $\epsilon = 0$. By the lemma,

$$d_t \leq (1 - 2\epsilon)d_{t-1},$$

so $d_t \leq d_{t-1}$. Thus, the sequence d_t is nonincreasing. Since P is regular, there is a number k such that P^k has all entries positive. Let ϵ' be the smallest entry of P^k. Applying the first part of the proof to P^k, we conclude that

$$d_{rk} \leq (1 - 2\epsilon')^{r-1}d_k.$$

Thus, the subsequence $d_k, d_{2k}, d_{3k}, \ldots$ approaches 0. Since the sequence d_t is nonincreasing, it follows that d_t also approaches 0. The rest of the proof is similar to the proof for P positive.   Q.E.D.

EXERCISES

1. Which of the following stochastic matrices are regular?

(a) $\begin{pmatrix} \frac{1}{2} & \frac{1}{2} & 0 \\ 0 & \frac{1}{2} & \frac{1}{2} \\ \frac{1}{2} & \frac{1}{2} & 0 \end{pmatrix}$
(b) $\begin{pmatrix} 0 & \frac{1}{3} & \frac{2}{3} \\ 0 & 1 & 0 \\ \frac{1}{3} & \frac{2}{3} & 0 \end{pmatrix}$

(c) $\begin{pmatrix} 0 & 1 & 0 \\ 0 & 0 & 1 \\ 1 & 0 & 0 \end{pmatrix}$
(d) $\begin{pmatrix} 0 & 1 & 0 \\ \frac{1}{2} & 0 & \frac{1}{2} \\ 0 & 1 & 0 \end{pmatrix}.$

2. Which of the stochastic digraphs of Fig. 5.16 are regular?

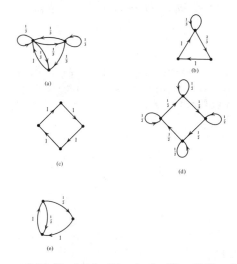

FIGURE 5.16. Stochastic digraphs for Exer. 2, Sec. 5.5.

3. Find the unique fixed point probability vector for the following matrices.

   (a)
   $$P = \begin{pmatrix} \frac{2}{3} & \frac{1}{3} \\ \frac{1}{3} & \frac{2}{3} \end{pmatrix}.$$

   (b)
   $$P = \begin{pmatrix} \frac{1}{2} & \frac{1}{2} \\ 1 & 0 \end{pmatrix}.$$

   (c)
   $$P = \begin{pmatrix} \frac{1}{2} & \frac{1}{2} & 0 \\ \frac{1}{3} & \frac{1}{3} & \frac{1}{3} \\ 0 & \frac{1}{2} & \frac{1}{2} \end{pmatrix}.$$

4. Let
   $$P = \begin{pmatrix} .1 & .9 \\ .5 & .5 \end{pmatrix}.$$

   (a) Compute P^2, P^3, and P^4 and guess at w.
   (b) Check your answer by computing w.
   (c) Compute W.
   (d) Check that $PW = WP = W$.

5. In the supermarket Markov chain of Exer. 4, Sec. 5.1, what is the long-run probability that the customer will be shopping Thrifty?

6. In the water supply Markov chain of Exer. 7 of Sec. 5.1, what is the long-run probability that the reservoir will be full at the beginning of the summer?

7. Referring to the Markov chain of Exer. 8, Sec. 5.2, suppose we pick out an individual at random whose great-grandfather was an Elk. How would you calculate the probability that he is also an Elk? Does this probability change if his great-grandfather was a Moose? What happens if we replace the great-grandfather with an ancestor who goes back sufficiently many generations? What does Markov chain theory say the ultimate proportion of Elks will be in the male population?

8. Referring to the marketing example of Exer. 11, Sec. 5.2, suppose a given individual starts out drinking brand A coffee. If we ask him many years from now, what is an estimate of the probability he will be drinking brand A coffee at the time we ask him?

9. Referring to the inheritance assumptions of Exer. 9, Sec. 5.2, what can you predict about the future proportions of tall, medium, and short men in the population if the current proportions are 20% tall, 20% short and 60% medium?

10. In the water supply situation of Exer. 7, Sec. 5.1, suppose the reservoir is observed to be full in a given year. On the average, how many years will it be before it is full again?

11. In the Tel Aviv weather (Example 6 of the text of Sec. 5.1 and Fig. 5.4), suppose the weather is observed to be dry on a given day. On the

average, how many days will it be before the weather will be dry again?
How many days will it be before the weather will be wet?

12. Is it possible for a Markov chain to be both regular and absorbing?
Why?

13. Does the matrix

$$P = \begin{pmatrix} 0 & 1 \\ 1 & 0 \end{pmatrix}$$

have a unique fixed point probability vector w? Does P^t approach W?
Use these conclusions to determine if P is regular.

14. Show that

$$P = \begin{pmatrix} 1 & 0 \\ 0 & 1 \end{pmatrix}$$

has more than one fixed point probability vector. *Use this conclusion*
to determine if P is regular.

15. If a stochastic matrix P has the property that every column of P is a
probability vector, P is called *doubly stochastic*.
  *(a) Show that if the transition matrix of a regular Markov chain is
doubly stochastic, then all components of w are equal.
  (b) If P is the transition matrix of a regular Markov chain and all
components of w are equal, is P necessarily doubly stochastic?
(Give proof or counterexample.)

16. Suppose P is the transition matrix of a 2-state regular Markov chain.
We may write

$$P = \begin{pmatrix} 1-a & a \\ b & 1-b \end{pmatrix}.$$

  (a) Prove that 1 and $1 - a - b$ are the eigenvalues of P.
  (b) Assume that $a + b \neq 0$. Since P then has distinct eigenvalues,
a well-known theorem in linear algebra implies that P is similar to
a diagonal matrix, i.e., there is a matrix Q so that

$$P = Q\begin{pmatrix} 1 & 0 \\ 0 & 1-a-b \end{pmatrix}Q^{-1}.$$

Show that the matrix

$$Q = \begin{pmatrix} 1 & a \\ 1 & -b \end{pmatrix}$$

satisfies this condition and calculate Q^{-1}.
  (c) Conclude that

$$P^t = Q\begin{pmatrix} 1 & 0 \\ 0 & (1-a-b)^t \end{pmatrix}Q^{-1}.$$

(d) The formula in part (c) may be used to calculate t-step transition probabilities. Use it on the Markov chain

$$P = \begin{pmatrix} \frac{1}{2} & \frac{1}{2} \\ \frac{1}{4} & \frac{3}{4} \end{pmatrix}$$

to calculate P^2 and P^3.

*(e) Show that in general if $a + b \neq 0$, then $|1 - a - b| < 1$.

(f) By part (e), $(1 - a - b)^t \to 0$. Use this observation and the results of parts (b) and (c) to show that

$$P^t \longrightarrow \begin{pmatrix} \dfrac{b}{a+b} & \dfrac{a}{a+b} \\ \dfrac{b}{a+b} & \dfrac{a}{a+b} \end{pmatrix}.$$

(g) Use this result to calculate w for P of part (d).

17. Show that every stochastic matrix P (and hence every Markov chain) has unity as an eigenvalue, by proving the following:

(a) Let $u = (1, 1, \ldots, 1)$. Show that $uP' = u$, where P' is the transpose of P.

(b) Show that λ is an eigenvalue of P iff it is an eigenvalue of P'.

18. Suppose P is the transition matrix of a Markov chain. It can be shown (cf. Cox and Miller [1965], Karlin [1969a], etc.) that if 1 is an eigenvalue of P of multiplicity k, then the chain has k ergodic sets. In particular, it has one ergodic set iff $k = 1$. Moreover, it can be shown that if P is known to be an ergodic chain, then P is regular if and only if 1 is the only eigenvalue of P of magnitude unity. Use these results to decide which of the following matrices correspond to chains with only one ergodic set. For these, check if the chains are ergodic. (This is not done using eigenvalues. Why?). If they are ergodic, use the eigenvalue results to decide if they are regular. (Suggestion: By Exer. 17, $\lambda - 1$ can always be factored out of the characteristic polynomial $C(\lambda)$.)

(a) $\begin{pmatrix} 0 & 1 & 0 \\ \frac{1}{2} & 0 & \frac{1}{2} \\ 0 & 0 & 1 \end{pmatrix}.$
(b) $\begin{pmatrix} 1 & 0 & 0 \\ 0 & \frac{1}{2} & \frac{1}{2} \\ 0 & \frac{1}{2} & \frac{1}{2} \end{pmatrix}.$

(c) $\begin{pmatrix} 0 & 1 & 0 \\ 0 & 0 & 1 \\ 1 & 0 & 0 \end{pmatrix}.$
(d) $\begin{pmatrix} \frac{1}{2} & \frac{1}{2} & 0 \\ 0 & \frac{1}{2} & \frac{1}{2} \\ \frac{1}{2} & 0 & \frac{1}{2} \end{pmatrix}.$

(e) $\begin{pmatrix} 0 & \frac{1}{2} & \frac{1}{2} \\ 1 & 0 & 0 \\ 1 & 0 & 0 \end{pmatrix}.$

19. Suppose P is the transition matrix of a regular Markov chain and P has no multiple eigenvalues. Using the results of Exer. 18, generalize the procedure in Exer. 16 for finding P^t and $\lim_{t \to \infty} P^t$.

20. Consider the stochastic matrix

$$P = \begin{pmatrix} \frac{1}{2} & \frac{1}{2} \\ \frac{1}{4} & \frac{3}{4} \end{pmatrix}.$$

Using the notation of the proof of Theorem 5.8, calculate $M_1, M_2, M_3, M_4, m_1, m_2, m_3$, and m_4 for $P^t e_1$. Use these calculations to guess at w_1 and check your guess by calculation.

*21. Suppose P is the transition matrix of a regular Markov chain and x is a fixed point column vector, i.e., $Px = x$. Show that x has all entries the same.

22. Consider a communication network. Make this into a Markov chain by assuming that if an individual has a message, he will pass it on to exactly one of the individuals (including himself) he can reach directly, and he does this with a certain probability. Assuming the resulting Markov chain is regular, Kemeny and Snell [1962, p. 102] suggest measuring importance of an individual i in such a communication network by using the component w_i of the fixed point probability vector w. Calculate this measure of importance for a communication network (with probabilities) of your choice. (Don't forget to include loops where they are appropriate.) Discuss whether or not w_i is a good measure of importance.

5.6. Ergodic Chains[18]

5.6.1. THE PERIODIC BEHAVIOR OF ERGODIC CHAINS

The stochastic digraph of Fig. 5.15 defines an ergodic Markov chain which is not regular. The passage from state to state is periodic, with the chain going from state u_1 to state u_2 to state u_1 to state u_2, and so on. It returns to state u_1 with period 2. A similar situation is true in general: The vertices of an ergodic Markov chain can be partitioned into d classes $C_0, C_1, \ldots, C_{d-1}$ such that starting from a vertex of class C_0, all paths of length 1 in the transition digraph lead to a vertex of C_1, all paths of length 2 lead to a vertex of $C_2, \ldots$, and in general all paths of length t lead to a vertex of class C_j where $t \equiv j \pmod{d}$. The Markov chain will be called *periodic of*

---

[18]This section may be omitted without loss of continuity. In particular, it should be omitted by the reader unfamiliar with the notion of congruence. However, Corollary 2 to Theorem 5.12 should be noted.

period d. If $d = 1$, the Markov chain will turn out to be regular. The partition of vertices turns out to be independent of transition probabilities and depends only on the digraph structure.

To illustrate this result, consider the digraph of Fig. 5.17. This digraph is strongly connected, so the result applies. Starting at u_1, let us list what vertices are reachable from this vertex in t steps, for several t:

Vertices reachable
from u_1 in

1 step	u_2
2 steps	u_1, u_3
3 steps	u_2, u_4
4 steps	u_1, u_3, u_5
5 steps	u_2, u_4, u_6
6 steps	u_1, u_3, u_5, u_8
7 steps	u_2, u_4, u_6, u_9
8 steps	u_1, u_3, u_5, u_7, u_8
9 steps	$u_2, u_4, u_6, u_9.$

The result about periodicity suggests that $d = 2$ and that the two classes are $C_0 = \{u_1, u_3, u_5, u_7, u_8\}$ and $C_1 = \{u_2, u_4, u_6, u_9\}$. Indeed, it is easy to see that all paths from a vertex of class C_0 to a vertex of class C_1 are of

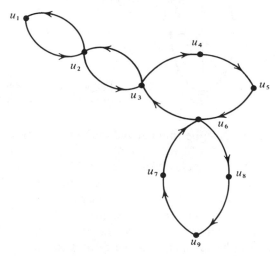

FIGURE 5.17. A digraph coming from an ergodic Markov chain.

odd length, all paths from a vertex of class C_0 to a vertex of class C_0 are of even length, etc. The theorems below will give us a simple way of calculating periodicity.

Let us now consider the digraph of Fig. 5.18. This is also strongly connected. Performing the same analysis, we see:

Vertices reachable
from u_1 in

1 step	u_2
2 steps	u_1, u_3, u_5
3 steps	u_2, u_4, u_6
4 steps	u_1, u_2, u_3, u_7
5 steps	u_1, u_2, u_3, u_4, u_5
6 steps	$u_1, u_2, u_3, u_4, u_5, u_6$
7 steps	$u_1, u_2, u_3, u_4, u_5, u_6, u_7.$

This analysis suggests that all vertices are reachable from u_1 in 7 steps and hence $d = 1$: There is just one class. It will follow that D is regular.

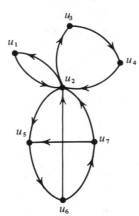

FIGURE 5.18. Another digraph coming from an ergodic Markov chain.

The main results can be summarized in the following theorem.

Theorem 5.12.   The states of an ergodic Markov chain can be partitioned into d periodic classes $C_0, C_1, \ldots, C_{d-1}$ such that if the chain starts in a state of class C_i, after t steps it can only be in a state of class $C_{i \oplus t}$, where $\oplus$ stands for addition modulo d. For t sufficiently large, if the chain starts in any state of class C_i, it can be in any state of class $C_{i \oplus t}$ at time t. Moreover, for any i, d can be calculated as the greatest common divisor[19] of the set of numbers

$$N_{ii} = \{t > 0: \text{ state } u_i \text{ can reach state } u_i \text{ in } t \text{ steps}\}.$$

---

[19]The *greatest common divisor* (g.c.d.) of a set of numbers is the largest number d which divides every number in the set. The theorem asserts that for all i and j, the g.c.d.'s of N_{ii} and N_{jj} are the same, and so *any* N_{ii} may be used to calculate d.

We have the following corollaries.

Corollary 1.   If $d = 1$, the ergodic chain is regular.

Proof. By the second part of the theorem, for t sufficiently large, there is a positive probability of going from any vertex (of C_0 = the set of all states) to any other vertex (of C_0) in t steps.   Q.E.D.

Corollary 2.   If in an ergodic Markov chain there is a state u_i so that transition probability p_{ii} is positive, then the chain is regular.

Proof.   1 is in N_{ii}, so d divides 1, i.e., $d = 1$.

For a proof of Theorem 5.12, we refer the reader to Kemeny and Snell [1960, p. 5 ff]. We shall, however, present a direct proof of Corollary 1 below. It is perhaps easier for the reader to understand Theorem 5.12 if he translates it into a theorem about digraphs. This second theorem, stated below, holds true without consideration of weights, and shows why periodicity depends only on the digraph structure, and not on the specific numerical probabilities.

Theorem 5.13.   The vertices of any strongly connected digraph can be partitioned into d periodic classes $C_0, C_1, \ldots, C_{d-1}$, such that starting from a vertex of class C_i, all paths of length t lead to a vertex of class $C_{i \oplus t}$. For t sufficiently large, there are paths of length t from any vertex of C_i to any vertex of $C_{i \oplus t}$. Moreover, for any i, d can be calculated as the greatest common divisor of the set of numbers

$$N_{ii} = \{t > 0 : \text{there is a path from } u_i \text{ to } u_i \text{ of length } t\}.$$

Theorem 5.13 can be illustrated on the digraphs of Figs. 5.17 and 5.18. In the digraph of Fig. 5.18, there are cycles through u_2 of lengths 2 and 3 (as well as other lengths). The greatest common divisor of 2 and 3 is 1, so d is 1. Thus, the digraph consists of one periodic class. In the digraph of Fig. 5.17, there are two periodic classes ($C_0 = \{u_1, u_3, u_5, u_7, u_8\}$ and $C_1 = \{u_2, u_4, u_6, u_9\}$). For, it is easy to see that all cycles and hence all closed paths have even length, and there is one of length 2. Thus, 2 is the greatest common divisor of each N_{ii}.

We now present a proof of Corollary 1 of Theorem 5.12. Specifically, we prove the following digraph statement: If D is a strongly connected digraph and

$$N_{ii} = \{t > 0 : \text{there is a path in } D \text{ from } u_i \text{ to } u_i \text{ of length } t\},$$

then if N_{ii} has greatest common divisor equal to 1 for some i, there is some k so that there is a path of length k from any vertex to any other vertex. Our proof follows closely that of Kemeny and Snell [1960, p. 5 ff], and gives a flavor of the proof of the full Theorem 5.13.[20]

Lemma 1.   For each i, N_{ii} is closed under addition.

Proof.   Suppose t and t' are in N_{ii}. If P and P' are paths of lengths t and t' respectively from u_i to u_i, then P followed by P' is a path of length $t + t'$ from u_i to u_i.   Q.E.D.

Lemma 2.   A set N of positive integers that is closed under addition contains all but a finite number of multiples of its greatest common divisor.

Proof.   Let d be the g.c.d. We may assume that $d = 1$, for if $d \neq 1$, we divide all elements of N by d. Now if 1 is the g.c.d. of the elements of N, it must be the g.c.d. of a finite subset of N, $\{n_1, n_2, \ldots, n_m\}$. If any n_i is 1, then by closure under addition, N contains every integer. Thus, suppose no n_i is 1. By a well-known theorem of number theory, there are integers α_i such that $\sum \alpha_i n_i = 1$. Let

$$s = \sum_{\alpha_i \text{ pos.}} \alpha_i n_i \quad \text{and} \quad t = - \sum_{\alpha_i \text{ neg.}} \alpha_i n_i.$$

Now s and t are in N since N is closed under addition. We have $s - t = 1$. Moreover, since no n_i is 1, t must be at least 2. Hence if q is at least $t(t - 1)$, then q is positive. Let l be the remainder when q is divided by t. We have $q = kt + l$, where $k \geq (t - 1)$ and $0 \leq l \leq t - 1$. It follows that $q = kt + l(s - t) = (k - l)t + ls$. Since $k \geq l$, since s and t are in N, and since N is closed under addition, q is in N. Thus, any number q greater than or equal to $t(t - 1)$ is in N.   Q.E.D.

Lemmas 1 and 2 together give us our desired result. For by Lemma 2, if $d = 1$, there is a number m so that for every $t \geq m$, there is a path from u_i to u_i of length t. We shall show that there is a number k so that for all r and s, there is a path from u_r to u_s of length k. Pick $k = m + 2n$, where n is the number of vertices of D. Now given any u_r and u_s, since D is strongly connected, there is a path P_{ri} from u_r to u_i of length at most n, and a path P_{is} from u_i to u_s of length at most n. Thus, there is a path of length $k = m + 2n$ from u_r to u_s. We start with P_{ri}, follow with a path from u_i to u_i of length $m + (2n - a - b)$, and then add P_{is}, where a is the length of P_{ri} and b is the length of P_{is}. This completes the proof of Corollary 1 of Theorem 5.12.

5.6.2.   Generalization from Regular Chains[21]

Most of the theorems for regular chains generalize fairly easily to ergodic chains. Naturally, P^t doesn't converge if $d > 1$. Indeed, the i, j entry

---

[20]The reader may wish to skip the proof, which uses elementary number theory.

[21]This subsection is fairly technical and may be omitted.

of P^t is 0 in $d - 1$ out of every d steps. The sequence P^t does converge in a certain sense. To make this sense precise, let us say that a sequence of matrices $M_1, M_2, \ldots$ is *Cesaro-summable* to a matrix M if the sequence

$$\frac{\sum_{s=1}^{t} M_s}{t}$$

converges to M. Thus, for example, suppose

$$P = \begin{pmatrix} 0 & 1 \\ 1 & 0 \end{pmatrix}.$$

Then $M_s = P^s$ is

$$\begin{pmatrix} 0 & 1 \\ 1 & 0 \end{pmatrix}$$

if s is odd, and it is

$$\begin{pmatrix} 1 & 0 \\ 0 & 1 \end{pmatrix}$$

if s is even. Now

$$\frac{\sum_{s=1}^{t} M_s}{t}$$

is

$$\begin{pmatrix} \frac{1}{2} & \frac{1}{2} \\ \frac{1}{2} & \frac{1}{2} \end{pmatrix}$$

if $t = 2k$ for some k and it is

$$\begin{pmatrix} \dfrac{k}{2k + 1} & \dfrac{k + 1}{2k + 1} \\ \dfrac{k + 1}{2k + 1} & \dfrac{k}{2k + 1} \end{pmatrix}$$

if $t = 2k + 1$. Thus the sequence $P, P^2, \ldots$, is Cesaro-summable to the matrix

$$\begin{pmatrix} \frac{1}{2} & \frac{1}{2} \\ \frac{1}{2} & \frac{1}{2} \end{pmatrix}.$$

The matrix

$$P = \begin{pmatrix} 0 & 1 \\ 1 & 0 \end{pmatrix}$$

is the simplest example of the transition matrix of a periodic (irregular) ergodic Markov chain. But Cesaro-summability of the sequence $P, P^2, \ldots$ holds in general for ergodic chains. The following theorem, which is proved in Kemeny and Snell [1960, Ch. V], summarizes this result and the other basic results about ergodic chains. We apply this theorem to the study of diffusion and Brownian motion in Sec. 5.11.

Theorem 5.14. Suppose P is the transition matrix of an ergodic Markov chain. Then

(a) P^t is Cesaro-summable to a stochastic matrix W all of whose rows are the same probability vector $w = (w_1, w_2, \ldots, w_n)$.

(b)   Each w_i is positive.

(c)   If p is any probability vector, then pP^t is Cesaro-summable to w.

(d) w is the unique fixed point probability vector of P.

(e) $PW = WP = W$.

(f) $Z = [I - (P - W)]^{-1}$ exists.

(g) $I + \sum_{s=1}^{\infty} (P^s - W)$ is Cesaro-summable to Z.

(h)   The mean first passage matrix E is given by Eq. (14) of Theorem 5.11.

For a further discussion of ergodic chains, as well as examples illustrating this theorem, the reader is referred to Kemeny and Snell.

EXERCISES

1.   Find the greatest common divisor of each of the following sets of integers.

(a) $\{2, 4, 6, 8, \ldots\}$.   (b) $\{1, 3, 5, 7, \ldots\}$.   (c) $\{2, 3, 4, \ldots\}$.

(d) $\{3, 6, 9, 12, \ldots\}$.   (e) $\{3, 4, 6, 8, 10, \ldots\}$.

2.   Suppose the sets in Exer. 1 are thought of as the set N_{11} for an ergodic Markov chain. In which cases is the chain regular?

3.   Of the ergodic Markov chains whose underlying digraphs are shown in Fig. 5.19, determine the periodicity d of each, using Theorem 5.13. Also exhibit the periodic classes of vertices.

4.   Use the results of this section to show that the following Markov chains are regular.

(a)   Coin tossing and die throwing (Fig. 5.1).

(b)   Coin tossing (Fig. 5.3.)

(c)   Weather in Tel Aviv (Fig. 5.4).

(d)   Message passage (Fig. 5.5).

(e)   Summer weather (Fig. 5.7).

(f)   Intergenerational occupational mobility (Fig. 5.8).

5.   (a)   Give an example of a Markov chain with three states which is ergodic and periodic with period 3.

(b)   For all $d > 0$, give an example of a Markov chain which is ergodic and periodic with period d.

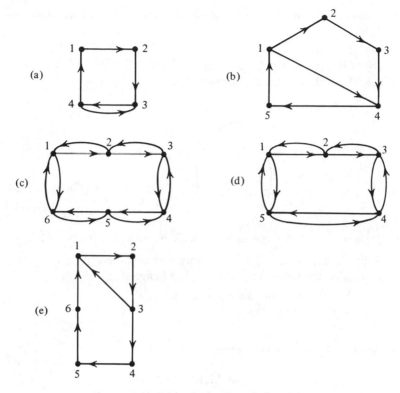

FIGURE 5.19. Digraphs for Exer. 3, Sec. 5.6.

6. Suppose a particle moves on the real line between the origin (0) and the point N. If the particle is at 0, it must move to 1. If it is at N, it must move to $N - 1$. From each other location, the particle moves left one step with probability p, with $0 < p < 1$, and moves right one step with probability $1 - p$. If $N = 100$, show that this *random walk* process defines an ergodic Markov chain and determine its periodicity. (This is a random walk with *reflecting barriers*, as opposed to the gambler's ruin (Example 11, Sec. 5.2) which is a random walk with *absorbing barriers*. Here, if the particle reaches one of the barriers 0 or N, it is reflected back into the system. A different random walk with reflecting barriers is studied in Sec. 5.11.)

7. Repeat Exer. 6 for the random walk with arbitrary $N \geqq 2$.

8. Suppose P is the transition matrix of an ergodic Markov chain. It can be shown (cf. Cox and Miller [1965] or Karlin [1969a]) that P has period d if and only if there are exactly d eigenvalues of magnitude unity. Use this result to calculate the periodicity of each of the Markov chains of Exer. 18, Sec. 5.5, which turned out to be ergodic.

9. Consider the Markov chain whose transition digraph is shown in Fig. 5.20.

   (a) Find w.
   (b) Show that P^t is Cesaro-summable to W.
   (c) Find the mean first passage matrix.

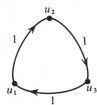

FIGURE 5.20. Transition digraph for Exer. 9, Sec. 5.6.

10. Repeat Exer. 9 for the random walk of Exer. 6 with $N = 2$.

11. Suppose a Markov chain has only one ergodic set. How would you determine the long-run probability of being in

    (a) A given state of this ergodic set.
    (b) A given transient set.

5.7.   Applications to Genetics

5.7.1.   THE MENDELIAN THEORY

Markov chains have a number of interesting applications in genetics. In this section, we present several of them. The discussion is based on a theory of inheritance which goes back to the Augustinian monk, Gregor Mendel, who published his work in 1866. In Mendel's experiments, he worked with pea plants. Such plants have pods of two different colors, green and yellow. Some stocks of plants are "pure" green and others "pure" yellow. The pure green are those which when crossed with each other give only green offspring, and similarly for the yellow. Mendel discovered that any time a plant from the pure green stock and another from the pure yellow stock are crossed, all of the offspring are green-podded. If two of these offspring are mated, *about*[22] $\frac{3}{4}$ of their offspring are green-podded and *about* $\frac{1}{4}$ of them are yellow-podded.

Such experiments led Mendel to the following theory. (This theory covers the situation known as *complete dominance*. For a discussion of incomplete dominance, see Exers. 4 and 6.) A trait such as color of pod is governed by two *genes*,[23] each of which can have one of two types, *domi-*

---

[22] The fractions may not come out exactly $\frac{3}{4}$, $\frac{1}{4}$. We discuss this point below.

[23] Mendel did not use the terminology "gene."

nant (*d*) or *recessive* (*r*). If two dominant genes are present, the dominant form of the trait appears. If two recessive genes are present, the recessive form of the trait appears. If one dominant and one recessive gene are present, the dominant form of the trait appears. In the case of peas, green-poddedness is dominant and yellow-poddedness is recessive. An individual with two dominant genes, a *dd* individual, has green pods, as does an individual with one dominant and one recessive gene, a *dr* or *rd* individual. The two individuals are said to have the same *phenotype* or appearance but different *genotype* or genetic makeup. The former is called *pure dominant* and the latter *hybrid*. An individual with two recessive genes, an *rr* individual, is called *pure recessive*, and has yellow pods. To give some other examples: In pea plants, smooth seeds dominate over wrinkled; in fruit flies, red eyes dominate over brown eyes; in humans, presence of pigment (in hair, eyes or skin) dominates over absence of pigment, and the pure recessive is called an albino.

If two individuals are mated, according to Mendel's theory, the offspring gets one gene from each parent. The choice of gene is random, and the choice in the two parents is independent. For example, suppose two pure dominant or *dd* individuals mate. Then the offspring is sure to get a *d* gene from each parent, and so the offspring is *dd*, or pure dominant, also. If a pure dominant *dd* and a pure recessive *rr* mate, the offspring gets one *d* gene from the pure dominant parent and one *r* gene from the pure recessive parent, and so is a hybrid *dr*. Its appearance is indistinguishable from that of a pure dominant. If a pure dominant *dd* and hybrid *dr* mate, the offspring gets a *d* gene from the *dd* parent and either a *d* or an *r* gene from the *dr* parent. The latter two possibilities occur with equal probability. Thus, the offspring is pure dominant *dd* with probability $\frac{1}{2}$ and hybrid *dr* with probability $\frac{1}{2}$. The reasoning behind this conclusion is easily illustrated in a tree diagram, such as that of Fig. 5.21. We shall be making more and more use of such tree diagrams in this chapter. If a hybrid *dr* mates with a second hybrid *dr*, the offspring gets a *d* or an *r* gene from the first parent, each with probability $\frac{1}{2}$, and similarly from the second parent. Thus, the offspring can be a *dd*, a *dr*, an *rd* (which is the same genetically as a *dr*), or an *rr*. The probability that the offspring is pure dominant (*dd*) is $\frac{1}{4}$, the probability it is hybrid (*dr* or *rd*) is $\frac{1}{2}$, and the probability it is pure recessive (*rr*) is $\frac{1}{4}$. (Once again, see Fig. 5.21.) In sum, if two hybrids mate, the probability is $\frac{1}{4}$ that an offspring will exhibit the recessive trait and $\frac{3}{4}$ that it will exhibit the dominant trait. (This explains the results about pod color in the second generation which we mentioned at the beginning of this subsection. For if a pure dominant (green-podded) individual and a pure recessive (yellow-podded) individual are mated, all offspring are hybrid, and so the second generation offspring are the results of a hybrid-hybrid mating.) Table 5.1 summarizes the results of various matings. The reader can easily verify the entries we have not discussed.

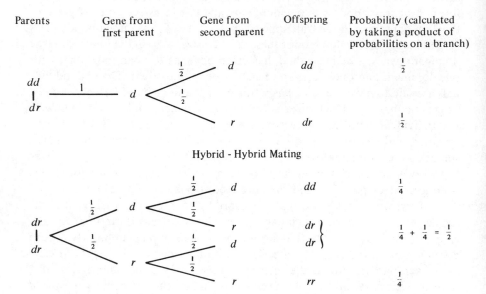

FIGURE 5.21. Results of a pure dominant-hybrid mating and of a hybrid-hybrid mating.

5.7.2. PREDICTIONS FROM THE MENDELIAN MODEL

The theory of Mendel may be thought of as a mathematical model, which was developed on the basis of certain data, specifically, Mendel's experi-

TABLE 5.1

Results of various matings

Parents	Offspring
Pure dominant–pure dominant	All pure dominant
Pure dominant–hybrid	Pure dominant, probability $\frac{1}{2}$; Hybrid, probability $\frac{1}{2}$
Pure dominant–pure recessive	All hybrid
Hybrid–hybrid	Pure dominant, probability $\frac{1}{4}$; Hybrid, probability $\frac{1}{2}$; Pure recessive, probability $\frac{1}{4}$
Hybrid–pure recessive	Pure recessive, probability $\frac{1}{2}$; Hybrid, probability $\frac{1}{2}$
Pure recessive–pure recessive	All pure recessive

ments on pea plants. As any mathematical model, it should be tested by deriving predictions. The specific predictions which are a direct result of the model seem to be approximately satisfied. For example, if a large number of pure dominants and pure recessives are mated, *approximately* $\frac{3}{4}$ of the second generation offspring exhibit the dominant trait. (It should be emphasized that usually not exactly $\frac{3}{4}$ will come out dominant-appearing. Indeed, on a single trial, it is possible that none of the second generation offspring exhibit the dominant trait. There is variation[24] in natural phenomena and that is why we have made a probabilistic prediction, not a deterministic one. A similar situation occurs with coin tossing, where we predict that with a fair coin, $\frac{1}{2}$ of the outcomes will be heads. However, we don't expect *exactly* $\frac{1}{2}$ of the outcomes to be heads every time we toss a coin a number of times. Our prediction is considered "verified" if this usually happens over a long period of time. Of course, a detailed verification of a prediction like the $\frac{3}{4}$ prediction requires a significant sophistication of the Mendelian model. This sophistication allows estimation of the expected departures from the predicted fraction $\frac{3}{4}$, and then the application of statistical methods to analyze the significance of these departures. See for example Elandt-Johnson [1971], Ewens [1969], or Karlin [1969b] for details of this statistical theory of mathematical genetics.)

Aside from testing the direct predictions made from the Mendelian model, we would like to test the model by deriving several more complicated predictions.[25] First, let us imagine starting with an individual of unknown genetic character. Cross the individual with a known hybrid. Pick one offspring from this crossing at random and cross again with a known hybrid. Pick one offspring from this crossing at random and cross again with a known hybrid. And so on. We might ask, in the long run, what is the probability that the individual chosen at a given generation will exhibit the dominant trait? That is, for example for peas, what is the probability that in the long run, the pea plant chosen at the tth generation will be green-podded? The process we have just described is called *continued crossings with a hybrid*, and the answer to our question can be obtained by using Markov chain theory. Thus, the mathematical tools of Markov chain theory are used in deriving predictions from a model.

Build a Markov chain where the state represents the genotype of the individual chosen at the tth generation. Thus, there are three states, D (for pure dominant), H (for hybrid), and R (for pure recessive). The transition

---

[24]There is a difference of opinion about this variation. Is it there inherently or does it only appear to be there because we don't have powerful enough models or explanations to understand the deterministic process which is going on?

[25]The rest of this section follows closely Kemeny, Snell, and Thompson [1956].

probabilities for this chain are given by the following matrix

$$P = \begin{array}{c} \\ D \\ H \\ R \end{array} \begin{array}{ccc} D & H & R \\ \begin{pmatrix} \frac{1}{2} & \frac{1}{2} & 0 \\ \frac{1}{4} & \frac{1}{2} & \frac{1}{4} \\ 0 & \frac{1}{2} & \frac{1}{2} \end{pmatrix} \end{array}.$$

The i, j entry of P represents the probability that, if a given individual of genetic type i is crossed with a known hybrid, and an offspring is chosen at random, the offspring will have genetic type j. Thus, for example, the DD entry is $\frac{1}{2}$ because if a pure dominant is crossed with a hybrid, the probability of obtaining a pure dominant offspring is $\frac{1}{2}$. (See Table 5.1.) The transition digraph corresponding to P is shown in Fig. 5.22.

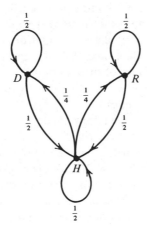

FIGURE 5.22. Transition digraph for process of continued crossings with a hybrid.

What kind of Markov chain does the matrix P define? It is certainly an ergodic chain, since the transition digraph is strongly connected. Since there is a loop from D to D, Corollary 2 to Theorem 5.12 implies that the chain is regular. Indeed, an easy calculation shows that P^2 has all entries positive. A further simple calculation shows that the fixed point probability vector w is given by $w = (\frac{1}{4}, \frac{1}{2}, \frac{1}{4})$. Thus, it is predicted that in the long run, independent of the genotype of the starting individual, the probability that a given individual at the tth generation will be pure dominant is $\frac{1}{4}$, hybrid $\frac{1}{2}$, and pure recessive $\frac{1}{4}$. Thus, the individuals will appear dominant with probability $\frac{3}{4}$. In the case of peas, it is predicted that in the long run, if there are continued crossings with a hybrid, $\frac{3}{4}$ of the peas at a given generation will have green pods. This probabilistic prediction, which is certainly easily testable, appears to be approximately[26] verified by data, not just for color

---

[26]Once again, the fractions are not expected to come out exactly $\frac{3}{4}$, $\frac{1}{4}$, etc.

of pods in peas, but for a large variety of traits. (The prediction that $\frac{1}{4}$ of the offspring are pure dominant is not as easily testable, because we cannot visibly distinguish pure dominants from hybrids. However, pure dominants can be identified if one can observe many generations of their offspring. It is always important in studying a mathematical model to derive a prediction which is testable.)

As a second test of the Mendelian model, let us imagine a process of *continued crossings with a dominant*, where the crossings are always with a known dominant. We again have a Markov chain, with states D, H, and R as before, but now the transition matrix (obtained using Table 5.1) is

$$P = \begin{array}{c} \\ D \\ H \\ R \end{array} \begin{array}{ccc} D & H & R \\ \begin{pmatrix} 1 & 0 & 0 \\ \frac{1}{2} & \frac{1}{2} & 0 \\ 0 & 1 & 0 \end{pmatrix} \end{array}.$$

This is the transition matrix of an absorbing Markov chain with one absorbing state, D. Thus, the theory of absorbing chains predicts that eventually all offspring will be pure dominant, and will appear dominant. This prediction too is borne out by data. (It is a deterministic prediction.) P is already in canonical form, and

$$Q = \begin{pmatrix} \frac{1}{2} & 0 \\ 1 & 0 \end{pmatrix}.$$

By calculation,

$$N = (I - Q)^{-1} = \begin{array}{c} \\ H \\ R \end{array} \begin{array}{cc} H & R \\ \begin{pmatrix} 2 & 0 \\ 2 & 1 \end{pmatrix} \end{array}.$$

The average number of generations before absorption is given by the row sums in N, i.e., it is two generations if the initial individual is hybrid and three generations if the initial individual is pure recessive. In the latter case, on the average, there will be one pure recessive and two hybrids before a pure dominant is produced.[27] These predictions are testable if we can observe many generations of offspring of a given individual. (For then we can distinguish pure dominants from hybrids.)

For our last prediction, let us imagine a situation of *inbreeding*. We cross two individuals of opposite sex. Two of their offspring are selected at random and crossed. Two of their offspring are in turn selected at random and crossed. And so on. We assume that the trait is independent of sex and that there are many offspring. Here, we are interested in the "gene pool," and whether a given gene (d or r) will disappear. We define a Markov chain by

---

[27]There can only be one pure recessive. Why?

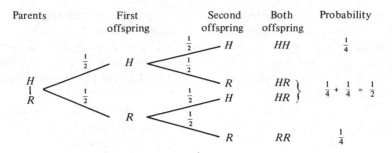

FIGURE 5.23. Tree diagram for mating of a hybrid and a recessive in the inbreeding situation.

taking as state the genotypes of the pair of offspring in the tth generation. Thus, the six states are DD, DH, DR, HH, HR and RR. For example, the state DD represents the situation where both individuals chosen are pure dominant. (Since the trait is independent of sex, we may treat DR and RD alike, and so on.) Let us calculate transition probabilities. Suppose for example that we start in state HR. Then, using Table 5.1, we find that $\frac{1}{2}$ the offspring are hybrid and $\frac{1}{2}$ pure recessive. Picking two offspring at random and assuming the number of offspring is large enough to assume that we are sampling with replacement, we determine that the offspring are both hybrid with probability $\frac{1}{4}$, both recessive with probability $\frac{1}{4}$, and one hybrid and one recessive with probability $\frac{1}{2}$. Hence, the probability of going from state HR to state HH is $\frac{1}{4}$, to state RR is $\frac{1}{4}$, and to state HR is $\frac{1}{2}$. The computation is summarized in the tree diagram of Fig. 5.23. Other transition probabilities are calculated similarly. Noting that DD and RR are absorbing states, we list them first and obtain the following transition matrix:

$$P = \begin{array}{c} \\ DD \\ RR \\ \\ DH \\ DR \\ HH \\ HR \end{array} \begin{array}{c} \begin{array}{cccccc} DD & RR & DH & DR & HH & HR \end{array} \\ \left(\begin{array}{cc|cccc} 1 & 0 & 0 & 0 & 0 & 0 \\ 0 & 1 & 0 & 0 & 0 & 0 \\ \hline \frac{1}{4} & 0 & \frac{1}{2} & 0 & \frac{1}{4} & 0 \\ 0 & 0 & 0 & 0 & 1 & 0 \\ \frac{1}{16} & \frac{1}{16} & \frac{1}{4} & \frac{1}{8} & \frac{1}{4} & \frac{1}{4} \\ 0 & \frac{1}{4} & 0 & 0 & \frac{1}{4} & \frac{1}{2} \end{array} \right). \end{array}$$

This matrix defines an absorbing chain, as is easy to show by drawing the transition digraph. From each nonabsorbing state, it is possible to reach one of the two absorbing states. (Indeed, it is easy to see that the four nonabsorbing states form a single transient set.) It is predicted that after a while, either the d gene or the r gene disappears from the gene pool, and all the offspring are pure dominant or pure recessive. Since P is already in canonical form, we find that

$$Q = \begin{pmatrix} \frac{1}{2} & 0 & \frac{1}{4} & 0 \\ 0 & 0 & 1 & 0 \\ \frac{1}{4} & \frac{1}{8} & \frac{1}{4} & \frac{1}{4} \\ 0 & 0 & \frac{1}{4} & \frac{1}{2} \end{pmatrix}$$

and, by calculation, that

$$
N = \begin{array}{c} \\ DH \\ DR \\ HH \\ HR \end{array}
\begin{array}{cccc} DH & DR & HH & HR & \text{row sum} \end{array}
\begin{pmatrix} \frac{8}{3} & \frac{1}{6} & \frac{4}{3} & \frac{2}{3} & 4\frac{5}{6} \\ \frac{4}{3} & \frac{4}{3} & \frac{8}{3} & \frac{4}{3} & 6\frac{2}{3} \\ \frac{4}{3} & \frac{1}{3} & \frac{8}{3} & \frac{4}{3} & 5\frac{2}{3} \\ \frac{2}{3} & \frac{1}{6} & \frac{4}{3} & \frac{8}{3} & 4\frac{5}{6} \end{pmatrix}
$$

Starting with one pure dominant and one pure recessive, one expects that on the average, the absorption will take place in $6\frac{2}{3}$ generations. Using

$$R = \begin{pmatrix} \frac{1}{4} & 0 \\ 0 & 0 \\ \frac{1}{16} & \frac{1}{16} \\ 0 & \frac{1}{4} \end{pmatrix},$$

we calculate

$$
B = NR = \begin{array}{c} \\ DH \\ DR \\ HH \\ HR \end{array}
\begin{array}{cc} DD & RR \end{array}
\begin{pmatrix} \frac{3}{4} & \frac{1}{4} \\ \frac{1}{2} & \frac{1}{2} \\ \frac{1}{2} & \frac{1}{2} \\ \frac{1}{4} & \frac{3}{4} \end{pmatrix}.
$$

Entry b_{ij} of the matrix B gives the probability of absorption in state j given that the process starts in state i. Thus, the probability of ending with a pure dominant depends on the starting state. For example, if the starting state is HH, then the probability of ending with a pure dominant is $\frac{1}{2}$. The reader should note that in the starting state HH, there are two d genes and two r genes. The probability of ending with a pure dominant is equal to the proportion of d genes in the starting state. This is true for all of the starting states. This very interesting prediction has a very simple ring to it, and it too seems, to be upheld by data.

The Mendelian theory, forerunner of much interesting work in mathematical genetics, is considered further in the exercises. For further work on mathematical genetics, see for example Elandt-Johnson [1971], Ewens [1969], or Karlin [1969b].

EXERCISES

1. Explain the following on the basis of the Mendelian theory. (Your explanation should involve some guesses as to genotype, etc.)
    (a) A man and woman, neither of whom appear to have sickle cell anemia, have a child who has the disease.
    (b) When a tall pea plant is crossed with a short pea plant, there are many offspring, all tall.
    (c) A fruitfly with long wings is mated with another whose wings are barely developed (vestigial). There are 150 offspring, all with long wings. Two of the offspring are chosen at random and mated. There are 200 offspring of this mating, 155 of which have long wings.
    (d) A guinea pig with rough hair mates one with smooth hair. Over a period of time, they have 20 offspring, 11 with smooth hair and 9 with rough.
    (e) An albino horse is mated with a normal-appearing horse, and some of the offspring are albinos.

2. A pure dominant and pure recessive are mated. Two offspring of the mating are picked at random and mated. If one of their offspring is picked at random, what is the probability it is a pure dominant?

3. The first offspring from a hybrid-hybrid mating is mated with the first offspring from a pure dominant-pure recessive mating. Is it possible to have a pure recessive offspring from this mating?

4. (Mosimann [1968].) In *incomplete dominance*, a hybrid appears (phenotypically) different from a pure dominant. For example, in plants, pure dominant might appear tall, hybrid medium, and pure recessive short. In such a situation of incomplete dominance, imagine a breed of cows which can appear red, roan, or white. If two red cows are mated, all offspring are red. If two white cows are mated, all offspring are white. If a red and white cow are mated, all offspring are roan. And so on. What do the offspring look like if two roan cows are mated?

5. In a process of continued crossings with a recessive, what is the expected number of generations before absorption given that we start with a pure dominant?

6. In Exer. 4, imagine that an inbreeding scheme is carried out, beginning with two roan cows. In the long run, what is the probability that a cow chosen in the tth generation will appear roan?

7. It is often reasonable to assume that inheritance of two different traits takes place independently. Thus, for example, in guinea pigs black hair dominates over white hair, and rough hair over smooth, and we can think of the genotype as a pair consisting of the genotype with

respect to color and the genotype with respect to texture. Thus, *DH* would stand for an individual which is pure dominant with respect to color and hybrid with respect to texture. Such an individual would appear as black- and rough-haired. There are thus nine genotypes and four phenotypes.

(a) Name them.
(b) Show that if a *dihybrid HH* mates a dihybrid *HH*, then the phenotypes appear in the offspring in the ratio $9:3:3:1$. (In the case of guinea pigs, that would say that black, rough hair appears in $\frac{9}{16}$ of the offspring.)
(c) Using the genotypes (genotype pairs) as states, calculate the transition probabilities for the Markov chain of continued crossings with a dihybrid.

*8. (Kemeny, Snell, and Thompson [1956].) In the inbreeding situation, suppose that hybrids have a high mortality rate. Assume that in a given mating, half of the hybrid offspring die out before reaching maturity, while only a negligible number of pure dominant or pure recessive offspring do. The two mates for the next generation are chosen only from the surviving mature offspring. Thus, for example, if a pure dominant is mated with a hybrid, $\frac{1}{2}$ of the offspring are hybrid and $\frac{1}{2}$ pure dominant. But only $\frac{1}{2}$ of the hybrids reach maturity, so only $\frac{1}{3}$ of the mature surviving offspring are hybrid. (This is the conditional probability that an offspring chosen at random is hybrid given that it has survived.) If we pick two of these offspring at random they are both hybrid with probability $\frac{1}{9}$, both pure dominant with probability $\frac{4}{9}$, and one hybrid and one pure dominant with probability $\frac{4}{9}$.

(a) Construct a Markov chain for this process by choosing the states and calculating transition probabilities.
(b) Show that the chain is absorbing.
(c) Find N.
(d) Find the expected number of generations before absorption for each starting state.
(e) Find B and interpret.

9. (Kemeny, Snell, and Thompson [1956]). Color blindness is a sex-linked characteristic. There is a pair of genes C and S, of which the C gene produces color-blindness and the S gene does not. The S gene is dominant. A male has only one gene and a female two. Inheritance takes place with a male offspring inheriting one of his mother's genes (chosen at random) and a female offspring inheriting one of her mother's genes and one of her father's. Thus, a man may have type C or type S, while a woman has type CC, CS, or SS. Consider again an inbreeding situation, where at the tth generation, one male offspring and one female

offspring are chosen at random and mated. Model this situation using a Markov chain whose state is the pair of genotypes for the tth generation. Thus, there are six states, of which (C, CS) is a typical one.

(a) List the remaining states, and derive the transition matrix.
(b) Show that the Markov chain is absorbing, and interpret the absorbing states.
(c) Find N.
(d) Find the expected number of generations before absorption for each starting state.
(e) Show that the probability of absorption in the state having only C genes is equal to the proportion of C genes in the starting state.

10. In the process of continued crossings with a hybrid, suppose the current generation exhibits the recessive trait. On the average, how many generations will there be before the recessive trait is exhibited again?

5.8. Flow Models

In this section we present several applications of the theory of Markov chains which are interesting because they apply to models which, as formulated, do not seem to have a probabilistic element to them. The first deals with the flow of pollutants between different locations and the second with money flow between cities. It will be very interesting to see that once the first problem is translated into mathematical terms and mathematical tools for dealing with it are derived, then these same tools can be used to handle the second problem, which on the surface seems quite unrelated to the first.

5.8.1. An Air Pollution Model

The following is a very simple mathematical model of the flow of air pollution. There are n locations, $u_1, u_2, \ldots, u_n$. At each discrete time $t = 0, 1, 2, \ldots$ there is a concentration $c_i(t)$ of a certain pollutant at each location u_i. In each unit time period, a certain fraction q_{ij} of the pollutant at location u_i travels or flows to location u_j. The numbers q_{ij} depend on geographical, meteorological, and other such data, and in principle should vary in each time period, though we treat them as fixed. The numbers q_{ij} satisfy

$$\sum_{j=1}^{n} q_{ij} \leq 1.$$

That is, some pollutant at each location can be dissipated or can flow to other locations than the ones considered. However, once outside the n locations u_i, it can never (for the purposes of the model) return.

It is convenient to introduce an imaginary 0th location, u_0, into which all pollutant which is dissipated flows, and to assume that pollutant once in location u_0 stays there. Then we can extend the $n \times n$ matrix Q to an

$(n + 1) \times (n + 1)$ stochastic matrix P by adding a 0th row of the form $(1, 0, 0, \ldots, 0)$ and a 0th column with entries chosen so that row sums are 1. For example, if Q is the matrix

$$\begin{pmatrix} \frac{1}{2} & \frac{1}{4} & 0 \\ \frac{1}{2} & \frac{1}{2} & 0 \\ 0 & \frac{1}{3} & \frac{1}{3} \end{pmatrix},$$

then P is the matrix

$$\begin{pmatrix} 1 & 0 & 0 & 0 \\ \frac{1}{4} & \frac{1}{2} & \frac{1}{4} & 0 \\ 0 & \frac{1}{2} & \frac{1}{2} & 0 \\ \frac{1}{3} & 0 & \frac{1}{3} & \frac{1}{3} \end{pmatrix}.$$

Even though nothing probabilistic is occurring, we can think of the matrix P as defining a Markov chain. In this chain, u_0 is an absorbing state. We shall assume that P defines an absorbing chain, with u_0 as the only absorbing state. One way this assumption is satisfied is if, for all i,

$$\sum_{j=1}^{n} q_{ij} < 1. \tag{17}$$

Under Eq. (17), some of the pollutant in each location is dissipated in each time period. (Why does Eq. (17) guarantee that the Markov chain is absorbing with single absorbing state u_0?)

Let us assume that some of the locations $u_1, u_2, \ldots, u_n$ have fixed sources of the pollutant in question. In each time period, the source at u_i puts out an amount or flow $f_i \geq 0$ of the pollutant. (We can take $f_0 = 0$ and $f_i = 0$ if there is no source at u_i. Again it is a simplification to assume that f_i is independent of time.)

The problem is the following. Suppose we have certain upper levels of pollutant concentration deemed "reasonable" or acceptable at the various locations. These might vary depending on whether the locations are industrial, residential, open space, etc. What restrictions or upper bounds on the amounts f_i allowed to be produced at each source will force concentration within the desired limits? These restrictions might in principle depend on the amount of pollutant present when the standards go into effect, though we shall show below that under our model, they do not. In any case, let m_i be the amount of pollutant in location u_i at time 0.

We express the acceptable upper level of pollution concentration at location u_i as a value g_i (the goal). We want to choose f_i so that after awhile, $c_i(t) \leq g_i$ for all i. Let $f = (f_1, f_2, \ldots, f_n)$, $g = (g_1, g_2, \ldots, g_n)$, $m = (m_1, m_2, \ldots, m_n)$, and $c(t) = (c_1(t), c_2(t), \ldots, c_n(t))$. Let us trace out what happens to pollutants present at time 0. Of the amount of pollutant m_i at location u_i, $i \neq 0$, at time 0, a portion $m_i p_{ij} = m_i q_{ij}$ ends up at location

$u_j, j \neq 0$, at time 1. Thus, if no additional pollutant is added, the distribution of pollutant at time 1 is given by the vector mQ. That is, $c(1) = mQ$. Similarly, the distribution of pollutant at time 2, $c(2)$, is given by the vector $mQ^2, \ldots$, and the distribution at time t, $c(t)$, is given by the vector mQ^t. Since Q comes from an absorbing Markov chain, Theorem 5.5 implies that $Q^t \rightarrow 0$. Thus, $mQ^t \rightarrow 0$. We conclude that if no additional pollutant is added, the initial concentration eventually dissipates. (All the pollutant is eventually "absorbed" into imaginary location u_0.)

To give an example, let us consider three locations. At each time period, $\frac{1}{3}$ of the pollutant at location 1 stays there, $\frac{1}{3}$ goes to location 3, and $\frac{1}{3}$ is dissipated. Also, $\frac{1}{3}$ of the pollutant at location 2 goes to location 1, $\frac{1}{3}$ stays there, and $\frac{1}{3}$ goes to location 3. Finally, $\frac{1}{3}$ of the pollutant at location 3 stays there, and the rest goes to location 2. Thus, the matrix Q is given by

$$Q = \begin{matrix} & \begin{matrix} 1 & 2 & 3 \end{matrix} \\ \begin{matrix} 1 \\ 2 \\ 3 \end{matrix} & \begin{pmatrix} \frac{1}{3} & 0 & \frac{1}{3} \\ \frac{1}{3} & \frac{1}{3} & \frac{1}{3} \\ 0 & \frac{2}{3} & \frac{1}{3} \end{pmatrix} \end{matrix}. \tag{18}$$

The matrix P is given by

$$P = \begin{matrix} & \begin{matrix} 0 & 1 & 2 & 3 \end{matrix} \\ \begin{matrix} 0 \\ 1 \\ 2 \\ 3 \end{matrix} & \begin{pmatrix} 1 & 0 & 0 & 0 \\ \frac{1}{3} & \frac{1}{3} & 0 & \frac{1}{3} \\ 0 & \frac{1}{3} & \frac{1}{3} & \frac{1}{3} \\ 0 & 0 & \frac{2}{3} & \frac{1}{3} \end{pmatrix} \end{matrix}. \tag{19}$$

Notice that Q does not satisfy Eq. (17). But P does define an absorbing Markov chain with single absorbing state the imaginary $0th$ location. Suppose that initially there are three units of pollutant at location 1, six units at location 2, and nine units at location 3. Then $m = (3, 6, 9)$. At time 1, the distribution of pollutant is given by

$$c(1) = mQ = (3, 8, 6).$$

Thus, there are still three units of pollutant at location 1, eight units at location 2, and six units at location 3. At time 2, the distribution is given by

$$c(2) = mQ^2 = (\tfrac{11}{3}, \tfrac{20}{3}, \tfrac{17}{3}).$$

Eventually, if no more pollutant is added, the pollutant present at any location approaches 0.

Next, let us consider what happens to pollutant f_i introduced at location u_i at time $t - s$. At time $t - s + 1$, a certain portion $f_i p_{ij} = f_i q_{ij}$ of that produced in location u_i has flowed to location u_j. Thus, the distribution among locations at time $t - s + 1$ of pollutant introduced at time $t - s$

is given by the vector fQ. The distribution at time t of pollutant introduced at time $t - s$ is given by fQ^s. In sum, the distribution of pollutant at locations at time t if all sources are operating at all times is given by

$$c(t) = mQ^t + \sum_{s=0}^{t} fQ^s. \tag{20}$$

Returning to our example, suppose locations 1 and 3 each emit one unit of pollutant at each time, and location 2 does not emit any pollutant. Then $f = (1, 0, 1)$ and

$$
\begin{aligned}
c(1) &= mQ + fQ^0 + fQ \\
&= mQ + f + fQ \\
&= (3, 8, 6) + (1, 0, 1) + (\tfrac{1}{3}, \tfrac{2}{3}, \tfrac{2}{3}) \\
&= (\tfrac{13}{3}, \tfrac{26}{3}, \tfrac{23}{3}).
\end{aligned}
$$

Suppose in our example that acceptable limits for pollutant concentration are given by the vector $g = (25, 25, 25)$. To meet these limits, we must set standards f_1, f_2, f_3 for emission levels at locations 1, 2, and 3 so that after awhile, $c(t) \leqq g$.[28] We shall show that, if $N = (I - Q)^{-1}$, then the limits are achieved for all practical purposes if

$$fN \leqq g. \tag{21}$$

To see this, consider Eq. (20). Since Q comes from an absorbing Markov chain, Theorem 5.5 implies that $\sum_{s=0}^{t} fQ^s$ approaches $f(I - Q)^{-1} = fN$. Thus, since mQ^t approaches 0, $c(t)$ approaches fN. It follows that for t sufficiently large, $c(t)$ is for all practical purposes equal to fN. If fN is chosen $\leqq g$, then for sufficiently large t, $c(t)$ is for all practical purposes $\leqq g$. Equation (21) gives rise to a system of n linear inequalities in n unknowns. Any solution to the system gives rise to a set of standards which will bring pollutant levels within desirable limits. Choice among possible sets of standards involves information which we have not built into our model.

In our example, we calculate

$$N = \begin{pmatrix} 3 & 3 & 3 \\ 3 & 6 & \tfrac{9}{2} \\ 3 & 6 & 6 \end{pmatrix}. \tag{22}$$

Hence, from $fN \leqq g$ we obtain the linear inequalities

$$
\begin{aligned}
3f_1 + 3f_2 + 3f_3 &\leqq 25 \\
3f_1 + 6f_2 + 6f_3 &\leqq 25 \\
3f_1 + \tfrac{9}{2}f_2 + 6f_3 &\leqq 25.
\end{aligned}
\tag{23}
$$

[28]If v and w are vectors or matrices, then $v \leqq w$ ($v < w$) means that each component of v is $\leqq$ ($<$) the corresponding component of w.

One solution to the system of inequalities (23) is the vector $f = (4, 1, 1)$. This amounts to allowing up to four units of pollutant to be introduced at location 1 in each time period, and up to one unit each at locations 2 and 3. Still another solution to the system of inequalities is the vector $f = (4, 2, 0)$. This amounts to closing down any sources of pollution at location 3. That a given vector f will lead to the attainment of desired limits turns out to be independent of the initial distribution of pollutant. The standards are thus independent of the initial distributions. As we have said, choice among two such sets of standards must be made on grounds which we have not built into our model; for example, what is the economic and social cost of meeting one set rather than another?

The model we have considered is unrealistic for various reasons. For example, it assumes that q_{ij} is independent of time t, which is unrealistic, and also that f_i is independent of t. But the model gives a hint of the potential application of Markov chain theory in the area of air pollution. Some general references on air pollution modelling are Briggs [1969], Slade [1968], and Stern, et al [1973].

5.8.2. A MONEY FLOW MODEL

A quite similar mathematical model, with some additional complications, is discussed by Kemeny and Snell [1962] in connection with the flow of money (currency) between cities. Here, the u_i are certain cities and q_{ij} is the fraction of the currency in city u_i which goes to city u_j each time period (say one year). As Kemeny and Snell point out, this could be estimated by marking a sample of bills. Once again this model is unrealistic because, among other things, q_{ij} should in reality depend on t. We again assume either that the q_{ij} satisfy Eq. (17) or more generally that the matrix P defines an absorbing Markov chain with one absorbing state u_0. Here, u_0 represents an imaginary additional city into which all currency which disappears from the other cities goes. It is assumed that once currency has disappeared from one of these cities, it never returns.[29] The vector m is assumed nonnegative and it gives the initial distribution of currency among cities. The government has in mind some ideal (long-run) distribution of currency, given by the vector g. It hopes to achieve this (in the long run) by placing an amount of currency f_i in city u_i at each time t. The complication arises by letting f_i be negative (equivalent to taking currency out of the city), but requiring each $c_i(t)$, the total amount of currency in city u_i at time t, to be nonnegative (no city can have a negative amount of currency). The

---

[29]The situation is somewhat different if no money can disappear, i.e., if $\sum_{j=1}^{n} q_{ij} = 1$. Then, the Markov chain is ergodic (it can be assumed regular). We refer the reader to Kemeny and Snell [1962, Ch. VI] and to Exers. 15 through 22 below for a discussion of this situation.

problem is to find f so that $c(t)$ will approach g. (We are demanding more than in the air pollution situation.) Once again, the model, as formulated, does not seem to have any probabilistic element in it, but the theory of Markov chains will provide a solution.

We have shown that, given f, $c(t)$ is given by Eq. (20) and $c(t)$ tends to fN. Thus, fN must be g, and f must be $g(I - Q)$.[30] This is the only f which could achieve the given goal. The problem arises from the requirement that each $c_i(t)$ be nonnegative, while this f may lead to negative $c_i(t)$'s. Thus, we are in the situation where, given g, there is at most one strategy f for achieving g. Whether this strategy does indeed meet the requirement that each $c_i(t)$ be nonnegative may depend on the initial distribution of currency m. (In our model of flow of pollutants, the "standards" f_i turned out to be independent of m.) If, relative to a given m, the strategy of setting $f = g(I - Q)$ meets the requirement that all $c_i(t)$ be nonnegative, we say that g is *feasible relative to m*.

To check for feasibility in general, let us use the value $f = g(I - Q)$ and substitute this into Eq. (20). We obtain

$$c(t) = mQ^t + g(I - Q) \sum_{s=0}^{t} Q^s \qquad (24)$$

$$= mQ^t + (g - gQ) \sum_{s=0}^{t} Q^s.$$

Now $m \geqq 0$ and similarly $Q \geqq 0$, so $Q^s \geqq 0$ and $\sum_{s=0}^{t} Q^s \geqq 0$. Thus, if $g - gQ \geqq 0$, i.e., if $g \geqq gQ$, we conclude that $c(t) \geqq 0$, all t, so $c_i(t) \geqq 0$, all i and t. This conclusion is independent of m. We summarize the result as a theorem.

Theorem 5.15. If $gQ \leqq g$, then g is feasible relative to all m.

To give an example illustrating our results so far, let us suppose there are three cities and the currency flow among them is given by the matrix Q of Eq. (18). Thus, $\frac{1}{3}$ of the currency from city 1 disappears each time period. The matrix P is again given by Eq. (19). Let us first consider the goal $g = (10, 10, 10)$, an equal distribution of currency in the cities. We have $gQ = (\frac{20}{3}, 10, 10)$. Since $gQ \leqq g$, Theorem 5.15 implies that g is feasible independent of the initial distribution of currency m. Moreover, the correct amounts to put into each city at each time period are given by the vector $f = g(I - Q) = (\frac{10}{3}, 0, 0)$. The government leaves cities 2 and 3 alone, and introduces $\frac{10}{3}$ units of currency at city 1 each time period. As a check, we note that N is again given by Eq. (22), and by a simple calculation, we have $fN = (\frac{10}{3}, 0, 0)N = (10, 10, 10) = g$.

---

[30]From the inequality $fN \leqq g$ which we encountered in the air pollution model, it would not have been correct to conclude $f \leqq g(I - Q)$. Why?

If it is not true that $g \geqq gQ$, we must derive other tests for feasibility. Let us return to Eq. (24), and observe that, by a simple calculation, it follows that
$$c(t) = mQ^t + g(I - Q^{t+1}).$$
It is desired that $c(t) \geqq 0$ for all t, i.e., that
$$mQ^t + g - gQQ^t \geqq 0$$
for all $t \geqq 0$, i.e., that
$$g \geqq (gQ - m)Q^t, \tag{25}$$
all $t \geqq 0$. Thus, g is feasible relative to m if and only if Eq. (25) holds for $t = 0, 1, 2, \ldots$. We shall say that *Condition* C_t is satisfied if (25) holds for t. Thus, g is feasible relative to m if and only if Condition C_t is satisfied for $t = 0, 1, 2, \ldots$.

For example, let us again use the matrix Q of Eq. (18), but study a different goal from that studied before, namely the new goal $g = (12, 6, 3)$. Here, $gQ = (6, 4, 7)$. Since $7 > 3$, Theorem 5.15 does not apply. Thus feasibility can depend on the initial distribution m. Let us consider first the distribution $m = (1, 1, 4)$. Then $(gQ - m) = (5, 3, 3)$. This is $\leqq g$, so Condition C_t holds for $t = 0$. However, $(gQ - m)Q = (\frac{8}{3}, 3, \frac{11}{3})$ and $\frac{11}{3} > 3$, so $(gQ - m)Q$ is not $\leqq g$. Thus, Condition C_t is violated for $t = 1$ and we conclude that g is infeasible relative to $m = (1, 1, 4)$. Below we shall see that g is feasible relative to the initial distribution $m = (6, 5, 4)$.

The trouble with using Condition C_t as a test for feasibility is that for any m, it can take an infinite number of tests to make sure that g is feasible relative to m. Thus, this is not practical. Following Kemeny and Snell [1962], we state a sufficient condition for feasibility which avoids this problem.

Theorem 5.16. Suppose each component g_i of g is positive.[31] Suppose that for some s, the following *Condition* D_s holds: The sum of the absolute values of the components of $(gQ - m)Q^s$ is at most the least entry of g. Then Condition C_t holds for all $t \geqq s$. Moreover, D_s holds for some s.

Theorem 5.16 can be used to test for feasibility as follows. First test Condition D_s for $s = 0$. If this holds, then we know that Condition C_t holds for all $t \geqq 0$ and so g is feasible relative to m. If it does not hold, then test C_t for $t = 0$. If this is violated, g is infeasible relative to m. If it is satisfied, then test D_s for $s = 1$. If this is satisfied, then g is feasible relative to m. (Why?) If it is violated, test C_t for $t = 1$. If this is violated, g is infeasible relative to m. If it is satisfied, then test D_s for $s = 2$. If this is satisfied, then g is feasible relative to m. If it is violated, then test C_t for $t = 2$. And so on. We know that this procedure eventually ends, since eventually either Condition C_t is violated for some t or Condition D_s is satisfied for some s.

[31] This assumption limits the theorem to a special situation, but quite a reasonable one.

To illustrate this test, let us return to our matrix Q of Eq. (18) and the goal $g = (12, 6, 3)$ which we have been studying, but let us change the initial distribution to $m = (6, 5, 4)$. Here, $gQ - m = (0, -1, 3)$. Now $0 + 1 + 3 = 4$, which is greater than 3, the least entry of g. Thus, Condition D_s fails for $s = 0$. Equation (25) is satisfied for $t = 0$, so Condition C_t holds for $t = 0$. Next, we calculate $(gQ - m)Q = (-\frac{1}{3}, \frac{5}{3}, \frac{2}{3})$. Now $\frac{1}{3} + \frac{5}{3} + \frac{2}{3} = \frac{8}{3}$, which is less than 3, the least entry of g. Thus, Condition D_s holds for $s = 1$. We conclude, by Theorem 5.16, that g is feasible relative to initial distribution m. Once again f is given by $g(I - Q)$, which is $(6, 2, -4)$. The government achieves its ideal distribution by, in each time period, adding six units of currency to city 1 and two units to city 2, and taking four units from city 3.

We conclude this subsection by proving Theorem 5.16. Let $h = h(s) = (gQ - m)Q^s$. Then for $t \geqq s$,

$$(gQ - m)Q^t = hQ^{t-s}. \tag{26}$$

Note that if Condition D_s holds, then $\sum_i |h_i| \leqq \min_i g_i$, the least entry of g. By (26), Condition C_t (Eq. (25)) for $t \geqq s$ can be written as $hQ^{t-s} \leqq g$. Assuming D_s, we demonstrate this inequality for $t \geqq s$. Note that in hQ^{t-s}, the jth component is $\sum_i h_i q_{ij}^{(t-s)}$, where $Q^{t-s} = (q_{ij}^{(t-s)})$. Since Q^{t-s} is a power of a matrix whose row sums are at most 1 and all of whose entries are nonnegative, we have each $q_{ij}^{(t-s)} \leqq 1$. Thus, the jth component of hQ^{t-s} is $\leqq \sum_i |h_i| q_{ij}^{(t-s)} \leqq \sum_i |h_i|$, which by assumption is $\leqq \min_i g_i$. We conclude that $hQ^{t-s} \leqq g$, which is Eq. (25) (Condition C_t) for t. This completes the proof of the first part of Theorem 5.16. To prove the second part, note that by Theorem 5.5, $Q^s \to 0$ as $s \to \infty$. Thus $h = h(s) = (gQ - m)Q^s \to 0$ as well. It follows that $\sum_i |h_i(s)| \to 0$, and so, since g has all positive entries, $\sum_i |h_i(s)| \leqq \min_i g_i$, for s sufficiently large. This completes the proof of Theorem 5.16.

EXERCISES

1. For the following matrices Q, find the corresponding matrices P.

(a) $Q = \begin{pmatrix} \frac{1}{3} & \frac{1}{3} & \frac{1}{3} \\ \frac{1}{2} & 0 & \frac{1}{4} \\ \frac{1}{4} & \frac{1}{4} & \frac{1}{4} \end{pmatrix}$. (b) $Q = \begin{pmatrix} \frac{1}{3} & \frac{1}{3} & \frac{1}{6} \\ 1 & 0 & 0 \\ 0 & 1 & 0 \end{pmatrix}$. (c) $Q = \begin{pmatrix} \frac{1}{2} & \frac{1}{2} & 0 \\ 0 & 1 & 0 \\ \frac{1}{6} & \frac{1}{3} & \frac{1}{3} \end{pmatrix}$.

2. In which of the cases in Exer. 1 does P define an absorbing Markov chain with u_0 as the only absorbing state?

3. In the air pollution example, suppose there are three locations. In each time period, $\frac{1}{3}$ of the pollutants present at each location pass to each of the other two locations, and the remaining $\frac{1}{3}$ of the pollutants present are dissipated.

(a) Find the matrices Q and P corresponding to this situation.

(b) Does our theory apply to this situation? Why?

(c) If initially there are two units of pollutant at locations 1 and 3 and five units of pollutant at location 2, and no additional pollutant is introduced, what is the distribution of pollutant after two time periods?

(d) Continuing with part (c), what is the pollutant concentration at location 1 after three time periods?

(e) Continuing with part (c), in the long run, what is the distribution of pollutant?

(f) Suppose in addition to the initial pollutant present, one unit of additional pollutant is introduced at location 1 each time period. What now is the pollutant concentration after two time periods?

(g) Continuing with part (f), how much pollutant is present at location 3 after three time periods?

(h) Continuing with part (f), in the long run, what is the distribution of pollutant?

4. In the air pollution example, suppose there are two locations. In each time period, $\frac{3}{4}$ of the pollutant in location 1 stays there, the rest going to location 2. Also, $\frac{1}{4}$ of the pollutant in location 2 stays there, $\frac{1}{2}$ goes to location 1, and the rest is dissipated. Suppose the initial distribution and reasonable limits are expressed by the vectors $m = (10, 10)$ and $g = (20, 20)$, respectively. Find standards on emissions at each location which will allow attainment of these limits. Discuss the standards.

5. Repeat Exer. 4 for $m = (10, 10)$, $g = (20, 20)$, and each of the following Q:

(a) $Q = \begin{pmatrix} \frac{1}{3} & \frac{1}{3} \\ 1 & 0 \end{pmatrix}$. (b) $Q = \begin{pmatrix} \frac{1}{3} & \frac{1}{3} \\ \frac{1}{3} & \frac{1}{3} \end{pmatrix}$. (c) $Q = \begin{pmatrix} \frac{1}{6} & \frac{1}{6} \\ \frac{1}{6} & \frac{1}{6} \end{pmatrix}$.

6. In the air pollution example, suppose a given solution f to the system of linear inequalities $fN \le g$ has at least one negative component. Does this make sense? Is it possible that every solution has at least one negative component?

7. Suppose $m = (3, 3, 3)$ and $g = (10, 10, 10)$. For each of the matrices Q of Exer. 1, determine if Condition C_t is satisfied for $t = 0$.

8. Repeat Exer. 7 for $t = 1$.

9. Repeat Exer. 7 for Condition D_s for $s = 0$.

10. Repeat Exer. 7 for Condition D_s for $s = 1$.

11. In the money flow example, suppose there are two cities. City 1 keeps $\frac{1}{4}$ of its currency each time period, and sends the rest overseas. City 2 keeps $\frac{1}{2}$ of its currency and sends the rest overseas.

(a)   Find the matrices Q and P corresponding to this situation and show that P is the transition matrix of an absorbing Markov chain with one absorbing state.

(b)   Show that in this situation, every goal g is feasible relative to any initial distribution m.

12.   In the money flow example, suppose there are two cities. City 1 keeps $\frac{1}{10}$ of its currency each time period and sends the rest to city 2; city 2 keeps $\frac{1}{10}$ of its currency, sends $\frac{1}{2}$ to city 1, and sends the rest overseas. The initial distribution of currency is given by the vector $m = (8, 16)$. For each of the following goals, determine if it is feasible relative to m and if it is, what distribution strategy f will attain the desired goal. Check your answer in the feasible cases by calculating N and showing that $fN = g$.

(a) $g = (10, 10)$.
(b) $g = (20, 10)$.
(c) $g = (100, 10)$.
(d) $g = (30, 20)$.

13.   In the money flow example, suppose Q is the matrix

$$\begin{pmatrix} \frac{1}{3} & \frac{1}{3} \\ \frac{1}{3} & \frac{1}{3} \end{pmatrix}.$$

For each of the following goals g and initial distributions m, determine if g is feasible relative to m, and if it is, what distribution strategy f will attain the desired goal.

(a) $g = (10, 10)$, $m = (10, 10)$.
(b) $g = (10, 50)$, $m = (10, 10)$.
(c) $g = (10, 50)$, $m = (5, 5)$.

14.   In the money flow situation of Exer. 12, the government aims at the goal $(5, 6)$. But it cannot introduce or remove currency from city 2. Is the goal feasible? If so, what is the appropriate amount of currency f_1 to put into and take out of city 1 at each time in order to achieve it? What about the goal $(5, 5)$? *Caveat*: What would feasibility mean here?

15.   Exercises 15 through 22 refer to the situation where the money flow matrix Q is ergodic, i.e., currency cannot disappear from the system. The results are due to Kemeny and Snell [1962]. We shall let $P = Q$ and assume that some p_{ii} is positive, so that P is a regular matrix. Let w be the fixed point probability vector of P and W the matrix each of whose rows is w. It is easy to see that $c(t)$ is still given by

$$c(t) = mP^t + \sum_{s=0}^{t} fP^s. \qquad (27)$$

(You do not have to prove this.) What does the first term of (27) converge to?

16. The second term of (27) does not necessarily converge; if it does for a given f, then $fP^s \to 0$. Since w has all positive entries, show that $fP^s \to 0$ implies that $\sum_i f_i = 0$. Interpret the result.

*17. By Theorem 5.10, $Z = (I - P + W)^{-1}$ exists and $I + \sum_{s=1}^{t} (P^s - W)$ converges to Z as $t \to \infty$. Assuming that $\sum_i f_i = 0$, show that $c(t)$ converges to h, where

$$h = (\sum_i m_i)w + fZ. \qquad (28)$$

*18. It follows from Exers. 16 and 17 that if $c(t)$ converges to g for a given f, then

$$g = (\sum_i m_i)w + fZ.$$

Show from this that $\sum_i g_i = \sum_i m_i$.

*19. If g is any vector at all and f is defined to be $g(I - P)$, show that $\sum_i f_i = 0$. (*Hint:* Let v be a column vector of all 1's and note that $fv = g(I - P)v$.)

*20. Suppose that $\sum_i g_i = \sum_i m_i$. Let $f = g(I - P)$. Show that if h satisfies Eq. (28), then $h = g$. (*Hint:* Multiply (28) by $I - P + W$ and use $wP = w$ and $wW = w$.)

21. Conclude from Exers. 17, 18, 19, and 20 that given a vector g, $c(t)$ converges to g for some f if and only if $\sum_i g_i = \sum_i m_i$ Moreover, if $c(t)$ converges to g for some f, then it converges to g for $f = g(I - P)$. (Even if $c(t)$ converges to g, $c(t)$ may become negative, in which case as before we say g is *infeasible* relative to m. Conditions for feasibility are analogous to those in the absorbing case.)

22. Apply the results of Exers. 15 through 21 to the following money flow situation. There are three cities. In each time period, $\frac{1}{2}$ of the currency in city 1 goes to city 2, the rest stays in city 1; $\frac{1}{2}$ the currency in city 2 goes to city 1 and the rest to city 3; and $\frac{1}{2}$ the currency in city 3 goes to city 2, the rest staying in city 3. Suppose that the initial distribution of money is given by the vector $m = (5, 15, 10)$.

(a) What is the matrix $P = Q$ for this situation?
(b) Is P the transition matrix of a regular Markov chain?
(c) Could the goal $g = (10, 12, 14)$ be feasible?
(d) Consider the goal $g = (10, 10, 10)$. Use Exer. 21 and a theorem analogous to Theorem 5.16 to show that g is feasible relative to m.
(e) Find a distribution vector f which achieves this goal g. Interpret the results.
(f) Check that with f as obtained in part (e), $c(t)$ does converge to g, by using the formula of Exer. 17.

(g)   Consider the new goal $g = (8, 10, 12)$. Is g feasible relative to m? If so, what f achieves the goal g?

23.   Formulate a model for the flow of electricity through a power grid in a period of allocation, when the government wants to restructure the system according to a newly worked out set of priorities (a new allocation scheme).[32]

24.   What are some simplifications in our flow models not mentioned in the text? How might some of the simplifications in these models be eliminated?

5.9.   Mathematical Models of Learning

Attempts to model the process of learning apparently go back as early as the year 1907 (Gulliksen [1934]). Most of the investigators, up to the present day, restrict themselves to specific learning situations, and develop models appropriate to describe learning in those situations. In this section, we shall take such an approach. The type of experimental situation we have in mind involves a T-maze, such as that shown in Fig. 5.24. A rat entering the maze

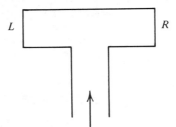

FIGURE 5.24. A T-maze.

can go either right (R) or left (L). Let us assume that we want the rat to "learn" to go right, and we try to accomplish this by rewarding him with food if he goes right, and not rewarding him if he goes left. By placing him in the maze in a sequence of trials, we can study his sequence of responses, R or L, and try to make predictions about this sequence. A similar experiment, involving humans, might place the subject before a panel consisting of two lights. A sequence of trials is run. Before each trial, the subject predicts which light, the right or left, will light up. Then the appropriate light does get lit. At the start, we will assume we are always lighting the right-hand light, i.e., always rewarding the response R. Again, we are interested in predicting the sequence of responses, or its properties.

---

[32]The author thanks K. Bogart (personal communication) for the idea behind this problem.

5.9.1. The Linear Model

Perhaps the simplest model of learning is the so-called *linear model.* This model assumes that learning is a gradual process, and we never completely learn. To state this model, let us define p_t as the probability of making a correct response R on trial t. Then $q_t = 1 - p_t$ is the probability of making an incorrect response on trial t. *Trials will be numbered* 0, 1, 2, ... That is, *the first trial is* 0. The linear model asserts that q_t decreases by the same proportionate amount at each trial. That is, there is a constant α, $0 < \alpha < 1$, so that

$$q_{t+1} = \alpha q_t.$$

Stated in terms of p, we have

$$p_{t+1} = 1 - \alpha(1 - p_t) = \alpha p_t + (1 - \alpha),$$

hence the terminology "linear." The constant α is a *parameter of the model,* a number whose specific value is needed to apply the model. The value of the parameter α will have to be estimated from data. The linear model predicts that $q_t = \alpha^t q_0$, or that

$$p_t = 1 - \alpha^t q_0. \tag{29}$$

The constant q_0 is a second parameter of the model, and will also have to be estimated from data. We may plot p_t against t, obtaining the so-called *learning curve.* (See Fig. 5.25.) The model predicts that we get closer and closer to

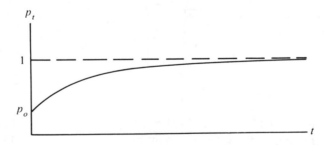

Figure 5.25. Learning curve in the linear model and the (Bower) 1-element all-or-none model.

perfection, and approach probability 1 of making a correct response as t gets large. The model can be tested by calculating the expected number of errors on trial t over a population of k subjects. This expectation is kq_t; it can be compared to data.

5.9.2. The 1-element All-or-none Model

To introduce an alternative learning model, let us recall the famous Pavlovian experiments with dogs. Normal dogs will salivate when presented with food. We say that food is a *stimulus* which when *sampled* elicits the *response* salivation. (In a psychological sense, the dog samples the stimulus food by seeing or smelling it, not by tasting it!) In our everyday life, we sample many stimuli from our environment and respond to some of them. These stimuli can be lights, smells, sounds, etc. Other stimuli present in the environment are disregarded—for example we sometimes shut out annoying sounds if we are trying to read. In his dog experiments, Pavlov was able, by repeated trials, to replace the stimulus food with another stimulus, bell-ringing, which elicited eventually the same response, salivation. We say that the new stimulus became *conditioned* to the response salivation.

The learning model we shall present as an alternative to the linear model is a *stimulus-sampling model* called the *(Bower) all-or-none model*, after G. H. Bower. It thinks of a subject sampling stimuli and eventually having them conditioned to responses he is trying to learn. Most stimulus-sampling models think of learning as taking place in stages, as shown in Fig. 5.26.

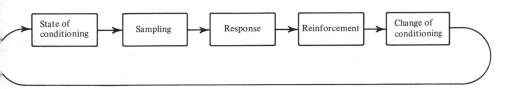

FIGURE 5.26. Stages of learning.

Before a given trial, each stimulus element is in some state of conditioning. Then, some or all of the available stimuli are sampled. On the basis of the sampled stimuli and their present state of conditioning, the subject makes a response. For example, if the dog samples the ringing bell and the ring is conditioned to the response salivation, he salivates. Next, the experimenter chooses a reward or *reinforcement*. Finally, the sampled stimuli can have their conditioning changed, i.e., they are reconditioned. Thus, if the bell-ringing was not yet conditioned to the response salivation, it can become conditioned if the dog is rewarded by food after the bell is rung. After reconditioning, things are ready for a new trial. A learning model of the stimulus-sampling type can be described by making assumptions about the different stages of learning.

In the simplest all-or-none model, the 1-*element all-or-none model*, we postulate the existence of a single, hypothetical stimulus element s. The element s is called hypothetical because we do not try to identify it with a specific item in the environment, unless perhaps it is the signal that the trial is to begin. We now make the following assumptions.

Assumptions for 1-Element All-or-None Model

State of Conditioning: Before each trial, the stimulus element s is or is not conditioned to the correct response R. Before the first trial, trial 0, s is not conditioned to R.

Sampling: On each trial, the subject begins by sampling the stimulus element s.

Response: The response on each trial is either R or L. If s is conditioned to the response R, the subject makes the response R, i.e., goes right. If s is not conditioned to the response R, he makes R with a certain probability, which we shall call g. (The number g is a parameter of the model.)

Reinforcement: The experimenter always rewards the response R and never the response L.

Change of Conditioning: If s was conditioned to the response R at the beginning of a trial, it remains conditioned afterwards. If s was not conditioned, it becomes conditioned at the end of the trial with a certain probability c, independent of the trial number, the response, etc. We assume that $0 < c \leq 1$. (The number c is a second parameter of the model.)

Using the all-or-none model, we can make various predictions. To use the predictions, we shall have to estimate the parameters g and c. The parameter g can be estimated by looking at the frequency of errors on the first trial, over a large number of subjects. (This assumes that g does not vary from subject to subject.) We shall see how to use the value of g to estimate the value of c.

Some simple predictions can be made if we form a Markov chain. The states will be called 1 (for conditioned) and 0 (for unconditioned), and stand for the state of conditioning just before a given trial. Thus, the transition matrix is:

$$P = \begin{matrix} & 1 & 0 \\ 1 & \\ 0 & \end{matrix}\begin{pmatrix} 1 & 0 \\ c & 1-c \end{pmatrix}.$$

Since $c > 0$, this is the transition matrix of an absorbing Markov chain with one absorbing state, state 1, which is the state we have called "conditioned." Our theory predicts that absorption takes place with probability 1, and so the model predicts that the subject eventually will become conditioned to or "learn" the correct response. Note that this is quite a different model from the linear model, which predicts that learning never completely takes place, whereas here learning is an all-or-none thing.

We may calculate the expected number of trials before learning by calculating the fundamental matrix N. Here, $Q = (1 - c)$, $I - Q = (c)$, and $N = (I - Q)^{-1} = (1/c)$. Thus, $1/c$ is the expected number of trials before learning, assuming that we start in the unconditioned state 0. For example, if $c = \frac{1}{4}$ then the expected number of trials before learning is 4; that is, we expect to enter state 1 after trial 3 (the trials are numbered 0, 1, 2, ...). If $c = \frac{1}{8}$, then the expected number of trials before learning is 8. We can also calculate the expected number of "errors" before learning, which turns out to be $(1/c)(1 - g)$, since $1/c$ is the expected number of trials before learning and $1 - g$ is the probability of making an error in an unconditioned state. (There are no more errors after conditioning.) If we know g, we may use this result to estimate c, by observing the average number A of errors before learning in a number of experiments. Then, setting $A = (1/c)(1 - g)$, we obtain $c = (1 - g)/A$. In this way, we may estimate c from the data of certain subjects and use it to predict the behavior of others. (This procedure assumes that c and g don't vary from subject to subject.)

Comparing our model to the linear model, we shall calculate q_t, the probability of an incorrect response on trial t. Specifically, in the all-or-none model, we obtain

$$q_t = (1 - c)^t (1 - g),$$

since $(1 - c)^t$ is the probability of being in state 0 after trial $t - 1$. Thus,

$$p_t = 1 - (1 - c)^t (1 - g). \tag{30}$$

This gives rise to the same shape learning curve as does the linear model, for we may obtain Eq. (29) from Eq. (30) by setting $\alpha = 1 - c$ and $q_0 = 1 - g$.

To obtain a prediction which is different for the linear model and for the 1-element all-or-none model, let us consider the (conditional) probability of an error following an error. In the linear model, the probability of an error depends only on the trial number, and so the probability r_t of making an error on trial t given that there has been an error on trial $t - 1$ is

$$r_t = q_t = \alpha^t q_0,$$

which decreases as t increases. In the all-or-none model, if an error was made on trial $t - 1$, this means that the subject was in state 0 before trial $t - 1$. Thus, his probability of making an error on trial t is

$$r_t = (1 - c)(1 - g),$$

the probability of not being conditioned after trial $t - 1$ and making an error on trial t. It follows that r_t is independent of t. By observing for a large number of subjects the number of errors which follow errors as t increases, one can thus compare the two models, and choose the one which accounts more for the data.

Dealing with a somewhat different experiment from the one we have discussed, Bower [1961] considers 20 predictions that can be derived for both the models which we have described so far. In Bower's experiment, the subject was presented with ten items, each consisting of a pair of consonants, and he was to learn to associate an integer, 1 or 2, with each such item. For five of the items, the integer 1 was the "correct" answer, while for the others 2 was correct. The learning of the ith item was considered one experiment, but Bower presented different items at different times, in effect performing ten experiments at once. Trial t for item i was the $(t + 1)$st time it was presented. After each presentation and response, Bower told the subject the correct response. For Bower's experiment, the data favored the all-or-none model on 18 of the 20 comparisons. Other experimenters have had less favorable results for the simple all-or-none model, though experiments similar to Bower's usually favor the all-or-none model. However, the less favorable results have led to a wide proliferation of learning models. We shall discuss some of these below, beginning with some generalizations of the all-or-none model.

5.9.3.   THE 2-ELEMENT ALL-OR-NONE MODEL

The above all-or-none model is called a 1-*element* all-or-none model since it postulates the existence of only one stimulus element. Let us generalize to two hypothetical stimulus elements, s_1 and s_2. In this 2-*element all-or-none model*, we introduce a sampling procedure, based on the observation that in everyday activity, an individual samples or attends to only a certain portion of the stimulus elements present.

Assumptions for 2-Element All-or-None Model

State of Conditioning: Before each trial, the stimulus element s_1 is or is not conditioned to the correct response R, and similarly for the stimulus element s_2. Before the first trial, trial 0, neither s_1 nor s_2 is conditioned to R.

Sampling: On each trial, the subject begins by sampling at random exactly one stimulus element.

Response: The response on each trial is either R or L. If the sampled stimulus is conditioned to the response R, the subject makes the response R. If it is not, the subject responds R with probability g. (The number g is a parameter of the model. It is a reasonable simplification to assume that g is the same for each stimulus element.)

Reinforcement: The experimenter always rewards the response R and never the response L.

Change of Conditioning: If a stimulus element s_i was sampled on a given trial and it was conditioned to R at the beginning of the trial, it remains conditioned afterwards. If it was not conditioned, it becomes conditioned at the end of the trial with probability c. The same probability of conditioning applies to each stimulus element. The nonsampled stimulus element retains its previous state of conditioning. (The number c is a parameter of the model, and it is assumed that $0 < c \leqq 1$.)

To analyze the 2-element model, we build a Markov chain with three states, 2, 1, and 0, by taking the state before a given trial to be the number of stimulus elements which are conditioned before that trial begins. Under the assumptions, we may calculate transition probabilities. For example, starting in state 2, the process must remain in state 2 with probability 1. Starting in state 1, exactly one stimulus element is conditioned. The process cannot go to state 0. (Why?) To go to state 2, the unconditioned stimulus element must be sampled (which happens with probability $\frac{1}{2}$) and conditioning must take place (which happens with probability c). Hence, $(\frac{1}{2})c$ is the probability of passing from state 1 to state 2. Similarly, we can calculate all the entries of the transition matrix, obtaining:

$$P = \begin{array}{c} 2 \\ 1 \\ 0 \end{array} \begin{pmatrix} \begin{array}{ccc} 2 & 1 & 0 \end{array} \\ 1 & 0 & 0 \\ \frac{c}{2} & 1 - \frac{c}{2} & 0 \\ 0 & c & 1 - c \end{pmatrix}.$$

These probabilities are very easily calculated using a tree diagram such as that in Fig. 5.27. The reader is urged to use such diagrams throughout this section. The matrix P is the transition matrix of an absorbing Markov chain

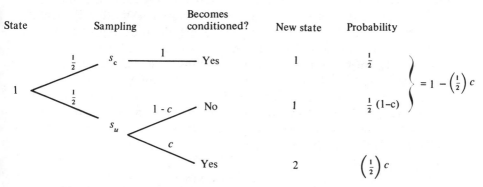

FIGURE 5.27. Typical tree diagram for calculating transition probabilities for the 2-element all-or-none model. s_c is the conditioned element and s_u the unconditioned element.

with one absorbing state, 2. Thus, as in the 1-element model, the probability of learning is 1. To get other information about the model, we calculate the fundamental matrix N. We have

$$Q = \begin{pmatrix} 1 - \dfrac{c}{2} & 0 \\ c & 1 - c \end{pmatrix}$$

$$(I - Q) = \begin{pmatrix} \dfrac{c}{2} & 0 \\ -c & c \end{pmatrix}$$

$$N = (1 - Q)^{-1} = \begin{array}{cc} & \begin{array}{cc} 1 & 0 \end{array} \\ \begin{array}{c} 1 \\ 0 \end{array} & \begin{pmatrix} \dfrac{2}{c} & 0 \\ \dfrac{2}{c} & \dfrac{1}{c} \end{pmatrix} \end{array}.$$

Thus, if n_{ij} is the i, j entry of N, then

$$n_{01} + n_{00} = \frac{2}{c} + \frac{1}{c} = \frac{3}{c}.$$

The prediction is that if we start in the unconditioned state 0, then it will take on the average $3/c$ trials before learning. Again, c can be estimated from the data of certain subjects, using g, and used to predict the behavior of others. If c is $\frac{1}{4}$, then we expect $3/c = 12$ trials before learning. If $c = \frac{1}{8}$, we expect 24 trials before learning.

One interesting way in which the 2-element all-or-none model differs from the 1-element all-or-none model is the following. In the latter, the probability of an error on any trial t known to be before the last error remains a constant, independent of t. Suppes and Ginsburg [1963] call this property *stationarity*. In the 2-element all-or-none model, the probability of an error on a trial t known to be before the last error decreases with t.

The 2-element all-or-none model naturally generalizes to an *n-element all-or-none model*. We leave the formulation of the 3-element model to the reader, as Exer. 21. A more detailed discussion of this all-or-none model can be found in Theios [1963]. (As n gets large, a number of the predictions from the n-element model reduce to the linear model. The reader is referred to Norman [1972] for details.)

5.9.4. THE ESTES MODEL

Other stimulus-sampling models apply to more complex experiments, evolving from the kind we have been discussing. Specifically in the T-maze experiment or the light prediction experiment, we might consider not rewarding the response R every time. Rather, we might reward it only a certain fraction of the times, or we might reward R a certain fraction of the times and L a certain fraction of the times. This fraction could be determined in practice by using some probabilistic device, and indeed the probability could depend, if the experimenter wanted to be quite subtle, on the subject's response. Let us modify the all-or-none stimulus-sampling model to apply to the situation where we have such probabilistic reward or reinforcement taking place. In the simplest case, let us assume that we reinforce the response R

with probability p, independent of response, and we reinforce the response L with probability $1 - p$. Thus, in the light prediction experiment, we reinforce by flashing the right light with probability p and the left with probability $1 - p$. In the maze situation, reinforcement is a little trickier. We might leave food in one side of the maze, and allow the rat, if he didn't find food on the side he chose, to immediately go to the other side and get the food.

We can no longer speak of trying to "learn" a specific response, so our conditioning assumptions must change. Let us go back to the situation of one stimulus element s, and make the following modified assumptions,[33] which are due to W. K. Estes. (Thus, we call the model the *Estes Model*.)

Assumptions for the Estes 1-Element Model

State of Conditioning: Before each trial the stimulus element s is conditioned to either the response R or the response L. Before the first trial, it is assumed that conditioning is determined at random.

Sampling: On each trial, the subject begins by sampling or not sampling the stimulus element s. He samples it with probability θ, $0 < \theta \leq 1$. (The number θ is a parameter of the model.)

Response: The response on each trial is either R or L. If the element s has been sampled on a given trial, and it is conditioned to the response R, the subject then responds R. Similarly, if s has been sampled and it is conditioned to the response L, the subject then responds L. If s is not sampled, then the subject responds R if and only if s is conditioned to R. Thus, independent of sampling, the subject responds R if and only if s is conditioned to R.

Reinforcement: The experimenter reinforces the response R with probability p, independent of response, and the response L with probability $1 - p$. (The number p is a parameter of the model.)

Change of Conditioning: Conditioning can change only if the stimulus element is sampled. If the stimulus element is conditioned to a given response and that response is reinforced, the conditioning does not change. If the stimulus element is sampled and if it is conditioned to a given response and the opposite response is reinforced, then conditioning changes with a certain probability c. It is assumed that $0 < c \leq 1$. (The number c is a second parameter of the model.)

Contrary to the all-or-none model, the Estes 1-element model does not allow any choice in the response. Response is determined entirely by conditioning. Variability in behavior enters from the sampling process and from whether or not change in conditioning takes place. The sampling process is somewhat different from that in the all-or-none model. It allows for the possibility that no stimulus element is sampled (or in the case of more than one

---

[33]The linear model must also be modified in order to apply to this situation, but we shall not do that here.

stimulus element, which we shall encounter below, it will allow for the possibility that more than one stimulus element is sampled). However, for simplicity, the response does not depend on the sampling. (It will in the case of more than one stimulus element.) We may again analyze this model by building a Markov chain. The states of this chain are 0 and 1, the number of stimulus elements conditioned to the response R. We shall assume, also for simplicity, that $c = 1$. Then, the transition matrix is given by

$$P = \begin{array}{cc} & \begin{array}{cc} 0 & \quad\quad 1 \end{array} \\ \begin{array}{c} 0 \\ 1 \end{array} & \begin{pmatrix} 1 - \theta p & \theta p \\ \theta(1 - p) & 1 - \theta(1 - p) \end{pmatrix}. \end{array}$$

For, since $c = 1$, conditioning changes from state 0 to state 1 if and only if the stimulus s is sampled (which happens with probability θ) and the response R is reinforced (which happens with probability p). And so on. See Fig. 5.28 for the calculation of some transition probabilities.

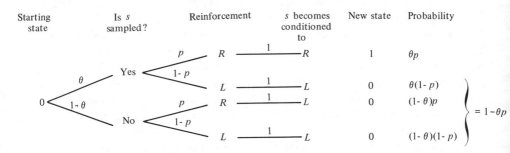

FIGURE 5.28. Some transition probabilities for the Estes 1-element model, with $c = 1$.

If $0 < p < 1$ and $0 < \theta \leq 1$, then P is the transition matrix of a regular Markov chain. A simple calculation shows that its fixed point probability vector w is given by $w = (1 - p, p)$. Thus, in the long run, the probability of being in state 1 is p. This implies that the probability of giving a response R equals the probability p of reinforcing an R response. The prediction from the model is that the subject will ultimately make a response R with the same probability that the experimenter reinforces R. This prediction is called *probability matching*. Let us ask if this is "optimal" behavior. Suppose for example that $p = \frac{3}{4}$. Then if you respond R with probability $\frac{3}{4}$, your probability of getting a correct response is $(\frac{3}{4})(\frac{3}{4}) + (\frac{1}{4})(\frac{1}{4}) = \frac{10}{16} = \frac{5}{8}$. However, suppose you always respond R. Then you will give the correct response $\frac{3}{4}$ of the time, a better performance. In short, the Estes 1-element Model predicts nonoptimal behavior! In tests of this prediction, it has been discovered that certain animals probability match, while others do not. For example, rats do not, while fish do, in similar sorts of experiments. Humans sometimes do, though they are less likely to if monetary rewards are offered.

To formulate a more general model, the *Estes n-element model*, we assume that there are n stimulus elements, $s_1, s_2, \ldots, s_n$. On each trial, each element is conditioned to one or another of the responses R and L. Reinforcement is done probabilistically, with a response R reinforced with probability p and a response L with probability $1 - p$. We introduce modified sampling, response, and change of conditioning assumptions.

Additional Assumptions for the Estes *n*-Element Model

Sampling: On any trial, any particular stimulus element is sampled with probability θ, $0 < \theta \leq 1$. Each element is treated independently.

Response: The response on each trial is either R or L. On a given trial, the subject responds R with probability equal to the proportion of *sampled* stimuli conditioned to the response R. That is, if this proportion is π, then we assume that the subject responds R with probability π. If *no stimuli are sampled*, it is assumed that he responds R with probability equal to the proportion of all stimuli which are conditioned to this response.

Change of Conditioning: If a given response is reinforced on trial t, then all stimulus elements conditioned to that response before trial t remain conditioned afterward. With probability c, $0 < c \leq 1$, *all sampled* elements conditioned to the unreinforced response have their conditioning changed. With probability $1 - c$, none do.

To exhibit the sampling, response, and change of conditioning assumptions, suppose that $n = 6$ and that before a given trial, stimulus elements s_1, s_2, and s_3 are conditioned to the response L while s_4, s_5, and s_6 are conditioned to the response R. Suppose s_1, s_2, and s_5 are sampled. This happens with probability $\theta^3(1 - \theta)^3$. The subject responds R with probability $\frac{1}{3}$, for exactly 1 of the 3 sampled stimulus elements is conditioned to R. If R is reinforced, then s_5 remains conditioned to the response R (and so do s_4 and s_6, since they were not sampled). Since s_3 was not sampled, it remains conditioned to L. Finally, s_1 and s_2 have a chance of becoming conditioned to R. With probability c, they both become conditioned to R, with probability $1 - c$, neither becomes conditioned to R.

The assumptions lead to an $(n + 1)$-state Markov chain, with states $0, 1, \ldots, n$, the number of stimulus elements conditioned to the response R at any given time. In Exer. 25, the reader is asked to analyze this Markov chain for the 2-element case (assuming $c = 1$) and to show that probability matching is once again predicted.

Further generalizations of the Estes model can be obtained by refining the reinforcement assumption. We might want to make reinforcement depend on response. A typical reinforcement assumption which does this is introduced in Exer. 23.

The reader who is interested in a more detailed introduction to mathematical learning theory can consult one of the following references: Atkinson, Bower, and Crothers [1965], Coombs, Dawes, and Tversky [1970], or Norman [1972].

EXERCISES

1. In each of the following experimental situations, state which of the three models (linear, all-or-none, Estes) seems most appropriate.

    (a) A monkey is being taught to distinguish red from blue. He is presented with red and blue balls and is reinforced for choosing the red one. After a while, he chooses the red one every time.

    (b) A pigeon has two food bins in her cage. Food is presented in the left bin 75% of the time, independent of which bin the pigeon went to on the previous trial.

    (c) A child is asked to memorize a given list of nonsense syllables. He is rewarded if he repeats the list correctly from memory. His performance gradually improves, but even once he has gotten the list right a few times, he tends to make a mistake.

2. In the linear model, with one subject, if $\alpha = \frac{1}{2}$ and $q_0 = 1$, what is the probability of an error on trial 3 (this is the fourth trial)?

3. In the 1-element all-or-none model, what is the probability that after trial 5, the stimulus element s is not conditioned to R?

4. In the 1-element all-or-none model, suppose it has been estimated that $c = \frac{1}{10}$. A given subject repeatedly learns by the third trial. Would you be justified in concluding that c has been estimated incorrectly? Why?

5. In the 1-element all-or-none model, suppose c is $\frac{1}{2}$ and g is $\frac{1}{2}$.

    (a) What is the expected number of "errors" before learning?

    (b) What is the expected number of "errors" after learning?

    (c) Plot the learning curve.

    (d) What is the probability of an error following a correct response?

    (e) What is the expected number of "errors" in N trials?

6. In the linear model, what is the probability of an error following a correct response?

7. In the 2-element all-or-none model, suppose before a given trial that s_1 is conditioned to R and s_2 is not.

    (a) What is the probability of response R on that trial?

    (b) What is the probability that s_2 will be conditioned to R before the next trial?

8. In the 2-element all-or-none model, draw a tree diagram to calculate the transition probabilities starting from state 0.

9. In the 2-element all-or-none model, suppose we start in state 1. What is the expected number of trials before learning?

10. In the linear model, let T be the first trial on which the probability of making a correct response is greater than the probability of making an incorrect response. Show that T is the least integer greater than
$$\frac{-\ln 2 - \ln q_0}{\ln \alpha}.$$

11. Could the concept of stationarity make sense for the linear model? Why?

12. In the linear model, calculate the probability $r_{t,t+1}$ of making two consecutive errors on trials t and $t + 1$ given that the subject made an error on trial $t - 1$.

13. Calculate $r_{t,t+1}$ (see Exer. 12) in the 1-element all-or-none model.

14. In the Estes 1-element model, what is the probability of sampling s on two consecutive trials?

15. In the Estes 1-element model (with $c = 1$), use tree diagrams to calculate the probability of going from state 1 to state 0 and from state 1 to state 1.

16. In the Estes 1-element model, suppose $c \neq 1$. What is the probability of going from state 0 to state 1?

17. In the Estes 1-element model with $c \neq 1$, if the element s is conditioned to R before trial t, what is the probability of responding R on trial $t + 1$?

18. In the Estes 1-element model with $c = 1$, what is the long-run probability of making a response on a given trial different from the response which is reinforced on that trial?

19. In the Estes 4-element model:
    (a) What is the probability that s_1 and s_2 will be sampled on a given trial and s_3 and s_4 will not?
    (b) What is the probability that no elements will be sampled?

20. In the Estes 4-element model, suppose s_1 and s_2 are conditioned to R before a given trial t and s_3 and s_4 are conditioned to L.
    (a) If s_1 and s_3 are sampled on trial t, what is the probability of responding R?
    (b) What if s_1 and s_2 are sampled?
    (c) What if s_1, s_2, and s_3 are sampled?
    (d) What if no stimulus elements are sampled?

(e)   If s_1 and s_3 are sampled on trial t and R is reinforced, what is the probability that $s_1, s_2,$ and s_3 will all be conditioned to R after trial t? (Do not assume that $c = 1$.)

(f)   Continuing part (e), what is the probability that $s_1, s_2,$ and s_4 will be conditioned to R after trial t? (Do not assume that $c = 1$.)

(g)   If s_1 and s_3 are sampled, what is the probability that $s_1, s_2,$ and s_3 will all be conditioned to R after trial t? (Do not assume that $c = 1$, and do not assume that R is reinforced.)

21.   (a)   Formulate assumptions for a 3-element all-or-none model.

(b)   Using the usual definition of states, construct a transition matrix for the corresponding Markov chain.

(c)   If $c = \frac{1}{2}$, what is the expected number of steps before learning?

(d)   If $c = \frac{1}{2}$, what is the probability of making an incorrect response on trial 2 (this is the third trial)?

22.   In the Estes 1-element model, remove the assumption that $c = 1$, but include the assumption that $\theta = 1$. (This is not a very satisfactory assumption psychologically.)

(a)   Using the usual definition of states, derive a transition matrix for a Markov chain representing learning under this model.

(b)   Is probability matching still predicted?

*23.   Modify the Estes model to allow reinforcement to depend on response. Specifically, assume that the *reinforcement schedule* is given by the following matrix:

$$\begin{array}{c c} & \begin{array}{cc} R & L \end{array} \\ \begin{array}{c} R \\ L \end{array} & \begin{pmatrix} 1 - a & a \\ b & 1 - b \end{pmatrix}. \end{array}$$

This matrix is read as follows. After the response R is made, the response R is reinforced with probability $1 - a$ and the response L with probability a. After the response L is made, the response R is reinforced with probability b, the response L with probability $1 - b$. (In the response-independent case treated previously, $1 - a = b = p$.)

(a)   Using the usual definition of states, show that the following is the transition matrix for a Markov chain under the modified Estes 1-element model, if c is assumed to be 1.

$$\begin{array}{c c} & \begin{array}{cc} 0 & 1 \end{array} \\ P = \begin{array}{c} 0 \\ 1 \end{array} & \begin{pmatrix} 1 - b\theta & b\theta \\ a\theta & 1 - a\theta \end{pmatrix}. \end{array}$$

(b)   Show that the long-run probability that the experimenter will reinforce the response R is given by $bw_0 + (1 - a)w_1$, where $(w_0, w_1) = w$ is the fixed point probability vector corresponding to P.

(c)   Calculate w_0 and w_1 and show that the long-run probability that the subject will respond R is equal to the long-run probability that the experimenter will reinforce R. (This is a more general sense of probability matching.)

24.   In the previous exercise, remove the assumption $c = 1$, and add the assumption $\theta = 1$. Derive a transition matrix. Is probability matching (in the general sense of the previous exercise) predicted?

*25.   (a)   Construct a transition matrix P for the Estes 2-element model with *noncontingent reinforcement* as described by the reinforcement axiom in the text. Assume that $c = 1$. (In Fig. 5.29, we give a sample calculation of the 0, 2 entry.)

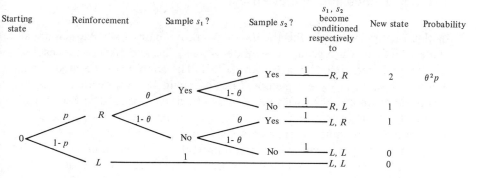

Starting state	Reinforcement	Sample s_1?	Sample s_2?	s_1, s_2 become conditioned respectively to	New state	Probability
			Yes $\xrightarrow{1}$	R, R	2	$\theta^2 p$
		Yes	No $\xrightarrow{1}$	R, L	1	
	R		Yes $\xrightarrow{1}$	L, R	1	
0		No	No $\xrightarrow{1}$	L, L	0	
	L	$\xrightarrow{1}$		L, L	0	

FIGURE 5.29. Calculation of the 0, 2 transition probability for the Estes 2-element model.

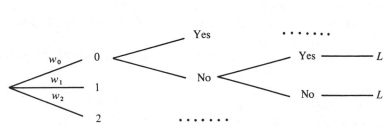

State	Sample s_1?	Sample s_2?	Response
	Yes	$\cdots\cdots$	
w_0 0		Yes	L
w_1 1	No		
w_2		No	L
2	$\cdots\cdots$		

FIGURE 5.30. Part of tree-diagram for determination of long-run probability of responding R under Estes 2-element model.

(b)   The following vector is a fixed point probability vector for P.

$$w = \left(\frac{(1-p)\theta + 2(1-p)^2(1-\theta)}{2-\theta},\ \frac{4p(1-p)(1-\theta)}{2-\theta},\ \frac{p\theta + 2p^2(1-\theta)}{2-\theta}\right).$$

Using this, show that probability matching is predicted. (*Hint:* The long-run probability of responding R can be determined using a diagram like that in Fig. 5.30.)

*26.[34] (a)   Construct a transition matrix P for the Estes 2-element model with noncontingent reinforcement if c is no longer assumed to be 1, but $\theta = 1$.

    (b)   Show that probability matching is still predicted.

27.   What are some of the simplifications in all of our models of learning? How could a more sophisticated model avoid these simplifications?

5.10. Influence and Social Power

In this section we apply the theory of Markov chains to a mathematical model of influence in a group. The model we describe, and most of the results, are due to French [1956] and Harary [1959]. The analysis of the model will illustrate how to analyze a general Markov chain by first dealing with an absorbing chain and then with an ergodic chain, as described in Sec. 5.3.

Suppose a group of n members, $u_1, u_2, \ldots, u_n$, has to make a decision, for example, how much money to spend on a particular project. At time $t = 0$, each member u_i has an opinion, represented by a real number a_i. Each member of the group has influence over certain other members. Let us assume that for each i and j, the nonnegative number α_{ij} denotes the influence of person u_i on person u_j. We shall assume that for every j,

$$\sum_{i=1}^{n} \alpha_{ij} = 1. \tag{31}$$

(This can be achieved by dividing each α_{ij} by $\sum_{i=1}^{n} \alpha_{ij}$.) Assuming (31), we can interpret α_{ij} as the *relative influence* of u_i on u_j, relative meaning compared to the influence of all other u_k's on u_j.

We can represent these influences by a weighted digraph D in the obvious way. The vertices of D are $u_1, u_2, \ldots, u_n$, we put an arc from u_i to u_j if $\alpha_{ij} \neq 0$, and we place a weight α_{ij} on the arc (u_i, u_j) if this arc appears. Let us call D the *influence digraph* of the group. We shall make the following additional assumptions about our model. First, the influence digraph never changes. Thus, our model is static. Second, power is exerted only at the discrete times $t = 1, 2, 3, \ldots$. The opinion of member u_j at time t is given by $a_j(t)$. Finally, we make the following assumption about how the exertion of power or influence affects opinions: At time $t + 1$, the opinion $a_j(t + 1)$ of member u_j is a weighted sum of the opinions $a_i(t)$ at time t of all members

---

[34]The reader should defer this exercise until after reading Sec. 5.10.

u_i having direct influence on u_j; in particular,

$$a_j(t+1) = \sum_{i=1}^{n} \alpha_{ij}a_i(t). \qquad (32)$$

We are interested in whether a particular member reaches a *stable final opinion*, i.e., in whether the sequence $a_i(t)$ approaches a limit as $t \longrightarrow \infty$, and whether each member of the group approaches the same final opinion, which if it exists we shall call the *final common opinion of the group*. (The model does not assert that attainment of a stable final opinion takes an infinite amount of time if $a_i(t)$ approaches a limit. Rather, the opinion for all practical purposes stabilizes once it is within a certain threshhold of the limit.) Using Markov chain theory, we shall be able to show that under certain assumptions about the structure of the influence digraph, the group does reach a final common opinion. Moreover, we shall be able to measure, in some sense, the relative power of each member toward the attainment of the final opinion. As in our study of flow models, it is rather surprising that, given the nonprobabilistic nature of our problem, the theory of Markov chains proves valuable.

Let us first define the *converse* of a weighted digraph D. This is a weighted digraph $C(D)$ which has an arc (u_i, u_j) if and only if D has an arc (u_j, u_i) and a weight p_{ij} equal to α_{ji} on the arc (u_i, u_j). In the converse, we simply reverse the directions of all arcs (except loops). Figure 5.31 shows an influence digraph D and its converse $C(D)$. By virtue of Eq. (31), if D is the influence digraph of a group, then $C(D)$ is a stochastic digraph. Hence, we may think of $C(D)$ as the transition digraph of a Markov chain.

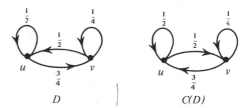

FIGURE 5.31. An influence digraph D and its converse $C(D)$, which is stochastic.

We shall assume that some α_{ii} is nonzero. This assumption, also made by French and Harary, is not too unreasonable, for it simply says that some person has influence on himself. (It is probably reasonable to assume that each person does.)

The theorems about Markov chains are now easily translated into theorems about influence. Let

$$A(t) = \begin{pmatrix} a_1(t) \\ a_2(t) \\ \cdot \\ \cdot \\ \cdot \\ a_n(t) \end{pmatrix}$$

be the vector of opinions at time t. Let P be the matrix (p_{ij}). Then Eq. (32) says that

$$A(t + 1) = PA(t).$$

Hence, we have the following equation:

$$A(t) = P^t A(0).^{35} \tag{33}$$

Theorem 5.17.   If the influence digraph D is strongly connected with at least one loop, then the members of the group attain a final common opinion. This opinion is a weighted average $\sum_i w_i a_i(0)$ of the initial opinions of the members, where the weights w_i are the components of the fixed point probability vector w of the Markov chain corresponding to the converse digraph $C(D)$.

Proof.   It is easy to prove that if D is strongly connected, then so is $C(D)$. Now $C(D)$ defines an ergodic chain and, since some p_{ii} is nonzero, Corollary 2 to Theorem 5.12 implies that the chain is regular. Thus P^t approaches a matrix W all of whose rows are the same vector $(w_1, w_2, \ldots, w_n)$. We conclude from Eq. (33) that for each j,

$$a_j(t) \longrightarrow \sum_{i=1}^{n} w_i a_i(0). \tag{34}$$

Thus, each $a_j(t)$ approaches the same final opinion.   Q.E.D.

Corollary.   If the influence digraph is strongly connected and has at least one loop, then the group reaches a final common opinion at the arithmetic mean of the initial opinions if and only if the matrix (α_{ij}) is stochastic.

Proof.   A matrix is called *doubly stochastic* if and only if every entry is nonnegative and both its row sums and column sums are 1. Thus, $P = (p_{ij})$ is doubly stochastic if and only if (α_{ij}) is stochastic. Finally, it is easy to prove that for regular Markov chains, the fixed vector $(w_1, w_2, \ldots, w_n)$ is the vector $(1/n, 1/n, \ldots, 1/n)$ if and only if P is doubly stochastic. The proof is left as an exercise (Exer. 15, Sec. 5.5).   Q.E.D.

To give an example, we note that if D is the influence digraph of Fig. 5.31, and if $u = u_1$ and $v = u_2$, then the converse $C(D)$ has fixed point probability vector $w = (\frac{3}{5}, \frac{2}{5})$. If u has initial opinion $a_u(0) = 5$ and v has initial opinion $a_v(0) = 10$, then the group reaches a final common opinion $w_1 a_u(0) + w_2 a_v(0) = (\frac{3}{5})5 + (\frac{2}{5})10 = 7$.

Let us recall from Chapter 2 that a vertex basis of a digraph is a minimal collection of vertices which can reach all other vertices. By Theorems 2.7

---

[35]The reader should compare this to the equation $P(t) = P(0)A^t$ for pulse processes on weighted digraphs studied in Chapter 4. He will then see that the present model is once again the pulse process model, except that the underlying matrix is stochastic.

and 2.8, if D is a digraph, then its condensation D^* has a unique vertex basis consisting of all strong components with no incoming arcs. Following Harary [1959], let us call a subgroup of the group $\{u_1, u_2, \ldots, u_n\}$ a *power subgroup* if it forms a strong component in the influence digraph D and this strong component is in the vertex basis of D^*. The power subgroups are exactly the ergodic sets in the Markov chain defined by the weighted digraph $C(D)$. In the influence digraph of Fig. 5.32, the power subgroups are the sets $\{u, v\}$ and $\{y, z\}$. Members of a power subgroup can always influence each other, at least indirectly. Moreover, the power subgroups, taken together, can influ-

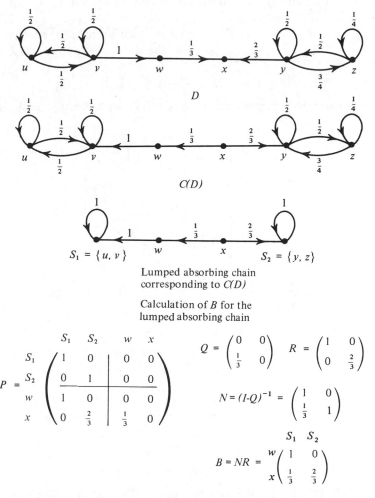

FIGURE 5.32. An influence digraph D, its converse $C(D)$, the lumped absorbing chain corresponding to $C(D)$, and the calculation of $B = NR$ for this absorbing chain.

ence at least indirectly all other individuals. We shall see that if at least one person in each power subgroup exerts influence on himself, then the members of each power subgroup reach a final common opinion, and each individual in the whole influence group reaches a stable opinion which is dependent only on the initial opinions of members of the power subgroups. Thus, the power subgroups themselves reach consensus and, moreover, they wield all the power.

Let us list the power subgroups as $S_1, S_2, \ldots, S_r$ and renumber the vertices so that those of S_1 come first, those of S_2 next, and so on up to those of S_r, with transient vertices (those not in any ergodic set) coming last. To be precise, let $S_1 = \{u_{11}, u_{12}, \ldots, u_{1n_1}\}$, $S_2 = \{u_{21}, u_{22}, \ldots, u_{2n_2}\}, \ldots, S_i = \{u_{i1}, u_{i2}, \ldots, u_{in_i}\}, \ldots$. Let T be the collection of transient vertices, and write $T = \{u_{r+1,1}, u_{r+1,2}, \ldots, u_{r+1,n_{r+1}}\}$. Thus, the states of the chain can be written as

$$\{u_{ij} : i = 1, 2, \ldots, r+1, j = 1, 2, \ldots, n_i\}.$$

We may write the transition matrix for $C(D)$ in the following canonical form:

$$
P = \begin{array}{c}
\\ S_1 \\ S_2 \\ \vdots \\ S_r \\ T
\end{array}
\overset{\displaystyle S_1 \quad S_2 \quad \cdots \quad S_r \qquad T}{
\left(\begin{array}{cccc|c}
\boxed{P_1} & & & & \\
 & \boxed{P_2} & & \mathbf{0} & \mathbf{0} \\
 & & \ddots & & \\
\mathbf{0} & & & \boxed{P_r} & \\
\hline
 & & R & & Q
\end{array}\right).}
$$

In the influence digraph D of Fig. 5.32, we have $S_1 = \{u, v\}$, $S_2 = \{y, z\}$, and $T = \{w, x\}$. Then

$$
P = \begin{array}{c}
u \\ v \\ y \\ z \\ w \\ x
\end{array}
\overset{\displaystyle u \quad\; v \quad\;\; y \quad z \quad\; w \;\; x}{
\left(\begin{array}{cc|cc|cc}
\frac{1}{2} & \frac{1}{2} & 0 & 0 & 0 & 0 \\
\frac{1}{2} & \frac{1}{2} & 0 & 0 & 0 & 0 \\
\hline
0 & 0 & \frac{1}{2} & \frac{1}{2} & 0 & 0 \\
0 & 0 & \frac{3}{4} & \frac{1}{4} & 0 & 0 \\
\hline
0 & 1 & 0 & 0 & 0 & 0 \\
0 & 0 & \frac{2}{3} & 0 & \frac{1}{3} & 0
\end{array}\right).}
$$

If the set of vertices forms just one ergodic set, we have already seen what happens. Let us assume then that the set of vertices does not form an ergodic set. We can define an absorbing Markov chain by lumping each ergodic set into one state. We know from Theorem 5.4 that a Markov chain starting in any transient state will eventually enter some ergodic set; so we might as well deal with this lumped absorbing chain to begin with. (This was the procedure suggested in our classification of Markov chains in Sec. 5.3.) In the example of Fig. 5.32, the lumped absorbing chain is shown.

If the chain starts in transient state $u_{r+1,j}$, its probability of entering ergodic set S_i is given by a number b_{ji}. (The numbers b_{ji} can be calculated according to the method of Theorem 5.7. We show the calculation for our example in Fig. 5.32.) Once the chain enters an ergodic set S_i, it never leaves, and we may study the chain within the ergodic set as a Markov chain in its own right. In particular, if $C(D)$ arises from an influence digraph with a loop at some vertex in each ergodic set, then each ergodic set S_i defines a regular Markov chain. We then know that the long-run probability of the chain being in a given state u_{ik} of S_i becomes independent of the starting state (the state in which the chain enters the set S_i) and, moreover, approaches a number $w_k^{(i)}$. Thus, these probabilities are given by a vector $w^{(i)} = (w_1^{(i)}, w_2^{(i)}, \ldots, w_{n_i}^{(i)})$, the fixed point probability vector of the transition matrix P_i of the chain restricted to the set S_i. In our example, these vectors are $w^{(1)} = (\frac{1}{2}, \frac{1}{2})$ and $w^{(2)} = (\frac{3}{5}, \frac{2}{5})$. To sum up, if the original chain starts in a transient state $u_{r+1,j}$, its long run probability of being in the state u_{ik} of ergodic set S_i is $b_{ji}w_k^{(i)}$. The long run probability of being in a nonergodic state is 0. If $W^{(i)}$ is the matrix with n_i rows $w^{(i)}$, we have shown that P^t approaches a matrix W of the following form:

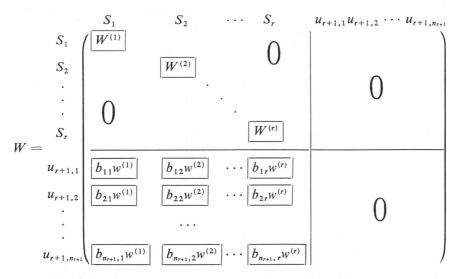

In our example, $b_{11}w^{(1)} = 1 \cdot (\frac{1}{2}, \frac{1}{2}) = (\frac{1}{2}, \frac{1}{2})$, $b_{12}w^{(2)} = 0 \cdot (\frac{3}{5}, \frac{2}{5}) = (0, 0)$, $b_{21}w^{(1)} = (\frac{1}{3}) \cdot (\frac{1}{2}, \frac{1}{2}) = (\frac{1}{6}, \frac{1}{6})$, and $b_{22}w^{(2)} = (\frac{2}{3}) \cdot (\frac{3}{5}, \frac{2}{5}) = (\frac{2}{5}, \frac{4}{15})$. Thus,

$$W = \begin{array}{c} \\ u \\ v \\ y \\ z \\ w \\ x \end{array}\begin{pmatrix} \begin{array}{cc} u & v \end{array} & \begin{array}{cc} y & z \end{array} & \begin{array}{cc} w & x \end{array} \\ \begin{array}{cc} \frac{1}{2} & \frac{1}{2} \end{array} & \begin{array}{cc} 0 & 0 \end{array} & \begin{array}{cc} 0 & 0 \end{array} \\ \begin{array}{cc} \frac{1}{2} & \frac{1}{2} \end{array} & \begin{array}{cc} 0 & 0 \end{array} & \begin{array}{cc} 0 & 0 \end{array} \\ \begin{array}{cc} 0 & 0 \end{array} & \begin{array}{cc} \frac{3}{5} & \frac{2}{5} \end{array} & \begin{array}{cc} 0 & 0 \end{array} \\ \begin{array}{cc} 0 & 0 \end{array} & \begin{array}{cc} \frac{3}{5} & \frac{2}{5} \end{array} & \begin{array}{cc} 0 & 0 \end{array} \\ \begin{array}{cc} \frac{1}{2} & \frac{1}{2} \end{array} & \begin{array}{cc} 0 & 0 \end{array} & \begin{array}{cc} 0 & 0 \end{array} \\ \begin{array}{cc} \frac{1}{6} & \frac{1}{6} \end{array} & \begin{array}{cc} \frac{2}{5} & \frac{4}{15} \end{array} & \begin{array}{cc} 0 & 0 \end{array} \end{pmatrix}.$$

Recall that $A(t) = P^t A(0)$. Thus, $A(t)$ approaches $WA(0)$. Let $a_{ij}(t)$ be the opinion of vertex u_{ij} at time t. If u_{ij} is in S_i, then $a_{ij}(t)$ approaches

$$o(i) = \sum_{k=1}^{n_i} w_k^{(i)} a_{ik}(0). \tag{35}$$

That is, the final opinion of member u_{ij} is dependent only on the initial opinions of other members of the same power subgroup. This opinion is the same for each j, so we may represent this as the power subgroup's final common opinion, using the notation $o(i)$. If $u_{r+1,j}$ is in T, then its final opinion approaches

$$\sum_{i=1}^{r} b_{ji}\left(\sum_{k=1}^{n_i} w_k^{(i)} a_{ik}(0)\right) = \sum_{i=1}^{r} b_{ji} o(i). \tag{36}$$

This depends on the initial opinion of all members of all power subgroups, but not on those of members not in power subgroups. It is a weighted average of the final common opinions of the different power subgroups. Thus, in any case, each member of the group reaches a final stable opinion.

In our example, suppose the initial opinions are given by the vector

$$\begin{pmatrix} 2 \\ 10 \\ 10 \\ 100 \\ 6 \\ 10 \end{pmatrix} = \begin{pmatrix} a_u(0) \\ a_v(0) \\ a_w(0) \\ a_x(0) \\ a_y(0) \\ a_z(0) \end{pmatrix}.$$

Then

$$o(1) = (\tfrac{1}{2})a_u(0) + (\tfrac{1}{2})a_v(0) = (\tfrac{1}{2})2 + (\tfrac{1}{2})10 = 6,$$

and

$$o(2) = (\tfrac{3}{5})a_y(0) + (\tfrac{2}{5})a_z(0) = (\tfrac{3}{5})6 + (\tfrac{2}{5})10 = \tfrac{38}{5}.$$

The final opinions of u and v approach 6, and those of y and z approach $\tfrac{38}{5}$. The final opinion of w approaches

$$b_{11}o(1) + b_{12}o(2) = 1 \cdot o(1) + 0 \cdot o(2) = 6,$$

and the final opinion of x approaches

$$b_{21}o(1) + b_{22}o(2) = (\tfrac{1}{3}) \cdot o(1) + (\tfrac{2}{3}) \cdot o(2) = \tfrac{106}{15}.$$

Each member of the subgroup reaches a final opinion, but it is not the same for each member.

If there is only one power subgroup, then $r = 1$ and $b_{ji} = 1$ for all j, so the right hand side of Eq. (36) reduces to $o(i) = o(1)$. It follows that the final common opinion of each member of the group is the same, the value given in Eq. (35). Here, the number $w_k^{(i)}$ can be interpreted as the *power* of the ergodic member $u_{ik} = u_{1k}$: It represents the relative weight attached to his initial opinion in calculating the group's final opinion. A transient member has power 0.

If there are several power subgroups, then each power subgroup S_i reaches a final common opinion $o(i)$. If each power subgroup reaches the same final opinion, then Eq. (35) gives the same number o for each $i = 1, 2, \ldots, r$. Each group member representing an ergodic state of $C(D)$ reaches this final opinion. It then follows from Eq. (36) that each group member representing a transient state also reaches this opinion o. For

$$\sum_{i=1}^{r} b_{ji}o(i) = \sum_{i=1}^{r} b_{ji}o$$

$$= o \sum_{i=1}^{r} b_{ji}$$

$$= o.$$

In this case, again all the power is in the hands of the ergodic members. In our example, if we change $a_y(0)$ to 10 and $a_z(0)$ to 0, then $o(2)$ is 6, which is the same as $o(1)$. It follows that every member of the group will attain a final common opinion of 6.

In sum, we have proved the following theorem.

Theorem 5.18.   In an influence group in which at least one person in each power subgroup has influence on himself:

(a)   Each person reaches a stable opinion, which is dependent only on the initial opinions of members of the power subgroups.

(b)   If there is only one power subgroup, then the whole group reaches a final common opinion which is the final common opinion of the power subgroup and depends only on the initial opinions of the members of the power subgroup.

(c)   If there are several power subgroups, each power subgroup reaches a final common opinion and the group reaches a final common opinion if and only if the final common opinion of each power subgroup is the same. The final common opinion of the group, if attained, depends only on the initial opinions of the members of the power subgroups.

EXERCISES

1. Suppose an influence group has four members and each member is influenced equally by the opinions of all other members, including his own. Draw an influence digraph D for this group and draw $C(D)$.

2. For each influence digraph D of Fig. 5.33,
   (a) Draw $C(D)$.
   (b) Determine if the group reaches a final common opinion.
   (c) Calculate this opinion if it exists.

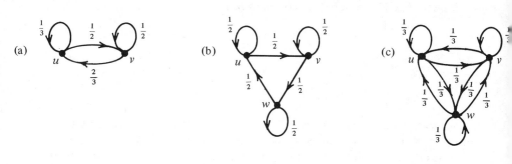

(a)

Initial opinions

$a_u(0) = 10$
$a_v(0) = 20$

(b)

Initial opinions

$a_u(0) = 20$
$a_v(0) = 30$
$a_w(0) = 40$

(c)

Initial opinions

$a_u(0) = 10$
$a_v(0) = 20$
$a_w(0) = 30$

FIGURE 5.33. Influence digraphs for Exer. 2, Sec. 5.10.

3. Consider the influence group of Exer. 1.
   (a) Does the group reach a final common opinion?
   (b) If so, what is it?

4. In the influence digraphs of Fig. 5.34, find all power subgroups.

5. In the influence digraphs of Fig. 5.34, determine if the group reaches a final common opinion. If so, what is it? In the situations with only one power subgroup, what is the power of each member of the group?

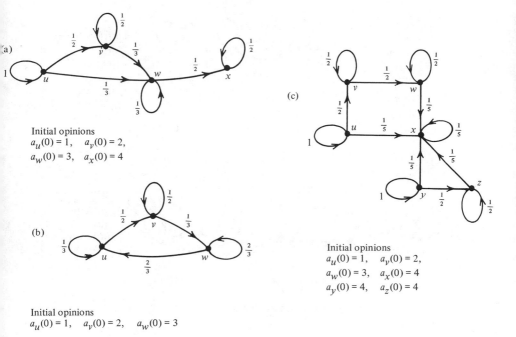

FIGURE 5.34. Influence digraphs for Exers. 4 and 5, Sec. 5.10.

6. For each influence digraph of Fig. 5.35:
   (a) Find all power subgroups.
   (b) Write the transition matrix P in the canonical form defined in this section.
   (c) Find the corresponding lumped absorbing chain.
   (d) Calculate the matrix B for this absorbing chain.
   (e) Calculate the vector $w^{(i)}$ for each power subgroup.
   (f) Calculate the limiting matrix W.
   (g) Calculate $o(i)$ for each power subgroup.
   (h) Calculate the final opinion of each member not in a power subgroup.
   (i) If the group does not reach a final common opinion, determine a different set of initial opinions which lead to one.

7. In the upper echelon of a company, the President's opinion is influenced equally by his own opinion and by that of his two First Vice Presidents.

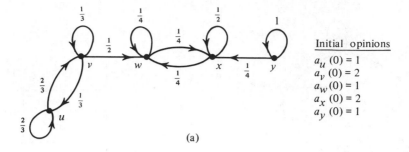

Initial opinions

$a_u(0) = 1$
$a_v(0) = 2$
$a_w(0) = 1$
$a_x(0) = 2$
$a_y(0) = 1$

(a)

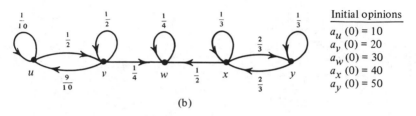

Initial opinions

$a_u(0) = 10$
$a_v(0) = 20$
$a_w(0) = 30$
$a_x(0) = 40$
$a_y(0) = 50$

(b)

FIGURE 5.35. Influence digraphs for Exer. 6, Sec. 5.10.

One of the First Vice Presidents (First Vice President 1) forms his opinion solely on the basis of the President's opinion. The other First Vice President gives equal weight to his own opinion and that of the two Second Vice Presidents. Finally, the two Second Vice Presidents are influenced only by their own opinions.

(a) Who holds the real power in this group, i.e., who really influences the group's final opinion?

(b) If the initial opinions are as follows, does the group reach a final common opinion? If so, what is it?

$$\text{President} = 10$$

$$\text{First Vice President } 1 = 20$$

$$\text{First Vice President } 2 = 30$$

$$\text{Second Vice President } 1 = 100$$

$$\text{Second Vice President } 2 = 100.$$

(c) What if the Second Vice President 2 changes his initial opinion to 200?

8. Consider the influence digraph shown in Fig. 5.36. What happens to the opinions of this group?

Initial opinions

$a_u(0) = 10$
$a_v(0) = 20$

FIGURE 5.36. An influence digraph.

9. Suppose a vertex u of an influence digraph D can reach all vertices of D (i.e., by paths) but there is no other vertex u with this property. Prove that the group attains a final common opinion which is u's initial opinion. (Why is it not necessary to explicitly assume that there is an arc from u to u?)

*10. (a) Prove that every group whose influence digraph is unilaterally connected reaches a final common opinion if at least one member of each power subgroup has influence on himself.

   (b) Show that this result is false without the assumption that at least one member of each power subgroup has influence on himself.

11. Consider the communication network of Fig. 5.37. A message is passed from person to person. When an individual receives the message, he can in the next time period either pass it on to one of the possible recipients

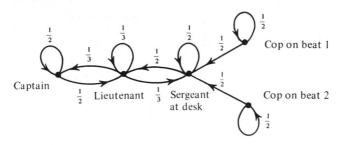

FIGURE 5.37. Communication network.

or keep it. It is assumed that his choice of whom to pass it on to is random, with himself considered equally with the other possible recipients. If he keeps it, he then makes another decision whether to pass it on or keep it in the following time period. Use the methods of this section to answer the following questions:

   (a) What is the probability that the captain will ever receive a message which starts with the cop on the beat 1?

   (b) In the long run, if you check with the captain, what is the probability he will just have received the message, supposing that it started initially with cop on beat 1?

   (c) If we treat this as the converse of an influence digraph, rather than as a communication network, then what are the power subgroups? What determines whether or not the group reaches a final common opinion? If one is reached, what is the power of the different members?

12. Repeat parts (a) and (b) of Exer. 11 for the communication networks of Fig. 5.38.

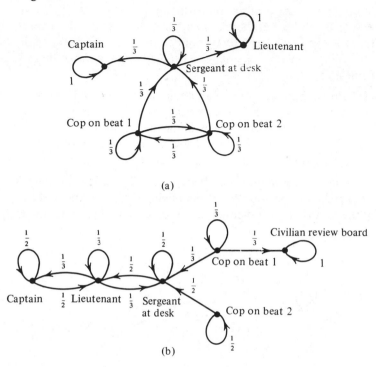

(a)

(b)

FIGURE 5.38. Communication networks for Exer. 13, Sec. 5.10.

*13. What do the results of Theorems 5.17 and 5.18 say about autonomous pulse processes on a weighted digraph D? Is pulse stability implied under certain conditions?

14. Do the results discussed in Exer. 13 hold for all weighted digraphs? (Give an example.)

15. What are some of the simplifications made in our model of influence? How might you modify the model?

5.11. Diffusion and Brownian Motion[36]

Most of the mathematical models of air flow and air pollution look at this flow as a diffusion process. Often they are stated in terms of infinite Markov

---

[36]The discussion in this section is based on Feller [1962], Kemeny and Snell [1960], and Kemeny, Mirkil, Snell, and Thompson [1959]. See also Kac [1947]. This section should be omitted if Sec. 5.6.2 has not been read.

processes under continuous time. But diffusion models stated in terms of finite Markov chains are conceptually simpler and predictions from these models do, in the limit, often agree with predictions from their infinite, continuous-time analogues. (Cf. Feller [1962, Sec. XIV. 6] for an illustration of this point.)

One of the simplest mathematical models of diffusion is that due to *P.* and *T.* Ehrenfest [1907]. (This model applies to various diffusion processes, including that in which particles suspended in a fluid move about under the rapid, successive, random impacts of their neighbors. The motion of these particles is called *Brownian motion.*) We imagine a number *a* of molecules of some substance (e.g., one pollutant) distributed in two regions *A* and *B* which are separated by a thin permeable barrier. In each time period, one molecule from one of the regions is chosen at random and moved from its region to the other one. We build a Markov chain to describe this system as follows. The state of the chain is given by the number of molecules in region *A*. Thus, there are $a + 1$ states, $0, 1, 2, \ldots, a$. To find the transition probabilities, suppose that we are in state i. If $i \neq 0$ and $i \neq a$, then we pick a molecule from region *A* with probability i/a and a molecule from region *B* with probability $1 - (i/a)$. In the former case, we end in state $i - 1$ and in the latter in state $i + 1$. Thus,

$$p_{i,i-1} = \frac{i}{a}$$

$$p_{i,i+1} = 1 - \frac{i}{a} \tag{37}$$

$$p_{i,k} = 0 \quad \text{for} \quad k \neq i - 1, i + 1.$$

If $i = 0$, we can only move to state 1, and do so with probability 1. Similarly, if $i = a$, we move to state $a - 1$ with probability 1. Thus, it is easy to verify that Eqs. (37) hold for all i. The Markov chain defined by the transition matrix $P = (p_{ij})$ is clearly ergodic. Moreover, it is easy to see that it is periodic of period 2.

The same transition matrix arises from a *random walk* on the line with locations $0, 1, 2, \ldots, a$ and states 0 and a acting as *reflecting barriers.* At each time, a particle is at some location i on the line. If i isn't 0 or a, the particle moves to the left one unit with probability i/a and to the right one unit with probability $1 - (i/a)$. The boundary points 0 and a "reflect" the particle back into the interior. (This random walk should be distinguished from the absorbing random walk studied in connection with the gambler's ruin problem, where the boundary points served to absorb any "particle" which reached them. It should also be distinguished from the reflecting random walk where the probability of moving left or right from location i is independent of i. See Sec. 5.6, Exer. 6, for an example of this type of random walk.)

In studying the diffusion process defined by the Markov chain with transition matrix (37), we are interested in the distribution of molecules between the two regions after a relatively long period of time has passed. This distribution is described by finding the fixed point probability vector w for the matrix P. It is easy to show that if $\binom{a}{i}$ is the binomial coefficient

$$\frac{a!}{i!(a-i)!}$$

and

$$w_i = \frac{\binom{a}{i}}{2^a},$$

then $w = (w_0, w_1, \ldots, w_a)$ defines a fixed point probability vector of P and hence, by Theorem 5.14(d), the unique such vector. The number w_i is the same as the probability of putting i molecules into region A if we choose a region for each molecule independently. In the long run, the probability of being in a state close to the "center" $a/2$ (or $(a \pm 1)/2$) if a is odd) is large (on every other trial); the probability of being in a state close to the boundary points 0 or a is small. This is true independent of starting point. In this sense, the Ehrenfest model exhibits a "central tendency."

It is of interest to calculate the expected number of steps e_{ij} required for going from state i to state j. If $i = j$, then Theorem 5.14(h) and a result analogous to the Corollary to Theorem 5.11 tell us that

$$e_{ii} = \frac{1}{w_i}.$$

Thus e_{ii} is small if i is close to the center $a/2$ (or $(a \pm 1)/2$) and large if i is close to the boundary. Similarly, using Theorem 5.14(h) one can calculate that e_{ij} is, relatively speaking, small if j is near the center and large if it is near the boundary. In particular, $e_{0,a/2}$ is quite small while $e_{a/2,0}$ is quite large. The reader will recall that $a/2$ represents the situation where the molecules are (essentially) evenly distributed and 0 represents the situation where all molecules are in region B. Our results say that the former is a sort of equilibrium point, and even if the process starts quite a bit away from it, the expected number of steps to return to this situation is small. Conversely, if the process starts in this evenly distributed situation, the expected number of steps before it reaches a very uneven distribution is large. Physically, this means that the distribution of molecules tends to become almost even between the two regions, even if all the molecules start in one region.

The numbers $e_{0,a/2}$ and $e_{a/2,0}$ are also interesting to consider in the limit as a gets large. For a detailed discussion of this situation, the reader is referred to Kemeny and Snell [1960, p. 173 ff].

EXERCISES

1. Show that the Ehrenfest model gives rise to an ergodic chain which is periodic with period 2.

2. For the Ehrenfest model with $a = 2$:
   (a) Write out P.
   (b) Calculate w.
   (c) Calculate the mean first passage matrix E.
   (d) Show that the results validate the discussion in the text.

3. For the Ehrenfest model with $a = 3$:
   (a) Write out P.
   (b) Calculate w using the formulas given in the text.
   (c) Check that $wP = w$.
   (d) Calculate the mean first passage times e_{ii} for every i and discuss the results.

4. For the Ehrenfest model with $a = 7$:
   (a) Calculate w using the formulas given in the text.
   (b) Calculate e_{ii} for each i and discuss the results.

5. One way of calculating the mean first passage time to a state i is to make i into an absorbing state and to calculate the expected number of steps to absorption. For example, to calculate the mean first passage time to state 0 in the Ehrenfest model with $a = 2$, we consider the Markov chain with

$$P = \begin{pmatrix} 1 & 0 & 0 \\ \frac{1}{2} & 0 & \frac{1}{2} \\ 0 & 1 & 0 \end{pmatrix}.$$

   Use this method to check your calculations in Exer. 2(c), for mean first passage times from states 1 and 2 to state 0.

6. Use the method of Exer. 5 to calculate mean first passage times in the Ehrenfest model with $a = 3$ as follows:
   (a) From the states 1, 2, and 3 to state 0.
   (b) From the states 0, 1, and 3 to state 2.

References

ATKINSON, R. C., BOWER, G. H., and CROTHERS, E. J., *An Introduction to Mathematical Learning Theory*, John Wiley & Sons, Inc., New York, 1965.

BHARUCHA-REID, A. T., *Elements of the Theory of Markov Processes and their Applications*, McGraw-Hill Book Company, New York, 1960.

BHAT, U. N., *Elements of Applied Stochastic Processes*, John Wiley & Sons, Inc., New York, 1972.

BOWER, G. H., "Application of a Model to Paired-associate Learning," *Psychometrika*, **26** (1961), 255–280.

BRIGGS, G. A., *Plume Rise*, AEC Critical Review Series, U.S. Atomic Energy Commission, Washington, D.C., 1969.

COOMBS, C. H., DAWES, R. M., and TVERSKY, A., *Mathematical Psychology*, Prentice-Hall, Inc., Englewood Cliffs, N.J., 1970.

COX, D. R. and MILLER, H. D., *The Theory of Stochastic Processes*, John Wiley & Sons, Inc., New York, 1965.

EHRENFEST, P. and EHRENFEST, T., "Über zwei bekannte Einwande gegen das Boltzmannsche H-Theorem," *Physikalische Zeitschrift*, **8** (1907), 311–314.

ELANDT-JOHNSON, R. C., *Probabilistic Models and Statistical Methods in Genetics*, John Wiley & Sons, Inc., New York, 1971.

EWENS, W. J., *Population Genetics*, Methuen, London, 1969.

FELLER, W., *An Introduction to Probability Theory and its Applications*, I, John Wiley & Sons, Inc., New York, 1962.

FRENCH, J. R. P., Jr., "A Formal Theory of Social Power," *Psychol. Rev.*, **63** (1956), 181–194.

GABRIEL, K. R. and Neumann, J., "A Markov Chain Model for Daily Rainfall Occurrence at Tel Aviv," *Quart. J. R. Met. Soc.*, **88** (1962), 90–95.

GLASS, D. V. and Hall, J. R., "Social Mobility in Great Britain: A Study of Integeneration Changes in Status," in *Social Mobility in Great Britain*, D. V. Glass (ed.), Routledge and Kegan Paul, London, 1954.

GOODMAN, A. W. and RATTI, J. S., *Finite Mathematics with Applications*, The Macmillan Company, New York, 1971, 1975.

GULLIKSEN, H., "A Rational Equation of the Learning Curve Based on Thorndike's Law of Effect," *J. of General Psychology*, **11** (1934), 395–434.

HARARY, F., "A Criterion for Unanimity in French's Theory of Social Power," in *Studies in Social Power*, D. Cartwright (ed.), Inst. Soc. Res., Ann Arbor, Mich., 1959, pp. 168–182.

HOWARD, R. A., "Stochastic Process Models for Consumer Behavior," *J. Advertising Res.*, **3** (1963), 35–42.

HOWARD, R. A., *Dynamic Probabilistic Systems I: Markov Models*, John Wiley, & Sons, Inc., New York, 1971.

KAC, M., "Random Walk and the Theory of Brownian Motion," *Amer. Math. Monthly*, **54** (1947), 369–391.

KARLIN, S., *A First Course in Stochastic Processes*, Academic Press, Inc., New York, 1969, a.

KARLIN, S., *Equilibrium Behavior of Population Genetic Models with Non-Random Mating*, Gordon and Breach, New York, 1969 b.

KEMENY, J. G., MIRKIL, H., SNELL, J. L., and THOMPSON, G. L., *Finite Mathematical Structures*, Prentice-Hall, Inc., Englewood Cliffs, N.J., 1959.

KEMENY, J. G., and SNELL, J. L., *Finite Markov Chains*, D. Van Nostrand Co., Inc., Princeton, N.J., 1960.

KEMENY, J. G. and SNELL, J. L., *Mathematical Models in the Social Sciences*, Blaisdell Publishing Co., New York, 1962; reprinted by M.I.T. Press, Cambridge, Mass., 1972.

KEMENY, J. G., SNELL, J. L., and THOMPSON, G. L., *Introduction to Finite Mathematics*, Prentice-Hall, Inc., Englewood Cliffs, N.J., 1956, 1957, 1966, 1974.

MOSIMANN, J., *Elementary Probability for the Biological Sciences*, Appleton-Century-Crofts, New York, 1968.

NORMAN, M. F., *Markov Processes and Learning Models*, Academic Press, Inc., New York, 1972.

PRAIS, S. J., "Measuring Social Mobility," *J. of the Royal Stat. Soc.*, **118** (1955), 56–66.

ROMANOVSKY, V. I., *Discrete Markov Chains*, translated by E. Seneta, Wolters-Noordhoff Publ., Gröningen, the Netherlands, 1970.

SLADE, D. H., *Meteorology and Atomic Energy*, AEC Critical Review Series, U.S. Atomic Energy Commission, Washington, D.C., 1968.

STERN, A. C., WOHLERS, H. C., BOWBEL, R. W., and LOWRY, W. P., *Fundamentals of Air Pollution*, Academic Press, New York, 1973.

SUPPES, P. and GINSBURG, R. A., "A Fundamental Property of All-or-None Models, Binomial Distribution Prior to Conditioning with Application to Concept Formation in Children," *Psychol. Rev.*, **70** (1963), 139–161.

THEIOS, J., "Simple Conditioning as Two-stage All-or-none Learning," *Psychol. Rev.*, **70** (1963), 403–417.

SIX

n-Person Games

6.1. Games in Characteristic Function Form

In this chapter we consider a situation, a "game," where n "players" try to reach agreement about possible outcomes. The situation might be a marketplace, a legislature, an international bargaining table, etc. A "solution" to such a game tells which of the set of possible outcomes will occur, or, in a weaker sense, predicts that one of a set of outcomes will occur. We shall be interested in developing various solution concepts for games, and applying the results.

Each of the games we shall study has a finite set $I = \{1, 2, \ldots, n\}$ of players. A subset S of I will be called a (potential) *coalition*. We allow players to cooperate and form coalitions, if it is to their benefit. We define the *value* $v(S)$ of a coalition S as the largest payoff the coalition can guarantee itself. The value v is a function assigning a real number to each subset S of the set I. We shall assume that v completely defines the game, and deal only with games in which this function v satisfies the following conditions:

$$v(\varnothing) = 0 \tag{1}$$

$$v(S \cup T) \geqq v(S) + v(T) \quad \text{if} \quad S \cap T = \varnothing. \tag{2}$$

Condition (1) is a convenience. Condition (2), called *superadditivity*, implies that in general coalition formation is not a bad thing. Two disjoint groups working together do at least as well as the two groups would do by working separately and then pooling their "winnings." A function v defined on the subsets of I and satisfying Eqs. (1) and (2) is called a *characteristic function*, and we shall speak of it as defining a game in *characteristic function form*. (Sometimes we shall even speak of v as a game.)

We begin by calculating characteristic functions for a number of games which we shall study throughout this chapter.[1]

Example 1: Pure Bargaining. A private foundation has offered n states a total of d dollars for development of water pollution abatement facilities provided that the states can agree on the distribution of the money. If there is no agreement, the foundation will withhold the money. In this game, the coalition $I = \{1, 2, \ldots, n\}$ consisting of all n players can guarantee itself d dollars. Any other coalition S of players cannot guarantee itself anything, since players outside the coalition may not agree to the coalition's plans for distribution. Thus,

$$v(S) = \begin{cases} 0 & (\text{if } |S| < n) \\ d & (\text{if } S = I), \end{cases}$$

where $|S|$ is the number of players in S. The reader should check here and in the following examples that v is indeed a characteristic function, i.e., satisfies conditions (1) and (2). In analyzing this game, we will want to determine whether or not the states will reach an agreement.

Example 2: Deterrence. Each of the n players has the means to destroy the wealth of any other player. If w_i is the wealth of player i, we have

$$v(S) = \begin{cases} -\sum_{i \in S} w_i & (\text{if } |S| < n) \\ 0 & (\text{if } S = I). \end{cases}$$

For only the grand coalition I can agree not to destroy anything and any smaller coalition is totally vulnerable from outside. (Note that $v(S)$ gives the change in wealth which results from the game; to obtain S's standing after the game, one should add the amount S started with.) In analyzing this game we want to determine whether or not there will be agreement or a battle.

Example 3: Land Development. A given tract of land is currently worth \$100,000 to its owner, player 1, who uses it for agricultural purposes. A prospective industrial user, player 2, considers the land worth \$200,000 and a prospective subdivider, player 3, considers it worth \$300,000. There are no

---

[1]The first two examples are due to L. Shapley and M. Shubik (unpublished) and the third is a Shapley-Shubik adaptation of a game, Example 3a below, which is studied in von Neumann and Morgenstern [1944]. The author thanks Dr. Shapley for introducing him to these games.

other prospective buyers. The characteristic function is given as follows:

$$v(\{1\}) = \$100,000$$
$$v(\{2\}) = \$0$$
$$v(\{3\}) = \$0$$
$$v(\{1, 2\}) = \$200,000$$
$$v(\{1, 3\}) = \$300,000$$
$$v(\{2, 3\}) = \$0$$
$$v(\{1, 2, 3\}) = \$300,000.$$

To explain, $v(\{1, 2\}) = \$200,000$ because the coalition could use the land for a plant site; $v(\{2, 3\}) = \$0$ because the industrial user and subdivider together cannot use the land without the assent of the present owner; etc. This example generalizes to any simple market with one seller and two buyers. (See the next example.) We would like to know whether or not player 1 makes a sale, and if so, to whom and at what price.

Example 3a: Two-Buyer Market.   Player 1 owns an object worth a units to him. Player 2 thinks the object is worth b units and player 3 thinks it is worth c units. Assuming $a < b < c$, the characteristic function of this game is given by

$$v(\{1\}) = a$$
$$v(\{2\}) = 0$$
$$v(\{3\}) = 0$$
$$v(\{1, 2\}) = b$$
$$v(\{1, 3\}) = c$$
$$v(\{2, 3\}) = 0$$
$$v(\{1, 2, 3\}) = c.$$

Example 4: Majority Rule.   A coalition "wins" if it has a majority of the n players and "loses" otherwise. Here, we may define v as follows:

$$v(S) = \begin{cases} 1 & \left(\text{if } |S| > \dfrac{n}{2}\right) \\ 0 & \text{(otherwise).} \end{cases}$$

We will want to know if a coalition is likely to form, and if so which one.

Example 5: Weighted Majority Games.   Player i has k_i votes. A coalition "wins" if the sum of its votes is at least as great as some quota Q. Thus, we may define v as follows:

$$v(S) = \begin{cases} 1 & (\text{if } \sum_{i \in S} k_i \geqq Q) \\ 0 & \text{(otherwise).} \end{cases}$$

We can represent a weighted majority game by giving an $(n + 1)$-tuple $(Q; k_1, k_2, \ldots, k_n)$, where n is the number of players. Weighted majority games arise in many situations. For example, the players could be stockholders in a company and the number of votes a player has might correspond to the percentage of shares he owns. Or the players could be states or election districts and the number of votes a district has in a legislature could correspond to its population. The number of votes a player has does not always correspond to his real power. For example, consider the game (51; 49, 48, 3). In this game, player 1 has 49 votes, player 2 has 48 votes, and player 3 has 3 votes; a simple majority of 51 votes is needed to win. The winning coalitions are {1, 2}, {1, 3}, {2, 3} and {1, 2, 3}. Thus, $v(S) = 1$ iff $|S| \geq 2$. The winning coalitions are exactly the same as in the game (2; 1, 1, 1). In the latter game, each player has equal power, since all players have the same number of votes. In terms of ability to enter into a winning coalition, the players in the game (51; 49, 48, 3) also have equal power. One interpretation of this result is the following: A small third party in a political system without a majority party can have as much real power as either majority party, if everyone votes along party lines. We return to this notion of power in Sec. 6.7. To give another example of a weighted majority game, in Nassau County, New York, in 1964, the Board of Supervisors consisted of six members with respectively 31, 31, 21, 28, 2, 2 votes. Needed to win were 59 votes. Thus, we have the game (59; 31, 31, 21, 28, 2, 2) (Banzhaf [1965]). In the Australian government, on some issues, each of the six states gets one vote and the federal government gets two votes. Five votes are needed to win. In case of a 4-4 tie, the federal government makes the decision. The winning coalitions are all coalitions of at least five states or at least two states and the federal government. These are the same as in the weighted majority game (5; 1, 1, 1, 1, 1, 1, 3), so we may think of the federal government as having three votes. The Security Council of the United Nations can also be thought of as a weighted majority game. Here, a measure wins if and only if it has the support of the five permanent members and at least four of the remaining ten members. It is easy to show that the Security Council can be thought of as the weighted majority game (39; 7, 7, 7, 7, 7, 1, 1, 1, 1, 1, 1, 1, 1, 1, 1), since this game has the same winning coalitions if players 1 through 5 are thought of as the permanent members.

Example 5a: Simple Games. A game in characteristic function form is called *simple* if each coalition S has value 0 or 1. Coalitions S with $v(S) = 1$ are called *winning*, all others *losing*. Examples of simple games are all weighted majority games. However, not every simple game is a weighted majority game. (See Exer. 10.)

Example 6: The Garbage Game.[2]   Each of n players has a bag of garbage which he *must* drop in someone else's yard. The utility or worth of b bags of

[2]This game and the next one are due to Shapley and Shubik [1969b].

garbage is $-b$. Then

$$v(S) = \begin{cases} 0 & (\text{if } s = 0) \\ -(n - s) & (\text{if } 0 < s < n) \\ -n & (\text{if } s = n), \end{cases}$$

where $s = |S|$. For if $s < n$, members of a coalition S of s players can agree not to drop garbage in each others' yards. Then the worst that can happen to S is that all other players drop garbage in the yards of members of S. For the coalition consisting of all n players, no such non-dumping agreement is possible. In studying this game, we will ask whether or not the players can reach some agreement on the disposal of garbage.

Example 7: The Lake.   There are n factories around a lake. We assume that it costs an amount B for a factory to treat its wastes before discharging them into the lake and that it costs an amount μC for a factory to purify its own water supply, if μ is the number of factories that *do not* treat their wastes. Assuming $C < B < nC$, we have

$$v(S) = \begin{cases} -snC & \left(\text{if } s \le \dfrac{B}{C}\right) \\ -snC + s(sC - B) & \left(\text{if } s \ge \dfrac{B}{C}\right), \end{cases} \tag{3}$$

where again $s = |S|$. (Proof that this is v is left as an exercise (Exer. 9).) We want to know whether or not the factories can reach some agreement on treatment of wastes.

EXERCISES

1.   Which of the following are characteristic functions? Why?
   (a) $n = 3$, $v(\varnothing) = 0$, $v(\{1\}) = v(\{2\}) = v(\{3\}) = -1$, $v(\{1, 2\}) = 3$, $v(\{1, 3\}) = 3$, $v(\{2, 3\}) = 4$, $v(\{1, 2, 3\}) = 2$.
   (b) $n = 3$, $v(\varnothing) = 0$, $v(\{1\}) = 0$, $v(\{2\}) = 0$, $v(\{3\}) = -1$, $v(\{1, 2\}) = 1$, $v(\{2, 3\}) = -1$, $v(\{1, 3\}) = 0$, $v(\{1, 2, 3\}) = 0$.
   (c)   Arbitrary n, $v(S) = -|S|$, all S.

2.   Show that the function v defined in the text in each of the following examples is indeed a characteristic function.
   (a)   Pure Bargaining.
   (b)   Deterrence.
   (c)   Land Development.
   (d)   Majority Rule.

3.   (L. Shapley and M. Shubik (unpublished).) In the game called Post Office, every player must mail one dollar to some *other* player. Calculate the characteristic function of this game.

4. For each of the following weighted majority games, describe all winning coalitions.

   (a)   (51; 50, 30, 20).
   (b)   (51; 60, 30, 10).
   (c)   The Board of Supervisors in Nassau County New York: (59; 31, 31, 21, 28, 2, 2).
   (d)   (201; 100, 100, 100, 100, 1).
   (e)   (151; 100, 100, 100, 1).

5. For the following weighted majority games, calculate the characteristic function and identify all *minimal* winning coalitions, i.e., winning coalitions with the property that a removal of any player results in a losing coalition.

   (a)   (7; 3, 3, 5, 6, 1).
   (b)   (15; 14, 1, 1, 1, 1, 5).
   (c)   All games of Exer. 4.

6. (A. W. Tucker).[3] Suppose $n = 4$, and define v as follows:

$$v(\varnothing) = 0$$

$$v(\{i\}) = 0 \qquad \text{(all } i)$$

$$v(\{i, j\}) = \frac{i + j}{10} \qquad \text{(all } i \neq j)$$

$$v(\{i, j, k\}) = \frac{i + j + k}{10} \qquad \text{(all distinct } i, j, k)$$

$$v(\{1, 2, 3, 4\}) = \frac{1 + 2 + 3 + 4}{10}.$$

   Show that v is a characteristic function.

7. (A. W. Tucker.) Generalizing the previous exercise, let $I = \{1, 2, \ldots, 2n\}$, and let

$$v(\varnothing) = 0$$

$$v(\{i\}) = 0 \qquad \text{(all } i)$$

$$v(S) = \sum_{i \in S} c_i \qquad \text{(for each } S \text{ with } |S| > 1),$$

   where the c_i are $2n$ positive constants with sum equal to one. Verify that v is the characteristic function of a $2n$-person game.

8. Suppose v is a characteristic function and $v(I) = 1$. Is it necessarily true that $v(S) \leq 1$ for all S? (Give proof or counterexample.)

---

[3]The author thanks Prof. Tucker for permission to use this and a number of other of his unpublished exercises in this chapter.

9. Show that (3) is the proper characteristic function for the Lake Game.

10. Consider a conference committee consisting of three senators x, y, and
    z, and three members of the House of Representatives, a, b, and c. A
    measure passes this committee if and only if it receives the support of at
    least two senators and at least two representatives.

    (a) Calculate the characteristic function of this simple game.
    (b) Show that this game is not a weighted majority game. That is, we
        cannot find votes $v(x)$, $v(y)$, $v(z)$, $v(a)$, $v(b)$, and $v(c)$ and a quota Q
        such that a measure passes if and only if the sum of the votes in
        favor of it is at least Q. (*Note:* A similar argument shows that, in
        general, a bicameral legislature cannot be thought of as a weighted
        majority game.)

11. Which of the following defines a weighted majority game in the sense
    that there is a weighted majority game with the same winning coalitions?
    Give a proof of your answer.

    (a) Three players, and a coalition wins if and only if player 1 is in it.
    (b) Four players, a, b, x, y; a coalition wins if and only if at least a or
        b and at least x or y is in it.
    (c) Four players and a coalition wins if and only if at least three
        players are in it.

12. A coalition S with $|S| \geq 2$ is called *essential* if for every partition T_1,
    $T_2, \ldots, T_k$ of S into (disjoint) smaller sets T_i, we have $\sum_{i=1}^{k} v(T_i) < v(S)$.
    Calculate all essential coalitions in the following games. (*Note:* If we
    know the value $v(S)$ for all 1-person coalitions and essential coalitions
    S, then we should be able to calculate $v(S)$ for all coalitions. Why?)

    (a) Pure Bargaining.
    (b) Deterrence.
    (c) Land Development.
    (d) Majority Rule.
    (e) The weighted majority game (3; 1, 1, 1, 1, 1).
    (f) The weighted majority game (6; 1, 2, 3, 4).
    (g) The Security Council.
    (h) The Garbage Game.

13. *Project:* Gather data on the voting distribution in some legislative
    bodies and discuss the relative power of the different parties. Lucas
    [1974] has data on the New York City Board of Estimates, the Israeli
    Knesset (in 1965), the Tokyo Metropolitan Assembly (in 1973), and
    others.

14. What are some of the omissions in treating a game as a characteristic
    function?

6.2.   The Core

6.2.1.   EFFECTIVE PREFERENCE

A game can have various possible outcomes. For us, a "solution" to a game tells which of these outcomes will occur, or at least predicts that the outcome will be one of a given group.

To define various solution concepts, we shall treat outcomes as vectors $(x_1, x_2, \ldots, x_n)$, with x_i interpreted as the "payoff" to player i, and n equal to the number of players. We assume that payoffs are measured in some quantity (for example, money) which is transferable and, in principle, infinitely divisible. Thus, suppose a coalition $S = \{i_1, i_2, \ldots, i_s\}$ receives a total payoff of u units. These u units can be split up in any way the individuals in S work out. That is, if $\sum_{j=1}^{s} x_{i_j} = u$, then the vector $(x_{i_1}, x_{i_2}, \ldots, x_{i_s})$ represents an allowable distribution of payoffs within S. Moreover, if $\sum_{j=1}^{s} y_{i_j} = 0$, then the vector

$$(x_{i_1} + y_{i_1}, x_{i_2} + y_{i_2}, \ldots, x_{i_s} + y_{i_s})$$

also represents an allowable distribution. We say that the players in S can give each other *side-payments*, i.e., payoffs or bribes, *after* the payoffs resulting from coalition behavior have been made. Assuming the existence of side payments, we may separate the question of the payoff which a coalition can guarantee itself from the question of how the payoffs in a game are split up among individuals.

If $x = (x_1, x_2, \ldots, x_n)$ is a possible outcome, it seems reasonable to assume that this vector x will not be accepted as a payoff by the ith player unless x_i is at least as great as what this player can guarantee himself. Thus, it is reasonable to assume that

$$x_i \geqq v(\{i\}) \qquad \text{(all } i\text{)}. \tag{4}$$

Condition (4) is called the condition of *individual rationality*. Similarly, it is reasonable to assume that

$$\sum_{i=1}^{n} x_i \geqq v(I). \tag{5}$$

Indeed, we will not lose anything by limiting ourselves to vectors x which satisfy the stronger property

$$\sum_{i=1}^{n} x_i = v(I). \tag{6}$$

Condition (5) is called *feasibility* and condition (6) is called *Pareto optimality*. Feasibility says that the total payoffs add up to at least what the grand coalition I can attain, and Pareto optimality says that they add up to exactly what the grand coalition can attain. The set of n-dimensional vectors $x = (x_1, x_2, \ldots, x_n)$ satisfying individual rationality and Pareto optimality is called the set of *imputations* for the game defined by the characteristic function *v*.

To give an example, let $n = 3$, and let

$$v(\varnothing) = 0$$
$$v(\{1\}) = v(\{2\}) = v(\{3\}) = 0$$
$$v(\{1, 2\}) = v(\{1, 3\}) = v(\{2, 3\}) = \tfrac{2}{3} \tag{7}$$
$$v(\{1, 2, 3\}) = 1.$$

Then $x = (x_1, x_2, x_3)$ is an imputation if and only if $x_i \geqq 0$, for $i = 1, 2, 3$, and $x_1 + x_2 + x_3 = 1$. Thus, $x = (\tfrac{1}{2}, \tfrac{1}{3}, \tfrac{1}{6})$ and $y = (\tfrac{1}{3}, \tfrac{5}{12}, \tfrac{1}{4})$ are examples of imputations.

The set $M = M(v)$ of imputations is the set of possible outcomes of the game. To determine which (individual or set of) imputations are realistic outcomes, we note that each coalition will have preferences among the various possible outcomes. However, only certain of these preferences are *effective* in the sense that the coalition S, if acting together, can enforce this preference on the whole group. We say a coalition S *effectively prefers* imputation $x = (x_1, x_2, \ldots, x_n)$ to imputation $y = (y_1, y_2, \ldots, y_n)$ if $S \neq \varnothing$,

$$x_i > y_i \qquad \text{(for all } i \in S), \tag{8}$$

and

$$\sum_{i \in S} x_i \leqq v(S). \tag{9}$$

Condition (8) says that S (strictly) prefers x to y and condition (9) says that S has the power to guarantee an outcome at least as good as $\sum_{i \in S} x_i$. We shall write $x \succ_S y$ if x is effectively preferred to y by the coalition S. If $x \succ_S y$ for some S, we say that imputation x is *effectively preferred* to imputation y and write $x \succ y$.

Continuing our example (7), let us note that for the imputations $x = (\tfrac{1}{2}, \tfrac{1}{3}, \tfrac{1}{6})$ and $y = (\tfrac{1}{3}, \tfrac{5}{12}, \tfrac{1}{4})$, we have $y \succ_{\{2, 3\}} x$ since $y_2 > x_2, y_3 > x_3$, and $y_2 + y_3 \leqq v(\{2, 3\})$. Next, let $z = (\tfrac{9}{24}, \tfrac{1}{3}, \tfrac{7}{24})$. Then $z \succ_{\{1, 3\}} y$. Thus, $z \succ y$ and $y \succ x$. However, it is not the case that $z \succ x$. For if $z \succ_S x$, then condition (8) implies that $S = \{3\}$. However, $z_3 > v(\{3\})$, which violates condition (9).[4] (Thus, the relation $\succ$ is not necessarily *transitive*, i.e., does not satisfy the condition that $z \succ x$ whenever $z \succ y$ and $y \succ x$. However, the relation $\succ_S$ is transitive (see Exer. 14).)

We could continue our example by finding an imputation w effectively preferred to $z = (\tfrac{9}{24}, \tfrac{1}{3}, \tfrac{7}{24})$, one effectively preferred to w, etc. However, there are some imputations which have no others effectively preferred to them. An example is the imputation $(\tfrac{1}{3}, \tfrac{1}{3}, \tfrac{1}{3})$. (The proof of this will be straightforward, using Theorem 6.1 below.)

---

[4]In general, effective preference cannot take place with respect to 1-element coalitions S. See Exer. 15.

6.2.2. COMPUTATION OF THE CORE

D. B. Gillies [1953] and L. S. Shapley (in Kuhn [1953]) introduced the name *core* for the set of all imputations x to which no other imputation is effectively preferred. (See also Gillies [1959].) Thus, $(\tfrac{1}{3}, \tfrac{1}{3}, \tfrac{1}{3})$ is in the core of the game (7). To gain insight into the notion of core, we build an *effective preference digraph* D by taking $M = M(v)$, the set of imputations, as the set of vertices of D, and drawing an arc from imputation x to imputation y if x is effectively preferred to y. It is reasonable to assume that D **has no loops. We shall make this assumption about all the digraphs in this chapter.** D may have an infinite number of vertices, in which case most of the results in graph theory which we have obtained in this book do not apply. The *core* C is the set of all vertices of D which have indegree 0, i.e., have no incoming arcs. It seems reasonable to suggest that a solution to the game will boil down to choice of some outcome in the core. For once an outcome in the core is proposed, no coalition has the strength to improve upon it. Thus, the core could be considered a "solution" to the game. Unfortunately, not every digraph has a core, i.e., a nonempty core.[5] For example, the directed cycle of length n does not. Moreover, the core may consist of several vertices, not just one; see the digraph of Fig. 6.1. Ideally, a solution to a game would specify

FIGURE 6.1. A digraph whose core consists of vertices 1 and 2.

a particular outcome, not a set. However, structural analysis of the set of outcomes may not necessarily be expected to yield such a single outcome, as we shall see.

In the weighted majority game $(51; 51, 48, 1)$, the characteristic function v is

$$v(S) = \begin{cases} 1 & (\text{if } 1 \in S) \\ 0 & (\text{if } 1 \notin S). \end{cases}$$

The imputations are all the vectors $x = (x_1, x_2, x_3)$ such that $x_1 \geq 1$, $x_2 \geq 0$, $x_3 \geq 0$, and $x_1 + x_2 + x_3 = 1$. Thus, $M(v)$, the set of imputations, consists of the single imputation $(1, 0, 0)$. The effective preference digraph consists of a single vertex. There is nothing effectively preferred to the imputation $(1, 0, 0)$, and so the core consists of this single imputation. It is easy to interpret this "solution" to the game. Clearly, player 1 is a "dictator." He wins all.

---

[5]It is a convention in game theory to say the *core exists* if it is nonempty.

To give another example, in the weighted majority game $(2; 1, 1, 1)$, $v(\{i\}) = 0$ for $i = 1, 2, 3$ and $v(\{1, 2, 3\}) = 1$. Thus, $M(v)$ is the set of all nonnegative vectors $x = (x_1, x_2, x_3)$ whose components sum to 1, an infinite set. The core here is the empty set. For given $y = (y_1, y_2, y_3)$ in $M(v)$, some pair of components will sum to less than 1. Thus, let us assume that $y_1 + y_2 < 1$ and define $x = (y_1 + (\epsilon/2), y_2 + (\epsilon/2), y_3 - \epsilon)$, where ϵ is chosen so that $y_1 + y_2 + \epsilon \leqq 1$. Then $x \succ_{\{1, 2\}} y$. For $x_1 > y_1$, $x_2 > y_2$ and $x_1 + x_2 \leqq 1 = v(\{1, 2\})$. The interpretation of this result is as follows. No agreement on split of the prize can satisfy every group of players to the point where they can be sure of not doing better.

In general, it is helpful to have a procedure for finding all imputations in the core. Our first theorem gives such a procedure.

Theorem 6.1. The core of a game with characteristic function v consists of all imputations $x = (x_1, x_2, \ldots, x_n)$ such that

$$\sum_{i \in S} x_i \geqq v(S) \qquad \text{(all } S \neq \varnothing\text{).}^{6} \tag{10}$$

Proof. Suppose x is an imputation. If x satisfies condition (10), then $y_i > x_i$ for all $i \in S$ implies $\sum_{i \in S} y_i > v(S)$. Thus, it is not the case that $y \succ_S x$, for any y and S, i.e., x is in the core. Next, suppose x violates condition (10). Then for some $S \neq \varnothing$,

$$\sum_{i \in S} x_i = v(S) - \epsilon,$$

$\epsilon > 0$. Let

$$\delta = v(I) - v(S) - \sum_{i \notin S} v(\{i\}).$$

By superadditivity (condition (2)), $\delta \geqq 0$. Let

$$y_i = \begin{cases} x_i + \dfrac{\epsilon}{s} & \text{(if } i \in S) \\ v(\{i\}) + \dfrac{\delta}{n - s} & \text{(if } i \notin S), \end{cases} \tag{11}$$

where $s = |S|$. It is relatively easy to check that $y \succ_S x$ and that y is an imputation. The proofs are left to the reader as an exercise (Exer. 21).

Q.E.D.

Theorem 6.1 has a very interesting interpretation. It says that an imputation is in the core if and only if it meets the minimal requirements of every coalition. A game with an empty core must always leave one coalition short of being satisfied.

---

[6]Condition (10) is often called *group rationality*. Some authors prefer to define the core as the set of all imputations x satisfying condition (10), and then prove that the core consists of all imputations to which no other imputation is effectively preferred. It should be noted that Theorem 6.1 does not hold without the assumption of superadditivity, or a somewhat weaker assumption described in Shapley and Shubik [1969a] (see Exer. 16).

In Section 6.4 we shall prove some general theorems on the existence of (nonempty) cores in games in characteristic function form. For now, let us calculate several more cores, using Theorem 6.1.

In Pure Bargaining (Example 1 of the previous section), the imputations are all the vectors x such that $x_i \geq 0$ for all i and $\sum_{i=1}^{n} x_i = d$. By the theorem, it is easy to see that every imputation is in the core. If we assume that some imputation in the core will be chosen as the solution to the game, the interpretation of this result is that the group will reach some bargain. But game theory does not tell us which bargain.

Let us return to the weighted majority game $(2; 1, 1, 1)$, which has an empty core, and demonstrate this result again using Theorem 6.1. By the theorem, an imputation $x = (x_1, x_2, x_3)$ is in the core if and only if

$$x_i \geq 0 \qquad (i = 1, 2, 3) \tag{12}$$

$$x_1 + x_2 \geq 1 \tag{13}$$

$$x_1 + x_3 \geq 1 \tag{14}$$

$$x_2 + x_3 \geq 1 \tag{15}$$

and

$$x_1 + x_2 + x_3 \geq 1. \tag{16}$$

Inequalities (12) and (16) are automatically satisfied for all imputations. In general, in applying Theorem 6.1, inequality (10) need be used only for coalitions S with $1 < |S| < n$. (Why?) Adding inequalities (13), (14), and (15), we get

$$x_1 + x_2 + x_3 \geq \tfrac{3}{2},$$

which contradicts $x_1 + x_2 + x_3 = 1$ (Pareto optimality). Thus, the core is empty.

In the Land Development Game (Example 3 of the previous section), the imputations are all vectors $x = (x_1, x_2, x_3)$ with $x_1 \geq \$100{,}000$, $x_2 \geq \$0$, $x_3 \geq \$0$, and $x_1 + x_2 + x_3 = \$300{,}000$. Using Theorem 6.1, we see that an imputation x is in the core if and only if it satisfies the following inequalities (17) through (23).

$$x_1 \geq \$100{,}000 \tag{17}$$

$$x_2 \geq \$0 \tag{18}$$

$$x_3 \geq \$0 \tag{19}$$

$$x_1 + x_2 \geq \$200{,}000 \tag{20}$$

$$x_1 + x_3 \geq \$300{,}000 \tag{21}$$

$$x_2 + x_3 \geq \$0 \tag{22}$$

$$x_1 + x_2 + x_3 \geq \$300{,}000. \tag{23}$$

As usual, for every imputation x, inequalities (17), (18), (19) and (23) are automatically satisfied. Now, if x is in the core, inequality (21) states that $x_1 + x_3 \geqq \$300,000$. Since x is an imputation, $x_1 + x_2 + x_3 = \$300,000$. Thus, $x_2 = \$0$. Now it follows from (20) that $x_1 \geqq \$200,000$. Moreover, $x_3 \leqq \$300,000$. (Why?) Thus, $x_2 = \$0$, $\$200,000 \leqq x_1 \leqq \$300,000$, and $x_3 = \$300,000 - x_1$. Conversely, if x satisfies these restrictions, then it is easy to verify inequalities (17) through (23). Thus, the core in the Land Development Game is the set

$$C = \{(x_1, 0, x_3): \$200,000 \leqq x_1 \leqq \$300,000, x_1 + x_3 = \$300,000\}. \quad (24)$$

An interpretation of this result is that the core consists of all those situations where the subdivider buys the land from the original owner, at a price between \$200,000 and \$300,000. Player 1 ends up with the money and player 3 with the land (which is worth \$300,000 to him) minus the money he paid for it. This "solution" makes sense, as Luce and Raiffa [1957, p. 208] point out, for clearly player 3 can always exclude player 2 from the bargaining by paying at least \$200,000. The main trouble with this "solution" is that it is not unique, it defines only a set of imputations. A standard reply of game theory to that complaint is that social standards or bargaining ability will determine the final price. The structure of the game itself does not determine it exactly, but only restricts it.[7]

This solution to the Land Development Game generalizes to the general Two-Buyer Market game, Example 3a of the previous section. In Exer. 22, the reader is asked to compute a core for this game.

In the Garbage Game, Example 6 of the previous section, the imputations are all the vectors $x = (x_1, x_2, \ldots, x_n)$ satisfying $x_i \geqq 1 - n$ and $\sum_{i=1}^{n} x_i = -n$. Suppose x is in the core. Then by Theorem 6.1, if $S \neq \varnothing$,

$$\sum_{i \in S} x_i \geqq v(S).$$

In particular, this is true for all $(n - 1)$-player coalitions S, and we have for such coalitions

$$\sum_{i \in S} x_i \geqq -1. \quad (25)$$

There are n such coalitions S. Adding inequalities (25) over all these sets, we obtain

$$(n - 1) \sum_{i=1}^{n} x_i \geqq -n. \quad (26)$$

But since x is an imputation, $\sum_{i=1}^{n} x_i = -n$. Thus (26) becomes

$$(n - 1)(-n) \geqq -n.$$

---

[7]The result does not explicitly bring in the inherent pride in ownership of land or other similar factors. In Chapter 8, we shall discuss how to measure the overall value or "utility" of an object such as land. Then, a number like 100,000 could be considered a utility or the number of dollars which would bring players a utility equal to that of owning the land. With this interpretation, the discussion could continue as above.

We conclude that $n - n^2 \geqq -n$, $n(n - 2) \leqq 0$, $n \leqq 2$. Thus, if $n > 2$, there is no core in the Garbage Game. There is no agreement on the disposal of garbage for which some group cannot do better by violating it!

We shall have more to say on the core in Secs. 6.4 and 6.6, where we calculate cores of some other games, for example, the Lake Game.

EXERCISES

1. If a coalition $S = \{1, 2, 3\}$ receives 10 units in payoff, which of the following vectors represent acceptable ways for the players in S to split up the 10 units?
   (a) $(3, 3, 4)$.
   (b) $(3.1, 3.1, 3.8)$.
   (c) $(0, 0, 10)$.
   (d) $(4, 4, 4)$.

2. In the game of Eqs. (7), which of the following payoff vectors satisfies individual rationality?
   (a) $(0, 0, 0)$.
   (b) $(1, 0, -1)$.
   (c) $(-1, -1, -1)$.
   (d) $(\frac{1}{3}, \frac{1}{3}, \frac{1}{3})$.
   (e) $(\frac{1}{2}, 0, \frac{1}{2})$.

3. In the game of Eqs. (7), which of the following vectors satisfies Pareto optimality and which satisfies feasibility?
   (a) $(0, 0, 0)$.
   (b) $(\frac{1}{3}, \frac{1}{3}, \frac{1}{3})$.
   (c) $(\frac{1}{2}, 0, \frac{1}{2})$.
   (d) $(1, 1, 1)$.

4. In the following games, find $M(v)$, the set of imputations.
   (a) $(51; 40, 30, 30)$.
   (b) $(51; 60, 30, 10)$.
   (c) Post Office (Exer. 3, Sec. 6.1).

5. In the game $(2; 1, 1, 1)$, find an imputation x which is effectively preferred to $(\frac{1}{2}, \frac{1}{2}, 0)$.

6. In the Land Development Game, find an imputation which is effectively preferred to the imputation
   $$(\$180{,}000, \$20{,}000, \$100{,}000).$$

7. In the game of Eqs. (7), find an imputation which is effectively preferred to the imputation $z = (\frac{9}{24}, \frac{1}{3}, \frac{7}{24})$.

8. In each of the digraphs of Fig. 6.2, find the core.

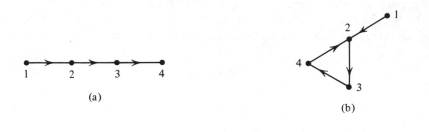

(a)

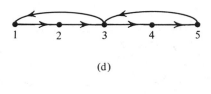

(b)

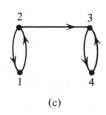

(c)

(d)

FIGURE 6.2. Digraphs for Exer. 8, Sec. 6.2 and Exer. 3, Sec. 6.3.

9. In the game of Eqs. (7), show that $(\frac{1}{3}, \frac{1}{3}, \frac{1}{3})$ is in the core.

10. Determine if the following weighted majority games have (nonempty) cores:
   (a) (51; 35, 35, 30).
   (b) (20; 10, 10, 1).
   (c) (51; 26, 26, 26, 22).

11. Let $n = 3$ and let v be defined as follows:

$$v(\{1, 2\}) = v(\{2, 3\}) = v(\{1, 2, 3\}) = 1,$$
$$v(S) = 0 \text{ otherwise.}$$

Find the core.

12. Let $n = 4$ and let v be defined as follows:

$$v(\{1, 2, 3\}) = v(\{1, 2, 3, 4\}) = 1 \quad \text{and} \quad v(S) = 0 \text{ otherwise.}$$

Find the core.

13. Consider the weighted majority game (6; 1, 1, 1, 2, 3, 3). Determine if the core exists, i.e., is nonempty.

14. Show that $\succ_s$ is transitive, i.e., that if $z \succ_s y$ and $y \succ_s x$, then $z \succ_s x$.

15. Show that for all i, and all imputations x and y, it is not the case that $x \succ_{\{i\}} y$.

16. Consider the following game on three players.

$$v(\varnothing) = 0$$
$$v(\{1\}) = v(\{2\}) = v(\{3\}) = v(\{2, 3\}) = v(\{1, 3\}) = 0$$
$$v(\{1, 2\}) = 2$$
$$v(\{1, 2, 3\}) = 1.$$

Show that Theorem 6.1 does not hold for this game. Why is this the case?

17. (A.W. Tucker.) Determine the core of the characteristic function v of Exer. 6, Sec. 6.1.

18. (A.W. Tucker.) Determine the core of the characteristic function v of Exer. 7, Sec. 6.1.

19. In the Security Council, $(39; 7, 7, 7, 7, 7, 1, 1, 1, 1, 1, 1, 1, 1, 1, 1)$, determine the core and discuss.

20. Determine the core in the following games and discuss.
    (a) Deterrence. (*Note:* If $x_i > 0$, how could player i have obtained extra wealth?)
    (b) The conference committee game of Exer. 10, Sec. 6.1.

21. Complete the proof of Theorem 6.1 by showing that if y is defined by Eq. (11), then y is an imputation and $y \succ_S x$.

22. Consider the general Two-Buyer Market (Example 3a, Sec. 6.1).
    (a) Determine $M(v)$.
    (b) Determine the core.

23. (A.W. Tucker.) Find the core for the weighted majority game $(12; 5, 5, 2, 2, 1)$.

24. Prove that if S is essential and $S \neq I$, then for some x and y, $x \succ_S y$. (See Exer. 12, Sec. 6.1, for definition of essential.)

25. Give proof or counterexample: If $x \succ_S y$ for some x, y, then S is essential.

26. Discuss critically the definition of effective preference given here. Can you think of other definitions?

27. Is it necessarily bad to have a core with more than one element in it?

6.3. Stable Sets

The core is not a totally satisfactory solution concept for a game, since sometimes it is empty and other times it can have more than one member. Von Neumann and Morgenstern, in their fundamental book [1944], anticipated some of the problems with the core and introduced the following ideas in a game theory context. A set B of vertices in a digraph D is said to be *internally stable* if no two vertices in B are joined by arcs; in terms of the effective preference digraph, for all x and y in B, neither x nor y is effectively preferred to the other. For example, the set $\{1, 3\}$ is internally stable in the digraph of Fig. 6.3. A set of vertices B is called *externally stable* if for every vertex $x \notin B$, there is some vertex $y \in B$ with an arc from y to x; in terms of effective preference, for every vertex not in B there is a vertex in B which is effectively preferred to it. The set $\{1, 3\}$ is also externally stable in the digraph of Fig. 6.3. A set of outcomes which is both internally and externally stable is called *stable*.

FIGURE 6.3. A digraph with two stable sets.

Von Neumann and Morgenstern proposed that the solution of a game should be limited to outcomes in a set of imputations which formed a stable set in the effective preference digraph. Such a set would be a reasonable one to limit consideration to. No outcome in the set is effectively preferred to any other outcome in the set, and every outcome outside is "dominated" by some outcome inside. The idea of stable set can be thought of as a generalization of the idea of vertex basis studied in Sec. 2.3. A stable set might be called a 1-basis. (Cf. Exer. 26.) Note that the core is always contained in a stable set. (Why?)

As a solution concept for *n*-person games, the idea of stable set has several drawbacks. First, there can be more than one stable set in a digraph. The digraph of Fig. 6.3 has two stable sets, $\{1, 3\}$ and $\{2, 4\}$. Indeed, any even cycle has two stable sets. Second, a digraph need not have any stable sets. The digraph of Fig. 6.4 is an example of such a digraph. Indeed, any odd cycle is an example. Finally, as with the core, there can be more than one vertex in a stable set. Ideally, a solution would specify a particular outcome rather than a set of outcomes.

Another problem with the concept of stable set is a practical one: There is no known general procedure, such as the one in Theorem 6.1, for calculating stable sets. Indeed, stable sets can take on very pathological forms.

FIGURE 6.4. A digraph with no stable sets.

In Sec. 6.5, we give some general theorems on the existence of stable sets. Here, we give several examples.

In Pure Bargaining, the core is the entire set of imputations. Since the core must be in every stable set, there is only one stable set here, the set of all imputations.

Turning next to the weighted majority game $(51; 51, 48, 1)$, we have a similar situation. For here, as we observed in Sec. 6.2.2., $M(v)$ consists of the single imputation $(1, 0, 0)$, which makes up the core. Since the core is in all stable sets, $\{(1, 0, 0)\}$ is also the one and only stable set.

In the weighted majority game $(2; 1, 1, 1)$, things are a bit different. Here, the core was the empty set. We shall show that the following set B of imputations is a stable set:

$$B = \{(\tfrac{1}{2}, \tfrac{1}{2}, 0), (\tfrac{1}{2}, 0, \tfrac{1}{2}), (0, \tfrac{1}{2}, \tfrac{1}{2})\}. \tag{27}$$

An interpretation of this result is as follows. Some two-player coalition forms, and splits the winnings entirely (and evenly) among the players. The stable set solution does not predict which coalition will form. Given the symmetry of the situation, it is not reasonable to expect it will.

To show that B of (27) is internally stable,[8] note that no imputation in B is effectively preferred to any other imputation in B. For example, if $y = (\tfrac{1}{2}, \tfrac{1}{2}, 0)$ were effectively preferred to $x = (\tfrac{1}{2}, 0, \tfrac{1}{2})$ via S, then S could only be the coalition $\{2\}$. However, $\sum_{i \in \{2\}} y_i = \tfrac{1}{2} > v(\{2\})$. (In general, effective preference cannot take place through 1-element coalitions.) To show that B is externally stable, suppose $x = (x_1, x_2, x_3)$ is an imputation not in B. We know that $x_i \geqq 0$ and $x_1 + x_2 + x_3 = 1$. Let x_i be the largest entry. By symmetry, we may take $i = 1$. If $x_1 > \tfrac{1}{2}$, then x_2 and x_3 are each less than $\tfrac{1}{2}$, since $x_1 + x_2 + x_3 = 1$. It follows that $(0, \tfrac{1}{2}, \tfrac{1}{2}) \succ_{\{2,3\}} x$. If $x_1 \leqq \tfrac{1}{2}$, then $x_2 \leqq \tfrac{1}{2}$ and $x_3 \leqq \tfrac{1}{2}$. If x_2 or x_3 is $\tfrac{1}{2}$, then $x = (\tfrac{1}{2}, \tfrac{1}{2}, 0)$ or $(\tfrac{1}{2}, 0, \tfrac{1}{2})$, and x is in B. Thus, $x_2 < \tfrac{1}{2}$ and $x_3 < \tfrac{1}{2}$, and again we conclude that $(0, \tfrac{1}{2}, \tfrac{1}{2}) \succ_{\{2,3\}} x$. This completes the proof that B is stable.

It should be remarked that even if an imputation x is in a stable set B, there can be imputations y outside B which are effectively preferred to x. For example, here we know this to be the case for $(\tfrac{1}{2}, \tfrac{1}{2}, 0)$, for otherwise this imputation would be in the core, which we know to be empty. It is easy to construct an imputation effectively preferred to $(\tfrac{1}{2}, \tfrac{1}{2}, 0)$, and we leave that to the reader (Exer. 5, Sec. 6.2).

It turns out that the set B of Eq. (27) is not the only stable set for the game $(2; 1, 1, 1)$. As pointed out in von Neumann and Morgenstern [1944], suppose we pick c in the interval $[0, \tfrac{1}{2})$. Then the set

$$B_c = \{(x_1, 1 - c - x_1, c): \quad 0 \leqq x_1 \leqq 1 - c\} \tag{28}$$

---

[8]In the remainder of this section, the reader may wish to skip the proofs, which are getting more technical.

is also stable. (Such a stable set in which a given player receives a fixed amount is called *discriminatory*.) To verify internal stability,[9] suppose $x = (x_1, 1 - c - x_1, c)$ and $y = (y_1, 1 - c - y_1, c)$ are both in B_c. Then $x_1 > y_1$, say, and so $x_2 < y_2$. Effective preference could only take place through a 1-player coalition, which is never possible (Exer. 15, Sec. 6.2). Thus, B_c is internally stable. To verify external stability, suppose $y = (y_1, y_2, y_3)$ is not in B_c. Then $y_3 > c$ or $y_3 < c$. If $y_3 > c$, let $y_3 = c + \epsilon$ and define $x = (y_1 + \frac{\epsilon}{2}, y_2 + \frac{\epsilon}{2}, c)$. It is easy to show that x is an imputation and that $x \succ_{\{1,2\}} y$. If $y_3 < c$, then either $y_1 \leq \frac{1}{2}$ or $y_2 \leq \frac{1}{2}$. If $y_1 \leq \frac{1}{2}$, let $x = (1 - c, 0, c)$. Then $x \succ_{\{1,3\}} y$. The proof is similar if $y_2 \leq \frac{1}{2}$.

It is more difficult to interpret the meaning of the stable set B_c of Eq. (28) than the set B of Eq. (27). One interpretation is that players 1 and 2 agree to give player 3 a side payment of c and split the pot between themselves. This stable set is a solution to the game again, in some sense. Which stable set among the many possibilities is chosen and which imputation in that stable set is decided upon is not settled by this analysis. The choice of stable set, if there is more than one, is made, according to von Neumann and Morgenstern, on the basis of standards of behavior in society. If society has a tradition that in such a voting situation, two players get together and split the proceeds, then B would be a reasonable "solution" to the game. The existence of many stable sets even for this simple game illustrates what Shapley and Shubik [1973, p. 40] call the "inherent indeterminacy of multiperson games."

Finally, let us consider the Land Development Game. The reader will recall that the core is given by the set

$$C = \{(x_1, 0, x_3): \quad \$200,000 \leq x_1 \leq \$300,000, \ x_1 + x_3 = \$300,000\}. \quad (24)$$

An interpretation of this result was that player 1 sells the land to player 3 for a selling price between $200,000 and $300,000. C is a subset of every stable set, but unfortunately C is not a stable set, for there is no imputation in C which is effectively preferred to ($100,000, $100,000, $100,000). To find a stable set, we need to add new imputations. One set of imputations which suffices consists of those where the selling price goes as low as $100,000. That is, the following set B is stable:

$$B = \{(x_1, 0, x_3): \quad \$100,000 \leq x_1 \leq \$300,000, \ x_1 + x_3 = \$300,000\}. \quad (29)$$

B is the set of all imputations with $x_2 = 0$. To prove that B is stable, observe that internal stability follows since effective preference cannot occur with respect to a 1-element coalition (Exer. 15, Sec. 6.2) and since $y_1 > x_1$ and $y_1 + y_3 = \$300,000$ implies $y_3 < x_3$. To prove external stability, suppose an imputation $x = (x_1, x_2, x_3)$ is not in B. Then $x_2 \neq 0$. Let $y_1 = x_1 + (x_2/2)$, $y_2 = 0$, and $y_3 = x_3 + (x_2/2)$. Then y is an imputation

---

[9] Once again, the verification may be omitted.

and $y \succ_{\{1,3\}} x$. Since $y_2 = 0$, y is in B. Thus, B is a stable set. This result deserves comment. Let x be the imputation ($100,000, $0, $200,000). Certainly the imputation $y = ($101,000, $1000, $198,000) is effectively preferred to x. This doesn't violate internal stability of B. (Why?) However, the imputation ($101,500, $0, $198,500) is, in turn, effectively preferred to y. Thus, we know that player 2 can be kept out of the bidding. Knowing this, all of the imputations in B are plausible solutions, since no one is effectively preferred to any other.

As in the weighted majority game $(2; 1, 1, 1)$, there is more than one stable set for the Land Development Game. Let

$$D = \{(x_1, x_2, x_3): \$100,000 \leq x_1 \leq \$200,000, x_2 = \$100,000 - \tfrac{1}{2}x_1,$$
$$x_3 = \$200,000 - \tfrac{1}{2}x_1\}.$$

Then $B' = C \cup D$, where C is the core, is also stable. Notice that imputations in D have the following interpretation. The land is sold by player 1 to player 3 for a price x_1 between $100,000 and $200,000. Player 3 then pays player 2 an amount of money equal to $100,000 less half the selling price for having stayed out of the bidding. If a society has a tradition of not condoning such "under the table" payoffs, then this stable set will not be a reasonable solution for that society.

There are many other stable sets in the Land Development Game. Suppose t is a real number in $(0, 1)$. Let

$$D_t = \{(x_1, x_2, x_3): \$100,000 \leq x_1 \leq \$200,000, x_2 = t(\$200,000 - x_1),$$
$$x_3 = \$100,000 + (1 - t)(\$200,000 - x_1)\}. \tag{30}$$

Then $B_t = C \cup D_t$ is stable. The set B' we have just discussed is of course $B_{1/2}$. The proof that B_t is stable is left to the reader as an exercise (Exer. 17). There are other stable sets in addition to these sets B_t.

We shall return to stable sets in Secs. 6.5 and 6.6. In Sec. 6.7 we shall introduce one additional solution concept for games, the Shapley value, which avoids some of the problems with cores and stable sets. It always exists and defines a single imputation solution.

EXERCISES

1. Give an example of a digraph with an internally stable set which is not externally stable.

2. Give an example of a digraph with an externally stable set which is not internally stable.

3. In each of the digraphs of Fig. 6.2, find
   (a) All maximal internally stable sets, i.e., internally stable sets not contained in any larger internally stable sets.
   (b) All minimal externally stable sets.
   (c) All stable sets.

4. Is it possible for one stable set to be a proper subset of another stable set? Why?

5. In the weighted majority game (51; 60, 30, 10), is the core a stable set?

6. The notion of internally stable set has applications in communications. When S is a set of signals, we draw a graph G with $V(G) = S$ and two signals joined by an edge if it is possible for the two signals to be confused. Think of G as a digraph, by drawing arcs in both directions between x and y if there is an edge joining x and y. Using this idea, find the maximum number of signals which can be used so that there is no way to confuse any signal, if $S = \{a, b, c, d, e\}$, and a can be read as a or b, b can be read as b or c, c can be read as c or d, d can be read as d or e, and e can be read as e or a. What does the solution have to do with internal stability? (*Note:* Additional discussion of the notion of internal stability in communications can be found in Berge [1962, p. 38] and in Shannon [1956].)

7. In a nuclear power plant, there are warning lights placed at various locations. Suppose from location 1, it is possible to see lights at locations 1, 2, and 3. From location 2 it is possible to see lights at locations 2 and 3. From location 3 it is possible to see lights at locations 3 and 5. From location 4 it is possible to see lights at locations 4 and 5. From location 5 it is possible to see only the light at location 5. What is the smallest number of fixed monitoring stations which can be set up to continuously monitor all the lights? (*Hint:* Draw a digraph and use external or internal stability.)

8. An old problem posed by Gauss asked whether or not it is possible to place eight queens on a chessboard so that no queen can be taken by any other. Can this problem be translated into one of internal stability? How? (For a solution, see Berge [1962, p. 35].)

9. A similar problem is to find the smallest number of queens which can capture a piece at any position of a chessboard. Can either internal or external stability help with this problem? (Incidentally, what is the solution? See Berge [1962, p. 41].)

10. In the game of Exer. 11, Sec. 6.2, is the core a stable set?

11. In the game of Exer. 12, Sec. 6.2, is the core a stable set?

12. In Deterrence, is the core a stable set? (Cf. Exer. 20(a), Sec. 6.2.)

13. In the weighted majority game $(51; 26, 26, 26, 22)$, show that the following set is stable:

$$B = \{(\tfrac{1}{2}, \tfrac{1}{2}, 0, 0), (\tfrac{1}{2}, 0, \tfrac{1}{2}, 0), (0, \tfrac{1}{2}, \tfrac{1}{2}, 0)\}.$$

14. If the core of a game is all of $M(v)$, show that the core is a stable set.

15. Suppose the core of a game is a stable set. Show that it is the unique stable set.

16. Find all stable sets in the 3-player simple game in which the only winning 2-player coalition is $\{2, 3\}$.

*17. In the Land Development Game, let C be the core and let D_t be the set defined by Eq. (30). Show that $B_t = C \cup D_t$ is a stable set.

*18. In the general Two-Buyer Market (Example 3a, Sec. 6.1), let C be the core (see Exer. 22, Sec. 6.2). Let t be in $(0, 1)$ and let

$$D_t = \{(x_1, x_2, x_3): a \leqq x_1 \leqq b, x_2 = t(b - x_1),$$

$$x_3 = c - b + (1 - t)(b - x_1)\}.$$

Show that $B_t = C \cup D_t$ is stable.[10]

19. (A.W. Tucker.) Find a stable set in the weighted majority game $(12; 5, 5, 2, 2, 1)$ (Exer. 23, Sec. 6.2).

20. Since the core is in all stable sets, it is a subset of the intersection of all stable sets. It was conjectured by Shapley in Kuhn [1953] that the intersection of stable sets of a game is equal to the core. The following counterexample is due to W. F. Lucas [1967]. Let $I = \{1, 2, 3, 4, 5\}$ and let $v(I) = 1, v(\{1, 2\}) = v(\{4, 5\}) = v(\{2, 3, 4\}) = v(\{1, 2, 4, 5\}) = \tfrac{1}{2}, v(S) = \tfrac{1}{2}$ for all supersets of $\{1, 2\}$ and $\{4, 5\}$, and $v(S) = 0$ for all other S.

   (a) Identify the set $M(v)$ of imputations.

   (b) Show that the core is given by

$$C = \{x \in M(v): x_1 + x_2 = \tfrac{1}{2}, x_4 + x_5 = \tfrac{1}{2}, x_2 + x_3 + x_4 \geqq \tfrac{1}{2}\}.$$

   (c) Show that the only stable set is given by

$$K = M(v) - \{x: y \succ x \text{ for some } y \in C\}$$

$$= \{x \in M(v): x_1 + x_2 = \tfrac{1}{2}, x_3 = 0, x_4 + x_5 = \tfrac{1}{2}\}.$$

*21. (A.W. Tucker.) Consider the characteristic function v of Exer. 6, Sec. 6.1. Show that the union of the following two sets is a stable set for v:

$$\{(x_1, x_2, 0.3, 0.4): x_1 + x_2 = 0.3, x_1 \geqq 0, x_2 \geqq 0\}$$

and

$$\{(0.1, 0.2, x_3, x_4): x_3 + x_4 = 0.7, x_3 \geqq 0, x_4 \geqq 0\}.$$

---

[10]The reader should beware: In the literature of game theory, this stable set is incorrectly defined by various authors.

22. (Berge [1962].) Show that if S is a stable set in a digraph D, then S is a maximal internally stable set.[11]

23. Show that if S is a stable set in a digraph D, then S is a minimal externally stable set.

24. (Berge [1962].) Show that the converse of the result of Exer. 22 is false in general.

25. (Berge [1962].) Prove the converse of the result of Exer. 22 for a symmetric digraph, i.e., a digraph with an arc from y to x whenever there is an arc from x to y.

26. (Harary, Norman, and Cartwright [1965].) An *n-cover* B of a digraph D is a set of vertices such that every vertex of D is reachable in at most n steps from a vertex of B. An *n-basis* is an n-cover B such that no vertex of B is reachable from any other vertex in B in less than or equal to n steps. Thus, an externally stable set is a 1-cover and a stable set is a 1-basis. Prove that every n-basis is a minimal n-cover. This generalizes Exer. 23.[12]

27. (Harary, Norman, and Cartwright [1965].) Show by example that it is possible for a digraph to have an n-basis and not an $(n + 1)$-basis.

28. (Berge [1962].) Let $\alpha(G)$, the *coefficient of internal stability* of the graph G, be defined as the size of the largest internally stable set of G if G is thought of as a digraph with arcs from x to y and y to x replacing each edge $\{x, y\}$. If $\chi(G)$ is the chromatic number of G (Sec. 3.6), show that $\alpha(G)\chi(G) \geqq n$, where n is the number of vertices.[13]

6.4. The Existence of a Nonempty Core

The core, if it is nonempty, is a reasonable solution concept for a game, and one often can limit discussion to imputations in the core. Unfortunately, as we have seen, the core may be empty. One of the goals of n-person game theory has been to prove theorems about the existence or nonexistence of the core. (By the convention we have been using, the core is said to *exist* if it is nonempty.)

The notion of core may be defined for general digraphs, not just effective preference digraphs arising from games in characteristic function form. A core is just the set of all vertices of the digraph with indegree equal to 0. Thus,

---

[11]**All digraphs in the exercises of this section can be assumed to be finite.**

[12]See footnote 11.

[13]See footnote 11.

one might look to digraph theory for theorems on the existence of a core. (Of course, it will often be most appropriate to look at digraphs on infinite sets, unless the set of outcomes is first limited by other means.) We know some theorems about the existence of vertices with indegree equal to 0. In particular, the Corollary to Theorem 2.8 says that every (finite) acyclic digraph has such vertices, and hence has a core. Unfortunately, the effective preference digraph D often does have cycles. In fact, it is even possible to have $x \succ y$ and $y \succ x$. For example, if $n = 5$, if $v(\{1, 2\}) = v(\{3, 4\}) = \frac{1}{2}$, and if $x = (\frac{1}{4}, \frac{1}{4}, \frac{1}{6}, \frac{1}{6}, \frac{1}{6})$ and $y = (\frac{1}{6}, \frac{1}{6}, \frac{1}{4}, \frac{1}{4}, \frac{1}{6})$, then $x \succ_{\{1,2\}} y$ and $y \succ_{\{3,4\}} x$. Theorems of digraph theory are more useful in proving the existence and uniqueness of stable sets than they are in proving the existence of a core. We shall see this in the next section. Here, we shall seek other tools.

The reader will recall that a characteristic function v is required to be superadditive, i.e., to satisfy

$$v(S \cup T) \geq v(S) + v(T) \qquad (\text{if } S \cap T = \varnothing).$$

If this inequality is always an equality, we say v is *additive* and call the game *inessential*. Thus, v defines an *essential* game if and only if there are disjoint coalitions S and T such that $v(S \cup T) > v(S) + v(T)$.[14] The weighted majority game $(2; 1, 1, 1)$ is certainly essential, while the game $(51; 51, 48, 1)$ is inessential. A game v is called *constant-sum* if

$$v(S) + v(I - S) = v(I) \qquad (\text{all } S).$$

Thus, all inessential games are constant-sum, but not conversely. The game $(2; 1, 1, 1)$ is both essential and constant-sum. The first theorem generalizes our result that $(2; 1, 1, 1)$ has no core.

Theorem 6.2.   If v is an essential constant-sum game, then v has no core.

Proof.   Suppose $x = (x_1, x_2, \ldots, x_n)$ is an imputation. The reader can readily verify (Exer. 12) that in an essential game, $v(I) > \sum_{i=1}^{n} v(\{i\})$. Since $\sum_{i=1}^{n} x_i = v(I)$, it follows that for some j, $v(\{j\}) < x_j$. Now since the game is constant-sum,

$$v(I - \{j\}) = v(I) - v(\{j\})$$

$$= \left(\sum_{i=1}^{n} x_i \right) - v(\{j\}).$$

Thus,

$$v(I - \{j\}) > \sum_{i \in I - \{j\}} x_i,$$

and Theorem 6.1 implies that x is not in the core.   Q.E.D.

---

[14]The reader should distinguish an essential game from an essential coalition, which we defined in Exer. 12, Sec. 6.1.

One point should be made here about games without cores. A coalition might be forced to accept less than it can get in such a game. This is because individuals belong to several coalitions, and split their loyalties. In games with cores, there is a way of satisfying all coalitions at once. In games without cores, this cannot happen.

Let us turn next to the general game with three players.

Theorem 6.3.   In a 3-person essential game v, there is a core if and only if

$$v(\{1, 2\}) + v(\{1, 3\}) + v(\{2, 3\}) \leq 2v(\{1, 2, 3\}).$$

Proof.   We shall prove the theorem for the special case where $v(\{1\}) = v(\{2\}) = v(\{3\}) = 0$ and $v(\{1, 2, 3\}) = 1$. In Sec. 6.6 we shall see that every essential game can be reduced to this form (its so-called $(0, 1)$-normalization) and so the more general theorem will follow (see Exer. 23, Sec. 6.6). We now use Theorem 6.1. Inequality (10) of Theorem 6.1 is always satisfied by an imputation $x = (x_1, x_2, x_3)$ for those coalitions S with $|S| = 1$ or $|S| = n$. Thus, inequality (10) is satisfied by x for all S if and only if

$$\begin{aligned}
x_1 + x_2 &\geq v(\{1, 2\}) \\
x_1 + x_3 &\geq v(\{1, 3\}) \\
x_2 + x_3 &\geq v(\{2, 3\}).
\end{aligned} \tag{31}$$

Adding these three inequalities and using the fact that $x_1 + x_2 + x_3 = 1$, we find that if x is in the core, then

$$2 \geq v(\{1, 2\}) + v(\{1, 3\}) + v(\{2, 3\}). \tag{32}$$

Conversely, suppose inequality (32) holds. We show that the core is non-empty. Suppose first that at least two of $v(\{1, 2\})$, $v(\{1, 3\})$ and $v(\{2, 3\})$ are less than one-half, say $v(\{1, 2\}) < \frac{1}{2}$ and $v(\{1, 3\}) < \frac{1}{2}$. Then, since $v(\{2, 3\}) \leq 1$ (why?), $(0, \frac{1}{2}, \frac{1}{2})$ satisfies the inequalities (31) and so is in the core. Suppose next that at most one of $v(\{1, 2\})$, $v(\{1, 3\})$, and $v(\{2, 3\})$, is less than one-half, and say that $v(\{1, 2\}) \geq \frac{1}{2}$ and $v(\{1, 3\}) \geq \frac{1}{2}$. Then $v(\{1, 2\}) + v(\{1, 3\}) \geq 1$. We can put

$$\begin{aligned}
x_1 &= v(\{1, 2\}) + v(\{1, 3\}) - 1 \\
x_2 &= 1 - v(\{1, 3\}) \\
x_3 &= 1 - v(\{1, 2\}).
\end{aligned}$$

Then, $x = (x_1, x_2, x_3)$ is an imputation (why?) and, using inequality (32), we see that x satisfies inequalities (31). Hence, x is in the core.   Q.E.D.

Theorem 6.3 gives us another proof that the weighted majority game $(2; 1, 1, 1)$, and indeed any essential, constant-sum 3-player game, has no core.

Conditions necessary and sufficient for the existence of a core which generalize those of Theorem 6.3 have been given by Bondareva [1962, 1963] and by Shapley [1967]. We shall not discuss them here.

A set K of vectors in $\mathfrak{R}^n$ is called *convex* if whenever $x \in K$ and $y \in K$ and α is in the interval $[0, 1]$, then the vector $\alpha x + (1 - \alpha)y \in K$. It is easy to show that the core is a convex set. For by Theorem 6.1, if x and y are in the core, then for all $S \neq \varnothing$,

$$\sum_{i \in S} \alpha x_i + \sum_{i \in S} (1 - \alpha)y_i = \alpha \sum_{i \in S} x_i + (1 - \alpha) \sum_{i \in S} y_i$$

$$\geqq \alpha v(S) + (1 - \alpha)v(S)$$

$$= v(S).$$

Thus, $\alpha x + (1 - \alpha)y$ is in the core. This observation helps us state conditions for the existence of cores in *symmetric games*, games where $v(S)$ is a function only of $|S|$. (The games Pure Bargaining, Garbage, the Lake, $(2; 1, 1, 1)$ and $(51; 49, 48, 3)$ are all symmetric.) Suppose v is the characteristic function of a symmetric game and let $s = |S|$. Let f be a function from $\{0, 1, 2, \ldots, n\}$ into the reals, with $f(s)$ defined to be $v(S)$. If the core exists, it is clearly symmetric, that is, $(x_1, x_2, \ldots, x_n)$ is in the core if and only if $(x_{\pi(1)}, x_{\pi(2)}, \ldots, x_{\pi(n)})$ is in the core, for all permutations π of $\{1, 2, \ldots, n\}$.[15] Since the core is symmetric and convex, it must contain the imputation $x = (f(n)/n, f(n)/n, \ldots, f(n)/n)$ if it contains any imputations. Thus, the core exists if and only if this imputation x belongs to it. By Theorem 6.1, x belongs to the core if and only if for all $s > 0$,

$$\frac{s}{n} f(n) \geqq f(s). \tag{33}$$

We summarize this result as the theorem.

Theorem 6.4.[16]   Let v be the characteristic function of a symmetric game and let $f(s) = v(S)$ for S a coalition with $|S| = s$. Then the core of v exists if and only if for all $s = 1, 2, \ldots, n$,

$$\frac{s}{n} f(n) \geqq f(s). \tag{33}$$

To give some examples, in Pure Bargaining,

$$f(s) = \begin{cases} d & (\text{if } s = n) \\ 0 & (\text{if } s < n). \end{cases}$$

Then (33) is trivially satisfied for all $s > 0$. In the weighted majority game

---

[15]A *permutation* is a one-to-one function from a set onto itself. It can be thought of as a rearrangement or relabelling of the set.

[16]This theorem is due to Shapley (unpublished) and will appear in a forthcoming Rand report.

$(2; 1, 1, 1)$, we have

$$f(s) = \begin{cases} 1 & \text{(if } s \geqq 2) \\ 0 & \text{(if } s < 2). \end{cases}$$

Then (33) is violated for $s = 2$. In the Garbage Game,

$$f(s) = \begin{cases} -(n - s) & \text{(if } s < n) \\ -n & \text{(if } s = n). \end{cases}$$

If $n > 2$, then (33) is violated for any s such that $n/2 < s < n$. If $n = 2$, (33) holds for every $s > 0$. Finally, in the Lake Game, (33) is satisfied for every $s > 0$. The proof is left as an exercise (Exer. 7). Thus, unlike the Garbage Game, there is a core for the Lake Game. There is a way for the factories to reach agreement on the treatment of wastes so that no coalition can be sure of doing better by violating the agreement. Unfortunately, the core here is so large that it gives little information about the proper agreement.[17]

EXERCISES

1. Give an example of a digraph with 10 vertices which has no core.

2. Which of the following games are essential?
    (a) $(51; 60, 30, 10)$.
    (b) Pure Bargaining (Example 1, Sec. 6.1).
    (c) The Land Development Game (Example 3, Sec. 6.1).
    (d) $(51; 40, 30, 30)$.

3. Which of the games of Exer. 2 are constant-sum?

4. For arbitrary n, give an example of a game with n players which is
    (a) Inessential.
    (b) Essential.

5. Give an example of a game in characteristic function form whose effective preference digraph is acyclic and has at least two vertices.

6. Suppose $n = 3$ and v is defined as follows. Does v have a core?
    (a) v of Eqs. (7).
    (b) $v(\varnothing) = 0$.
         $v(\{1\}) = v(\{2\}) = v(\{3\}) = 0$.
         $v(\{1, 2\}) = v(\{1, 3\}) = v(\{2, 3\}) = 1$.
         $v(\{1, 2, 3\}) = 2$.
    (c) Same as part (b) but
         $v(\{1, 2\}) = v(\{1, 3\}) = v(\{2, 3\}) = \frac{3}{2}$.

---

[17]See Shapley and Shubik [1969b] for further discussion.

7. Prove that in the Lake Game, inequality (33) is satisfied for every $s > 0$.

8. Does the game (51; 49, 48, 3) have a core?

9. Does Post Office (Exer. 3, Sec. 6.1) have a core?

10. Suppose $v(S) = [s/2]$, where $s = |S|$ and $[x]$ is the greatest integer less than or equal to x. Is there a core? Does the answer depend on n? If so, how?

11. Describe all possible 3-person simple games and use Theorem 6.3 to determine which have a core.

12. Prove that v is essential if and only if

$$v(I) > \sum_{i=1}^{n} v(\{i\}).$$

13. Give an example of an inessential constant-sum game which has a core.

*14. (Shapley [1953b].) A game v is a *quota game* if there is a vector $k = (k_1, k_2, \ldots, k_n)$ so that for all $i \neq j$, $k_i + k_j = v(\{i, j\})$, and so that $\sum_{i=1}^{n} k_i = v(I)$. Show that every 4-person constant-sum game can be represented as a quota game.

*15. Show that if the core of a quota game exists, it consists of just the imputation k.

6.5. Existence and Uniqueness of Stable Sets

We have seen in Sec. 6.3 that there may be more than one stable set and that, moreover, a stable set may have more than one imputation in it. For a long time it was believed that, contrary to the case of the core, some stable set always existed for a game in characteristic function form. This belief turned out to be wrong, as was demonstrated by Lucas [1968, 1969].

In this section, we present certain theorems on the existence and uniqueness of stable sets. The first result is basically negative, saying that the hope of finding a single imputation making up a stable set is only realized in very dull games, the inessential ones. Such an imputation, if it were to exist, could have been selected as *the* solution to the game.

Theorem 6.5 (von Neumann and Morgenstern [1944]). A game v in characteristic function form possesses a singleton stable set $B = \{x\}$ if and only if v is inessential. In such games, B is the only stable set.

Proof. The reader should observe (see Exer. 12, Sec. 6.4) that v is inessential if and only if $\sum_{i=1}^{n} v(\{i\}) = v(I)$. Suppose that v is inessential. Then $M(v)$ consists of the single imputation $(v(\{1\}), v(\{2\}), \ldots, v(\{n\}))$. This imputation makes up a stable set, and of course it is the unique stable set.

Conversely, suppose that v is essential, and suppose that there is a singleton stable set $B = \{x\}$. Now $\sum_{i=1}^{n} x_i = v(I) > \sum_{i=1}^{n} v(\{i\})$. Thus, $x_j > v(\{j\})$ for some j. Define y as follows:

$$y_j = v(\{j\})$$

$$y_i = x_i + \frac{x_j - v(\{j\})}{n-1} \qquad \text{(if } i \neq j\text{)}.$$

Then y is an imputation and $y \neq x$, so by external stability of $B = \{x\}$, $x \succ y$. However, suppose S is any coalition so that $x \succ_S y$. Then $x_i > y_i$ for all $i \in S$, so S must be $\{j\}$. By Exer. 15, Sec. 6.2, effective preference by 1-element coalitions is impossible and we have reached a contradiction.

Q.E.D.

Some theorems on the existence and uniqueness of stable sets can be proved by passing to the more general setting of a digraph, thought of as the effective preference digraph. (Of course, it should be remembered that these theorems have only limited applicability to games in characteristic function form unless they are generalized to infinite digraphs, or unless the theorems are applied to finite sets of imputations which have been chosen ahead of time by some other means.)

Theorem 6.6 (von Neumann and Morgenstern). Every (finite) acyclic digraph D has a unique stable set.

Proof. The argument is by induction on $|V|$, where V is the vertex set of D. The result is trivial if $|V| = 1$. Let $|V| > 1$. By the Corollary to Theorem 2.8, an acyclic digraph has a vertex u of indegree 0. Let

$$R_1(u) = \{u\} \cup \{x: \text{there is an arc from } u \text{ to } x\}.$$

Then the subgraph D' of D generated by $V - R_1(u)$ is also acyclic and has fewer vertices than D. By inductive assumption, D' has a unique stable set B. Then $B' = B \cup \{u\}$ is stable in D. To prove this, note that B' is internally stable since u has indegree 0, since B is not in $R_1(u)$, and since B is internally stable in D'. Next, B' is externally stable since any $x \notin B'$ is either reached in one step from u or from a vertex in B.

Finally, we show that B' is the unique stable set of D. Suppose B'' is also stable in D. Now u has indegree 0, so $u \in B''$. By internal stability, $B'' - \{u\} \subseteq V - R_1(u)$, so $B'' - \{u\}$ is a set of vertices in D'. It is certainly internally stable in D'. It is also externally stable in D', since $x \in V(D') - (B'' - \{u\})$ is reachable in one step from a vertex of B'', and that vertex could not be u. We conclude that $B'' - \{u\}$ is stable in D', and so by inductive assumption $B'' - \{u\}$ is B.   Q.E.D.

The proof suggests a constructive procedure for finding the unique stable set B in an acyclic digraph. The procedure is illustrated in Fig. 6.5. At

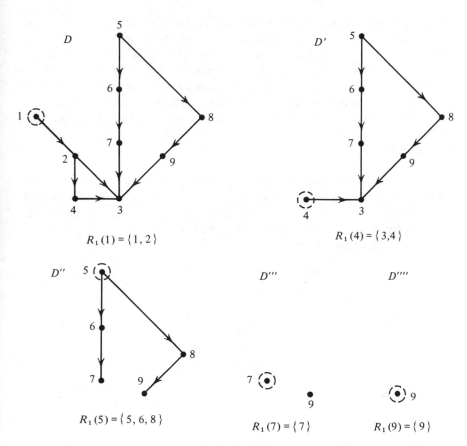

FIGURE 6.5. Procedure for finding a stable set B in an acyclic digraph. Vertex chosen at each stage is surrounded by dotted boundary. $B = \{1, 4, 5, 7, 9\}$.

the first stage, we choose a vertex of indegree 0, here vertex 1, and put it in B. Let D' be the digraph generated by vertices in $V(D) - R_1(1)$. Pick a vertex of indegree 0 in D', say vertex 4, and add it to B. Let D'' be the digraph generated by $V(D') - R_1(4)$. Pick a vertex of indegree 0 in D'', say vertex 5. And so on. This procedure results in the unique stable set $B = \{1, 4, 5, 7, 9\}$ for D.

Theorem 6.7 (Harary, Norman, and Cartwright [1965]). Every (finite) strongly connected digraph D consisting of more than one vertex and having no odd cycles has at least two stable sets.

Proof. Theorem 3.13 (König's Theorem on 2-colorable graphs) says that if an undirected graph has no odd circuits, then the vertices can be divided into two classes such that every edge of the graph joins vertices of different classes. An analogous proof verifies the fact that if a strongly connected digraph D has no odd cycles, its vertices can be partitioned into two classes V_1 and V_2 such that every arc of D joins vertices of two different classes. (To define V_1 and V_2, pick any vertex x and put it in V_1. Then V_1 is the set of all vertices y such that there is a path of even length from x to y and V_2 is the set of all remaining vertices. Proof that this construction works is left to the reader (Exer. 12).) Since D is strongly connected and has more than one vertex, V_1 and V_2 are both nonempty. It is easy to show that both V_1 and V_2 are stable sets.   Q.E.D.

The proof of Theorem 6.7 again gives us a constructive procedure for finding a stable set, if we have a strongly connected digraph D with no odd cycles. Simply use V_1 or V_2 as constructed in the proof. In the digraph of Fig. 6.6, for example, we choose x to be vertex 1. Then $V_1 = \{1, 2, 3\}$ and $V_2 = \{4, 5\}$. Both are stable sets.

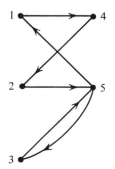

FIGURE 6.6. A strongly connected digraph with no odd cycles. $V_1 = \{1, 2, 3\}$ and $V_2 = \{4, 5\}$ are two stable sets.

Theorem 6.8 (Richardson [1946]).   Every (finite) digraph D with no odd cycles has a stable set.

Proof.   We shall define step-by-step two sets of vertices, B and C, the first of which will be a stable set and the second of which is its complement. Let D^* be the condensation by strong components of D. (See Sec. 2.3.) We know D^* is acyclic (Theorem 2.7) and so there is a strong component K_1 of D with indegree 0 in D^*. Since K_1 has no odd cycles, Theorems 6.6 and 6.7 imply that K_1 has a stable set B_1. Let all vertices of B_1 be put in B and let all vertices reachable from a vertex of B_1 by an arc of D be put in C. Let D_1 be the subgraph of D generated by all vertices of D not yet in B or C. Repeat the process with D_1 and continue it until all vertices of D are in B or in C.

We show that B is a stable set. Any vertex in C is placed in C because it is reached by an arc from B. Thus, B is externally stable. Next, no vertex in B is adjacent to any other vertex of B, because we always selected for B only vertices of an internally stable set not reachable by arcs from vertices previously in B. Thus, B is also internally stable.   Q.E.D.

The procedure for constructing a stable set B discussed in the proof of Theorem 6.8 is easily illustrated. Figure 6.7 shows successive choice of sets K_i, B_i, and C_i, and digraphs D_i. D_0 is D. K_{i+1} is a strong component with indegree 0 in the condensation of D_i. B_{i+1} is a stable set of K_{i+1}. C_{i+1} consists of all vertices of D_i reachable by an arc of D from a vertex of B_{i+1}. Finally, D_{i+1} is the subgraph of D generated by vertices of D_i not in B_{i+1} or C_{i+1}. $B = \bigcup_i B_i$ is a stable set.

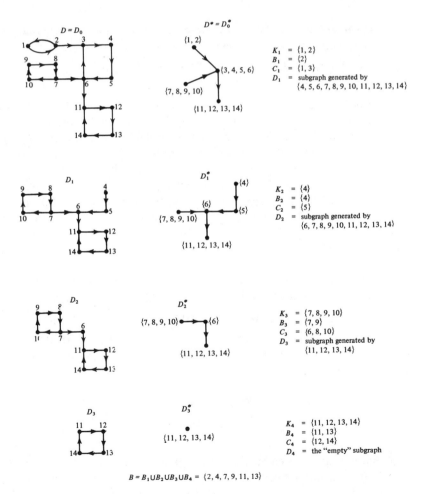

$$B = B_1 \cup B_2 \cup B_3 \cup B_4 = \{2, 4, 7, 9, 11, 13\}$$

FIGURE 6.7. Construction of a stable set B in a digraph with no odd cycles.

As we observed before, to be applicable to many games in character-istic function form, theorems like Richardson's must be applied to pre-selected finite sets of imputations or, alternatively, must be proved for infinite digraphs. Richardson's Theorem, it turns out, can be proved for certain classes of infinite digraphs, the so-called locally finite ones, and the so-called totally inductive ones. See Berge [1962, p. 50].

Theorems like the ones we have proved apply to digraphs with certain specialized structural properties. Other theorems about existence and unique-ness of stable sets can be proved by making special assumptions about the characteristic function v. We illustrate this by proving a theorem about those v which define simple games, games in which every coalition has value 1 (winning) or 0 (losing).

Theorem 6.9.   Every simple game has a stable set. In particular, if S is a minimal winning coalition, then

$$B_S = \{x \in M(v): x_i = 0 \text{ if } i \notin S\} \tag{34}$$

is stable.

Proof.   If there is no winning coalition, then $v(S) = 0$, all S, and so $(0, 0, \ldots, 0)$ is the only imputation. It forms a 1-element stable set. If there is a winning coalition, let S be a minimal such coalition, and define B_S by Eq. (34). If x and y are in B_S, then any set T of i for which $x_i > y_i$ must be a sub-set of S. Indeed, since $\sum_{i \in S} x_i = \sum_{i \in S} y_i = 1$, T must be a proper subset of S. By minimality of S, $v(T) = 0$ and no effective preference via T can occur. This proves that B_S is internally stable. Next, let x be an imputation outside B_S. Then $\sum_{i \in S} x_i < 1$. Split up $\sum_{j \notin S} x_j$ among the x_i for $i \in S$ to obtain an imputation y in B_S which is effectively preferred to x via S. Thus, B_S is externally stable.   Q.E.D.

Let us apply this theorem to the weighted majority game $(2; 1, 1, 1)$. We find for example that one stable set is given by

$$B_{\{1,2\}} = \{(x_1, x_2, 0): x_1 \geqq 0, x_2 \geqq 0, x_1 + x_2 = 1\}.$$

(The reader will note that this is the same as the set B_c defined in Eq. (28) when $c = 0$.) Stable sets of the form B_S are called *exclusive solutions;* certain players are excluded.

EXERCISES

1.   Show that the weighted majority game $(2; 1, 1, 1)$ does not possess a singleton stable set.

2.   Give an example of a game which possesses a singleton stable set.

3.   Find the unique stable set for each of the digraphs of Fig. 6.8.

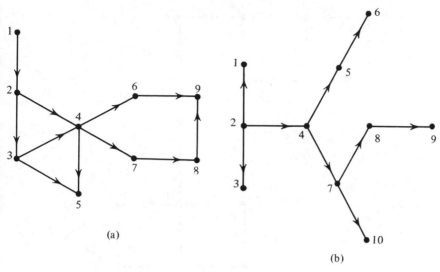

(a)

(b)

FIGURE 6.8. Digraphs for Exer. 3, Sec. 6.5.

4.   Use the procedure of the proof of Theorem 6.7 to find a stable set for each of the digraphs of Fig. 6.9.

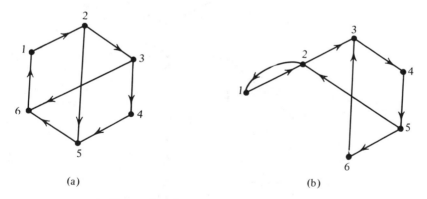

(a)   (b)

FIGURE 6.9. Digraphs for Exer. 4, Sec. 6.5.

5.   Find a stable set for each of the digraphs of Fig. 6.10.

6.   Find an exclusive solution for the weighted majority game (51; 51, 48, 1).

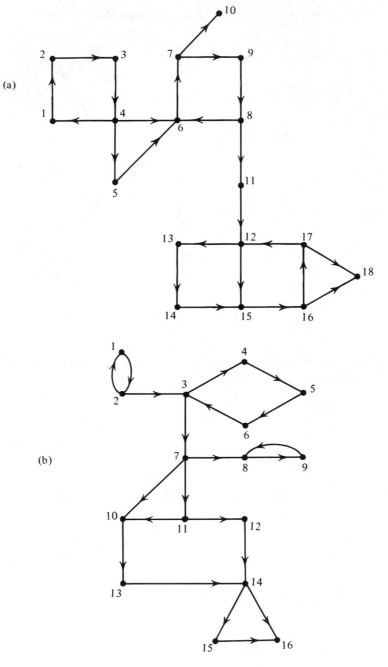

FIGURE 6.10. Digraphs for Exer. 5, Sec. 6.5.

7. Find an exclusive solution for the weighted majority game (51; 50, 30, 20).

8. Find an exclusive solution for the conference committee game of Exer. 10, Sec. 6.1. Interpret the results.

9. Find all exclusive solutions for the weighted majority game (5; 1, 1, 1, 2, 3).

10. Show that two stable sets in a digraph need not have the same number of vertices.

11. Give proof or counterexample: If a (finite) digraph D has no odd cycles, then either D is acyclic or it has at least two stable sets.

12. Prove that in a (finite) strongly connected digraph D with no odd cycles, the vertices can be partitioned into two classes so that every arc of D joins vertices of different classes.

13. (Harary, Norman, and Cartwright [1965].) Show that no more than one pair of sets (and hence stable sets) can partition the set of vertices of a (finite) strongly connected digraph so that if vertices u and v are joined by an arc, then u and v are in different sets.

14. (Harary, Norman, and Cartwright [1965].) Prove that if B is a stable set in a (finite) strongly connected digraph D with vertex set $V(D)$, then $V(D) - B$ is externally stable. Is this true without the hypothesis that D be strongly connected?

·15. (von Neumann and Morgenstern [1944].) Let v define a constant-sum simple game. Suppose $x^* = (x_1^*, x_2^*, \ldots, x_n^*)$ satisfies

$$x_i^* \geq 0 \qquad \text{(all } i)$$

and

$$\sum_{i \in S} x_i^* = 1 \qquad \text{(all minimal winning coalitions } S).$$

For each minimal winning coalition S, let $x(S) = (x_1, x_2, \ldots, x_n)$ be the imputation with

$$x_i = \begin{cases} x_i^* & \text{(if } i \in S) \\ 0 & \text{(if } i \notin S). \end{cases}$$

Then

$$\{x(S) : S \text{ is a minimal winning coalition}\}$$

forms a *main simple solution* to the game. Find a main simple solution to:

(a) The 3-person essential, constant-sum simple game.
(b) The weighted majority game (3; 1, 1, 1, 2).
(c) The weighted majority game (6; 1, 1, 1, 2, 3, 3).

(d) (A.W. Tucker.) The 5-person constant-sum game in which the minimal winning coalitions are all 3-element coalitions.

16. (von Neumann and Morgenstern [1944].) Prove that a main simple solution, if it exists, is always a stable set.

17. (von Neumann and Morgenstern [1944].) A weighted majority game $(Q; k_1, k_2, \ldots, k_n)$ is *homogeneous* if $\sum_{i=1}^{n} k_i = 2Q - 1$ and each minimal winning coalition S has $\sum_{i \in S} k_i = Q$. Show that every homogeneous weighted majority game has a main simple solution.

6.6. *S*-equivalence and the (0, 1)-Normalization

6.6.1. ISOMORPHISM AND *S*-EQUIVALENCE

Let us take two characteristic functions v and v' on the same set of players $I = \{1, 2, \ldots, n\}$. We shall say that the game (function) v is *isomorphic* to the game (function) v' if there exists a one-to-one function f from $M(v)$ onto $M(v')$ such that for all $x, y \in M(v)$ and all coalitions S,

$$x \succ_S y \quad \text{iff} \quad f(x) \succ_S f(y). \tag{35}$$

We call the function f an *isomorphism* from v to v'. It is easy to prove that if v is isomorphic to v' then v' is isomorphic to v. An isomorphism is given by f^{-1}. Thus, we can say that v and v' are isomorphic, without mentioning the direction of the isomorphism. To give an example, suppose v' on $\{1, 2, 3\}$ is defined as follows:

$$v'(S) = \begin{cases} 0 & (\text{if } |S| = 0 \text{ or } 1) \\ 2 & (\text{if } |S| = 2 \text{ or } 3). \end{cases} \tag{36}$$

Then the characteristic function v of the weighted majority game $(2; 1, 1, 1)$ is isomorphic to v'. For

$$M(v) = \{x : x_1 \geqq 0, x_2 \geqq 0, x_3 \geqq 0, x_1 + x_2 + x_3 = 1\}$$

and

$$M(v') = \{x : x_1 \geqq 0, x_2 \geqq 0, x_3 \geqq 0, x_1 + x_2 + x_3 = 2\}.$$

If $f(x) = 2x$, i.e., $f(x_1, x_2, x_3) = (2x_1, 2x_2, 2x_3)$, then f is a function from $M(v)$ onto $M(v')$. Moreover, $(x_1, x_2, x_3) \succ_S (y_1, y_2, y_3)$ holds (in v) if and only if $(2x_1, 2x_2, 2x_3) \succ_S (2y_1, 2y_2, 2y_3)$ holds (in v'). Thus, f is an isomorphism.

Clearly, isomorphic games have isomorphic cores and families of stable sets. That is, if C is a core or a stable set in v, and f is an isomorphism from v to v', then $f(C) = \{f(x) : x \in C\}$ is respectively a core or a stable set in v'. For example, by the discussion in Sec. 6.3,

$$B = \{(\tfrac{1}{2}, \tfrac{1}{2}, 0), (\tfrac{1}{2}, 0, \tfrac{1}{2}), (0, \tfrac{1}{2}, \tfrac{1}{2})\} \tag{37}$$

is a stable set in the game $(2; 1, 1, 1)$. Then

$$f(B) = \{(1, 1, 0), (1, 0, 1), (0, 1, 1)\}$$

is a stable set in the game v' of Eq. (36).

From a game-theoretic point of view, isomorphic games are interchangeable in many ways.[18] In this section, we shall show that all "interesting" games in characteristic function form are isomorphic to games whose characteristic functions satisfy very simple conditions, namely

$$v(\{i\}) = 0 \qquad (\text{all } i) \tag{38}$$

and

$$v(I) = 1. \tag{39}$$

A characteristic function v satisfying Eq. (38) is said to be in 0-*normalization* and one satisfying both Eqs. (38) and (39) is said to be in (0, 1)-*normalization*. Characteristic functions in 0-normalization have a number of useful properties.

Theorem 6.10. If v is a characteristic function in 0-normalization, then

(a) $v(S) \geqq 0$ \qquad (all S)

and

(b) $S \subseteq T \Longrightarrow v(S) \leqq v(T)$.

Proof. Using superadditivity (Eq. (2)) and the 0-normalization, we have

$$v(S \cup \{i\}) \geqq v(S) + v(\{i\}) = v(S).$$

Part (a) can be proved by induction on $|S|$, using $v(\varnothing) = 0$. Part (b) follows from part (a), since $v(T - S) \geqq 0$, so

$$v(T) \geqq v(S) + v(T - S) \geqq v(S). \qquad \text{Q.E.D.}$$

Let us recall that a game v is essential if there exist disjoint coalitions S and T such that $v(S \cup T) > v(S) + v(T)$. We shall show that every essential game is isomorphic to a game in (0, 1)-normalization. To do this, let us introduce the notion that two characteristic functions v and v' on the same set of players I are *strategically equivalent* or *S-equivalent*. (The reader should not confuse S with a coalition.) We say that v' is *S-equivalent* to v if there are

---

[18]They are not totally interchangeable, however. In Shapley and Shubik [1969a], it is shown that there are isomorphic games one of which is "totally balanced" but the other of which is not. (A game is *totally balanced* if every subgame, i.e., restriction of the game to a subset of I, has a core.) It is also the case that isomorphisms do not necessarily preserve the Shapley value, the solution concept we introduce in the next section. For Shapley and Shubik [1969a] show that if v is a so-called "balanced game," then the so-called "cover" $\bar{v}$ of v is isomorphic to v via an isomorphism f. However, the Shapley value of $\bar{v}$ is not necessarily f of the Shapley value of v.

real numbers $a_1, a_2, \ldots, a_n$ and a positive real number α such that

$$v'(S) = \alpha v(S) + \sum_{i \in S} a_i \quad \text{(all } S\text{)}. \tag{40}$$

If v' is S-equivalent to v, one can think of getting v' from v as follows: Before the game, each player is paid an amount a_i and the units in which payoffs are measured are then changed by multiplying all payoffs by α. To give an example, suppose v is the following characteristic function:

$$v(\varnothing) = 0$$
$$v(\{1\}) = 1, \quad v(\{2\}) = 1, \quad v(\{3\}) = 2$$
$$v(\{1, 2\}) = 3, \quad v(\{1, 3\}) = 3, \quad v(\{2, 3\}) = 4 \tag{41}$$
$$v(\{1, 2, 3\}) = 6.$$

Pick $a_1 = 1, a_2 = 0, a_3 = 2$ and $\alpha = 2$. Then v' satisfying Eq. (40) is given by

$$v'(\varnothing) = 0$$
$$v'(\{1\}) = 2v(\{1\}) + a_1 = 3$$
$$v'(\{2\}) = 2v(\{2\}) + a_2 = 2$$
$$v'(\{3\}) = 2v(\{3\}) + a_3 = 6$$
$$v'(\{1, 2\}) = 2v(\{1, 2\}) + a_1 + a_2 = 7$$
$$v'(\{1, 3\}) = 2v(\{1, 3\}) + a_1 + a_3 = 9$$
$$v'(\{2, 3\}) = 2v(\{2, 3\}) + a_2 + a_3 = 10$$
$$v'(\{1, 2, 3\}) = 2v(\{1, 2, 3\}) + a_1 + a_2 + a_3 = 15.$$

To give another example, if $d = \sum_{i \in I} w_i$, then Pure Bargaining (Example 1, Sec. 6.1) is S-equivalent to Deterrence (Example 2, Sec. 6.1). Why?

S-equivalence has the following properties:

$$v \text{ is } S\text{-equivalent to itself.} \tag{42}$$

$$\text{If } v' \text{ is } S\text{-equivalent to } v, \text{ then } v \text{ is } S\text{-equivalent to } v'. \tag{43}$$

$$\text{If } v \text{ is } S\text{-equivalent to } v' \text{ and } v' \text{ is } S\text{-equivalent to } v'', \atop \text{then } v \text{ is } S\text{-equivalent to } v''. \tag{44}$$

In the terminology of relation theory (Sec. 8.2, Exer. 20), S-equivalence is an equivalence relation. Verification of properties (42) to (44) is left as an exercise (Exer. 16).

Theorem 6.11. Every essential game is S-equivalent to a game in $(0, 1)$-normalization.

Proof. Let v be the characteristic function of an essential game. Define $a_i = -v(\{i\})$. Then define v' by

$$v'(S) = v(S) + \sum_{i \in S} a_i.$$

Now v' is S-equivalent to v and it is 0-normalized. Moreover, $v'(I) > 0$. (This follows from Exers. 18 and 19.) Finally,

$$v''(S) = \frac{1}{v'(I)} v'(S)$$

defines a characteristic function which is S-equivalent to v and in $(0, 1)$-normalization. Q.E.D.

Corollary 1. Every game is S-equivalent to a game in 0-normalization.

Proof. This is a corollary of the proof.

Corollary 2. If v is essential, then v is S-equivalent to exactly one game in $(0, 1)$-normalization, the game v'' given by

$$v''(S) = \frac{1}{v(I) - \sum_{i=1}^{n} v(\{i\})} [v(S) - \sum_{i \in S} v(\{i\})]. \tag{45}$$

Proof. Left to the reader (Exer. 20).

If v is essential, the unique $(0, 1)$-normalized game S-equivalent to v is called *the $(0, 1)$-normalization* of v. To give an example, the game v of Eqs. (41) is essential. Its $(0, 1)$-normalization can be obtained by using Corollary 2. It is

$$v''(S) = 0 \quad (\text{if } |S| = 0 \text{ or } 1)$$
$$v''(\{1, 2\}) = \tfrac{1}{2}, \quad v''(\{1, 3\}) = 0, \quad v''(\{2, 3\}) = \tfrac{1}{2}$$
$$v''(\{1, 2, 3\}) = 1.$$

The next theorem completes the proof that every essential game is isomorphic to a game in $(0, 1)$-normalization.

Theorem 6.12. Suppose v and v' are the characteristic functions of two S-equivalent games. Then the games are isomorphic.

Proof. Suppose v and v' are S-equivalent and satisfy Eq. (40). Let $a = (a_1, a_2, \ldots, a_n)$. Define f on $M(v)$ by $f(x) = \alpha x + a$. The verification that f is an isomorphism is straightforward. (See Exer. 22.) Q.E.D.

The converse of Theorem 6.12 has been proved by McKinsey [1950] for a special class of essential games called *zero-sum*. Such games satisfy

$$v(S) + v(I - S) = 0 \quad (\text{all } S).$$

(Zero-sum games are special kinds of constant-sum games.)

Theorem 6.13 (McKinsey). Suppose v and v' are isomorphic characteristic functions of essential zero-sum games. Then v and v' are S-equivalent.

The proof of Theorem 6.13 is considerably more difficult than that of Theorem 6.12. A nice presentation of the proof can be found in Berge [1957]. Apparently Theorem 6.13 is true[19] without the assumption that the games are essential and zero-sum, but the author knows of no published proof of this fact.

Using the $(0, 1)$-normalization, it is possible to calculate cores and stable sets for many games. We have already accomplished this for the game v' of Eq. (36). For $(2; 1, 1, 1)$ is the $(0, 1)$-normalization. (Why?) We showed how to find a stable set for v' from the stable set B for $(2; 1, 1, 1)$ of Eq. (37). To give a second example, consider the 3-player Garbage Game v. Here, if $|S| = 1$, $v(S) = -2$. By Corollary 2 to Theorem 6.11, the $(0, 1)$-normalization is given by

$$v''(S) = \tfrac{1}{3}[v(S) + \sum_{i \in S} 2] = \tfrac{1}{3}v(S) + \sum_{i \in S} (\tfrac{2}{3}).$$

Thus,

$$v''(S) = \begin{cases} 0 & (\text{if } |S| \leq 1) \\ 1 & (\text{if } |S| \geq 2). \end{cases}$$

The reader will recognize v'' as the weighted majority game $(2; 1, 1, 1)$ once again. One stable set for v'' is given by set B of Eq. (37). It is easy to translate B into a stable set for the 3-player garbage game v. Let $a = (a_1, a_2, a_3)$ be $(\tfrac{2}{3}, \tfrac{2}{3}, \tfrac{2}{3})$ and let $f(y) = (\tfrac{1}{3})y + a$, where $y = (y_1, y_2, y_3)$. Then by Exer. 22, f is an isomorphism from game v to game v''. Hence, $B' = \{y : f(y) \in B\} = f^{-1}(B)$ defines a stable set in v. Note that

$$(\tfrac{1}{3})y + a = z \qquad \text{iff} \qquad y = 3z - 3a,$$

so $f^{-1}(z) = 3z - 3a$. Now

$$f^{-1}[(\tfrac{1}{2}, \tfrac{1}{2}, 0)] = 3(\tfrac{1}{2}, \tfrac{1}{2}, 0) - 3(\tfrac{2}{3}, \tfrac{2}{3}, \tfrac{2}{3}) = (-\tfrac{1}{2}, -\tfrac{1}{2}, -2).$$

A similar calculation using $(\tfrac{1}{2}, 0, \tfrac{1}{2})$ and $(0, \tfrac{1}{2}, \tfrac{1}{2})$ gives us

$$B' = \{(-\tfrac{1}{2}, -\tfrac{1}{2}, -2), (-\tfrac{1}{2}, -2, -\tfrac{1}{2}), (-2, -\tfrac{1}{2}, -\tfrac{1}{2})\}.$$

This stable set is interpreted as follows. Some 2-player coalition forms. The two players agree to each drop their bags of garbage in the third player's yard, and to split the bag of garbage which the third player is expected to drop in one of their yards.

6.6.2. THE GEOMETRIC POINT OF VIEW

The $(0, 1)$-normalization gives rise to a very useful geometric representation of various sets of imputations. Suppose v is a game in $(0, 1)$-normaliza-

---

[19] L. S. Shapley (personal communication).

tion. Then the set of imputations is given as follows:

$$M(v) = \{(x_1, x_2, \ldots, x_n): x_i \geqq 0, \text{ all } i, \text{ and } \sum_{i=1}^{n} x_i = 1\}.$$

If the set $M(v)$ is drawn as a set of points in Euclidean n-space, then this set determines an *n-simplex*, an $(n-1)$-dimensional polyhedron or regular n-gon with vertices the n vectors $(1, 0, 0, \ldots, 0)$, $(0, 1, 0, \ldots, 0)$, $\ldots$, $(0, 0, \ldots, 0, 1)$. This is an equilateral triangle if $n = 3$, a regular tetrahedron if $n = 4$, etc. The space of imputations consists of all vertices on the boundary and in the interior of the regular n-gon in n-space. It is convenient to draw this n-gon in $n-1$ dimensions. Figure 6.11 shows the space of imputations in a 3-player game in $(0, 1)$-normalization and locates several imputations. To locate an imputation $(x_1, x_2, \ldots, x_n)$, the following technique works: Find the center of gravity if for each i, an object of mass x_i is placed at the ith vertex, the one corresponding to the vector $(0, 0, \ldots, 0, 1, 0, \ldots, 0)$, with a 1 in the ith place.

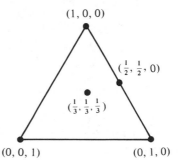

FIGURE 6.11. The space of imputations for a 3-player game in $(0, 1)$-normalization.

It is instructive to locate cores and stable sets geometrically. Consider the weighted majority game $(2; 1, 1, 1)$, which is already in $(0, 1)$-normalization. Figure 6.12 shows the stable set B of Eq. (37) and Fig. 6.13 shows the

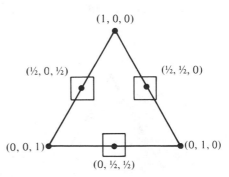

FIGURE 6.12. The boxes enclose a stable set B for the weighted majority game $(2; 1, 1, 1)$.

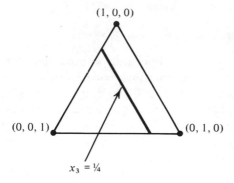

FIGURE 6.13. The line $x_3 = \frac{1}{4}$ is a stable set $B_{1/4}$ for the weighted majority game $(2; 1, 1, 1)$.

stable set B_c of Eq. (28), with $c = \frac{1}{4}$. The latter is a straight line consisting of all points (x_1, x_2, x_3) in the triangle such that $x_3 = \frac{1}{4}$.

In the Land Development Game, the $(0, 1)$-normalization gives us

$$v(\varnothing) = 0$$
$$v(\{1\}) = v(\{2\}) = v(\{3\}) = 0$$
$$v(\{1, 2\}) = \tfrac{1}{2}, \; v(\{1, 3\}) = 1, \; v(\{2, 3\}) = 0$$
$$v(\{1, 2, 3\}) = 1.$$

The core in this normalization consists of a line segment corresponding to

$$C' = \{(x_1, 0, x_3): \tfrac{1}{2} \leq x_1 \leq 1, \; x_1 + x_3 = 1\}. \tag{46}$$

This is shown in Fig. 6.14. As for stable sets, the stable set B of Eq. (29) corresponds to the set

$$B' = \{(x_1, 0, x_3): 0 \leq x_1 \leq 1, \; x_1 + x_3 = 1\}$$

and the stable set $B_{1/2}$ corresponds to the set $B'_{1/2}$ which is the union of the set representing the core plus the set

$$D'_{1/2} = \{(x_1, x_2, x_3): 0 \leq x_1 \leq \tfrac{1}{2}, \; x_2 = (\tfrac{1}{2})(\tfrac{1}{2} - x_1), \; x_3 = \tfrac{1}{2} + (\tfrac{1}{2})(\tfrac{1}{2} - x_1)\},$$

which corresponds to the set $D_{1/2}$ of Eq. (30). Now, B' is a single line seg-

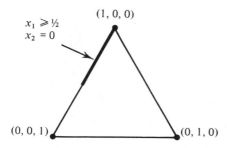

FIGURE 6.14. Core of the Land Development Game in $(0, 1)$-normalization.

ment and $B'_{1/2}$ is the union of two line segments, as shown in Fig. 6.15.[20] More generally, stable sets for the Land Development Game can be constructed by adding certain kinds of continuous curves to the core. A sample of such a stable set is given in Fig. 6.16. The reader is referred to Shapley and Shubik [1973, p. 61] for further details.

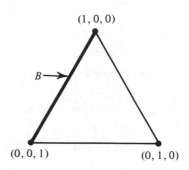

FIGURE 6.15. Stable sets B and $B_{1/2}$ for the Land Development Game in (0, 1)-normalization.

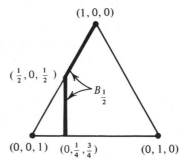

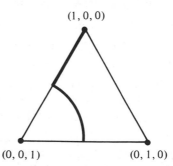

FIGURE 6.16. A stable set for the Land Development Game in (0, 1)-normalization.

[20]To see that $D'_{1/2}$ is a line segment, observe that it can be written as: $x_1 = t$, $x_2 - \frac{1}{4} = (-\frac{1}{2})t$, $x_3 - \frac{3}{4} = (-\frac{1}{2})t$, for $0 \leq t \leq \frac{1}{2}$. The reader will recognize these as the parametric equations for a straight line in 3-space.

EXERCISES

1. Find a game v'' which is isomorphic to the game $(2; 1, 1, 1)$ and such that $v''(I) = 3$.

2. For the game constructed in Exer. 1, find a stable set.

3. In which of the following games is the characteristic function 0-normalized and in which is it $(0, 1)$-normalized?
   (a)   Pure Bargaining.
   (b)   Land Development.
   (c) $(51; 50, 30, 20)$.

4. Give an example of a game violating conditions (a) and (b) of Theorem 6.10.

5. Find a game in $(0, 1)$-normalization which is S-equivalent to the game v defined as follows:

   $$v(\varnothing) = 0$$
   $$v(\{1\}) = -3, \; v(\{2\}) = -1, \; v(\{3\}) = 0$$
   $$v(\{1, 2\}) = v(\{2, 3\}) = v(\{1, 3\}) = 0$$
   $$v(\{1, 2, 3\}) = 0.$$

6. Find a game in $(0, 1)$-normalization which is S-equivalent to the 4-player Lake Game if:
   (a) $B = 3, \, C = 2$.
   (b) $B = 2, \, C = 1$.

7. Show that $(2; 1, 1, 1)$ is the $(0, 1)$-normalization of the game defined by Eq. (36).

8. Find a 0-normalization for the game defined as follows: $v(S) = |S|$. Does this game have a $(0, 1)$-normalization? Why?

9. Find the $(0, 1)$-normalization for the 4-player Garbage Game.

10. Prove that the 3-player Garbage Game is isomorphic to the weighted majority game $(2; 1, 1, 1)$.

11. (a)   Prove that Deterrence is S-equivalent to Pure Bargaining.
    (b)   Find the core and all stable sets for Deterrence from the corresponding solutions for Pure Bargaining.

12. Suppose $I = \{1, 2, 3\}$ and consider the game $v(S) = 10$ iff $\{1, 2\} \subseteq S$, $v(S) = 0$ otherwise. Can you find the core and a stable set for v by passing to the $(0, 1)$-normalization?

13. Using the core C' given in Eq. (46), rederive an expression for the core of the Land Development Game.

14. For the 3-player game in (0, 1)-normalization, locate the following imputations geometrically.

    (a) $(\frac{1}{2}, 0, \frac{1}{2})$.
    (b) $(.99, .01, 0)$.
    (c) $(\frac{1}{4}, \frac{1}{4}, \frac{1}{2})$.

15. In the 3-player game in (0, 1)-normalization, locate the following subsets of $M(v)$ geometrically:

    (a) $\{(x_1, x_2, x_3) \in M(v): x_1 + x_2 + x_3 = 1\}$.
    (b) $\{(x_1, x_2, x_3) \in M(v): x_1 = 0, x_2 = 0, \text{ or } x_3 = 0\}$.
    (c) The stable set B_c of Eq. (28) with $c = 0$.
    (d) $\{(x_1, x_2, x_3) \in M(v): x_1 \geqq \frac{1}{2}\}$.

16. Prove that S-equivalence satisfies properties (42), (43), and (44).

17. (A. W. Tucker.) Suppose u is a 0-normalized characteristic function on a set of individuals H and v is a 0-normalized characteristic function on a set of individuals I, with $H \cap I = \varnothing$. Define w on $H \cup I$ as follows: $w(R \cup S) = u(R)v(S)$ for each $R \subseteq H$ and each $S \subseteq I$.

    (a) Show that w is a characteristic function on $H \cup I$.
    (b) Is w necessarily 0-normalized? (Give proof or counterexample.)
    (c) Is w necessarily essential if u and v are?
    (d) Is w necessarily simple if u and v are?
    (e) Can w necessarily be represented as a weighted majority game if u and v can?
    (f) If u and v have nonempty cores, does w?
    (g) Is part (a) true if u and v are not both 0-normalized?

18. Prove that if v' is S-equivalent to v, then v' is essential if and only if v is.

19. Prove that in an essential game v in 0-normalization, $v(I) > 0$.

20. Prove Corollary 2 of Theorem 6.11.

21. Suppose v and v' are characteristic functions and for all coalitions S,

$$v'(S) = \alpha v(S) + \sum_{i \in S} a_i,$$

$\alpha > 0$. Let $a = (a_1, a_2, \ldots, a_n)$. If $x = (x_1, x_2, \ldots, x_n)$ is a real vector, let $f(x)$ be the vector $\alpha x + a$. Prove that $x \in M(v)$ if and only if $f(x) \in M(v')$.

22. Prove that if v and v' are S-equivalent and satisfy Eq. (40) with $\alpha > 0$ and if $a = (a_1, a_2, \ldots, a_n)$ and $f(x) = \alpha x + a$, then f is an isomorphism from v to v'.

23. Theorem 6.3 has been proved only for the (0, 1)-normalization. Use the result of Exer. 22 to complete the proof of the theorem.

6.7. The Shapley Value

6.7.1. SHAPLEY'S AXIOMS

Ideally, an analysis of a game in characteristic function form would result in the singling out of a specific imputation which would be regarded as the solution of the game. Since the core may be empty, since a stable set may not exist, and since the core or a stable set may have more than one imputation, we are led to search for yet another solution concept if a single imputation solution is sought. That is the approach we take in this section. Specifically, our approach is axiomatic, comparable to the approach we used to axiomatize a measure of trophic status in Sec. 3.5.2 and comparable to the axiomatizations for group decisionmaking to be studied in Chapter 7. We list reasonable axioms which the solution concept should satisfy. A solution concept is considered satisfactory if it satisfies these axioms. The axioms we use are slightly different from those originally presented by Shapley [1953a], and are based on those in Shapley and Shubik [to appear a].

Suppose v is a game in characteristic function form. We seek a single imputation $\phi[v]$, to be called the *Shapley value*, which will serve as a solution to the game. (The value only makes sense for games in characteristic function form, and it is assumed that it depends only on the characteristic function v.) Let $\phi_i[v]$ be the ith component of the vector $\phi[v]$, i.e., the payoff to player i. We shall assume that the function $\phi[v] = (\phi_1[v], \phi_2[v], \ldots, \phi_n[v])$ satisfies serveral axioms. To state the first, let us suppose that π is a permutation of $I = \{1, 2, \ldots, n\}$.[21] Then for every coalition $S = \{i_1, i_2, \ldots, i_s\}$ contained in I, πS is the coalition $\{\pi(i_1), \pi(i_2), \ldots, \pi(i_s)\}$.

Axiom 1. If v is a characteristic function on a set of players I, if π is a permutation of I, and if the characteristic function[22] w is defined on I by $w(S) = v(\pi S)$, then for all $i \in I$,

$$\phi_i[w] = \phi_{\pi(i)}[v].$$

Axiom 1 states that the value assigned to a player is independent of the "label" given him. In particular, it implies that in a symmetric game, all players are assigned equal value.

The second axiom is needed to make $\phi[v]$ an imputation.

Axiom 2. $\sum_{i=1}^{n} \phi_i[v] = v(I).$

We do not explicitly assume that $\phi_i[v] \geqq v(\{i\})$, the second requirement needed for $\phi[v]$ to be an imputation. This condition follows from the other

---

[21]Recall that a permutation is just a rearrangement or relabelling of the elements of the set.

[22]One needs to prove that w is a characteristic function. This is straightforward.

axioms. In listing axioms, we try to list as few axioms as possible, and omit conditions which follow from other assumptions.

Axiom 3 says that if a player i adds nothing to any coalition, then his value is 0. Such a player is called a *dummy*.

Axiom 3. If $v(S - \{i\}) = v(S)$ for all S, then

$$\phi_i[v] = 0.$$

The final axiom is an additivity axiom.

Axiom 4. If v and v' are characteristic functions defined on the same set of players I and w is the characteristic function[23] $v + v'$ on I, then

$$\phi[w] = \phi[v] + \phi[v'].$$

Axiom 4 at first reading is aesthetically pleasing. However, its justification requires some understanding of the meaning of the sum of two games v and v'. If these games are played one after another, using the same set of players, it is not clear that the outcomes add: The first game might affect the play of the second or change the psychological feeling of players toward each other. A more plausible justification would be to think of the games v and v' as being played simultaneously, but with each player i appointing a "double" to play in the second game v'.

As we have discussed in Sec. 3.5.2, a set of axioms can have several properties. It might give rise to several $\phi[v]$; alternatively, there might be no $\phi[v]$ satisfying all of the axioms; or finally there might be exactly one $\phi[v]$ satisfying them. In this case, the latter very pleasing situation occurs. (In such a situation, we say that the axioms are *categorical*.)

Theorem 6.14 (Shapley). There is a unique function $\phi[v]$, defined on all characteristic functions v, and satisfying Axioms 1–4. Specifically, $\phi_i[v]$ is given by

$$\phi_i[v] = \sum \{\gamma(s)[v(S) - v(S - \{i\})] : S \text{ such that } i \in S\}, \qquad (47)$$

where $s = |S|$ and

$$\gamma(s) = \frac{(s-1)!\,(n-s)!}{n!}. \qquad (48)$$

We defer the proof of Theorem 6.14 to the end of this section.

6.7.2. EXAMPLES

Using the theorem, let us calculate the Shapley value $\phi[v]$ for certain games. First as we have already observed, in a symmetric game the Shapley value divides $v(I)$ into n equal parts. Thus, in the weighted majority game

---

[23]Again, one needs to prove that w is a characteristic function.

$(2; 1, 1, 1)$, $\phi[v]$ is the imputation $(\frac{1}{3}, \frac{1}{3}, \frac{1}{3})$. Indeed, $\phi[v]$ is $(\frac{1}{3}, \frac{1}{3}, \frac{1}{3})$ also for the game $(51; 49, 48, 3)$, since that is also symmetric. We observed earlier that in $(51; 49, 48, 3)$, the three players have essentially equal power, and the Shapley value captures this idea. In some sense, the number $\phi_i[v]$ may be looked at as a measure of the "power" of the ith player. In Pure Bargaining, another symmetric game, the Shapley value is $(d/n, d/n, \ldots, d/n)$. In the Garbage Game it is $(-1, -1, \ldots, -1)$ for any n. Similar calculations can be made for other symmetric games such as the Lake.

The Land Development Game will be our first interesting example. Here, using the convention $0! = 1$, we find

$$\gamma(1) = \tfrac{1}{3}$$
$$\gamma(2) = \tfrac{1}{6}$$
$$\gamma(3) = \tfrac{1}{3}.$$

Thus, since 1 is in the coalitions $\{1\}$, $\{1, 2\}$, $\{1, 3\}$, and $\{1, 2, 3\}$,

$$\phi_1[v] = \gamma(1)[v(\{1\}) - v(\varnothing)] + \gamma(2)[v(\{1, 2\}) - v(\{2\})] +$$
$$\gamma(2)[v(\{1, 3\}) - v(\{3\})] + \gamma(3)[v(\{1, 2, 3\}) - v(\{2, 3\})]$$
$$= \$100{,}000\gamma(1) + \$200{,}000\gamma(2) + \$300{,}000\gamma(2) + \$300{,}000\gamma(3)$$
$$\approx \$216{,}667.$$

Similarly,
$$\phi_2[v] \approx \$16{,}667$$
and
$$\phi_3[v] \approx \$66{,}667.$$

One interpretation of this result is that the owner pays the industrialist $16,667 for not bidding, and then sells his land to the subdivider for $233,334. The reader will notice that the value is not in the core. Indeed, it can be shown that it is not in any stable set. This, of course, is disturbing, and it suggests that value again may not be the "right" solution concept for a game.

6.7.3. SIMPLE GAMES

Suppose the game v is simple. Then $v(S) - v(S - \{i\})$ is always 0 or 1, and it is 1 if and only if S is winning and $S - \{i\}$ is losing. Then we have

$$\phi_i[v] = \sum_{S \in \mathcal{S}_i} \gamma(s) = \sum_{S \in \mathcal{S}_i} \frac{(s-1)!(n-s)!}{n!}, \tag{49}$$

where $s = |S|$ and the summation in (49) is carried out over the set $\mathcal{S}_i$ of all winning coalitions S containing i and such that $S - \{i\}$ is not winning. Thus, for example, it is easy to calculate the value for weighted majority games. Let us take the game $(51; 51, 48, 1)$. Here, a coalition S is winning if and only if player 1 is in S. We have $v(S) - v(S - \{i\}) = 1$ iff $i = 1$. Thus,

$\phi_i[v] = 0$ for $i \neq 1$ and, since $\sum_{i=1}^{n} \phi_i[v] = 1$, $\phi_1[v] = 1$. Thus, $\phi[v] = (1, 0, 0)$. Player 1 has all the power, as is clear since a coalition is winning if and only if 1 is in it. A player who himself forms winning coalition is called a *dictator*.

To give another example, let us consider the game $(51; 40, 30, 20, 10)$. The winning coalitions are

$$\{1, 2\}, \{1, 3\}, \{1, 2, 3\}, \{1, 2, 4\}, \{1, 3, 4\}, \{2, 3, 4\}, \{1, 2, 3, 4\}.$$

We have $\gamma(2) = \frac{1}{12}$ and $\gamma(3) = \frac{1}{12}$. The winning coalitions containing player 1 which are not winning after player 1 is removed are

$$\{1, 2\}, \{1, 3\}, \{1, 2, 3\}, \{1, 2, 4\}, \{1, 3, 4\}.$$

Hence,

$$\phi_1[v] = 2\gamma(2) + 3\gamma(3) = \tfrac{5}{12}.$$

The winning coalitions containing player 2 which are not winning after player 2 is removed are

$$\{1, 2\}, \{1, 2, 4\}, \{2, 3, 4\}.$$

Hence,

$$\phi_2[v] = \gamma(2) + 2\gamma(3) = \tfrac{1}{4}.$$

Similarly, $\phi_3[v] = \frac{1}{4}$ and $\phi_4[v] = \frac{1}{12}$. Thus, $\phi[v] = (\frac{5}{12}, \frac{1}{4}, \frac{1}{4}, \frac{1}{12})$. In the sense of the Shapley value, players 2 and 3 have equal power and player 1 has five times the power of player 4. Note that power is not necessarily proportional to percentage of votes held. Here, player 1 has four times as many votes as player 4, but five times the power; in the game $(51; 49, 48, 3)$, player 1 has more than 16 times the votes of player 3, but no more power. Power, under our interpretation, has to do with ability to maneuver into "winning" coalitions, and the proportion of votes is clearly not a good measure.

With this interpretation of power, let us calculate the power in the U.N. Security Council. The reader will recall that the Security Council is the game $(39; 7, 7, 7, 7, 7, 1, 1, 1, 1, 1, 1, 1, 1, 1, 1)$. A coalition is winning if and only if it consists of all five permanent members (players 1 through 5) and at least four nonpermanent members (players 6 through 15). If player i is not one of the permanent members, and i is in S, then S is winning while $S - \{i\}$ is not if and only if S consists of nine players, including all the permanent members. Thus, choice of S amounts to choosing three nonpermanent members out of the nine other than i. It follows that there are $\binom{9}{3} = \dfrac{9!}{3!\,6!}$ such coalitions. Now $n = 15$. Hence,

$$\gamma(9) = \frac{(9 - 1)!\,(15 - 9)!}{15!} = \frac{8!\,6!}{15!}.$$

It follows that

$$\phi_i[v] = \frac{9!}{3!\,6!} \times \frac{8!\,6!}{15!} \approx .001865.$$

Now, since there are 10 nonpermanent members (with equal power),

$$\sum_{\substack{i \\ \text{nonpermanent}}} \phi_i[v] \approx 10 \times .001865 = .01865.$$

Thus,

$$\sum_{\substack{i \\ \text{permanent}}} \phi_i[v] \approx 1 - .01865 = .98135.$$

By symmetry, each permanent member has power

$$\phi_i[v] \approx \frac{1}{5}(.98135) \approx .1963.$$

The conclusion is that the permanent members together carry almost all, namely more than 98%, of the power in the Security Council. The nonpermanent members, although not powerless, are virtually so. The idea of calculating power in the Security Council and other legislative organs in this manner was first introduced in Shapley and Shubik [1954].

6.7.4. THE PROBABILISTIC INTERPRETATION AND APPLICATIONS TO LEGISLATURES

Frequently the Shapley value can be calculated in a different way, which gives it a probabilistic interpretation.[24] Imagine that the players in a game are chosen in random order. If $o = (i_1, i_2, \ldots, i_n)$ is such an order, then let $S_i(o)$ denote the set of players who precede i in o, and let S be the set $S_i(o) \cup \{i\}$. It is easy to see that the numerator of $\gamma(s)$, $(s-1)!(n-s)!$, gives the number of orderings of $\{1, 2, \ldots, n\}$ in which the players of $S - \{i\}$ come first, then i, and then the rest of the players. The denominator of $\gamma(s)$, $n!$, gives the number of orderings of $\{1, 2, \ldots, n\}$. Thus $\gamma(s)$ is the probability that exactly the players of $S - \{i\}$ will precede i if an ordering o is chosen at random. It follows from Theorem 6.14 that $\phi_i[v]$ is the expected value of the expression

$$\lambda_i(o) = v(S_i(o) \cup \{i\}) - v(S_i(o)), \tag{50}$$

with the expectation calculated over all orderings o. We formulate this result as a theorem.

Theorem 6.15 (Shapley [1953a]). If $S_i(o)$ represents the set of players who precede player i in the ordering o of I, then $\phi_i[v]$ is the expected value over all orderings o of

$$\lambda_i(o) = v(S_i(o) \cup \{i\}) - v(S_i(o)). \tag{50}$$

In a simple game, this result has a very nice interpretation. Here, $\lambda_i(o)$ is 1 if $S_i(o)$ is losing and $S_i(o) \cup \{i\}$ is winning, while $\lambda_i(o)$ is 0 otherwise. We imagine coalitions expanding with players added one at a time at random.

---

[24]The reader may wish to skip directly to the Corollary to Theorem 6.15.

The ordering of all the players obtained is o. Then $\phi_i[v]$ is the probability that player i is exactly the player whose addition makes the coalition a winning one. Such a player is called *pivotal* for the ordering o. He is the player whose addition throws the coalition "over the top." This interpretation is particularly useful in calculating the value. Since all orderings are considered equally likely, we may restate the result as follows:

Corollary. In a simple game, $\phi_i[v]$ is the probability that player i is pivotal, i.e., $\phi_i[v]$ is the number of orderings in which player i is pivotal divided by $n!$, the number of orderings of n players.

Using the Corollary, let us return to the game (2; 1, 1, 1). The possible orderings of the players 1, 2, 3 are

Order Number	Ordering
1	1 2 3
2	1 3 2
3	2 1 3
4	2 3 1
5	3 1 2
6	3 2 1

In the third and fifth, player 1 is pivotal, so $\phi_1[v] = 2/3! = \frac{1}{3}$, which agrees with our earlier calculation. In the game (51; 40, 30, 20, 10), the orderings are

No.	Ordering	No.	Ordering	No.	Ordering
1	1 2 3 4	9	2 3 1 4	17	3 4 1 2
2	1 2 4 3	10	2 3 4 1	18	3 4 2 1
3	1 3 2 4	11	2 4 1 3	19	4 1 2 3
4	1 3 4 2	12	2 4 3 1	20	4 1 3 2
5	1 4 2 3	13	3 1 2 4	21	4 2 1 3
6	1 4 3 2	14	3 1 4 2	22	4 2 3 1
7	2 1 3 4	15	3 2 1 4	23	4 3 1 2
8	2 1 4 3	16	3 2 4 1	24	4 3 2 1

Player 1 is pivotal in orderings 7, 8, 9, 11, 13, 14, 15, 17, 21 and 23, so $\phi_1[v] = \frac{10}{24} = \frac{5}{12}$, which agrees with our earlier calculations. To give yet another example, let us recall that for certain decisions, the Australian government acts like the game (5; 1, 1, 1, 1, 1, 1, 3), with the federal government acting as player 7 and the states as the remaining six players. The federal government is pivotal in a given ordering if and only if it is third, fourth, or fifth. The probability that the federal government is picked third is $(\frac{6}{7})(\frac{5}{6})(\frac{1}{5}) = \frac{1}{7}$; the probability it is picked fourth is $(\frac{6}{7})(\frac{5}{6})(\frac{4}{5})(\frac{1}{4}) = \frac{1}{7}$; and the probability that it is picked fifth is $(\frac{6}{7})(\frac{5}{6})(\frac{4}{5})(\frac{3}{4})(\frac{1}{3}) = \frac{1}{7}$. Thus, $\phi_7[v] = \frac{3}{7} = \frac{9}{21}$. Now, by symmetry, $\phi_1[v] = \phi_2[v] = \cdots = \phi_6[v]$. Since $\phi_1[v] + \phi_2[v] + \cdots + \phi_6[v] = \frac{4}{7}$, we have $\phi_1[v] = \phi_2[v] = \cdots = \phi_6[v] = (\frac{1}{6})(\frac{4}{7}) = \frac{2}{21}$. Thus, according to the Shapley value, the federal government has $4\frac{1}{2}$ times as much power as any state.

To give a more complicated example, suppose we have a bicameral legislature with n_1 members in the first house and n_2 members in the second house.[25] Suppose a measure can pass only if it has a majority in each house of the legislature, and suppose for simplicity that n_1 and n_2 are both odd. Let I be the union of the sets of members of both houses, and let o be any ordering of I. A player i is pivotal in o if he is the $[(n_j + 1)/2]$-th player of his house in the ordering and a majority of players in the other house precede him. However, for every ordering o in which a player in the first house is pivotal, the reverse ordering makes a player in the second house pivotal. (Why?) Moreover, every ordering has some player as pivotal. Thus, the probability that a player in house number 1 will be pivotal is $\frac{1}{2}$. Since all players in house number 1 are treated equally, the probability that any one of these players will be pivotal is $1/(2n_1)$. Similarly, the probability that any player of house number 2 will be pivotal is $1/(2n_2)$. Thus, each player of house number 1 has power $1/(2n_1)$ and each player of house number 2 has power $1/(2n_2)$. In the United States House and Senate, $n_1 = 435$ and $n_2 = 101$, including the Vice President, who votes in case of a tie. According to our calculation, each representative has power $1/870 \approx .0011$ and each senator (including the Vice President) has power $1/202 \approx .005$. Thus, a senator has about five times as much power as a representative.

Next, let us add an executive (governor, president, etc.) who can veto the vote in the two houses, but let us assume that there is no possibility of overriding the veto. Now there are $n_1 + n_2 + 1$ players and a coalition is winning if and only if it contains the executive and a majority from each house. Assuming n_1 and n_2 are large, Shapley and Shubik [1954] argue that the probability that the executive will be pivotal is approximately $\frac{1}{2}$. (This argument is a bit complicated and we omit it.) The two houses divide the remaining power almost equally. Finally, if the possibility of overriding the veto with a $\frac{2}{3}$ majority of both houses is added, a similar discussion implies that the executive has power approximately $\frac{1}{6}$, and the two houses divide the remaining power almost equally. The reader is referred to Shapley and Shubik's paper for details.

Similar calculations can be made for the relative power various states wield in the electoral college. Mann and Shapley [1960, 1962] calculated this, using the distribution of electoral votes as of 1961. New York had 43 out of the total of 538 electoral votes, and had a power of .0841. This compared to a power of .0054 for states like Alaska, which had three electoral votes.

---

[25]This example is again due to Shapley and Shubik [1954].

According to the Shapley value, the power of New York exceeded its percentage of the vote, while that of Alaska lagged behind its percentage.

Similar results for the new distribution of electoral votes as of 1972 were obtained by Boyce and Cross [unpublished]. In the new situation, New York had a total of 41 electoral votes (the total was still 538) and a power of .0797, while Alaska still had three electoral votes and a power of .0054.

In the exercises, we consider an alternative value concept, due to Banzhaf [1965].[26] Both this concept and Shapley's are considered plausible and experience has shown that they usually agree fairly well. However, Dubey and Shapley [to appear] show that this is not always the case. Banzhaf's value especially is coming to be used in the courts, in cases involving proper representation of voters. (For references, see Banzhaf [1968, p. 306] or Johnson [1969].)

6.7.5. Proof of the Value Formula[27]

We close this section by proving Theorem 6.14. We first prove several lemmas.

Lemma 1.   Let R be a coalition on a set of individuals I and define a characteristic function v_R on I as follows:

$$v_R(S) = \begin{cases} 1 & (\text{if } R \subseteq S) \\ 0 & (\text{otherwise}). \end{cases}$$

If a function ϕ satisfying Axioms 1–4 exists, c is a nonnegative real number, and $r = |R|$, then

$$\phi_i[cv_R] = \begin{cases} \dfrac{c}{r} & (\text{if } i \in R) \\ 0 & (\text{if } i \notin R). \end{cases}$$

Proof.   By Axiom 3, $\phi_i[cv_R] = 0$ if $i \notin R$. By Axiom 1, $\phi_i[cv_R] = \phi_j[cv_R]$ if $i, j \in R$. Thus, by Axiom 2, if $i \in R$, $r\phi_i[cv_R] = (cv_R)(I) = cv_R(I) = c$. (Note: $c \geqq 0$ is needed cv_R to be a characteristic function.)

Q.E.D.

Lemma 2.   Any characteristic function v is a linear combination

$$v = \sum_{R \subseteq I} c_R v_R$$

of the characteristic functions v_R defined in Lemma 1.

---

[26]The Banzhaf value has recently been put on an axiomatic foundation by Dubey and Shapley [to appear].

[27]This subsection can be omitted.

Proof.[28]   If R is a coalition on a set of individuals I, define the function v_R' as follows:

$$v_R'(S) = \begin{cases} 1 & \text{(if } R = S) \\ 0 & \text{(otherwise).} \end{cases}$$

Note that v_R' is not a characteristic function if $R \neq I$. However, every characteristic function v on I is a linear combination of the functions v_R', since

$$v = \sum_{R \subseteq I} v(R) v_R'.$$

Finally, we show that each v_R' is a linear combination of the functions v_R of Lemma 1. We prove this by induction on $k = n - |R|$, where $n = |I|$. If $k = 0$, then $v_R' = v_I' = v_I$. Assume the statement is true for k and prove it for $k + 1$. Note that

$$v_R' = v_R - v_{R_1}' - v_{R_2}' - \cdots - v_{R_j}',$$

where $R_1, R_2, \ldots, R_j$ are the proper supersets of R. By inductive assumption, each v_{R_i}' is a linear combination of functions v_R, and hence so is v_R'.   Q.E.D.

To complete the proof of Theorem 6.14, we observe that if the function ϕ exists, then if 0 is the characteristic function which is identically 0, we have $\phi[0] = 0$. (Why?) By Lemma 1, ϕ is uniquely defined for games of the form $c v_R$, $c \geqq 0$. By Axiom 4, if u, v, and u-v are all characteristic functions,

$$\phi[u\text{-}v] = \phi[u] - \phi[v].$$

Using this result, the result of Lemma 1, and the result of Lemma 2, we see that ϕ is uniquely determined for any v, since

$$\phi[v] = \sum_{R \subseteq I} \phi[c_R v_R].$$

Thus, we have proved that if a function ϕ satisfying Axioms 1–4 exists, ϕ is uniquely determined. It remains to prove that such a ϕ does exist. This is accomplished by showing that the function ϕ defined by Eq. (47) does indeed satisfy the axioms. Proof is left as an exercise (Exer. 17).

One thing which we have omitted to consider is the fact that $\phi[v]$ might not be an imputation. By Axiom 2, $\phi[v]$ satisfies Pareto optimality (Eq. (6)). But the second requirement, individual rationality (Eq. (4)) was omitted from our axioms because it follows from them. To show that a function ϕ satisfying Axioms 1–4 also satisfies individual rationality, we want to show that $\phi_i[v] \geqq v(\{i\})$ for all i. Note that $v(S) - v(S - \{i\}) \geqq v(\{i\})$, by superadditivity. Thus, using Eq. (47), it suffices to show that

$$\sum \{\gamma(s): S \text{ such that } i \in S\} = 1.$$

[28]This proof is due to Dubey [1974].

Now

$$\sum \{\gamma(s)\colon S \text{ such that } i \in S\}$$

$$= \frac{1}{n!} \sum \{(s-1)!\,(n-s)!\colon S \text{ such that } i \in S\}.$$

The number $n!$ is the number of permutations of the set I. For each S, as we have seen, the number $(s-1)!(n-s)!$ gives the number of permutations of I in which elements of $S - \{i\}$ come first (in some order), then i, then the rest of the elements (in some order). Since each permutation is realized in this fashion for some S containing i and since $n!$ is the total number of permutations,

$$\sum \{(s-1)!(n-s)!\colon S \text{ such that } i \in S\} = n!$$

This concludes the proof.

EXERCISES

1. According to one of the axioms for the Shapley value, $\phi_i[v] = 0$ if i is the third player in the game (51; 60, 30, 10). Which axiom and why?

2. According to one of the axioms for the Shapley value, $\phi[v] = (\frac{1}{4}, \frac{1}{4}, \frac{1}{4}, \frac{1}{4})$ if v is the weighted majority game (3; 1, 1, 1, 1). Which axiom and why?

3. According to one of the axioms for the Shapley value, $\phi[v] = (1, 1, 1)$ could not be the Shapley value for a weighted majority game. Which axiom and why?

4. Calculate the Shapley value for the Lake with $B = 2$, $C = 1$, $n = 4$.

5. For each of the following weighted majority games:
   (a) List the winning coalitions.
   (b) Comment (prior to making calculations) on the relative power of each player. In particular, indicate dummies (players who are in no minimal winning coalitions) and dictators (players who are in every winning coalition.)
   (c) Calculate the Shapley value and compare with your intuition. (You may use either formula (49) or the Corollary to Theorem 6.15.)
       (i)   (51; 49, 47,4).
       (ii) (201; 100, 100, 100, 100, 1).
       (iii) (151; 100, 100, 100, 1)
       (iv) (51; 26, 26, 26, 22).
       (v)   (16; 9, 9, 7, 3, 1, 1) (this game arose in the Nassau County, New York, Board of Supervisors in 1958; cf. Banzhaf [1965]).

(vi)   (59; 31, 31, 21, 28, 2, 2)   (this   game   arose   in   the   Nassau
        County, New York, Board of Supervisors in 1964; again see
        Banzhaf [1965]).

6.   Calculate the Shapley value for the conference committee (Exer. 10,
     Sec. 6.1).

7.   Calculate the Shapley value for the Two-Buyer Market Game (Example
     3a, Sec. 6.1).

8.   Calculate the Shapley value for the game defined in Exer. 6, Sec. 6.1.

9.   Calculate the Shapley value for the game defined in Exer. 7, Sec. 6.1.

10. Calculate the Shapley value for the weighted majority game
     (12; 5, 5, 2, 1). (*Hint:* Is player 4 ever pivotal?)

11. Calculate the Shapley value for the weighted majority game
     (12; 5, 5, 2, 2, 1).

12. Calculate the Shapley value for the weighted majority game
     (6; 1, 1, 1, 2, 3, 3).

13. (Lucas [1974].) In the original Security Council, there were five perma-
     nent members and only six nonpermanent members. The winning coali-
     tions consisted of all five permanent members plus at least two
     nonpermanent members. Formulate this as a weighted majority game
     and calculate the Shapley value.

14. (Lucas [1974].) It has been suggested that Japan be added as a sixth
     permanent member of the Security Council. If this were the case, assume
     that there would still be 10 nonpermanent members and winning coali-
     tions would consist of all six permanent members plus at least four non-
     permanent members. Formulate this as a weighted majority game and
     calculate the Shapley value.

15. One problem with the Shapley value when calculated for simple games
     is that we bring in a notion of order, and not all orders are equally
     likely. Banzhaf [1965] has introduced the following value concept, also
     a power index, which is very similar. Consider a simple game and let P
     be a partition of the players into two sets. Let us assume that we are
     dealing with a constant-sum game, i.e., that $v(S) + v(I - S) = v(I)$, all
     S. Then, in each partition P, one of the two groups is winning and
     one is losing. We say that player i is *marginal* with respect to the parti-
     tion P if by switching his vote he changes the winning group. (Several
     players could be marginal for a given partition.) For example, in the
     game (2; 1, 1, 1), one partition is {1, 2} vs. {3}. Here, the first group
     wins, but both players 1 and 2 are marginal with respect to this parti-
     tion. If we let b_i be the number of partitions in which i is marginal, then

the *Banzhaf index* is given by

$$\beta_i = \frac{b_i}{\sum_{j=1}^{n} b_j}.$$

For example, in the game $(2; 1, 1, 1)$, we can make a table like Table 6.1, where an x in column i in a given row means that player i is marginal with respect to that row's partition. From this table, we see that $b_1 = b_2 = b_3 = 4$ and $\beta_1 = \beta_2 = \beta_3 = \frac{4}{12} = \frac{1}{3}$. All three players are equally

TABLE 6.1

Calculation of the Banzhaf Index for the Game $(2; 1, 1, 1)$

Partition		Marginal players		
		1	2	3
$\varnothing$	$\{1, 2, 3\}$			
$\{1\}$	$\{2, 3\}$		x	x
$\{2\}$	$\{1, 3\}$	x		x
$\{3\}$	$\{1, 2\}$	x	x	
$\{1, 2\}$	$\{3\}$	x	x	
$\{1, 3\}$	$\{2\}$	x		x
$\{2, 3\}$	$\{1\}$		x	x
$\{1, 2, 3\}$	$\varnothing$			

powerful. Calculate the Banzhaf index for the following games, and compare to the Shapley value:

(a) $(51; 51, 48, 1)$.
(b) $(20; 1, 10, 10, 10)$.
(c) $(102; 80, 40, 80, 20)$.
(d)   The Australian government $(5; 1, 1, 1, 1, 1, 1, 3)$.

*16.   Show that S-equivalence preserves the Shapley value, in the sense that if v' satisfies Eq. (40), then $\phi_i[v'] = \alpha\phi_i[v] + a_i$. (*Note:* isomorphism does not necessarily preserve the Shapley value: if v is isomorphic to v' via an isomorphism f, then $\phi[v']$ is not necessarily $f(\phi[v])$. See footnote 18 in Sec. 6.6.)

*17.   Prove that the function $\phi[v]$ defined in Eq. (47) satisfies Axioms 1–4. (*Hint:* For Axiom 2, note that if v'_R is defined as in the proof of Lemma 2, then $\phi[v'_R]$ is defined even though $v'_{R|}$ is not a characteristic function. Also,

$$\sum_{i=1}^{n} \phi_i[v'_R] = \begin{cases} 1 & \text{(if } R = I) \\ 0 & \text{(if } R \neq I). \end{cases}$$

Finally,

$$\phi[v] = \sum_{R \subseteq I} v(R)\phi[v'_{R_i}].)$$

18. Show that if ϕ defines the Shapley value, and v and w are characteristic functions on the same set of players I, then for all α in $(0, 1)$,

$$\phi[\alpha v + (1 - \alpha)w] = \alpha\phi[v] + (1 - \alpha)\phi[w].$$

19. (Shapley [1953a].) A *carrier* for a game v is a coalition T such that for all S, $v(S) = v(S \cap T)$. Show directly from Axioms 1–4 that if T is a carrier, then

$$\sum_{i \in T} \phi_i[v] = v(T).$$

20. Prove that in a symmetric game in which the core exists, the Shapley value is in the core.

*21. (Shapiro and Shapley [1960].) Show that in the weighted majority game with n players $(Q; k, 1, 1, \ldots, 1)$, if $k < Q \leqq n - 1$, then ϕ is $(k/n, (n - k)/n(n - 1), (n - k)/n(n - 1), \ldots, (n - k)/n(n - 1))$.

22. Discuss the results of our calculation of the Shapley value for some of the games in this section (Security Council, U.S. Congress, etc.). Do they agree with your intuition? What oversimplifications have been made in our analysis?

23. Try to list your own set of axioms for a power measure.

References

BANZHAF, J. F., III, "Weighted Voting Doesn't Work: A Mathematical Analysis," *Rutgers Law Review*, **19** (1965), 317–343.

BANZHAF, J. F., III, "One Man, 3.312 Votes: A Mathematical Analysis of the Electoral College," *Villanova Law Review*, **13** (1968), 304–332.

BERGE, C., *Théorie générale des jeux à n personnes*, Mem. des Sciences Math., **138**, Gauthier-Villars, Paris, 1957.

BERGE, C., *The Theory of Graphs and its Applications*, John Wiley & Sons, Inc., New York, 1962.

BONDAREVA, O. N., "Teoriya yadra v igre n lits" ("Theory of the Core in the *n*-person Game"), *Vestnik Lening. Univ.*, **13** (1962), 141–142.

BONDAREVA, O. N., "Nekotorye primeneniya metodor linejnogo programmirovaniya k teorii kooperativnykh igr" ("Some Applications of Linear Programming Methods to the Theory of Cooperative Games"), *Problemy Kibernetiki*, **10** (1963), 119–139.

BOYCE, W. M., and CROSS, M. J., "An Algorithm for the Shapley-Shubik Voting Power Index for Weighted Voting," unpublished Bell Telephone Laboratories manuscript.

DUBEY, P., "On the Uniqueness of the Shapley Value," Technical Report, Applied Mathematics Center, Cornell University, Ithaca, New York, June 1974.

DUBEY, P., and SHAPLEY, L. S., "Some Properties of the Banzhaf Power Index," to appear as a Rand Corporation Report, 1975.

GILLIES, D. B., *Some Theorems on n-Person Games*, Ph.D. Thesis, Department of Mathematics, Princeton University, Princeton, N.J., 1953.

GILLIES, D. B., "Solutions to General Non-zero-sum Games," *Annals of Mathematics Studies* No. 40, Princeton University Press, Princeton, N.J., 1959, pp. 47–85.

HARARY, F., NORMAN, R. Z., and CARTWRIGHT, D., *Structural Models*, John Wiley, & Sons, Inc., New York, 1965.

JOHNSON, R. E., "An Analysis of Weighted Voting as Used in Reapportionment of County Governments in New York State," *Albany Law Review*, **34** (1969), 1–45.

KUHN, H. W. (ed.), *Report of an Informal Conference on the Theory of n-Person Games Held at Princeton University*, March 20–21, 1953, mimeographed. (Out of print.)

LUCAS, W. F., "A Counterexample in Game Theory," *Management Science*, **13** (1967), 766–767.

LUCAS, W. F., "A Game with no Solution," *Bull. Amer. Math. Soc.*, **74** (1968), 237–239.

LUCAS, W. F., "The Proof that a Game may not have a Solution," *Trans. Amer. Math. Soc.*, **137** (1969), 219–229.

LUCAS, W. F., "Measuring Power in Weighted Voting Systems," Tech. Report No. 227, Department of Operations Research, Cornell University, September 1974. (To be published by CUPM Applied Mathematics Modules Project, Berkeley, Calif.)

LUCE, R. D. and RAIFFA, H., *Games and Decisions*, John Wiley & Sons, Inc., New York, 1957.

MANN, I. and SHAPLEY, L. S., "Values of Large Games, IV: Evaluating the Electoral College by Monte Carlo Techniques," Rand Corporation Memorandum RM-2651, September 1960; reproduced in *Game Theory and Related Approaches to Social Behavior*, M. Shubik (ed.), John Wiley & Sons, Inc., New York, 1964.

MANN, I. and SHAPLEY, L. S., "Values of Large Games, VI: Evaluating the Electoral College Exactly," Rand Corporation Memorandum RM-3158-PR, May 1962; reproduced in part in *Game Theory and Related Approaches to Social Behavior*, M. Shubik (ed.), John Wiley & Sons, Inc., New York, 1964.

McKINSEY, J. C. C., "Isomorphism of Games and Strategic Equivalence," in *Contributions to the Theory of Games*, II, H. W. Kuhn and A. W. Tucker (eds.), Annals of Mathematics Studies No. 24, Princeton University Press, Princeton, N.J., 1950.

OWEN, G., *Game Theory*, W. B. Saunders, Philadelphia, 1968.

RAPOPORT, A., *N-Person Game Theory: Concepts and Applications*, University of Michigan Press, Ann Arbor, Mich., 1970.

RICHARDSON, M., "On Weakly Ordered Systems," *Bull. Amer. Math. Soc.*, **52** (1946), 113–116.

SHANNON, C. E., "The Zero-error Capacity of a Noisy Channel," *Comp. Information Theory*, *IRE Trans.*, **3** (1956), 3–15.

SHAPIRO, N. V., and SHAPLEY, L. S., "Values of Large Games I: A Limit Theorem," Rand Corporation Memorandum RM-2648, 1960.

SHAPLEY, L. S., "A Value for *n*-Person Games," in *Contributions to the Theory of Games, II*, H. W. Kuhn and A. W. Tucker (eds.), Annals of Mathematics Studies No. 28, Princeton University Press, Princeton, N.J., 1953, pp. 307–317, a.

SHAPLEY, L. S. "Quota Solutions of *n*-Person Games," in *Contributions to the Theory of Games II*, H. W. Kuhn and A. W. Tucker (eds.), Annals of Mathematics Studies No. 28, Princeton University Press, Princeton, N.J., 1953, pp. 343–359, b.

SHAPLEY, L. S., "On Balanced Sets and Cores," *Nav. Res. Logist. Quart.*, **14** (1967), 453–460.

SHAPLEY, L. S. and SHUBIK, M., "A Method for Evaluating the Distribution of Power in a Committee System," *The American Political Science Review*, **48** (1954), 787–792.

SHAPLEY, L. S. and SHUBIK, M., "On Market Games," *J. Econ. Theory*, **1** (1969), 9–25, a.

SHAPLEY, L. S. and SHUBIK, M., "On the Core of an Economic System with Externalities," *The American Economic Review*, **59** (1969), 678–684, b.

SHAPLEY, L. S. and SHUBIK, M., "Characteristic Function, Core, and Stable Sets," *Game Theory in Economics*, Chapter 6, Rand Corporation Report R-904/6-NSF, July 1973.

SHAPLEY, L. S. and SHUBIK, M., *Game Theory in Economics*, to appear. (Certain chapters are presently available as Rand Corporation Reports R-904/1-NSF, R-904/2-NSF, R-904/3-NSF, R-904/4-NSF, R-904/6-NSF), a.

SHAPLEY, L. S. and SHUBIK, M., "The Value," *Game Theory in Economics*, Chapter 7, Rand Corporation Report R-904/7-NSF, to appear, b.

VON NEUMANN, J. and MORGENSTERN, O., *Theory of Games and Economic Behavior*, Princeton University Press, Princeton, N.J., 1944, 1947, 1953.

SEVEN

Group Decisionmaking

7.1. Social Welfare Functions

In Chapter 6, we considered groups in a situation of conflict and cooperation. Coalitions formed and individuals used their power and influence to
extract results. In this chapter, we imagine a group trying to make a decision
on democratic grounds, where each individual is treated equally. The group
might be a legislative body; a panel of experts or judges choosing among
job candidates, among potential variables for a signed digraph, or among
rankings of the players in a tournament; or the group may be society as a
whole voting on some new law or some candidates for election. Each member
of the group has his preferences or opinions. We shall be interested in trying
to establish fair and uniform procedures for the group to combine the
individual opinions to reach some sort of consensus opinion.

One of the classic situations to which the subject matter of this chapter
applies is of course an election. Historically in elections, each voter has voted
for his first choice, and the winner is then declared as the candidate who
gets the most votes. If no candidate gets a majority, there is sometimes
provision for a runoff between the top two vote-getters. The result of such
an election is often not a very popular winner. For example, two regional
candidates for president might very well each win about half the primary
elections. Meanwhile, a popular national candidate who is ranked second in
each of the regions might win no primaries, and yet be a much more popular

choice for a majority of people than either of the other two. To take account of such information in elections, we could ask each voter to rank or list in order of his preference all possible candidates. (In practice, the voter would be presented with all possible rankings (orderings) of the candidates, and asked to vote for one such ranking.) Given this information, however, we will have to come up with a rule for choosing a winner. In this chapter, we shall discuss methods for determining this winner, or more generally, for obtaining a consensus ranking of all the candidates which somehow represents the group's opinion.

In many situations it is more useful to produce a ranking of all the candidates than it is to just produce a winner. This is the case, for example, if the candidates are candidates for a job and the rankings are provided by different judges, for we shall want to make an offer to a second choice if the top candidate does not accept the job, an offer to a third choice if the second candidate does not accept, and so on.

Formally, suppose we have a set of individuals I, to be labelled 1, 2, $\ldots, t$.[1] Let A be a set of objects, alternatives, factors, events, candidates, etc., among which the individuals make certain judgements. For each $i \leq t$, let P_i be the ith individual's ranking of the alternatives in A. We shall abbreviate by aP_ib the statement that the ith individual ranks a over b. Thus, aP_ib means that from the point of view of individual i, a is preferred to b, or a is more qualified than b, or a is more important than b, etc. We shall usually interpret P_i as a ranking in terms of preference, though the discussion of this chapter applies to other contexts as well.

In our rankings, it will be convenient to allow ties. If a and b are considered by individual i to be tied, we shall write aT_ib. (If ranking is in order of preference, then T_i has the interpretation "indifference.") It will be convenient to represent P_i by listing the elements of A in a column, with order of preference decreasing from top to bottom. Thus, if $A = \{x, y, u, v\}$, then one ranking P_i is given below:

$$P_i$$

$$x$$
$$y\text{-}v$$
$$u$$

The hyphen between y and v represents the fact that they are tied in the ranking, i.e., yT_iv. We also have element x ranked above both y and v and both y and v ranked above u.

We shall not define the term *ranking* any more precisely in this chapter than we have so far. The reader who goes on to Chapter 8 will note that a ranking is just a strict weak order. We shall make use of the fact that every

---

[1] The reader should note that in the previous chapter, we used n where we are now using t.

ranking P is *transitive*, i.e., if aPb and bPc, then aPc. Moreover, it is *asymmetric*, i.e., if aPb then not bPa.

Given a group, the rankings $(P_1, P_2, \ldots, P_t)$ of the members of the group define the group's *profile*. One profile for the set $A = \{x, y, u, v\}$ with a group of three individuals is given by

P_1	P_2	P_3
x	y	$x\text{-}u$
y	x	v
$u\text{-}v$	u	y
	v	

$\hat{\mathcal{P}} = \hat{\mathcal{P}}(A)$ will denote the collection of all possible rankings of A and $\hat{\mathcal{P}}_t(A)$ will be $\hat{\mathcal{P}} \times \hat{\mathcal{P}} \times \cdots \times \hat{\mathcal{P}}$, t times. Thus, $\hat{\mathcal{P}}_t(A)$ will denote the collection of all profiles of a group of t individuals over A.

Our problem is the following: Given a profile $(P_1, P_2, \ldots, P_t)$, find a "winning" alternative or, less generally, find a ranking P on A which represents the consensus group judgement. In case each P_i is a preference ranking, P is the group preference. Since the winner can be read off from a group ranking, we shall concentrate on the problem of finding such a ranking. However, the reader should observe that this may be more difficult than finding just a winner.[2] A rule for determining the group ranking from the group profile will be called a *group consensus function*. In case the rankings are preference rankings, and particularly where I stands for society as a whole, this is often called a *social welfare function*. The group consensus function can be thought of as a function $F\colon \hat{\mathcal{P}}_t(A) \longrightarrow \hat{\mathcal{P}}(A)$. We shall be concerned with describing which functions F are "reasonable." For most of the remainder of this chapter, we shall concentrate on the case of preference and use the term "social welfare function," primarily for conformity with the literature. The reader should keep in mind the other interpretations of the rankings P_i.

One obvious rule for obtaining a group ranking is the *simple majority rule*: Given a profile $(P_1, P_2, \ldots, P_t)$, let the group rank alternative a over alternative b if and only if a majority (more than half) of the individuals rank a over b. Suppose the set A is $\{x, y, z\}$, t is 3, and the group profile is

P_1	P_2	P_3
x	y	z
y	z	x
z	x	y

---

[2]In Exers. 5, 7, 9, 10 and 11, we shall mention several rules for determining just the group *winner*.

A majority of individuals rank x above y, a majority rank y above z, and a majority rank z above x. The simple majority rule would assign a group ranking P with xPy, yPz and zPx. But there is no way a ranking can have these three properties! (For since a ranking is asymmetric, zPx implies that xPz does not hold. Hence, the ranking wouldn't be transitive.) The above example is an illustration of *the voter's paradox* (*Condorcet's pardox*):[3] Individual preference rankings, combined under the simple majority rule, do not necessarily lead to group preference rankings. By our definition of group consensus function or social welfare function, the simple majority rule does not even define such a function, as it does not assign a ranking to each profile.

It is not hard to describe social welfare functions which do assign legitimate rankings to each profile. One example is the Borda count.[4] To describe this, let P_i be a ranking, let $B_i(a)$ be the number of alternatives b ranked *below a* in P_i, and let $B(a)$ be $\sum_{i=1}^{t} B_i(a)$. $B(a)$ is the *Borda count* of a. The Borda Count social welfare function is defined as follows: Rank a over b for the group if and only if $B(a)$ is bigger than $B(b)$. To give an example, suppose a group has the following profile:

P_1	P_2	P_3	P_4
x	y	v	x
y	u	x	y
u	x	y	u-v
v	v	u	

Then $B_1(x) = 3$, $B_2(x) = 1$, $B_3(x) = 2$, and $B_4(x) = 3$, so $B(x) = 9$. Similarly, $B(y) = 2 + 3 + 1 + 2 = 8$, $B(u) = 1 + 2 + 0 + 0 = 3$, and $B(v) = 0 + 0 + 3 + 0 = 3$. Thus, according to the Borda count, x is ranked over y, y is ranked over u and v, and u and v are tied. The group's ranking is

P
x
y
u-v

The Borda count always leads to a ranking, but this ranking does not always agree with our intuition of what is "fair." For example, consider the fol-

---

[3]The paradox is named after the philosopher and social scientist Marie Jean Antoine Nicolas Caritat, Marquis de Condorcet, who discovered it in the eighteenth century. The paradox was later discovered by Jean-Charles de Borda, an eighteenth century soldier and sailor. The paradox was rediscovered in the nineteenth century by Lewis Carroll.

[4]This is due to Jean-Charles de Borda. See the previous footnote.

lowing profile:

P_1	P_2	P_3	P_4	P_5
x	x	x	x	y
y	y	y	y	z
z	z	z	z	u
u	u	u	u	v
v	v	v	v	w
w	w	w	w	x

Here, $B(x) = 20$ and $B(y) = 21$, so y is the winner, even though four out of five individuals think x is the best candidate.

It is not hard to describe other social welfare functions. We list several in Exers. 13 and 14. However, as we ask the reader to show, each of these leads to some unpleasant examples much like the one with the Borda count. Thus, it is hard to decide, given two social welfare functions, which one is more "reasonable." Under the circumstances, a possible approach is to list conditions which a "reasonable" social welfare function should satisfy, and see if we can derive from them a particular social welfare function or class of such functions. This is the axiomatic approach, which we have encountered previously in our discussion of trophic status in Sec. 3.5.2 and in our discussion of the Shapley value in Sec. 6.7.1. It will be our approach in the next section. For a discussion of group decisionmaking more comprehensive than ours, the reader is referred to Luce and Raiffa [1957], Sen [1970], or Fishburn [1972]. A classic reference is Arrow [1951]. Other useful references on the theory of elections are Black [1958], Farquharson [1969], and Riker and Ordeshook [1973].

EXERCISES

1. (a) List all possible rankings of $\{x, y, z\}$.
    (b) If $A = \{u, v, w, x\}$, how many elements are there in $\hat{\mathcal{P}}(A)$?
    (c) If $A = \{u, v, w\}$, how many elements are there in $\hat{\mathcal{P}}_2(A)$?

2. If an individual prefers Buicks to Cadillacs, Cadillacs to Volkswagens, and Volkswagens to Buicks, is he transitive?

3. If an individual prefers vacations in Oregon to vacations in California, California to New Hampshire, New Hampshire to Maine, and Maine to Oregon, is he transitive? Why?

4. If an individual prefers Buicks to Cadillacs, Cadillacs to Volkswagens, and Cadillacs to Buicks, is he asymmetric?

5. Given a group profile, if we want to pick just a winner, rather than a group consensus ranking, the *plurality method* works as follows: Alternative a is a winner if the number of individuals ranking a first is larger than the number of individuals ranking any other alternative first. Suppose $A = \{x, y, z\}$ and consider all possible rankings of A without ties:

P_1	P_2	P_3	P_4	P_5	P_6
x	y	z	x	y	z
y	z	y	z	x	x
z	x	x	y	z	y

If 10 voters vote for ranking P_1, 31 for ranking P_2, 45 for ranking P_3, 30 for ranking P_4, and none for rankings P_5 or P_6, which alternative is the winner using the plurality method?

6. What are some of the problems with the plurality method of Exer. 5? (Illustrate with profiles.)

7. If we want to pick just a winner, rather than a group consensus ranking, the *Condorcet criterion* is the following: Alternative a wins if for every other alternative b, a is ranked over b by a majority of individuals.
    (a) Is there a winner if the Condorcet criterion is used in the election of Exer. 5?
    (b) What are some of the problems with the Condorcet criterion? (Illustrate with profiles.)

8. (Riker and Ordeshook [1973].) Use the following example to compare the plurality method (Exer. 5) to the Condorcet criterion (Exer. 7). In the 1912 Presidential election, it is estimated that voters had the following preferences for the three candidates x (Wilson, Democrat), y (Roosevelt, Progressive), and z (Taft, Republican):

$$\begin{array}{ccc} x & y & z \\ 45\% \text{ ranked } y, & 30\% \text{ ranked } z, & \text{and } 25\% \text{ ranked } y. \\ z & x & x \end{array}$$

(Wilson was the plurality winner when everyone just voted for his first place candidate.) Discuss your conclusions.

9. If we want to pick just a winner, rather than a group consensus ranking, the *runoff procedure* is the following: Use the plurality method of Exer. 5 unless no alternative is ranked first by a majority of the individuals; in the latter case, pick the two alternatives a and b which receive the two highest number of first place votes. Then choose a over b if more

individuals rank *a* over *b* than rank *b* over *a*. If there are more than two alternatives with the two highest first place vote counts, the procedure does not determine a winner and a new election is held.

(a) Who is the winner if the runoff procedure is used in the election of Exer. 5?

(b) Give an example to show that it is possible for *a* to be a winner under this method even though he gets a smaller number of first place votes than *b*.

10. (Malkevitch and Meyer [1974, p. 405].) Assume that *A* has four elements. If no alternative gets a majority of first place votes, one could use the following procedure to determine the winner: Compare the top and third best first place vote-getters by the runoff procedure of Exer. 9, compare the second and fourth best first place vote-getters by the same method, and then compare the two "winners" of these comparisons. Comment on the possible problems with this procedure. (Can you name some sports where a similar procedure is used to decide a winner?)

11. The simple majority rule can be used to determine a winner, if it applies, that is, if it gives a ranking. Is it possible for the simple majority rule to apply and to give a different winner from the runoff procedure of Exer. 9?

12. (Malkevitch and Meyer [1974, p. 411].) This exercise compares the Borda count to other group decisionmaking procedures. In the following problems, give examples with no ties in any of the rankings P_i.

(a) Show by example that it is possible for the winner by the Borda count method to be different from the winner by the plurality method of Exer. 5.

(b) Show by example that it is possible for the winner by the Borda count method to be different from the winner by the Condorcet criterion of Exer. 7.

(c) Show by example that if the simple majority rule does give a ranking, the *ranking* may be different from that given by the Borda count.

(d) Show that it is possible for the Borda count, the plurality procedure of Exer. 5, and the runoff procedure of Exer. 9 all to yield different winners.

(e) Give an example in which the Borda count and the plurality method give the same winner and the simple majority rule gives a ranking which has the same winner as the runoff procedure, but the two winners are different.

(f) Give an example in which the simple majority rule gives a ranking and therefore a winner and the four methods, Borda count, simple majority rule, plurality, and runoff, give different winners.

13. The following is a social welfare function called the *lexicographic group preference*. We rank alternative a over alternative b for the group if and only if aP_1b or $(aT_1b$ and $aP_2b)$ or $(aT_1b$ and aT_2b and $aP_3b)$ or $\cdots$ or $(aT_1b$ and aT_2b and $\cdots$ and $aT_{i-1}b$ and $aP_ib)$. That is, we consider individual 1's preference first, unless he is indifferent, in which case we consider individual 2's preference; if both are indifferent, we consider individual 3's preference; and so on. If all are indifferent, the group is also said to be indifferent.

    (a) Give an example to show that even if the simple majority rule gives a ranking, this ranking may be different from that given by the lexicographic group preference.

    (b) What are some of the weaknesses of this social welfare function?

14. The *plurality system* can be used to define another social welfare function. Rank a over b for the group if and only if a receives more first place votes than b. Show that the Borda count and the plurality system may give different rankings for the same group profile.

15. In each of the following situations, discuss whether or not obtaining individual rankings is (i) feasible and (ii) desirable, and discuss whether or not it is reasonable to try to obtain a group ranking rather than just a winner.

    (a) Members of a congressional committee are considering several alternative versions of a piece of legislation.

    (b) The organizers of a post-season college football bowl game are considering which of 20 teams to invite.

    (c) A wire service (A.P., U.P.I.) wants to produce a list of the top twenty college football teams. (Incidentally: What is the method presently used?)

16. In Chapter 4 we discussed the problem of choosing variables for an energy demand signed digraph. One method being used asks each of a group of experts to rank alternative variables in order of importance. Discuss possible procedures for the group to use in choosing say 10 of the variables on the basis of this information. Does the group want a ranking?

17. In Chapter 3, we discussed the problem of finding a ranking of the players in a tournament. We saw that various rankings can be defined, using complete simple paths through the tournament. Do any of the methods of this section help with the problem of choosing a consensus ranking?

18. If each individual chooses his ranking at random, it is possible to calculate the probability that the voter's paradox will arise if the simple majority rule is used. Show that if there are $t = 3$ individuals ranking $n = 3$ alternatives, and no ties are allowed, there are 216 possible profiles and 12 of these lead to the voter's paradox. Hence, the probability of the paradox is $\frac{12}{216} = .0556$, a small number. (Niemi and Weisberg [1968] have performed a similar calculation for various sizes of t and n, and found that the probability is under .31 if the number of alternatives is under 6, no matter how many individuals there are making the choice.)

19. One reason that the calculation in Exer. 18 may underestimate the probability of the voter's paradox is that individuals do not decide on rankings at random. Riker and Ordeshook [1973] speak of *sophisticated voting*, where a voter "acts other than in accord with his true tastes in order to achieve a higher goal." To illustrate the effect of sophisticated voting, consider the 1912 Presidential election (Exer. 8). Suppose each voter votes only for his first choice and the winner is chosen by the plurality method. Here, Wilson is the winner. Suppose everyone knows everyone else's preferences. Show that if people in the 25% group vote in a sophisticated way, then it is possible to have a different winner. If it were possible to prevent sophisticated voting, should this be done? Can you imagine how one would be able to prevent sophisticated voting?

7.2. Arrow's Impossibility Theorem

7.2.1. ARROW'S AXIOMS

In this section, we discuss the consequences of a set of axioms for a social welfare function originally listed in somewhat different form by Nobel Prize winner Kenneth Arrow [1951]. These axioms are intended to list conditions which a "reasonable" social welfare function should satisfy. Our discussion here follows closely that of Luce and Raiffa [1957, Ch. 14].

Axiom 1 (Positive Association of Social and Individual Values). If the social welfare function asserts that a is ranked over b for a given profile, then it asserts the same thing when the profile is modified as follows:

(a) Each individual's preferences between alternatives other than a are not changed;

and

(b) Each individual's preference between a and any other alternative either remains unchanged or is modified in a's favor.

To illustrate this axiom, suppose we imagine a group of four individuals they have choosing among cars, and suppose they have limited the choice to the set $A = \{$Ford, Chevy, Plymouth$\}$. The group's profile is as follows:

P_1	P_2	P_3	P_4
Ford	Chevy	Chevy	Plymouth
Chevy	Ford	Ford	Ford-Chevy
Plymouth	Plymouth	Plymouth	

Suppose the social welfare function chooses the ranking

Ford
Chevy
Plymouth

Consider now a second group with the profile

P_1'	P_2'	P_3'	P_4'
Ford	Ford	Chevy	Plymouth
Chevy	Chevy	Ford	Ford
Plymouth	Plymouth	Plymouth	Chevy

Now $P_1' = P_1$ and $P_3' = P_3$. In P_2', Chevy has been moved below Ford, though it is still above Plymouth. In P_4', Chevy has been moved below Ford and it remains below Plymouth. Thus, by Axiom 1 with $a =$ Ford, the social welfare function applied to the second group should still rank Ford over Chevy.

Axiom 2 (Independence of Irrelevant Alternatives). Let A_1 be any subset of the set of alternatives A. If a profile is modified in such a way that each individual's ranking among elements in A_1 is unchanged, then the group's ranking resulting from the original and the modified profiles should be the same for alternatives in A_1.

Axiom 2 says that the ranking among alternatives outside of A_1 and the ranking of these alternatives relative to those of A_1 is irrelevant to the group's ranking of the alternatives in A_1. To illustrate this axiom, let us take $A = \{$Ford, Chevy, Plymouth, Volkswagen, Datsun$\}$. Let $A_1 = \{$Ford, Chevy, Plymouth$\}$ and let $t = 3$. Consider the profiles

P_1	P_2	P_3
Ford	Volkswagen	Ford-Chevy-Plymouth-Volkswagen
Volkswagen	Plymouth-Datsun	Datsun
Datsun	Chevy	
Chevy-Plymouth	Ford	

and

P_1'	P_2'	P_3'
Datsun	Plymouth	Ford-Chevy-Plymouth
Ford	Chevy	Volkswagen-Datsun
Chevy-Plymouth	Volkswagen	
Volkswagen	Datsun	
	Ford	

In both P_1 and P_1', we have on A_1 a ranking

> Ford
> Chevy-Plymouth

In both P_2 and P_2' we have a ranking

> Plymouth
> Chevy
> Ford

And in both P_3 and P_3' we have a ranking

> Ford-Chevy-Plymouth

Thus, according to Axiom 2, whatever group rankings are imposed on A from the first and second profiles, these rankings should be the same on A_1. Adding two new alternatives, Datsun and Volkswagen, and bringing in comparisons between Datsun and Volkswagen and comparisons of these new alternatives to Ford, Chevy, and Plymouth, should not change the group's consensus ranking among Ford, Chevy, and Plymouth unless addition of the new alternatives brings about change in some individual's comparisons *among* Ford, Chevy, and Plymouth.

Axiom 3 (Citizens' Sovereignty).   For each pair of alternatives a and b, there is some profile for which the group ranks a above b.

If this axiom were not satisfied, there would be a pair of alternatives a and b so that regardless of individual preferences, even if each individual prefers a to b, the group never prefers a to b. In this case, society's (the group's) ordering between a and b is *imposed* on the group and the citizens' or individuals' preferences do not carry any weight. Axiom 3 is satisfied if the following reasonable condition holds: Whenever everyone in the group ranks a first and b last, then the group ranks a over b.

Axiom 4 (Nondictatorship).   There is no individual with the property that, for all $a, b \in A$, whenever he ranks a over b, then the group does the same, regardless of the rankings supplied by the other individuals.

If Axiom 4 is violated and individual j is such an individual, it is reasonable to call him a *dictator*.[5] The rest of society (the group) can have an impact on the final ranking between a and b only if j is indifferent.

Let us investigate Arrow's axioms for various choices of t and of $n = |A|$. In the case $t = 1$, where there is only one individual, Arrow's axioms are not particularly interesting. Similarly, if $n = 1$, things are not particularly interesting. If $n = 2$, then for all $t \geq 2$, the simple majority rule defines a social welfare function and this social welfare function satisfies Arrow's axioms. To check Axiom 1, note that if a is ranked over b by a majority of people, then the same is true if b is moved lower in some of the rankings. Axiom 2 is only interesting for $A_1 \neq A$, and here A_1 can only have one element. Since there is only one ranking of a 1-element set, Axiom 2 is trivially satisfied. Axiom 3 is satisfied because for the following profile, the simple majority rule ranks a over b:

P_1	P_2	$\cdots$	P_t
a	a	$\cdots$	a
b	b	$\cdots$	b

To show that Axiom 4 is satisfied, consider first the case $t > 2$. Then for the following profile, the simple majority rule ranks a over b, even though individual j does not.

P_1	P_2	$\cdots$	P_{j-1}	P_j	P_{j+1}	$\cdots$	P_t
a	a	$\cdots$	a	b	a	$\cdots$	a
b	b	$\cdots$	b	a	b	$\cdots$	b

It $t = 2$, then for the following profile, the group considers a and b tied even though individual j does not.

P_i	P_j
a	b
b	a

Let us now consider the case of three or more alternatives and two or more individuals, i.e., $n \geq 3$ and $t \geq 2$. Then we have the following theorem.

Theorem 7.1 (Arrow's Impossibility Theorem). Suppose A has at least three elements, the number t of individuals is at least two, and $\hat{\mathcal{P}}_t(A)$

---

[5]Compare the definition of dictator given in Chapter 6.

is the set of all profiles of a group of t individuals over A. Then there is no social welfare function defined on $\mathcal{P}_t(A)$ and satisfying Axioms 1–4.

This theorem is quite startling, as at first glance Axioms 1–4 are quite reasonable. Put another way, the theorem says that if a social welfare function satisfies Axioms 1 and 2, then it is either imposed or dictatorial. We defer a proof to the end of this section.

7.2.2. Discussion

Arrow's Theorem is considered by many people to say some very negative things about the possibility of truly democratic decisionmaking. Faced with a negative result like this, there are several approaches open to us. One is to re-evaluate the axioms, perhaps a little more critically. Another is to re-evaluate the setting in which the theorem is stated, and observe that Arrow's Theorem says that decisionmaking is impossible only in the sense that it is impossible to obtain a group ranking based on the input of individual rankings. Perhaps we could modify demands on either the input (individual rankings) or the output (a group ranking) and still be able to make some sort of rational decisionmaking. We shall try all of these different approaches. (For a more detailed discussion of the possibilities, see Luce and Raiffa [1957].)

Most writers have found Axiom 2 the easiest of Arrow's axioms to attack. Let us consider why. Suppose the group has only two individuals. Let us assume that society treats the two individuals equally and if one prefers a to b and the other b to a, then society is indifferent between a and b. Consider the situation where $A = \{$Ford, Chevy, Plymouth$\}$ and the group profile is the profile

P_1	P_2
Ford	Ford
Chevy	Chevy
Plymouth	Plymouth

Since P_1 and P_2 agree, it seems reasonable to assume that if P is the consensus ranking assigned to this profile, then $P = P_1 = P_2$. But now consider the profile

P_1'	P_2'
Ford	Plymouth
Chevy	Ford
Plymouth	Chevy

Suppose P' is the consensus ranking assigned to this profile. By Axiom 2, P' must agree with P on $A_1 = \{$Ford, Chevy$\}$. We have Ford P Chevy, so Ford P' Chevy, i.e. Ford ranks above Chevy in P'. Next consider the profile

P_1''	P_2''
Ford	Plymouth
Plymouth	Ford
Chevy	Chevy

If each individual's opinion is treated equally, it seems reasonable for society to be indifferent between Ford and Plymouth here. But now by Axiom 2 applied to (P_1', P_2') and $A_1 = \{$Ford, Plymouth$\}$, society would have to be indifferent between Ford and Plymouth in the situation (P_1', P_2') as well. Thus, Ford and Plymouth must be tied in P'. Since we observed before that Ford P' Chevy, P' must be the ranking

P'
Ford-Plymouth
Chevy

Similarly, using in place of (P_1'', P_2'') the profile

P_1'''	P_2'''
Chevy	Plymouth
Plymouth	Chevy
Ford	Ford

one can argue that Chevy and Plymouth are tied in P'. This contradicts what P' looks like, namely ranking Plymouth above Chevy. The above example seems to produce a strong argument against Axiom 2. However, it should be pointed out that the contradiction does not depend on Axiom 2 alone. It also depends on the assumption that, given the profile (P_1, P_2), society would choose $P_1 = P_2$ as the consensus ranking; and it depends on the assumption that given the profile (P_1'', P_2''), society would be indifferent between Ford and Plymouth, and that given the profile (P_1''', P_2'''), society would be indifferent between Chevy and Plymouth.

A second argument against Axiom 2 goes something like the following. Let $t = 2$. Suppose $A_1 = \{$Ford, Chevy$\}$, where the group is deciding whether to buy a Ford or a Chevy, or perhaps no automobile at all. Suppose individual 1 prefers Ford to Chevy and individual 2 prefers Chevy to Ford. Let us now ask each individual to compare the group's receiving Ford or Chevy to the group's receiving certain monetary sums. Suppose that the individual preference orderings are:

P_1	P_2
$6000	Chevy
$5000	$6000
$4000	$5000
Ford	$4000
Chevy	Ford

We would be tempted to choose Chevy over Ford for the group, since the first individual, though he prefers Ford to Chevy, thinks Ford and Chevy are not too valuable, while the second thinks Chevy is very valuable and Ford not so valuable. However, Axiom 2 applied to $A_1 = \{$Ford, Chevy$\}$ would then say that we would also choose Chevy over Ford given the preference orderings

P_1'	P_2'
Ford	$6000
$6000	$5000
$5000	$4000
$4000	Chevy
Chevy	Ford

Examples like this suggest that what causes Axiom 2 to go wrong is that we have not considered in the allowable input (a profile) the *strength of preference* for one alternative over another. Thus, we might consider changing the basic set-up which we have been studying, namely, being given only each individual's ranking, and use more information, specifically the strength of preference for various alternatives. For suggestions on how to introduce strength of preference into the group decisionmaking procedure, the reader is referred to Luce and Raiffa [1957], Coombs [1954], or Goodman and Markowitz [1952]. An alternative to this approach is of course to allow less information, rather than more. We could, as has been done historically, gather only information about each individual's first choice, rather than a ranking. We could also put restrictions on the possible rankings which can be given. This will be our approach in the next section. All of these approaches involve a change in the basic input for decisionmaking. Some students of group decisionmaking have chosen to keep the basic set-up, and to replace Arrow's axioms with others. But there is no universally accepted set of axioms which does not lead to the same sort of difficulties. Still another approach left to us is to deal with the same basic input information, but to modify demands on what the output of the social welfare function must be. One obvious modification is to require only that the social welfare function produce a winner, rather than a complete ranking. But the exercises of the previous section show that there are then many reasonable alternative social welfare functions, and it is hard to choose among them. Moreover, axiomatizing choice of a winner rather than choice of a ranking also leads to trouble. Another obvious modification is to allow the social welfare function to choose some sort of ordering which may not be a ranking. (This approach could be developed from some recent work of Bogart [1973, 1975], who shows how to find a so-called strict partial order from a set of rankings. In a strict partial order, we may find some alternatives incomparable.) Finally, we could allow the social welfare function to choose several possible consensus rankings, rather than just one. This will be the approach of Sec. 7.4.

7.2.3. PROOF OF ARROW'S THEOREM[6]

We conclude this section by presenting a proof of Arrow's Theorem. Let us first introduce some definitions. Suppose $\mathcal{P} = (P_1, P_2, \ldots, P_t)$ is a profile in $\hat{\mathcal{P}}_t(A)$ and F is a social welfare function on $\hat{\mathcal{P}}_t(A)$. Suppose J is a subset of the set of individuals I and $a \neq b$ are in A. We shall say that the profile $\mathcal{P}$ is *J-favored for* (a, b) if for all i in J, aP_ib. We shall say that $\mathcal{P}$ is *strictly J-favored for* (a, b) if it is J-favored for (a, b) and bP_ia for all i not in J. Thus, $\mathcal{P}$ is J-favored for (a, b) if every member of J ranks a over b and it is strictly J-favored for (a, b) if in addition every other member of I ranks b over a. Finally, we say that J is *decisive* for (a, b) if whenever $\mathcal{P}$ is a profile with aP_ib for all i in J, then $aF(\mathcal{P})b$. In other words, J is decisive for (a, b) if for every profile which is J-favored for (a, b), the social welfare function ranks a over b. To give an example, suppose the social welfare function is the simple majority rule (forgetting momentarily that this is not a social welfare function) and suppose J consists of $\frac{2}{3}$ of the members of I. Then J is decisive for every (a, b).

Lemma 1. Under Arrow's axioms, J is decisive for (a, b) if and only if there is a profile $\mathcal{P}$ which is strictly J-favored for (a, b) and for which we have $aF(\mathcal{P})b$.

Proof. If J is decisive for (a, b), then every profile $\mathcal{P}$ which is strictly J-favored for (a, b) has $aF(\mathcal{P})b$. It is left to assume the existence of such a $\mathcal{P}$ and prove that J is decisive for (a, b). To prove this, let $\mathcal{P}'$ be any profile which is J-favored for (a, b). We shall show that $aF(\mathcal{P}')b$. In the proof, we are of course free to use Arrow's axioms, and we shall do so.

---

[6]This subsection can be omitted.

We show that $aF(\mathcal{P}')b$ by successively changing $\mathcal{P}$ into $\mathcal{P}'$. The procedure is illustrated in Fig. 7.1. The reader is urged to follow along with this figure as he reads the proof. Let the profile $\mathcal{P}''$ be obtained from $\mathcal{P}$ as follows.

(a) For each i in J, replace P_i by P_i'.

(b) For each i not in J such that $bP_i'a$, replace P_i by P_i'.

(c) For each i not in J such that $aP_i'b$ or $aT_i'b$, modify P_i by moving b down to just above a.

Thus, in Fig. 7.1, by rule (a), $P_1'' = P_1'$, $P_2'' = P_2'$, and $P_3'' = P_3'$; by rule (b), $P_5'' = P_5'$; and by rule (c), P_4'' and P_6'' are obtained from P_4 and P_6 respectively by moving b down to just above a. The reader should verify that by Axiom 2, applied to the set $A_1 = \{a, b\}$, since $aF(\mathcal{P})b$, it follows that $aF(\mathcal{P}'')b$.

Next, assume that no element is tied with a in $\mathcal{P}'$. In case of such ties, the construction is left to the reader. Obtain $\mathcal{P}'''$ from $\mathcal{P}''$ as follows

(a) If i is not in J and $aP_i'b$, move b one step below a.

(b) If i is not in J and $aT_i'b$, move b into a tie with a.

(c) Otherwise, let P_i''' be P_i''.

Thus, in Fig. 7.1, P_6''' is obtained from P_6'' by rule (a) and P_4''' is obtained from P_4'' by rule (b). By Axiom 1, $aF(\mathcal{P}'')b$ implies $aF(\mathcal{P}''')b$. Finally, obtain $\mathcal{P}''''$

$\mathcal{P}'$:	P_1'	P_2'	P_3'	P_4'	P_5'	P_6'
	a	x	y	x	b	x
	b	a	a	a-b	z	a
	x	y	b	z	x	y
	y	z	x	y	a	b
	z	b	z		y	z

$\mathcal{P}$:	P_1	P_2	P_3	P_4	P_5	P_6
	a	x	y	b	z	b
	x	a	z	y	x	z
	y	z	x	a	b	x
	b	y	a	x	y	y
	z	b	b	z	a	a

$\mathcal{P}''$:	P_1''	P_2''	P_3''	P_4''	P_5''	P_6''
	a	x	y	y	b	z
	b	a	a	b	z	x
	x	y	b	a	x	y
	y	z	x	x	a	b
	z	b	z	z	y	a

$\mathcal{P}'''$:	P_1'''	P_2'''	P_3'''	P_4'''	P_5'''	P_6'''
	a	x	y	y	b	z
	b	a	a	a-b	z	x
	x	y	b	x	x	y
	y	z	x	z	a	a
	z	b	z		y	b

$\mathcal{P}''''$:	P_1''''	P_2''''	P_3''''	P_4''''	P_5''''	P_6''''
	a	x	y	x	b	x
	b	a	a	a-b	z	a
	x	y	b	z	x	y
	y	z	x	y	a	b
	z	b	z		y	z

FIGURE 7.1. Illustration of the proof of Theorem 7.1. Here, $J = \{1, 2, 3\}$.

from $\mathcal{P}'''$ as follows. For all i not in J such that aP'_ib or aT'_ib, replace P'''_i by P'_i. In Fig. 7.1, P''''_4 and P''''_6 are obtained from P'''_4 and P'''_6 respectively by use of this rule. By Axiom 2 applied to $\{a, b\}$, $aF(\mathcal{P}'''')b$. Moreover, it is easy to show that $\mathcal{P}''''$ is $\mathcal{P}'$.   Q.E.D.

Lemma 2.   Under Arrow's axioms, I is decisive for every (a, b).

Proof.   By Axiom 3, there is a profile $\mathcal{P} = (P_1, P_2, \ldots, P_t)$ such that $aF(\mathcal{P})b$. Obtain P'_i from P_i by moving a to the top of the list and let $\mathcal{P}' = (P'_1, P'_2, \ldots, P'_t)$. By Axiom 1, $aF(\mathcal{P}')b$. Note that $\mathcal{P}'$ is strictly I-favored for (a, b). Now by Lemma 1, it follows that I is decisive for (a, b).   Q.E.D.

To complete the proof of Arrow's Theorem, let J be a minimal decisive set, i.e., a set which is decisive for some (a, b) and such that no proper subset is decisive for any (c, d). There must be such a J since by Lemma 2, I is decisive for some (a, b). Keep removing elements from I until we no longer have a decisive set. The last decisive set obtained is J. The set J is nonempty, since otherwise, for some pair, the set $I - J = I$ is not decisive. Fix $j \in J$. We shall show that j is a dictator, which contradicts Axiom 4.

Suppose J is decisive for (a, b). Let $c \neq a, b$. (Here, the assumption that A has at least three elements is used, allowing us to conclude that there is an element c different from a, b.) Consider the following profile $\mathcal{P}$:

P_i for $i \in J - \{j\}$	P_i for $i \notin J$	P_j
c	b	a
a	c	b
b	a	c
$A - \{a, b, c\}$	$A - \{a, b, c\}$	$A - \{a, b, c\}$

In each ranking P_i $(i = 1, 2, \ldots, t)$, the elements of $A - \{a, b, c\}$ are all tied and ranked last. Note that aP_ib for all $i \in J$. Since J is decisive for (a, b), we have aPb, where $P = F(\mathcal{P})$. We also have $\sim cPb$. For $\mathcal{P}$ is strictly $(J - \{j\})$-favored for (c, b). Thus, if cPb, then by Lemma 1, $J - \{j\}$ is decisive for (c, b). This contradicts the minimality of J as a decisive set. Thus, we have aPb and $\sim cPb$. Among the elements $a, b,$ and c, P must look like one of the following rankings:

$$
\begin{array}{cc}
a & a \\
b\text{-}c & b \\
 & c
\end{array}
$$

In either case, we conclude aPc. Since aPc and j is the only individual who prefers a to c, Lemma 1 implies that $\{j\}$ is decisive for (a, c). Thus $\{j\}$ cannot be a proper subset of J, and we conclude that $\{j\} = J$.

We have now shown that $\{j\}$ is decisive for (a, c), any $c \neq a$. For the above proof demonstrates this for $c \neq a, b$ and $\{j\}$ is decisive for (a, b) since $\{j\} = J$. To show j is a dictator, it remains to show that $\{j\}$ is decisive for (d, a), any $d \neq a$, and also that $\{j\}$ is decisive for (d, c), any $d, c \neq a$. To demonstrate the latter, consider the profile $\mathcal{P}$:

P_j	$P_i, i \neq j$
d	c
a	d
c	a
$A - \{a, c, d\}$	$A - \{a, c, d\}$

Let $P = F(\mathcal{P})$. By Lemma 2, I is decisive for any (x, y), so dPa. Since $\{j\}$ is decisive for (a, c), we have aPc. Thus in P, d is ranked over a and a is ranked over c, so dPc. By Lemma 1 it follows that $\{j\}$ is decisive for (d, c), which is what we wanted to prove.

It remains to demonstrate that $\{j\}$ is decisive for (d, a) whenever $d \neq a$. To do this, consider the profile $\mathcal{P}$:

P_j	$P_i, i \neq j$
d	c
c	a
a	d
$A - \{a, c, d\}$	$A - \{a, c, d\}$

Let $P = F(\mathcal{P})$. Since $\{j\}$ is decisive for (d, c), we have dPc. By Lemma 2, we have cPa. Thus dPa follows. Finally, by Lemma 1, we conclude that $\{j\}$ is decisive for (d, a). This completes the proof of Theorem 7.1.

EXERCISES

1. Suppose three old friends are considering choices among alternative vacation sites, Oregon (O), New York (N), Florida (F), and California (C). Consider the two profiles (the first representing the friends' true preferences and the second suggested to them by their wives).

P_1	P_2	P_3		P_1'	P_2'	P_3'
F	F	F		F	O	F
O	O	O		O	F	O-C-N
C	C	C		C	C-N	
N	N	N		N		

Suppose we assume that for the first profile, the three friends' social welfare function ranks F over N.

(a) Does Arrow's first axiom allow them to conclude that this holds for the second profile as well?

(b) Does Arrow's second axiom?

2. Repeat Exer. 1 with the first profile as the second and the second as the first.

3. In the vacation choice situation of Exer. 1, suppose whenever friend 1 prefers Florida to Oregon, then the group prefers Florida to Oregon. Is friend 1 a dictator in the technical sense defined in this section? What if the group always goes where friend 1 chooses unless he is indifferent?

4. If $F(\mathcal{P})$ is a social welfare function (which does not necessarily satisfy Arrow's axioms), is it possible for there to be two dictators relative to $F(\mathcal{P})$? (Use the technical definition of dictator given in this section.)

5. Show that the lexicographic preference of Exer. 13, Sec. 7.1 violates at least one of Arrow's axioms.

6. Give an example to show that the Borda count violates at least one of Arrow's axioms. (Give an example with no ties in any of the P_i's.)

7. Does the plurality system of Exer. 14, Sec. 7.1 give a social welfare function which satisfies all of Arrow's axioms? (Give proof or counterexample.)

8. (Riker and Ordeshook [1973].) Suppose a President knows that some of his advisors are loyal and others are disloyal. He decides to discount the votes of his disloyal advisors by counting their votes negatively. Specifically, he lets the number of votes for alternative a be the number of loyal advisors who rank a first minus the number of disloyal advisors who rank a first. He ranks alternatives on the basis of the number of votes they receive under this procedure. Which of Arrows' axioms are violated by the procedure?

9. Use the following profile to make an argument against Arrow's second axiom.

P_1	P_2
Fish	Beer
Steak	Fish
Roast Beef	Steak
Scotch	Roast Beef
Beer	Scotch

10. Start with the following profile and make an argument against Arrow's second axiom.

P_1	P_2
Fish	Fish
Steak	Steak
Roast Beef	Roast Beef

11. Arrow's axioms also make sense under a somewhat different interpretation. Here, there is only one decisionmaker. The elements of the set I are considered to be different contingencies. P_i represents the single decisionmaker's ranking of the set of alternatives under contingency i. (For example, we might prefer as President a person with foreign affairs experience in a time of international tensions and we might rather have a person with a background in economics in a time of inflation or depression. Thus, our ranking changes under the different contingencies.) We treat the different contingencies as equally likely (or count the more likely ones more often). We seek a ranking of alternatives which is in some sense the best possible under all the different contingencies being considered. The whole problem might be called the *contingency ranking problem.*

    (a) Make up an argument to show why Axiom 2 may not be appropriate for the contingency ranking problem.

    (b) What about Axiom 1 in this situation?

12. Suppose $J = \{1, 2\}$.

    (a) Which of the profiles of Exer. 1 is J-favored for (F, N)?

    (b) Which of the profiles is J-favored for (F, O)?

    (c) Which of the profiles is strictly J-favored for (O, C)?

13. A social welfare function on $\mathcal{P}_t(A)$ is *Pareto optimal*[7] if for every a and b in A, whenever a is ranked over b in every ranking of a profile, then a is ranked over b in the corresponding group ranking.

    (a) Restate Pareto optimality in terms of J-favorability.

    (b) Assuming the simple majority rule is a social welfare function, is it Pareto optimal?

    (c) Is the Borda count Pareto optimal?

    (d) Is the lexicographic group preference of Exer. 13, Sec. 7.1 Pareto optimal?

    (e) Is the plurality system of Exer. 14, Sec. 7.1 Pareto optimal?

14. Suppose $t = 3$ and person 3 is a dictator in the technical sense. Do not assume Arrow's axioms. Name all decisive sets and all minimal decisive sets.

15. Illustrate the proof of Lemma 1 by using the following $\mathcal{P}$ and $\mathcal{P}'$ and taking $J = \{1\}$.

[7]We used the term Pareto optimal for a somewhat different notion in Chapter 6.

	P_1	P_2	P_3	P_4		P_1'	P_2'	P_3'	P_4'
	a	b	x	b		a	a	b	x
	b	a	b	x		x	y	a	a-b
$\mathcal{P} =$	x	x	a	z	$\mathcal{P}' =$	y	x	x	z
	y	y	z	y		b	z	y	y
	z	z	y	a		z	b	z	

16. Suppose I is decisive for all (a, b) and suppose J is decisive for (x, y). Show that for the following profile $\mathcal{P}$, $zF(\mathcal{P})y$.

P_i for $i \in J$	P_i for $i \notin J$
z	y
x	z
y	x
$A - \{x, y, z\}$	$A - \{x, y, z\}$

17. Suppose J is a minimal decisive set and J is decisive for (x, y). Let $k \in J$. Show that for the following profile $\mathcal{P}$, $xF(\mathcal{P})z$. (You may assume Lemma 1.)

P_i for $i \in J - \{k\}$	P_i for $i \in I - J$	P_k
z	y	x
x	z	y
y	x	z
$A - \{x, y, z\}$	$A - \{x, y, z\}$	$A - \{x, y, z\}$

18. Does the conclusion of Lemma 1 follow if only Axiom 1 is assumed? (Give proof or counterexample.)

19. Does the conclusion of Lemma 1 follow if only Axiom 2 is assumed? (Give proof or counterexample.)

20. Discuss whether or not Arrow's axioms seem appropriate for decisionmaking in the situations discussed in Exer. 15, Sec. 7.1.

21. Discuss Arrow's axioms in the context of ranking alternative job candidates.

22. Discuss Arrow's axioms in the situation of choosing variables for an energy demand signed digraph. (See Exer. 16, Sec. 7.1.)

23. Discuss Arrow's axioms in the situation of choosing a ranking of the players in a tournament (see Exer. 17, Sec. 7.1).

24. Compare Arrow's axioms for the situation where each voter is voting

his own feelings independently and the situation where sophisticated voting in the sense of Exer. 19, Sec. 7.1, can take place.

25. Invent some axioms of your own for a social welfare function.

7.3. Joint Scales and Single-Peakedness

In this section we consider the possibility of limiting the set of possible inputs or profiles in the group decisionmaking situation. We shall modify our definition of social welfare function and define it to be a procedure which chooses a group ranking for every profile in some set of profiles. We shall show that if a profile results from certain procedures, then the simple majority rule does not lead to the voter's paradox, i.e., it can be used to define a ranking. Hence, the simple majority rule is a social welfare function in the sense that it assigns a ranking to each of a selected set of profiles. This social welfare function satisfies all of Arrow's axioms, if we are careful to modify them to refer only to profiles in the set.

As a preliminary to this section, let us recall the notion of median. If S is a set of numbers (with repetitions allowed) with an odd number of elements, the *median* of S is the middle number when numbers of S are listed in order (including repetitions). That is, if S has $2k + 1$ members, the median of S is the $(k + 1)$st. For example, if S is $\{1, 5, 2\}$, then its members in order are 1, 2, 5, and 2 is the median. If S is the set $\{1, 4, 3, 3, 6\}$, then its members in order are 1, 3, 3, 4, 6, and 3 is the median.

One situation where the simple majority rule gives a ranking is the following. Suppose presidential candidates x, y, z, etc. are rated on a scale from 0 to 100 according to their Civil rights voting record, or their conservatism vs. liberalism, or their support of defense spending, etc. Suppose an individual in considering possible candidates locates his ideal rating on this scale, say 20, 70, or 100. Then he chooses candidate a over candidate b if and only if the distance from a to his ideal is less than the distance from b to his ideal. For example, in Fig. 7.2(a), individual i prefers candidate u to candidates x, y, and v, candidate v to candidates x and y, and candidate y

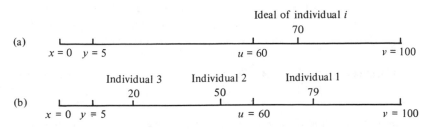

FIGURE 7.2. A joint quantitative scale.

to candidate x.[8] Thus, individual i's ranking is

P_i
u
v
y
x

Suppose next that each individual uses the same ratings of candidates, and we locate each individual (each individual's ideal rating) on this scale. Then we obtain a *joint quantitative scale of individuals and alternatives,* as shown in Fig. 7.2(b). Suppose the group's profile is obtained from this joint scale by the procedure we have described. Coombs [1954] has suggested the following social welfare function for this situation, if there is an odd number of individuals. We call it the *Coombs social welfare function.* Find that individual i whose ideal rating is the median of all the ideal ratings of all the individuals. Then use as the group's ranking individual i's ranking. For example, in Fig. 7.2(b), we have the following profile:

P_1	P_2	P_3
u	u	y
v	y	x
y	x-v	u
x		v

Individual 2 has the median position, 50, among the three individuals, for the set of ratings is $\{20, 50, 79\}$. Thus, using the Coombs social welfare function, we choose individual 1's ranking to be the group's. It is interesting to note that this ranking is the same as that obtained using the simple majority rule. (Check it.) This is no accident.

Theorem 7.2 (Goodman [1954]). Suppose A is a set of alternatives and $\hat{\mathcal{R}}_{2k+1}(A)$ is the set of all profiles of a group of $2k + 1$ individuals over A which are obtained from a joint quantitative scale. On $\hat{\mathcal{R}}_{2k+1}(A)$, the Coombs social welfare function is the same as the simple majority rule.

Corollary. On $\hat{\mathcal{R}}_{2k+1}(A)$, the simple majority rule gives rise to a ranking.

---

[8]The distance from i's ideal to candidate a can also be used to measure the strength of i's preference for a.

Proof of Theorem 7.2.   Let us first suppose that the median individual i prefers alternative a to alternative b. Then i's ideal rating is closer to a than to b. We shall show that since i's ideal is a median, at least k of the remaining ideal ratings are closer to a than to b, and so a is preferred to b by at least $k + 1$ individuals, a majority. The relative ordering of i, a, and b must be one of cases (a) to (f) of Fig. 7.3. In cases (a), (c), and (e), all individuals with ratings less than or equal to i's prefer a to b, and in cases (b), (d), and (f), all individuals with ratings greater than or equal to i's prefer a to b. Next, let us suppose that the median individual i rates alternatives a and b as tied. Then there are two possibilities, cases (g) and (h) of Fig. 7.3. In case (g), i is half-way between a and b. Only the k individuals with ratings less than or equal to i's could be closer to a than to b and similarly only the k individuals with ratings greater than or equal to i's could be closer to b than to a. Hence, neither a nor b is preferred by a majority. In case (h), a and b receive the same rating, and all individuals rate them tied. This proves Theorem 7.2.   Q.E.D.

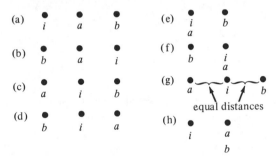

FIGURE 7.3. The median individual rating i and the ratings of two alternatives a and b in various cases.

Coombs has stated other conditions under which the simple majority rule gives rise to a ranking. Suppose we cannot rate each alternative or each individual's ideal on a quantitative numerical scale. However, suppose we can find a ranking or ordering which lists both alternatives and individuals. We shall suppose for simplicity that this ranking is *complete*, i.e., allows no ties. For example, for the four alternatives x, y, u, and v, and the three individuals 1, 2, and 3, one such ranking is

$$
\begin{array}{c}
x \\
2 \\
y \\
u \\
1 \\
v \\
3
\end{array}
\tag{1}
$$

We could interpret this as follows: The alternatives are rated in decreasing order as x, y, u, v; individual 2 considers himself somewhere between alternatives x and y on this scale, individual 1 considers himself between u and v, etc. Such a ranking is called a *joint qualitative scale of individuals and alternatives*, with the word qualitative replacing the word quantitative. In such a ranking, we can define the *distance* from one element to another as one more than the number of elements between them. (If there are ties in the joint scale, distance has to be defined a little differently.) In our example, the distance from 2 to 1 is 3, from y to u is 1, from y to v is 3, and so on. If there is a joint qualitative scale of individuals and alternatives, an individual i can define his ranking as follows: If i is equidistant from two alternatives a and b, then he chooses between a and b arbitrarily; otherwise, he prefers the alternative closer to himself. The resulting ranking has no ties.[9] In our example, individual 1 definitely prefers v to y and to x, u to y and to x, and y to x. He chooses arbitrarily between u and v. Let us say he chooses u. Then his ranking is

$$u$$
$$v$$
$$y$$
$$x$$

We shall assume that a group obtains its profile in this manner. If the set of individuals has an odd number of members, $2k + 1$, we can still pick a median from the set of individuals: The $(k + 1)$st individual counting from either end. In our example, the median individual is 1. A natural social welfare function, which we shall also call *Coombs' social welfare function*, can now be defined: Choose as the group's ranking the ranking of the median individual. A proof analogous to our earlier one now establishes the following result. (The proof uses the fact that there are no ties allowed in the complete joint qualitative scale. See Exer. 12.[10])

Theorem 7.3. Suppose A is a set of alternatives and $\hat{S}_{2k+1}(A)$ is the set of all profiles of a group of $2k + 1$ individuals over A obtained from a (complete) joint qualitative scale. On $\hat{S}_{2k+1}(A)$, the Coombs social welfare function is the same as the simple majority rule. Moreover, the simple majority rule gives rise to a ranking.

---

[9]An alternative method for i to define his ranking is for i to declare a and b tied if he is equidistant from a and b. However, in what follows, it will be convenient to think of breaking ties.

[10]If the ranking of individual i is obtained from the joint qualitative scale by declaring a and b tied if i is equidistant from a and b, then the proof goes over without the assumption that the joint qualitative scale is complete.

Theorem 7.4.   On $\hat{S}_{2k+1}(A)$, the Coombs social welfare function (the simple majority rule) satisfies all of Arrow's axioms.[11]

Proof.   Left to reader (Exer. 13).

Let us return to the (complete) joint qualitative scale (1) and let us choose a profile from it by breaking ties arbitrarily. We obtain the following profile

P_1	P_2	P_3
u	x	v
v	y	u
y	u	y
x	v	x

Let us plot each ranking as a curve in the following way. The x-axis shows the alternatives in the order shown in the joint qualitative scale. The y-axis shows $B(x)$, where $B(x)$ is the Borda count of x. We show the results for the three rankings in our profile in Fig. 7.4, first separately and then plotted on the same set of axes. The reader will note that each curve has at most one change of direction, and if there is a change, it is from up to down. That is,

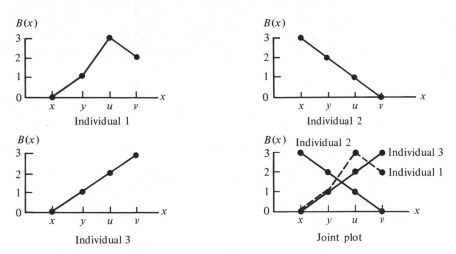

FIGURE 7.4. Single-peaked graphs resulting from individual and joint plots.

[11]We must be careful in our statement of the axioms. Axioms 1 and 2 are only required to hold for profiles in $\hat{S}_{2k+1}(A)$ and for modifications of these profiles which are also in $\hat{S}_{2k+1}(A)$. Similarly, Axiom 3 must be modified to require that there be some profile in $\hat{S}_{2k+1}(A)$ which has the desired property. And the definition of dictator for Axiom 4 must be modified to mention only profiles in $\hat{S}_{2k+1}(A)$.

the curve either goes always up or always down or it goes up for awhile, hits a high point, and then goes down. Such a curve is called *single-peaked*. A profile $\mathcal{P}$ satisfies the *single-peakedness condition* if we can find an ordering of the set of alternatives A so that under this ordering, the curve corresponding to every ranking in $\mathcal{P}$ is single-peaked.

Theorem 7.5 (Coombs [1954]). Suppose $\mathcal{P}$ is a profile of rankings without ties. Then we can find a complete joint qualitative scale from which $\mathcal{P}$ can be obtained if and only if $\mathcal{P}$ satisfies the single-peakedness condition.

Proof. Omitted.

Corollary. If there is an odd number of individuals and we limit ourselves to profiles without ties which satisfy the single-peakedness condition, then the simple majority rule satisfies all of Arrow's axioms.[12,13]

The advantage of the condition single-peakedness is that we do not have to restrict ourselves to a set procedure (the joint qualitative scale method) for obtaining rankings. Given a profile, we simply have to decide whether or not it satisfies single-peakedness. If it does, we can use the simple majority rule.

In testing for single-peakedness, the reader should be careful. Any single ranking without ties can be graphed as a single-peaked curve, simply by using on the x-axis the set of alternatives ranked in the order of the ranking itself. What is required here is that the same ordering of alternatives give single-peaked curves for every ranking in the profile. To give an example, suppose we consider the profile

P_1	P_2	P_3
x	y	z
y	z	x
z	x	y

We know that the simple majority rule will not give rise to a ranking, so we know that there is no ordering of the alternatives which will give rise to single-peaked curves for each ranking in the profile. A proof of this fact is left as an exercise (Exer. 9).

For a further discussion of single-peakedness, see Black [1958], Luce and Raiffa [1957], or Riker and Ordeshook [1973].

[12]See the previous footnote.

[13]The single-peakedness condition was introduced by Black [1948a, b], who proved that under single-peakedness, there is exactly one alternative which is favored by a majority over every other alternative. Arrow proved that single-peakedness implies that the simple majority rule gives a ranking.

EXERCISES

1. Find the median of each of the following sets:
   (a) $\{0, 5, 10, 11, 16\}$.
   (b) $\{5, 2, 7, 73, 21\}$.
   (c) $\{5, 5, 5, 1, 2\}$.
   (d) $\{5, 5, 5, 1, 1\}$.

2. Suppose that in the Americans for Democratic Action ratings, Senator A is rated 0, Senator B 20, Senator C 60, and Senator D 95. If your ideal is about 80, what is the ranking you might obtain from these ratings?

3. Suppose five individuals use the ratings of Exer. 2 and give their ideals as 0, 10, 25, 50, and 80 respectively.
   (a) Find the profile obtained from these five individuals.
   (b) Identify the median individual.
   (c) Check that his ranking is the same as that obtained from the simple majority rule.

4. Repeat Exer. 3 if the individuals' ideal ratings are 40, 50, 70, 90, and 100.

5. Repeat Exer. 3 given the joint quantitative scale shown in Fig. 7.5.

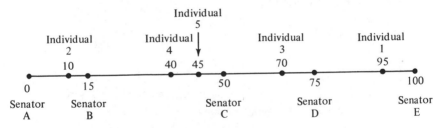

FIGURE 7.5. Joint quantitative scale for Exer. 5, Sec. 7.3.

6. Consider the joint qualitative scale for individuals 1, 2, and 3 and alternatives $x, y, u, v,$ and w given by

$$
\begin{array}{c}
x \\
y \\
1 \\
u \\
2 \\
v \\
w \\
3
\end{array}
$$

    (a) Find a ranking obtained for each individual.

    (b) Find the median individual.

    (c) Check that his ranking is the same as that obtained from all the rankings by the simple majority rule.

7. Repeat Exer. 6 for the following joint qualitative scale:

$$1$$
$$3$$
$$x$$
$$y$$
$$4$$
$$z$$
$$5$$
$$2$$

8. Plot the ranking

$$x$$
$$y$$
$$u$$
$$v$$

using each of the following orders on the x-axis and identify which curves are single-peaked.

    (a) x, y, u, v.

    (b) v, u, y, x.

    (c) x, u, y, v.

    (d) u, x, y, v.

9. By considering all six possible orderings on the x-axis, prove that the following profile is not single-peaked.

P_1	P_2	P_3
x	y	z
y	z	x
z	x	y

10. Determine if the following profiles are single-peaked:

    (a)

P_1	P_2	P_3
x	z	x
y	y	z
z	x	y

(b)

P_1	P_2	P_3	P_4
z	x	y	x
x	z	x	y
y	y	z	z

11. Determine if the following profile can be obtained from a (complete) joint qualitative scale of individuals and alternatives.

P_1	P_2	P_3
x	y	z
z	x	x
y	z	y

12. Show that in a joint qualitative scale which is incomplete, i.e., allows ties, the Coombs social welfare function and the simple majority rule may differ. *Hint*: Consider

$$3$$
$$a$$
$$1-2$$
$$b$$

13. Prove Theorem 7.4.

14. Under what circumstances is it reasonable to assume that we can find a joint quantitative or qualitative scale of individuals and alternatives?

7.4. Distance between Rankings

7.4.1. THE KEMENY-SNELL AXIOMS

An approach to determining a group consensus function which is quite different from Arrow's is that suggested by Kemeny and Snell [1962, Ch. 2]. Our discussion in this section follows that of Kemeny and Snell, though our axioms are slightly modified.

Let us consider the following rankings on the set
$A = \{$Ford, Chevy, Plymouth, Volkswagen, Datsun$\}$.

P	Q	R
Ford	Datsun	Chevy
Chevy	Volkswagen	Ford
Plymouth	Plymouth	Plymouth
Volkswagen	Chevy	Volkswagen
Datsun	Ford	Datsun

It seems reasonable to say that rankings P and Q are quite far apart and that ranking R is quite close to ranking P. Thus, perhaps it makes sense to speak of the *distance* between two rankings. Let $d(P, Q)$ represent the distance between the rankings P and Q. If we could measure distance precisely, a plausible rule for obtaining group consensus would be the following: Given a profile of rankings $(P_1, P_2, \ldots, P_t)$, find that ranking P whose distance is as small as possible from all of the P_i. One reasonable way to make this rule more precise is to choose P as a *median ranking*, a ranking P such that $\sum_{i=1}^{t} d(P, P_i)$ is a minimum.[14] We shall mention other ways to make this rule precise later on. Naturally, in light of Arrow's Theorem, we expect that this procedure will either violate one of his axioms or it will fail to pick out a *unique P*, or both. However, even if the procedure fails to produce a unique P, it might limit consideration to a set of reasonable consensus rankings P.

This new procedure is not very helpful unless we know how to measure $d(P, Q)$. Rather than write down ad hoc rules for measuring distance, let us take the axiomatic approach, much as we did in Sec. 7.2. Following Kemeny and Snell, we list axioms which our distance measure can be reasonably expected to satisfy. Given such a set of axioms, there are several possibilities. First, there could be a number of distance functions satisfying them. Second, there could be no distance function satisfying them; our experience with Arrow's Theorem suggests this as a strong possibility. Finally, there could be exactly one distance function satisfying the axioms. This is the best possible situation; the axioms determine exactly how to measure distance. In such a situation, we say the axioms are *categorical*.[15]

Before stating axioms for distance, let us note that it also seems to make sense to say that one ranking Q is *between* two others, P and R. This holds if for each pair of alternatives a and b in A, the judgement of relative ranking of a and b in Q is between those in P and R. Namely, if P and R agree on the ranking of a and b, then Q must agree with them. If P and R differ, then Q must agree with one or the other or, in the case that one ranks a higher and the other ranks b higher, then Q can declare a tie. For example, let $A = \{$Ford, Chevy, Plymouth, Volkswagen$\}$ and let P, Q, Q', R be the rankings

P	Q	Q'	R
Ford	Chevy	Chevy	Chevy
Chevy	Ford	Ford-Plymouth	Volkswagen
Plymouth	Volkswagen	Volkswagen	Plymouth
Volkswagen	Plymouth		Ford

[14]Why is this notion of median similar to that used for a set of numbers?

[15]In the study of relative balance in Exers. 21 through 26 of Sec. 3.1, we suggested that it would be nice to find axioms for a measure of balance. We could first axiomatize distance between two small groups and then axiomatize how to obtain a measure of balance from a measure of distance. Such an approach, based on the Kemeny-Snell axioms presented here, can be found in Norman and Roberts [1972].

Then Q is between P and R. For on the pair {Ford, Chevy}, Q agrees
with R; on the pair {Plymouth, Chevy}, Q agrees with both; on the pair
{Ford, Plymouth}, Q agrees with P; on the pair {Ford, Volkswagen}, Q agrees
with P; on the pair {Chevy, Volkswagen}, Q agrees with both; and on the
pair {Plymouth, Volkswagen}, Q agrees with R. Also, Q' is between P and R,
for on {Ford, Chevy}, Q' agrees with R; on {Chevy, Plymouth}, Q' agrees
with both; on {Ford, Plymouth}, Q' declares a tie while P ranks Ford ahead
and R ranks Plymouth ahead; and so on. If Q is between P and R, we shall
sometimes denote this fact by $B(P, Q, R)$.

To state our axioms for distance, let us note that a distance measure
is a real-valued function $d: \hat{\mathcal{P}}(A) \times \hat{\mathcal{P}}(A) \longrightarrow \mathcal{R}$, where $\hat{\mathcal{P}}(A)$ is the set of all
rankings over the set A. It seems reasonable that our distance function should
satisfy the following conditions for all P, Q, R in $\hat{\mathcal{P}}(A)$.

Axiom 1.1 $d(P, Q) \geqq 0$, with equality iff $P = Q$.

Axiom 1.2. $d(P, Q) = d(Q, P)$.

Axiom 1.3. $d(P, Q) + d(Q, R) \geqq d(P, R)$, with equality iff $B(P, Q, R)$.

The first part of Axiom 1.3 is the usual triangle inequality. If one thinks of
points in the plane, one point Q is between two others P and R if and only
if they lie on a line, i.e., if and only if to get from P to R by the shortest route
we must go through Q, i.e., if and only if $d(P, Q) + d(Q, R) = d(P, R)$. This
is the motivation for the second part of Axiom 1.3.

Our next axiom asserts that the measure of distance is independent
of the particular "names" we have given the elements of A. That is, if we
relabel the elements of A, distances should not change. In particular, this
implies that if P, Q, P', and Q' are the rankings

P	Q	P'	Q'
Ford	Plymouth	Chevy	Ford
Chevy	Chevy	Plymouth	Plymouth
Plymouth	Ford	Ford	Chevy

then $d(P, Q)$ should be the same as $d(P', Q')$. For P' and Q' can be obtained
from P and Q by relabelling Ford as Chevy, Chevy as Plymouth, and Plymouth as Ford. A relabelling of the objects of A is usually called a *permutation* of A, a one-to-one map from A into itself. The relabelling axiom is now
stated as follows:

Axiom 2. If ranking P' results from ranking P by a permutation of
the set A and ranking Q' results from ranking Q by the same permutation,
then $d(P, Q) = d(P', Q')$.

458   *Ch. 7 Group Decisionmaking*

Our next axiom can be informally stated as follows:

Axiom 3 (Informal Version). If two rankings P and Q are the same at the top and the bottom, and differ only on a set of elements in the middle, then the distance between P and Q depends only on the ranking of these middle objects.

Before making this axiom precise, let us illustrate it with an example. Suppose we are choosing among color television sets, and our set of alternatives is $A = \{$Zenith, RCA, Magnavox, Sylvania, Motorola, Philco$\}$. Consider the following rankings:

P	Q	P'	Q'
Zenith	Zenith	RCA	RCA
RCA	RCA	Magnavox-Sylvania	Magnavox
Magnavox-Sylvania	Magnavox	Motorola	Motorola
Motorola	Motorola	Philco	Sylvania
Philco	Sylvania	Zenith	Philco
	Philco		Zenith

(2)

Then P and Q differ only on the middle segment consisting of Magnavox, Sylvania and Motorola. They agree on the top segment consisting of Zenith and RCA and on the bottom segment consisting of Philco. Moreover, P' and Q' differ in the same way on the same middle segment consisting of Magnavox, Sylvania, and Motorola and agree with each other on their top segment consisting of RCA and their bottom segment consisting of Philco and Zenith. By Axiom 3, $d(P, Q) = d(P', Q')$. The reader should note the similarity of this axiom to Arrow's Axiom 2, the independence of irrelevant alternatives.

To make Axiom 3 precise,[16] let us say that a subset S of A is a *segment* in a ranking P if every element a in $A - S$ is either above every element of S or below every element of S. S is a *proper segment* if $S \neq A$. For example, in the ranking P of Eq. (2), the sets $\{$Magnavox, Sylvania, Motorola$\}$ and $\{$RCA, Magnavox, Sylvania, Motorola$\}$ are segments. But the set $\{$RCA, Motorola$\}$ is not, for Magnavox is between RCA and Motorola. The set $\{$RCA, Magnavox$\}$ is not a segment either, for Sylvania is neither above nor below Magnavox. If $S = S(P)$ is a segment in a ranking P, then the set $\bar{S} = \bar{S}(P)$ of all elements above S in P and the set $\underline{S} = \underline{S}(P)$ of all elements below S in P are also segments. $P(S)$ will be the ranking of S obtained from P. $P(\bar{S})$ and $P(\underline{S})$ have a similar interpretation. For example, for P of Eq. (2), if $S(P)$ is $\{$Magnavox, Sylvania, Motorola$\}$, then $\bar{S} = \bar{S}(P) = \{$Zenith, RCA$\}$ and $\underline{S} = \underline{S}(P) = \{$Philco$\}$. Moreover, $P(\bar{S}), P(S)$, and $P(\underline{S})$

[16]The reader may wish to skip the formal statement of Axiom 3 and pass directly to Axiom 4.

are given by

$P(\bar{S})$	$P(S)$	$P(\underline{S})$
Zenith	Magnavox-Sylvania	Philco
RCA	Motorola	

If we deal with two rankings P and Q, and S is a segment of both, then $\bar{S}(P)$ might be different from $\bar{S}(Q)$. (Why?) If S is a common segment of the rankings P and Q, let us say that P and Q *agree outside* S if $\bar{S}(P) = \bar{S}(Q) = \bar{S}$, $\underline{S}(P) = \underline{S}(Q) = \underline{S}$, $P(\bar{S}) = Q(\bar{S})$, and $P(\underline{S}) = Q(\underline{S})$. With these preliminaries, we can now state Axiom 3 precisely.

Axiom 3. Suppose P, Q, P' and Q' are rankings and S is a segment in each. Suppose moreover that P and Q agree outside S and P' and Q' agree outside S and that $P(S) = P'(S)$ and $Q(S) = Q'(S)$. Then $d(P, Q) = d(P', Q')$.

In our example of Eq. (2), if S is {Magnavox, Sylvania, Motorola}, the reader can easily check that P and Q agree outside S, P' and Q' agree outside S, and $P(S) = P'(S)$, $Q(S) = Q'(S)$. The appropriate rankings are shown below.

$P(\bar{S}) = Q(\bar{S})$	$P(\underline{S}) = Q(\underline{S})$	$P'(\bar{S}) = Q'(\bar{S})$	$P'(\underline{S}) = Q'(\underline{S})$
Zenith	Philco	RCA	Philco
RCA			Zenith

$P(S) = P'(S)$	$Q(S) = Q'(S)$
Magnavox-Sylvania	Magnavox
Motorola	Motorola
	Sylvania

Thus, by Axiom 3, $d(P, Q) = d(P', Q')$.

The final axiom is quite arbitrary and is simply a convention: It chooses a unit of measurement.

Axiom 4. The minimum positive distance between elements in $\hat{\mathcal{P}}(A)$ is 1, i.e., for all P and Q in $\hat{\mathcal{P}}(A)$, $d(P, Q) = 0$ or $d(P, Q) \geqq 1$, and for some P and Q in $\hat{\mathcal{P}}(A)$, $d(P, Q) = 1$.

7.4.2. CALCULATION OF DISTANCE

We now ask: Is there a distance function satisfying the Kemeny and Snell axioms, and if so, is it uniquely determined? The answers to both of these questions is yes.

Theorem 7.6 (Kemeny and Snell). For every set A with two or more members, there is a distance function d on $\hat{P}(A) \times \hat{P}(A)$ satisfying Axioms 1–4. Moreover, d is uniquely determined.

To prove the first part of Theorem 7.6, we construct an explicit distance function d which satisfies Axioms 1–4. The proof of uniqueness we omit. Suppose P and Q are rankings of A. If a and b are in A, let $\delta_{P,Q}(a, b)$ count 0 if P and Q agree on their ordering of a and b, let $\delta_{P,Q}(a, b)$ count 2 if one ranking puts a over b and the other puts b over a, and let $\delta_{P,Q}(a, b)$ count 1 if one ranking puts a over b or b over a and the other ranking has a and b tied. Then $d(P, Q)$ is the sum of $\delta_{P,Q}(a, b)$ over all (unordered) pairs $\{a, b\}$ from A. For example, suppose P and Q are given as follows:

P	Q
Ford	Ford-Plymouth
Chevy	Chevy
Plymouth	

Then

$$d(P, Q) = \delta_{P,Q}(\text{Ford, Chevy}) + \delta_{P,Q}(\text{Ford, Plymouth}) + \delta_{P,Q}(\text{Plymouth, Chevy})$$
$$= 0 + 1 + 2 = 3.$$

We now verify that the function d we have defined satisfies Axioms 1–4 for every A with two or more members.

Axiom 1.1. Each $\delta_{P,Q}(a, b)$ is nonnegative. Thus the sum of the terms $\delta_{P,Q}(a, b)$ is also nonnegative. Moreover, the sum is 0 if and only if each term is 0, which is true if and only if for all a, b in A, P and Q agree, which is true if and only if $P = Q$.

Axiom 1.2. This is true since $\delta_{P,Q}(a, b) = \delta_{Q,P}(a, b)$.

Axiom 1.3. Define

$$\delta^P(a, b) = \begin{cases} +1 & \text{(if } aPb) \\ -1 & \text{(if } bPa) \\ 0 & \text{(if } a \text{ and } b \text{ are tied in } P). \end{cases} \tag{3}$$

Then for all a, b in A,

$$\delta_{P,Q}(a, b) = |\delta^P(a, b) - \delta^Q(a, b)|.$$

Now

$$|\delta^P(a, b) - \delta^Q(a, b)| + |\delta^Q(a, b) - \delta^R(a, b)| \geqq |\delta^P(a, b) - \delta^R(a, b)|. \tag{4}$$

Adding the inequalities (4) over all $\{a, b\}$ proves the first part of Axiom 1.3, i.e.,

$$d(P, Q) + d(Q, R) \geqq d(P, R). \tag{5}$$

The only way to come out with equality in (5) is to have equality in (4) for each a, b in A, and equality in (4) means that $\delta^Q(a, b)$ is between $\delta^P(a, b)$ and $\delta^R(a, b)$. Thus, equality in (4) for each a, b is easily seen to mean that Q is between P and R (Exer. 21).

Axiom 2. A permutation of the elements of A has no effect on d, because it simply amounts to adding the summands $\delta_{P,Q}(a, b)$ in a different order.

Axiom 3. It is easy to verify that if P and Q are the same at the top and bottom and differ only on a middle segment S, then $\delta_{P,Q}(a, b) = 0$ whenever $a \notin S$ or $b \notin S$. Thus, $d(P, Q)$ is obtained by summing the terms $\delta_{P,Q}(a, b)$ for a and b in S, which verifies the informal version of Axiom 3. To verify the formal version, note that the same argument implies that if P' and Q' are the same at the top and bottom and differ only on the middle segment S, then $d(P', Q')$ is obtained by summing the terms $\delta_{P',Q'}(a, b)$ for a and b in S. Finally, if P and P' agree on S and Q and Q' agree on S, then $\delta_{P,Q}(a, b) = \delta_{P',Q'}(a, b)$ whenever a and b are in S. Thus, $d(P, Q) = d(P', Q')$.

Axiom 4. Since $\delta_{P,Q}(a, b)$ is always a nonnegative integer, it follows that $d(P, Q)$ is always a nonnegative integer. Now suppose that $a_0 \neq b_0$ (by assumption A has at least two elements) and P and Q are

P	Q
$a_0\text{-}b_0$	a_0
$A - \{a_0, b_0\}$	b_0
	$A - \{a_0, b_0\}$

Then $\delta_{P,Q}(a_0, b_0) = 1$ and $\delta_{P,Q}(a, b) = 0$ otherwise, so $d(P, Q) = 1$. Thus, 1 is the minimum positive distance.

The distances for the case where $|A| = 3$ are easily calculated. The rankings on a 3-element set are conveniently represented as a hexagon as shown in Fig. 7.6, which shows distances between some of the rankings. To find the distance between any two rankings P and Q, find a chain from P to Q in this figure, the sum of whose distances is minimal. The length of this chain is $d(P, Q)$. For example, the distance between

$$\begin{pmatrix} a \\ b \\ c \end{pmatrix} \quad \text{and} \quad \begin{pmatrix} c \\ a\text{-}b \end{pmatrix}$$

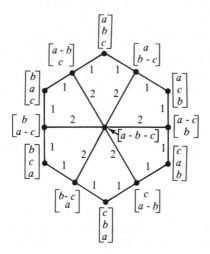

FIGURE 7.6. Distances between rankings of 3 objects. The distance between P and Q is the smallest sum of weights on a chain joining P and Q. Reprinted from Kemeny and Snell [1972] with permission of M.I.T. Press.

is 5. One chain the sum of whose distances is 5 is given by

$$\begin{pmatrix} a \\ b \\ c \end{pmatrix} \quad \text{to} \quad \begin{pmatrix} a \\ b\text{-}c \end{pmatrix} \quad \text{to} \quad (a\text{-}b\text{-}c) \quad \text{to} \quad \begin{pmatrix} c \\ a\text{-}b \end{pmatrix}.$$

7.4.3. MEDIANS AND MEANS

Once we have a way to measure distance between two rankings, we can use it to obtain a group consensus function. If $(P_1, P_2, \ldots, P_t)$ is a profile of rankings, let us define a *median ranking* for the profile as a ranking P such that $\sum_{i=1}^{t} d(P, P_i)$ is a minimum and a *mean ranking* as a ranking P such that $\sum_{i=1}^{t} d(P, P_i)^2$ is a minimum. It is tempting to define the group consensus as either a median or a mean. Both median and mean are measures of "centrality" or "central tendency" in statistics, and indeed any notion of centrality could be used instead of median or mean.

Let us give some examples. Suppose the profile is

P_1	P_2	P_3
Ford	Ford	Chevy
Chevy	Chevy	Ford
Plymouth	Plymouth	Plymouth

Using Fig. 7.6 to calculate distances, we see that a median is the ranking

P
Ford
Chevy
Plymouth

For $\sum d(P, P_i) = 0 + 0 + 2 = 2$, while from any other ranking Q, $\sum d(Q, P_i) \geqq 3$. To show this, observe first that if Q is P_3, then $\sum d(Q, P_i) = 2 + 2 + 0 = 4$. Second, if Q is not P_1, P_2, or P_3, then $d(Q, P_i) \geqq 1$ for all i, by Axiom 4. The median here gives as consensus the ranking which two of the three experts gave. In this same example, a mean is given by

$$Q$$

Ford-Chevy
Plymouth

For $\sum d(Q, P_i)^2 = 1^2 + 1^2 + 1^2 = 3$. This is easily seen to be minimal, for any other ranking Q must have either $d(Q, P_1) \geqq 2$ or $d(Q, P_3) \geqq 2$, and so $\sum d(Q, P_i)^2 \geqq 4$. Thus, the mean is different from the median, and simply gives a tie between Ford and Chevy while putting Plymouth least. Both procedures seem reasonable: The first accepts the majority and the second reports the evidence as inconclusive. Thus, there will be a problem in trying to decide which procedure gives the more reasonable group consensus function, and this is a question which is not decided by the axioms for distance.

There is another serious problem with using the mean or median to define a group consensus function. Namely, the mean or median need not be unique. Consider for example the case of the profile

P_1	P_2	P_3
Ford	Chevy	Plymouth
Chevy	Plymouth	Ford
Plymouth	Ford	Chevy

It turns out that there are three medians, the rankings P_1, P_2, and P_3. The mean is unique here, and is the ranking Ford-Chevy-Plymouth, i.e., a tie among all three elements. Again, both choice of median and choice of mean seem reasonable procedures: The first says "take your pick" of the experts and the second says the evidence suggests there is no difference among the elements. Unfortunately, if the median is not unique, then it cannot be used to define a group consensus function in our sense: We require choice of a unique consensus ranking. However, in this case where the profile is completely symmetric, it is just as natural to list several rankings as to list one.

Means also need not be unique, as can be seen by considering the profile

P_1	P_2
Ford	Plymouth
Chevy	Chevy
Plymouth	Ford

The means are the rankings

Chevy	Ford-Plymouth	Ford-Chevy-Plymouth
Ford-Plymouth	Chevy	

As mentioned in our discussion of Arrow's Theorem, a reasonable approach at this point is to relieve the requirements on a group consensus function. If we do not require a selection of a unique consensus ranking, then the median or mean give perfectly reasonable group consensus functions. However, the problem of deciding between these two remains.

7.4.4. REMARKS

1. One of the problems with applying the Kemeny-Snell method is that there is no efficient procedure known for finding a median or a mean from a given profile. It is always possible to make calculations of $\sum d(Q, P_i)$ or $\sum d(Q, P_i)^2$ for each possible Q. But this can become very time-consuming. Sometimes, as in our first example, it is clear that some Q's will give smaller values of $\sum d$ or $\sum d^2$ than others. This observation can then be used to eliminate many of the computations.

2. The calculation of $d(P, Q)$ can often be reduced to the calculation of distances between rankings which disagree on a small set of elements. Let O be the ranking in which all elements are tied, and suppose that O is between P and Q. Then $d(P, Q) = d(P, O) + d(O, Q) = d(P, O) + d(Q, O)$. Thus, it is sufficient to calculate $d(P, O)$ and $d(Q, O)$. We show how to calculate the former. If P has no tied elements, then $d(P, O) = n(n - 1)/2$, where n is the number of elements in A. (See Exer. 5.) Thus, let us assume that P has tied elements. Let $P_0 = P$, and define P_1 from P_0 by choosing the first set of tied elements and ordering them arbitrarily. Repeat the process to define P_2, P_3, etc., until a ranking P_r with no ties is obtained. Then it is easy to prove that

$$d(P, O) = \frac{n(n - 1)}{2} - d(P_0, P_1) - d(P_1, P_2) - \cdots - d(P_{r-1}, P_r). \quad (6)$$

(The proof is left as Exer. 22.) Calculation of $d(P_i, P_{i+1})$ is simple since P_i and P_{i+1} differ only on the set S of elements whose ties in P_i were broken to get P_{i+1}. Thus,

$$d(P_i, P_{i+1}) = \frac{s(s - 1)}{2},$$

where $s = |S|$. (Why?) We conclude that

$$d(P, O) = \frac{n(n - 1)}{2} - \sum_S \frac{s(s - 1)}{2}, \quad (7)$$

where the sum is taken over all sets S of tied elements in P.

To illustrate the procedure, suppose $P = P_0$ is

$$x\text{-}y\text{-}z$$
$$u\text{-}v$$
$$a$$
$$b\text{-}c\text{-}d$$

Then there are three sets of tied elements in P, $\{x, y, z\}$, $\{u, v\}$, and $\{b, c, d\}$. Hence,

$$d(P, O) = \frac{n(n-1)}{2} - \frac{3(3-1)}{2} - \frac{2(2-1)}{2} - \frac{3(3-1)}{2}$$

$$= \frac{9(8)}{2} - \frac{3(2)}{2} - \frac{2(1)}{2} - \frac{3(2)}{2}$$

$$= 29.$$

The rankings P_1, P_2, and P_3 are as follows:

P_1	P_2	P_3
x	x	x
y	y	y
z	z	z
$u\text{-}v$	u	u
a	v	v
$b\text{-}c\text{-}d$	a	a
	$b\text{-}c\text{-}d$	b
		c
		d

If Q is the ranking

$$a\text{-}b\text{-}c\text{-}d$$
$$u\text{-}v\text{-}x$$
$$y$$
$$z,$$

a similar computation shows that $d(Q, O) = 27$. Since O is between P and Q, we have $d(P, Q) = d(P, O) + d(Q, O) = 29 + 27 = 56$.

3. The axiomatization has really not put the entire procedure on firm axiomatic grounds. For, there is nothing in the axioms which has explained why we should use something like mean or median to find consensus.

4. There is a subtle assumption hidden in the axioms. Namely, the axioms for distance depend on our definition of betweenness. Given a notion of betweenness, the axioms seem reasonable. But why should this particular definition of betweenness be adopted? We have accepted something rather than putting it on a firm axiomatic foundation. (We ran across a similar problem in axiomatizing a measure of trophic status in Sec. 3.5.2.) It is often

possible to have such hidden assumptions when we use words with common connotations. Other definitions of betweenness could give rise to other ways of measuring distance.

5. If we worry about the uniqueness of means and medians, one approach is to weaken the restrictions on the type of order relations which the rankings are required to be. In this section, we have taken a ranking to be what we shall call in Chapter 8 a strict weak ordering. Bogart [1973] has recently shown that if rankings are only so-called strict partial orders, then a set of axioms analogous to Axioms 1–4 can be stated which again give rise to a unique distance function. The mean and median may now be a strict partial order as well. Bogart [1975] has obtained a similar uniqueness result if the rankings are only required to be asymmetric relations, i.e., only required to satisfy the condition that if aPb, then not bPa. Here, with respect to the distance function obtained, the median in the space of asymmetric relations is essentially unique (Bogart [1975, Theorem 4]). Moreover, it is given by the simple majority rule ordering, i.e., that asymmetric ordering M such that

$$aMb \quad \text{iff} \quad aP_ib \quad \text{by a majority of} \quad i.$$

This result specializes to the Kemeny-Snell context as follows. If the simple majority rule applied to a profile of rankings (rankings in the sense of this chapter) gives rise to a ranking, then this ranking is a median in the space of all asymmetric relations and it is the unique median unless for some a and b, the number of individuals preferring a to b is equal to the number preferring b to a.[17]

6.[18] A common notion in social science and statistics is that of correlation between two sets of data. A *correlation coefficient* is a function which assigns to two sets of data a number between -1 and $+1$. The larger the absolute value of this number, the more predictable one set of data is from the other. A positive correlation means that the data reflect each other, a negative correlation means that the data are opposite. A classical coefficient of correlation is *Kendall's tau*, which measures the correlation between two rankings P and Q of the same set of objects. Kendall's tau is given by

$$\tau(P, Q) = 1 - \frac{2d(P, Q)}{n(n-1)},$$

where d is the distance function we have been using and n is the cardinality of A. To understand this a little better, let us note that this is really just a "normalization" of $d(P, Q)$ to a number between 0 and 1 and then a shift to a scale $[-1, 1]$. For,

---

[17]The author thanks Professor Bogart for pointing this out to him.

[18]This remark is due to Bogart [1973].

$$\tau(P, Q) = 1 - 2\left[\frac{d(P, Q)}{d_{\max}}\right],$$

where $d_{\max}$ is the maximum distance between two rankings of A. It is left to the reader to prove that $d_{\max}$ is indeed $n(n - 1)$ (Exer. 17). Thus, Axioms 1–4 may be taken as a theoretical foundation of this standard rank correlation procedure.

EXERCISES

1. Suppose $A = \{\text{Ham, Bologna, Salami, Roast Beef}\}$. Consider the rankings

P	Q	R
Ham	Ham	Salami
Salami	Bologna-Roast Beef	Ham-Bologna-Roast Beef
Bologna	Salami	
Roast Beef		

   (a) Using the distance measure derived in this section, calculate $d(P, Q)$, $d(Q, R)$, and $d(P, R)$.
   (b) *Using Axiom 1.3*, conclude whether or not Q is between P and R.
   (c) Check your conclusion in part (b) by using the definition of betweenness.

2. Repeat Exer. 1 for the following rankings on
   $A = \{\text{Hemingway, Faulkner, Shakespeare, Milton}\}$.

P	Q	R
Hemingway	Hemingway-Faulkner	Milton
Faulkner	Shakespeare-Milton	Shakespeare
Shakespeare		Faulkner
Milton		Hemingway

3. Suppose $d(P_i, P_j)$ is given by the i, j entry of the following table.

$$\begin{array}{c} \begin{array}{cccc} P_1 & P_2 & P_3 & P_4 \end{array} \\ \begin{array}{c} P_1 \\ P_2 \\ P_3 \\ P_4 \end{array} \left(\begin{array}{cccc} 0 & 6 & 1 & 3 \\ 6 & 0 & 5 & 5 \\ 1 & 5 & 0 & 4 \\ 3 & 5 & 4 & 0 \end{array}\right). \end{array}$$

(a) Find a ranking P_i which is a median if the set of all possible rankings is $\{P_1, P_2, P_3, P_4\}$.
(b) Is the median unique?
(c) Find a ranking P_j which is a mean.
(d) Is the mean unique?

4. If P is a ranking of A without ties, let P^c be the reverse ranking. Show that $d(P, P^c) = n(n - 1)$, where n is the number of elements in A.

5. If P is a ranking of A without ties and O is the ranking of A in which all elements are tied, show that $d(P, O) = n(n - 1)/2$, where n is the number of elements of A.

6. Apply Axiom 2 to the following rankings.

P	Q	R	S
x	z	x-z	z-y
y	x	y	x
z	y		

(a) Can you conclude that $d(P, Q) = d(R, S)$? Why?
(b) Can you conclude that $d(P, R) = d(Q, S)$?
(c) Can you conclude that $d(P, S) = d(Q, R)$?

7. Repeat Exer. 6 with the following rankings. (*Note*: In a relabelling, not every element has to change its label.)

P	Q	R	S
x	x-y-z	y	x-y-z
y		x	
z		z	

8. Consider the following rankings. Use Axiom 3 to draw a conclusion.

P	Q	P'	Q'
Cadillac	Cadillac	Fiat	Fiat
Ford	Ford	Ford	Ford
Plymouth	Plymouth-Mercury	Plymouth-Mercury	Plymouth
Mercury	Fiat	Cadillac	Mercury
Fiat	Toyota	Toyota	Cadillac
Toyota			Toyota

9. For the ranking Q of Exer. 8, which of the following sets form segments?

  (a) $S = \{$Cadillac, Ford$\}$.
  (b) $S = \{$Ford, Plymouth$\}$.
  (c) $S = \{$Plymouth, Mercury, Fiat$\}$.

10. For the ranking P' of Exer. 8, find $\bar{S}(P')$, $\underline{S}(P')$, $P'(S)$, $P'(\bar{S})$, and $P'(\underline{S})$ for each of the following segments S.
  (a) $S = \{$Ford, Plymouth, Mercury$\}$.
  (b) $S = \{$Ford$\}$.
  (c) $S = \{$Plymouth, Mercury, Cadillac$\}$.

11. Suppose $A = \{x, y, z\}$. Using the distances calculated in Fig. 7.6, find all means and medians of the following profile:

P_1	P_2	P_3
x	x-y-z	z
y		y
z		x

12. Suppose $A = \{x, y, z\}$. Using the distances calculated in Fig. 7.6, find all means and medians of the following profile:

P_1	P_2	P_3	P_4	P_5
x-y-z	x-y-z	x-y-z	x-y-z	x
				y
				z

13. Suppose $A = \{$Ham, Bologna, Salami$\}$. Using the distances calculated in Fig. 7.6, find all means and medians of the following profile:

P_1	P_2	P_3
Ham	Ham	Ham
Bologna	Salami	Bologna-Salami
Salami	Bologna	

14. Give an example of a profile with $|A| = 4$ which does not have a unique median.

15. Give an example of a profile with $|A| = 4$ which does not have a unique mean.

16. Is it possible to find two rankings P and R so that the only rankings Q which are between P and R are P and R themselves? (Give proof or counterexample.)

17.   If $|A| = n$ and $d_{max} = \max d(P, Q)$, where the max is taken over all rankings P and Q of A, show that $d_{max} = n(n - 1)$.

18.   Let P and Q be the following rankings and let O be the ranking in which all elements are tied.

P	Q
a-b-c	h-i
d	g
e-f	e-f
g-h-i	d
	a-b-c

(a)   Show that $B(P, O, Q)$, i.e., O is between P and Q.
(b)   Calculate $d(P, Q)$ using the formula of Eq. (7).

19.   (Kemeny and Snell [1962].) Define a matrix (a_{ij}) by listing elements of A arbitrarily as $a_1, a_2, \ldots, a_n$ and taking $a_{ij} = \delta^P(a_i, a_j)$, where δ^P was defined in Eq. (3). The matrices (a_{ij}) satisfy the following conditions for all i, j:

   (i) $a_{ij} = +1, -1,$ or 0.
   (ii) $a_{ij} = -a_{ji}$.
   (iii) If $a_{ij} \geqq 0$ and $a_{jk} \geqq 0$, then $a_{ik} \geqq 0$; $a_{ik} = 0$ only if $a_{ij} = 0$ and $a_{jk} = 0$.

A matrix satisfying conditions (i)–(iii) is called an *ordering matrix*. Show that every ordering matrix satisfies the following conditions. (*Note*: You may use only conditions (i)–(iii) in your proof.)

(a)   If $a_{ij} = +1$ and $a_{jk} = +1$, then $a_{ik} = +1$.
(b) $a_{ii} = 0$.
(c)   If $a_{ij} = 0$, then $a_{ji} = 0$.
(d)   If $a_{ij} = +1$ and $a_{jk} = 0$, then $a_{ik} = +1$.
(e)   If $a_{ij} = 0$ and $a_{jk} = +1$, then $a_{ik} = +1$.

20.   We say that the rankings $P_1, P_2, \ldots, P_r$ *lie on a line* if for all $i < j < k$, $B(P_i, P_j, P_k)$, i.e., P_j is between P_i and P_k. Show *from the axioms* that if $P_1, P_2, \ldots, P_r$ lie on a line, then

$$d(P_1, P_r) = d(P_1, P_2) + d(P_2, P_3) + \cdots + d(P_{r-1}, P_r).$$

21.   (a)   Prove that if δ^P is defined as in Eq. (3), then Q is between P and R if and only if for all a, b in A,

$$\delta^P(a, b) \leqq \delta^Q(a, b) \leqq \delta^R(a, b)$$

       or

$$\delta^P(a, b) \geqq \delta^Q(a, b) \geqq \delta^R(a, b).$$

      (b)   Conclude that equality in Eq. (4) for each a, b holds if and only if Q is between P and R.

22. Prove that Eq. (6) holds (*Hint*: Show that $O, P_0, P_1, \ldots, P_r$ lie on a line in the sense of Exer. 20.)

23. (Kemeny and Snell [1962].) Draw a figure analogous to Fig. 7.6 for all rankings of a set of four alternatives. (*Hint*: O can be taken as the center of a semiregular solid, a solid bounded by two kinds of regular polygons, and the other points can be located on the surface of the solid. Draw only the surface.)

24. Give a proof or a counterexample for each of the following statements:
    (a) If $B(P_i, P_{i+1}, P_{i+2})$ for all $i = 1, 2, \ldots, r - 2$, then $P_1, P_2, \ldots, P_r$ lie on a line (Exer. 20).
    (b) If $B(P_i, P_{i+1}, P_{i+2})$ for all $i = 1, 2, \ldots, r - 2$, then $d(P_1, P_r) = d(P_1, P_2) + d(P_2, P_3) + \cdots + d(P_{r-1}, P_r)$.
    (c) If $d(P_1, P_r) = d(P_1, P_2) + d(P_2, P_3) + \cdots + d(P_{r-1}, P_r)$, then $B(P_i, P_j, P_k)$ for all $i < j < k$.

25. (Kemeny and Snell [1962].)
    (a) Suppose $\mathcal{P} = (P_1, P_2, P_3)$ and P_2 is between P_1 and P_3. Prove that P_2 is the unique median of $\mathcal{P}$.
    (b) Can you extend this result to the case of five individuals?
    (c) What about four individuals? (Is there a similar result?)

26. Suppose $R_1, R_2,$ and R_3 are the following rankings on $A = \{a, b\}$.

R_1	R_2	R_3
a	b	a-b
b	a	

    (a) Prove *from the axioms* that if d is the distance measure on $\hat{\mathcal{P}}(A)$ and $d(R_1, R_3) = \alpha$, then distance is given by the following matrix.

    $$\begin{array}{c} \\ R_1 \\ R_2 \\ R_3 \end{array} \begin{array}{ccc} R_1 & R_2 & R_3 \\ \begin{pmatrix} 0 & 2\alpha & \alpha \\ 2\alpha & 0 & \alpha \\ \alpha & \alpha & 0 \end{pmatrix} \end{array}.$$

    (b) Furthermore, conclude that $\alpha = 1$.

27. Illustrate how sophisticated voting (Exer. 19, Sec. 7.1) can change the median or mean of a profile.

28. Discuss possible situations in which you might not want Axiom 3 to hold.

29. Think of other ways of measuring distance between two rankings, and see if they satisfy Kemeny and Snell's axioms.

References

ARROW, K., *Social Choice and Individual Values,* Cowles Commission Monograph 12, John Wiley & Sons, Inc., New York, 1951; second edition, 1963.

BLACK, D., "On the Rationale of Group Decision-making," *J. of Political Economy,* **56** (1948), 23–24, a.

BLACK, D., "The Decisions of a Committee Using a Special Majority," *Econometrica,* **16** (1948), 245–261, b.

BLACK, D., *The Theory of Committees and Elections,* Cambridge University Press, London, 1958.

BOGART, K., "Preference Structures I," *J. Math. Sociology,* **3** (1973), 49–67.

BOGART, K., "Preference Structures II," *SIAM J. Appl. Math.,* **29** (1975), 254–262.

COOMBS, C. H., "Social Choice and Strength of Preference," in *Decision Processes,* R. M. Thrall, C. H. Coombs, and R. L. Davis (eds.), John Wiley & Sons, Inc., New York, 1954, pp. 69–86.

FARQUHARSON, R., *Theory of Voting,* Yale University Press, New Haven, Conn., 1969.

FISHBURN, P. C., *The Theory of Social Choice,* Princeton University Press, Princeton, N.J., 1972.

GOODMAN, L. A., "On Methods of Amalgamation," in *Decision Processes,* R. M. Thrall, C. H. Coombs, and R. L. Davis (eds.), John Wiley & Sons, Inc., New York, 1954, pp. 39–48.

GOODMAN, L. A., and MARKOWITZ, H., "Social Welfare Functions Based on Individual Rankings," *Amer. J. of Sociology,* **58** (1952), 257–262.

KEMENY, J. G., and Snell, J. L., *Mathematical Models in the Social Sciences,* Blaisdell, New York, 1962; reprinted by M.I.T. Press, Cambridge, Mass., 1972.

LUCE, R. D., and Raiffa, H., *Games and Decisions,* John Wiley & Sons, Inc., New York, 1957.

MALKEVITCH, J., and MEYER, W., *Graphs, Models, and Finite Mathematics,* Prentice-Hall, Inc., Englewood Cliffs, N.J., 1974.

NIEMI, R. G., and WEISBERG, H. F., "A Mathematical Solution for the Probability of the Paradox of Voting," *Behav. Sci.,* **13** (1968), 317–323.

NORMAN, R. Z., and ROBERTS, F. S., "A Derivation of a Measure of Relative Balance for Social Structures and a Characterization of Extensive Ratio Systems," *J. Math. Psychol.,* **9** (1972), 66–91.

RIKER, W. H., and ORDESHOOK, P. C., *Positive Political Theory,* Prentice-Hall, Inc. Englewood Cliffs, N.J., 1973.

SEN, A. K., *Collective Choice and Social Welfare,* Holden-Day, San Francisco, Calif., 1970.

EIGHT

Measurement
and Utility

8.1. Introduction

In the previous chapter, we studied a *group* of individuals, each of whom expressed preferences among a collection of alternatives. In this chapter, we consider the notion of preference from the *individual* point of view. Specifically, we shall study what it means to *measure* an individual's preference. We shall introduce the concept of a utility function as a means for measuring preference. For a more detailed treatment of utility theory, the reader is referred to Fishburn [1968, 1970c] or Luce and Suppes [1965], which also contain references to other surveys of the literature.

More generally, we shall study what measurement in many scientific contexts means. In Sec. 3.1 we discussed the idea that scientific progress often follows when imprecise concepts or relations are made precise. Frequently one can make such an imprecise relation precise by "measuring" it. In this chapter, we shall try to understand what measurement at such a fundamental level means. Indeed, we shall try to make precise a theory of measurement. We shall apply the results to measurement of preference, to measurement of loudness, to mental testing, to comparisons using the consumer price index, to analysis of data from pair comparison experiments, and to a variety of other problems.

One can distinguish two basic types of measurement, *fundamental measurement* and *derived measurement*. Fundamental measurement, as we shall

473

describe it, takes place at an early stage of scientific development, when several fundamental concepts are measured for the first time. Examples of such fundamental concepts in physics are mass and volume. Derived measurement takes place later, when some concepts have already been measured, and new measures are defined in terms of existing ones. For example, density is defined as mass divided by volume. We shall deal mostly with fundamental measurement,[1] though many of our ideas are applicable to derived measurement as well.

The reader should keep in mind that as a scientific discipline develops, fundamental measurement is not usually performed in as formalistic a way as we shall describe. Our discussion outlines an attempt to put measurement on a firm foundation, rather than to describe actual processes of measurement.

Our approach to fundamental measurement follows very closely that of Scott and Suppes [1958], Suppes and Zinnes [1963], Krantz, *et al.* [1971], and Roberts [to appear]. The reader might wish to consult some of the extensive literature on the nature of measurement, for other points of view. Some references are Adams [1965], Campbell [1920, 1928], Cohen and Nagel [1934], Ellis [1966], Helmholtz [1887], Pfanzagl [1968], Reese [1943], or Stevens [1946, 1951, 1959, 1968].

8.2. Relations

In this section, we shall present a mathematical topic which will be used freely in the remainder of the chapter. For a more thorough discussion of this topic, the reader is referred to Roberts [to appear] or Suppes [1957].

8.2.1. DEFINITION OF RELATION

Suppose D is a digraph, with A as the set of vertices and U as the set of arcs. Then U is a set of ordered pairs from A, i.e., a subset of the Cartesian product $A \times A$. Such a set of ordered pairs from a set A is often called a *binary relation*. Thus, every digraph defines a binary relation and every binary relation defines a digraph. The difference between the theory of digraphs and the theory of relations is primarily one of emphasis. The former often deals with structure, and in many ways can be classified as geometric. The latter might be classified as algebraic. Also, in digraphs, the set A is often assumed to be finite, while in binary relations, most of the concepts developed apply to infinite sets as well as finite ones. In addition to binary relations, one may also speak of *n*-ary relations, where n is a positive integer. An *n-ary*

---

[1]For a treatment of derived measurement, see Suppes and Zinnes [1963] or Roberts [to appear].

relation is a set of ordered *n*-tuples from a set *A*, i.e., a subset of the Cartesian product $A \times A \times \cdots \times A$, with *n* factors.

Usually, we shall denote relations by letters like *P*, *R*, *S*, *T*, etc. Examples of binary relations on the set $A = \{1, 2, 3, 4, 5\}$ are

$$R = \{(1, 1), (2, 2), (1, 2), (2, 1)\} \tag{1}$$

and

$$S = \{(1, 2), (2, 3), (3, 4), (4, 5), (1, 3), (1, 4), (1, 5), (2, 4), (2, 5), (3, 5)\}. \tag{2}$$

(Since binary relations correspond to digraphs, it is sometimes helpful to draw the corresponding digraph; we do so for these two binary relations in Figs. 8.1 and 8.2.) If $A = \{1, 2, 3, 4, 5\}$, a 3-ary or *ternary* relation *R* on *A* is given by

$$R = \{(1, 1, 1), (1, 2, 1), (2, 5, 5), (4, 5, 3)\} \tag{3}$$

and a 4-ary or *quaternary* relation *S* on *A* is given by

$$S = \{(1, 1, 1, 1), (2, 2, 2, 2), (3, 4, 5, 5), (3, 5, 5, 4), (1, 2, 2, 1)\}. \tag{4}$$

FIGURE 8.1. Digraph corresponding to the binary relation (A, R), where $A = \{1, 2, 3, 4, 5\}$ and *R* is defined by Eq. (1).

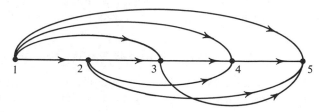

FIGURE 8.2. Digraph corresponding to the binary relation (A, S), where $A = \{1, 2, 3, 4, 5\}$ and *S* is defined by Eq. (2).

Everyday language is a rich source of relations. For example, if *A* is the set of people in the United States, then

$$S = \{(a, b): a \in A \text{ and } b \in A \text{ and } a \text{ is the sister of } b\} \tag{5}$$

defines a binary relation on *A*. We might call this relation "sister of." Similarly,

$$L = \{(a, b): a \in A \text{ and } b \in A \text{ and } a \text{ loves } b\}$$

defines a binary relation on A. If B is a set of alternative candidates in an election,

$$P = \{(a, b): a \in B \text{ and } b \in B \text{ and you strictly prefer } a \text{ to } b\}$$

defines a binary relation on B called preference. A quaternary relation S on B is given by

$$S = \{(a, b, c, d): a, b, c, d \in B \text{ and you prefer } a$$
$$\text{to } b \text{ at least as much as you prefer } c \text{ to } d\}.$$

Finally, if C is the set of all students in a given school, then a ternary relation R on C can be defined as follows:

$$R = \{(a, b, c): a, b, c \in C \text{ and } b\text{'s grade point}$$
$$\text{average is strictly between those of } a \text{ and } c\}.$$

The reader should be cautioned that a relation is not properly defined without giving its underlying set. Thus, if B is the set of all women in the United States, then

$$S' = \{(a, b): a \in B \text{ and } b \in B \text{ and } a \text{ is the sister of } b\}$$

is considered different from the relation S defined on the set A of all people in the United States in Eq. (5). To emphasize the dependence on the underlying set, we shall usually speak of a relation (A, R), rather than of a relation R.

Suppose (A, R) is a binary relation. We shall write aRb for the statement $(a, b) \in R$. Thus, if R is defined by Eq. (1), we have $1R1$ and $1R2$, but not $1R3$. If (A, R) is a ternary relation, we shall write $R(a, b, c)$ if $(a, b, c) \in R$. Analogous notation will be used for n-ary relations with $n > 3$.

8.2.2. PROPERTIES OF BINARY RELATIONS

Many binary relations share certain common properties, and we next mention some of these properties. For convenience, they are summarized in Table 8.1. A binary relation (A, R) is *reflexive* if for all $a \in A$, aRa. To give an example, if A is the set $\Re$ of real numbers and R is equality, then (A, R) is reflexive. If A is $\{1, 2, 3, 4, 5\}$ and R is defined by Eq. (1), then (A, R) is not reflexive, since it is not the case that $3R3$. On the other hand, if R is thought of as a relation defined on the set $\{1, 2\}$, then R is reflexive. This is one reason why it is important to speak of the underlying set.

If A is the set of people in the United States and S is the relation "sister of" on A, then (A, S) is not reflexive: A person is not her own sister. Indeed, for every a in A, we have $\sim aSa$, i.e., aSa does not hold. Such a binary relation

TABLE 8.1

Properties of Binary Relations.

A binary relation (A, R) is	provided that
reflexive	aRa, all $a \in A$
nonreflexive	it is not reflexive*
irreflexive	$\sim aRa$, all $a \in A$
symmetric	$aRb \Rightarrow bRa$, all $a, b \in A$
nonsymmetric	it is not symmetric
asymmetric	$aRb \Rightarrow \sim bRa$, all $a, b \in A$
antisymmetric	aRb & $bRa \Rightarrow a = b$, all $a, b \in A$
transitive	aRb & $bRc \Rightarrow aRc$, all $a, b, c \in A$
nontransitive	it is not transitive
negatively transitive	$\sim aRb$ & $\sim bRc \Rightarrow \sim aRc$, all $a, b, c \in A$; equivalently: $xRy \Rightarrow (xRz$ or $zRy)$, all $x, y, z \in A$.

*The reader should think of examples of nonreflexive binary relations which are not irreflexive.

is called *irreflexive*. The binary relation (A, S), where $A = \{1, 2, 3, 4, 5\}$ and S is defined by Eq. (2), is also irreflexive.

A binary relation (A, R) is called *symmetric* if whenever aRb, then bRa. The relation "equality" on the set $\Re$ is symmetric. So is the relation (A, R), where $A = \{1, 2, 3, 4, 5\}$ and R is defined by Eq. (1). The relation "sister of" on the set of all people in the United States is not symmetric, though the relation "sister of" on the set of all women in the United States is. Unfortunately, the relation "loves" is probably not symmetric.

The binary relation (A, S), where $A = \{1, 2, 3, 4, 5\}$ and S is given by Eq. (2), is highly nonsymmetric: For all a, b in A, if aSb, then $\sim bSa$, i.e., bSa does not hold. Such a binary relation is called *asymmetric*. Another example of an asymmetric binary relation is $(\Re, >)$. The rankings studied in Chapter 7 were assumed to be asymmetric. Probably, the relation of strict preference in general is also asymmetric.

The binary relation $(\Re, \geqq)$ is not asymmetric. Take $a = b = 2$ and observe that since $2 \geqq 2$, asymmetry would imply $\sim(2 \geqq 2)$. This binary relation is almost asymmetric, in the sense that whenever aRb and bRa, then a must equal b. A binary relation with this property is called *antisymmetric*. The binary relation $\subseteq$ on any collection of sets gives another example of an antisymmetric relation.

We have at various places in this book referred to transitive binary relations. (A, R) is *transitive* if whenever aRb and bRc hold, then aRc holds as well. The binary relation $(\Re, \geqq)$ is transitive, as is the binary relation $(\Re, =)$. So is the binary relation (A, S), where $A = \{1, 2, 3, 4, 5\}$ and S is defined by Eq. (2). (For S is simply $<$ on A.) It seems reasonable to assume that strict preference is transitive: If you prefer a to b and b to c, you should prefer a to c. However, we shall present arguments against this point later.

A binary relation is called *negatively transitive* if whenever $\sim aRb$ and $\sim bRc$ hold, then $\sim aRc$ holds as well. The binary relation $(\Re, >)$ is negatively transitive, for if $\sim (a > b)$ and $\sim (b > c)$, then $a \leqq b$ and $b \leqq c$, so $a \leqq c$, so $\sim (a > c)$. In general, (A, R) is negatively transitive if the relation "not in R," defined on the set A, is transitive. If A is $\{1, 2, 3, 4, 5\}$ and S is defined by Eq. (2), then (A, S) is negatively transitive. For the relation "not in S" is simply the relation $\geqq$ on A. It is easy to prove that (A, R) is negatively transitive if and only if whenever xRy, then for any z in A, either xRz or zRy. (This statement $\mathfrak{I}$ is just the contrapositive of the statement $\mathfrak{S}$ used to define negative transitivity. For $\mathfrak{S}$ is the statement "X implies Y" and $\mathfrak{I}$ is the statement "not Y implies not X.") Thus, we see that the binary relation $\subseteq$ on a collection of sets is not necessarily negatively transitive. For it might very well be that while $x \subseteq y$, there is a z so that $x \nsubseteq z$ and $z \nsubseteq y$.

8.2.3. OPERATIONS

Turning from binary relations to ternary relations, we note that operations like addition of numbers define ternary relations. The operation of addition on the set of real numbers can be thought of as corresponding to a ternary relation $\oplus$ on $\Re$. $\oplus$ is defined as

$$\{(a, b, c): a, b, c \in \Re \text{ and } c = a + b\}.$$

Thus, $(4, 5, 9) \in \oplus$ and $(6, 8, 14) \in \oplus$, but $(3, 5, 9) \notin \oplus$. Similarly, multiplication defines a ternary relation $\otimes$ on $\Re$ defined as

$$\{(a, b, c): a, b, c \in \Re \text{ and } c = a \times b\}.$$

The relation $\oplus$ has the following properties: For all $a, b \in \Re$, there is $c \in \Re$ such that $(a, b, c) \in \oplus$; and if $(a, b, c) \in \oplus$ and $(a, b, d) \in \oplus$, then $c = d$. The relation $\otimes$ has similar properties. Generalizing, we shall call a ternary relation (A, o) a *(binary) operation* if

    (a)   for all $a, b \in A$, there is $c \in A$ such that $(a, b, c) \in \mathrm{o}$

and if

    (b)   for all $a, b, c, d \in A$, if $(a, b, c) \in \mathrm{o}$ and $(a, b, d) \in \mathrm{o}$, then $c = d$.

If (A, o) is an operation and $(a, b, c) \in \mathrm{o}$, we usually write $c = a \, \mathrm{o} \, b$.

To give an example, suppose we define o on $\Re$ by

$$\mathrm{o}(a, b, c) \quad \text{iff} \quad c = \frac{a}{b}.$$

Then there is no c such that o$(1, 0, c)$, and so (A, o) is not an operation. If o is thought of as a ternary relation on the *positive* reals, however, then o is an operation. To give another example, suppose we define o on $\mathfrak{R}$ by

$$\text{o}(a, b, c) \quad \text{iff} \quad c^2 = a \times b.$$

Then for every $a, b \in \mathfrak{R}$, there is $c \in \mathfrak{R}$ such that o(a, b, c). However, $(1, 1, 1)$ and $(1, 1, -1)$ are both in o, which violates property (b) of operations. Next, suppose we define o on $\mathfrak{R}$ by

$$\text{o}(a, b, c) \quad \text{iff} \quad c = 2\,(a \times b).$$

Then o is an operation on $\mathfrak{R}$. We have $6 = 1 \text{ o } 3$ and $16 = 4 \text{ o } 2$. Finally, suppose $A = \{1, 2, 3\}$ and o is defined on A as follows:

$$\text{o} = \{(1, 1, 0), (1, 2, 1), (1, 3, 1), (2, 1, 1), (2, 2, 0), (2, 3, 1),$$
$$(3, 1, 1), (3, 2, 1), (3, 3, 0)\}.$$

Then o is an operation on A; $a \text{ o } b$ is 1 if $a \neq b$ and 0 if $a = b$. However, if o is defined as

$$\text{o} = \{(1, 1, 0), (1, 2, 1), (1, 3, 1), (2, 1, 1), (2, 2, 0), (2, 3, 1),$$
$$(3, 1, 1), (3, 2, 1), (3, 3, 0), (1, 1, 1)\},$$

then o is not an operation on A, for o$(1, 1, 0)$ and o$(1, 1, 1)$.

EXERCISES

1. Write out as a set of ordered pairs the binary relation $>$ on the set $A = \{1, 6, 8\}$.

2. Suppose $A = \{1, 2, 3, 4, 5\}$ and S is defined on A by Eq. (4). Write out as a set of ordered n-tuples the following relations:
   (a) The ternary relation S' on A defined by $(a, b, c) \in S'$ iff for some d in A, $(a, b, c, d) \in S$.
   (b) The binary relation S'' on A defined by $aS''b$ iff for some s and t in A, $(a, s, t, b) \in S$.
   (c) The quaternary relation S''' on A defined by $(a, b, s, t) \in S'''$ iff $(t, s, b, a) \in S$.

3. A binary relation on a set of real numbers, being a subset of the Cartesian product, may be plotted as a collection of points using Cartesian coordinates. Plot the binary relations (A, R) and (A, S), where $A = \{1, 2, 3, 4, 5\}$ and R and S are given by Eqs. (1) and (2) respectively.

4. If (A, R) is a binary relation, define the *converse* relation $\check{R}$ on A by

$$a\check{R}b \quad \text{iff} \quad bRa.$$

State in English or write out as a set of ordered pairs the converse of the following relations:

(a)   The relation (A, R) where $A = \{1, 2, 3, 4, 5\}$ and R is given by Eq. (1).

(b)   The relation (A, S) where $A = \{1, 2, 3, 4, 5\}$ and S is given by Eq. (2).

(c)   The relation "father of."

(d)   The relation "aunt."

(e)   The relation $>$ on a set of numbers.

5. (Thrall, *et al.* [1967].) In blood typing, there are four blood types, A, B, AB and O. In general, a person of type A can receive blood from persons of types A and O; a person of type B can receive blood from persons of types B and O; a person of type AB can receive blood from persons of all types; and a person of type O can receive blood only from persons of type O. Let $X = \{A, B, AB, O\}$ and define a binary relation S on X as follows: aSb iff a person of type a can receive blood from a person of type b.

(a)   Write out (X, S) as a set of ordered pairs.

(b)   What is the interpretation of the converse $(X, \check{S})$? (See Exer. 4.)

6. If (A, R) and (A, S) are binary relations, define $R \cap S$ on A as follows:

$$R \cap S = \{(a, b): aRb \text{ and } aSb\}.$$

This is the usual set-theoretical intersection. In the following, interpret $(A, R \cap S)$.

(a) (A, R) is "brother of" and (A, S) is "sibling of."

(b) (A, R) is "$\geq$" and (A, S) is "$\leq$."

(c) (A, R) is "parent of" and (A, S) is "ancestor of."

(d) aRb if and only if a is at least as qualified as b; and aSb if and only if a is no more qualified than b.

7. If (A, R) and (A, S) are binary relations, define $R \cup S$ on A as follows:

$$R \cup S = \{(a, b): aRb \text{ or } aSb\}.$$

This is the usual set-theoretical union. In the following, interpret $(A, R \cup S)$.

(a) (A, R) is "brother of" and (A, S) is "sister of."

(b) (A, R) is "father of" and (A, S) is "mother of."

(c) (A, R) is "$>$" and (A, S) is "$\geq$."

(d) (A, R) is "louder than" and (A, S) is "quieter than."

8. If (A, R) is "father of" and (A, S) is "mother of," what is $(A, (\overparen{R \cup S}))$, i.e., the converse of $(A, R \cup S)$? (See Exer. 4.)

9. If (A, R) and (A, S) are binary relations, define

$$R/S = \{(a, b): \text{for some } c \text{ in } A, aRc \text{ and } cSb\}.$$

$(A, R/S)$ is called the *relative product* of (A, R) with (A, S). For example, if (A, R) is "brother of" and (A, S) is "parent of," then $a(R/S)b$ holds if and only if for some c, a is a brother of c and c is a parent of b, i.e., $a(R/S)b$ holds if and only if a is an *uncle* of b.

(a) If R and S are defined as above, is $(A, S/R)$ the same relation as $(A, R/S)$? Why?

(b) If (A, S) is defined as above, what is $(A, S/S)$?

(c) If (A, R) is "sister of" and (A, S) is "mother of," what is $(A, R/S)$? What is $(A, S/R)$?

(d) If (A, R) is "father of," what is $(A, \breve{R}/R)$, where $\breve{R}$ is the converse of R?

(e) If (A, R) is "brother of," (A, S) is "father of" and (A, T) is "mother of," what is $(A, R/(S \cup T))$?

10. For each of the following binary relations, state which of the properties reflexivity, irreflexivity, symmetry, asymmetry, antisymmetry, transitivity, and negative transitivity it satisfies.

(a) The relation (A, R) where $A = [0, 1]$ and $R = \geqq$.

(b) The relation (A, R) where $A = \{a, b, c, d, e\}$ and R is defined as $\{(a, b), (b, b)\}$.

(c) The relation (A, R) where $A = \{$river, lake, ocean, estuary, stream$\}$ and R is defined as $\{($river, lake$)$, $($ocean, ocean$)$, $($lake, river$)\}$.

(d) The relation (A, R) where $A = \{$EPA, HEW, FAA$\}$ and R is defined as $\{($EPA, EPA$)$, $($HEW, HEW$)\}$.

(e) The relation (A, R) where A is a collection of electric power generating stations and aRb holds if and only if the average hourly nitrogen oxide output by station a is greater than the average hourly nitrogen oxide output by station b.

(f) The relation (A, R) where $A = \{1, 2, 3\}$ and aRb iff a divides b.

(g) The relation (A, R) where A is a set of automobiles and aRb iff a and b have the same horsepower.

(h) The relation (X, S) described in Exer. 5.

11. If (A, R) is a binary relation, let $(A, \breve{R})$ denote its converse. (See Exer. 4.) Which of the following properties holds for $\breve{R}$ whenever it holds for R?

(a) Reflexivity.

(b) Irreflexivity.

(c) Symmetry.

(d) Asymmetry.

(e) Antisymmetry.
(f) Transitivity.
(g) Negative transitivity.

12. If (A, R) and (A, S) are binary relations, let $(A, R \cap S)$ denote their intersection. (See Exer. 6.)

   (a) If (A, R) and (A, S) are both reflexive, is $(A, R \cap S)$?
   (b) Same question for irreflexive.
   (c) Same question for symmetric.
   (d) Same question for asymmetric.
   (e) Same question for antisymmetric.
   (f) Same question for transitive.
   (g) Same question for negatively transitive.

13. If (A, R) and (A, S) are binary relations, let $(A, R \cup S)$ be their union. (See Exer. 7.) Answer the questions of Exer. 12 for $(A, R \cup S)$.

14. If (A, R) and (A, S) are binary relations, let $(A, R/S)$ be the relative product of (A, R) with (A, S). (See Exer. 9.) Answer the questions of Exer. 12 for $(A, R/S)$.

15. Give an example of a binary relation which is both antisymmetric and transitive.

16. Give an example of a binary relation which is both symmetric and antisymmetric.

17. Give an example of a binary relation which is symmetric but not transitive.

18. Suppose (A, D) is a quaternary relation. We shall say it is *reflexive* if for all $a, b \in A$, $D(a, b, a, b)$. It is *symmetric* if for all $a, b, s, t \in A$, $D(a, b, s, t)$ implies $D(s, t, a, b)$. It is *transitive* if for all $a, b, s, t, p, q \in A$, $D(a, b, s, t)$ & $D(s, t, p, q)$ imply $D(a, b, p, q)$. Give examples of quaternary relations (A, D) which are

   (a) Reflexive.
   (b) Symmetric.
   (c) Transitive.
   (d) Reflexive and symmetric but not transitive.

19. Let $\hat{\mathcal{P}}$ be the collection of all rankings of the set A (see Chapter 7). Let $B(P, Q, R)$ be the ternary relation on $\hat{\mathcal{P}}$ defined as follows: (P, Q, R) is in B if and only if Q is between P and R in the sense of Sec. 7.4. Let B' be the binary relation defined on $\hat{\mathcal{P}}$ as follows: (P, Q) is in B' if and only if for some R in $\hat{\mathcal{P}}$ with $R \neq P$ or Q, $B(P, R, Q)$ holds. In general, is B'

   (a) Reflexive?
   (b) Irreflexive?
   (c) Symmetric?

(d)   Asymmetric?

(e)   Antisymmetric?

(f)   Transitive?

(g)   Negatively transitive?

20. A binary relation (A, R) is an *equivalence relation* if it is reflexive, symmetric, and transitive. Which of the following binary relations (A, R) are equivalence relations?

(a) $(\Re, =)$.

(b) $(\Re, \geqq)$.

(c) $(\Re, >)$.

(d) $A =$ a set of people, aRb iff a and b have the same weight.

(e) $A = \{0, 1, 2, \dots, 22\}$, aRb iff $a \equiv b \pmod{5}$.

(f) A is the collection of all finite sets of real numbers and aRb iff $a \cap b \neq \phi$.

(g) $A = \{(1, 1), (2, 3), (3, 8)\}$ and $R = \{\langle(1, 1), (1, 1)\rangle, \langle(2, 3), (2, 3)\rangle, \langle(3, 8), (3, 8)\rangle, \langle(1, 1), (2, 3)\rangle, \langle(2, 3), (1, 1)\rangle\}$.

21. If (A, R) and (A, S) are equivalence relations (Exer. 20), is

(a) $(A, R \cap S)$?   (b) $(A, R \cup S)$?   (c) $(A, R/S)$?

22. To show that all the properties in the definition of an equivalence relation are needed, give an example of a binary relation which is

(a)   Reflexive, symmetric, and not transitive.

(b)   Reflexive, transitive, and not symmetric.

(c)   Symmetric, transitive, and not reflexive.

23. If (A, R) is an equivalence relation (Exer. 20), let a^* be $\{b \in A: aRb\}$. This is called the *equivalence class containing a.* For example, if $A = \{1, 2, 3, 4, 5\}$ and $R = \{(1, 1), (1, 2), (2, 1), (2, 2), (3, 3), (4, 4), (5, 5)\}$, then (A, R), is an equivalence relation. The equivalence classes are $1^* = 2^* = \{1, 2\}$, $3^* = \{3\}$, $4^* = \{4\}$, and $5^* = \{5\}$.

(a)   Find all equivalence classes in the equivalence relation (A, R) of Exer. 20, part (e).

(b)   Show that two equivalence classes a^* and b^* are either disjoint or identical.

(c)   Give an example of an equivalence relation with three distinct equivalence classes.

(d)   Give an example of an equivalence relation with two distinct equivalence classes, one of which has three elements and the other two.

24. Which of the following relations (A, o) define operations?

(a) $A = \{SO_2, DDT, NO_x\}$,
$o = \{(SO_2, SO_2, NO_x), (SO_2, DDT, NO_x), (SO_2, NO_x, NO_x),$
$(DDT, SO_2, NO_x), (DDT, DDT, NO_x), (DDT, NO_x, NO_x),$
$(NO_x, SO_2, NO_x), (NO_x, DDT, NO_x), (NO_x, NO_x, NO_x)\}$.

(b) $A = \{SO_2, NO_x\}$,
        $o = \{(SO_2, NO_x), (NO_x, SO_2)\}$.
(c) $A = \{SO_2, NO_x\}$,
        $o = \{(SO_2, SO_2, NO_x), (SO_2, NO_x, SO_2), (NO_x, SO_2, SO_2),$
                $(SO_2, SO_2, SO_2), (NO_x, NO_x, SO_2)\}$.
(d) $A =$ the set of positive integers and $o(a, b, c)$ iff $a + b = c$.
(e) $A =$ the set of positive integers and $o(a, b, c)$ iff $a - b = c$.
(f) $A =$ the set of real numbers and $o(a, b, c)$ iff $a - b = c$.
(g) $A =$ the set of real numbers and $o(a, b, c)$ iff $a + b + c = 0$.

25.   Represent the operation of multiplication modulo 4 on $\{0, 1, 2, 3, 4\}$ as a ternary relation.

8.3. The Theory of Measurement

It seems almost redundant to say that measurement has something to do with assignment of numbers. (However, in Sec. 8.8 we shall argue that measurement without numbers is a perfectly legitimate and useful activity.) Looking at the paradigm examples of physics, such as measurement of temperature and measurement of mass, we see that measurement has something to do with assigning numbers which "preserve" certain observed relations. In the case of temperature, measurement is the assignment of numbers which preserves the observed relation "warmer than"; in the case of mass, the relation preserved is the relation "heavier than."

More precisely, suppose A is a set of days and the binary relation aWb holds if and only if you judge a to be warmer than b. Then we would like to assign a real number $f(a)$ to each $a \in A$ such that for all $a, b \in A$,

$$aWb \Longleftrightarrow f(a) > f(b). \qquad (6)$$

Similarly, if A is a set of objects which you lift and H is the judged relation "a is heavier than b," then we would like to assign a real number $f(a)$ to each $a \in A$ such that for all $a, b \in A$,

$$aHb \Longleftrightarrow f(a) > f(b). \qquad (7)$$

Measurement in the social sciences can be looked at in a similar manner. Thus, for example, measurement of preference is assignment of numbers preserving the observed binary relation "preferred to." If A is a set of alternatives and aPb holds on A if and only if you (strictly) prefer a to b, then we would like to assign a real number $u(a)$ to each $a \in A$ such that for all $a, b \in A$,

$$aPb \Longleftrightarrow u(a) > u(b). \qquad (8)$$

The function u is often called a *utility function* or an *ordinal utility function* and the value $u(a)$ is called the *utility* of a. If we can find a utility function, we can use it to make decisions: Always choose the alternative with highest utility. The case of measurement of loudness is analogous to the case of preference, and calls for an assignment of numbers preserving the relation "louder than." So is the case of measurement of air quality: We are trying to preserve the observed relation "the air quality on day a was better than the air quality on day b."

In the case of mass, we actually demand more of our "measure." We want it to be "additive" in the sense that the mass of the combination of two objects is the sum of their masses. Formally, we need to speak of a binary operation o of combination on the set A of objects (think of a o b as the object obtained by placing a next to b).[2] We want a real-valued function f on A which not only satisfies condition (7) but also "preserves" the binary operation o, in the sense that for all $a, b \in A$,

$$f(a \text{ o } b) = f(a) + f(b). \tag{9}$$

There is no comparable operation in the case of temperature. Whether there is one in the case of preference depends on the structure of the set of alternatives being considered, and on how demanding we want to be in our measurement. We might want to allow complex alternatives like paper and pencil $(a \text{ o } b)$ and we might want to require utility to be additive, i.e., to satisfy

$$u(a \text{ o } b) = u(a) + u(b). \tag{10}$$

A utility function which is also additive is often called a *cardinal utility function*.[3]

Abstracting from these examples, let us introduce the concept of a relational system. A *relational system* is an ordered n-tuple

$$\mathfrak{A} = (A, R_1, R_2, \ldots, R_p, \text{o}_1, \text{o}_2, \ldots, \text{o}_q),$$

where A is a set, $R_1, R_2, \ldots, R_p$ are relations on A and $\text{o}_1, \text{o}_2, \ldots, \text{o}_q$ are (binary) operations on A. (The number n is of course $p + q + 1$.) The *type* of the relational system is a sequence $(r_1, r_2, \ldots, r_p; q)$ of length $p + 1$, where r_i is m if R_i is an m-ary relation. For example, in the case of mass, we are dealing with a relational system (A, H, o) of type $(2; 1)$. In the case of temperature, we are dealing with a relational system (A, W) of type $(2; 0)$. The relational system $(\mathfrak{R}, >, \geqq, +)$ has type $(2, 2; 1)$. This is an example of what we shall call a *numerical relational system*, i.e., one where

[2]Certain formal difficulties are encountered in making combination into an operation in the precise sense of relation theory. For we need to speak of a o a, of $(a \text{ o } b)$ o a, etc. However, what interpretation do these combinations have? To get around this difficulty, we think of having an infinite number of "copies" of each element a. For details, see Roberts [to appear].

[3]In the literature, a utility function is often called cardinal if it gives rise to a scale at least as strong as what we shall call an interval scale.

A is the set of real numbers. A second example of a numerical relational system is $(\Re, >, +, \times)$, which has type $(2; 2)$.

In our examples, we have seen that in measurement, we start with an observed or empirical relational system $\mathfrak{A}$ and we seek a mapping to a numerical relational system $\mathfrak{B}$ which "preserves" all the relations and operations in $\mathfrak{A}$. For example, in measurement of mass, we seek a mapping from $\mathfrak{A} = (A, H, \text{o})$ to $\mathfrak{B} = (\Re, >, +)$ which "preserves" the relation H and the operation o. In measurement of temperature, we seek a mapping from $\mathfrak{A} = (A, W)$ to $\mathfrak{B} = (\Re, >)$ which "preserves" the relation W. A mapping f from one relational system $\mathfrak{A}$ to another $\mathfrak{B}$ which preserves all the relations and operations is called a *homomorphism*. (To make this precise, suppose $\mathfrak{B} = (B, R'_1, R'_2, \ldots, R'_p, \text{o}'_1, \text{o}'_2, \ldots, \text{o}'_q)$ is a second relational system of the same type as $\mathfrak{A}$. A function $f: A \longrightarrow B$ is called a *homomorphism* from $\mathfrak{A}$ *into* $\mathfrak{B}$ if for all $a_1, a_2, \ldots, a_{r_i} \in A$,

$$R_i(a_1, a_2, \ldots, a_{r_i}) \Longleftrightarrow R'_i(f(a_1), f(a_2), \ldots, f(a_{r_i})), \qquad (i = 1, 2, \ldots, p)$$

and for all $a, b \in A$,

$$f(a \text{ o}_i b) = f(a) \text{ o}'_i f(b) \qquad (i = 1, 2, \ldots, q).)$$

The mapping f need not be one-to-one or onto. If a homomorphism f exists, we say $\mathfrak{A}$ is *homomorphic* to $\mathfrak{B}$.

To give a concrete example, suppose $A = \{a, b, c\}$ and suppose $R = \{(a, b), (b, c), (a, c)\}$. Then a homomorphism from $\mathfrak{A} = (A, R)$ into $\mathfrak{B} = (\Re, >)$ is given by $f(a) = 3$, $f(b) = 2$, $f(c) = 1$. For the only ordered pairs in the relation $>$ among $f(a)$, $f(b)$ and $f(c)$ are $(f(a), f(b))$, $(f(b), f(c))$, and $(f(a), f(c))$, which correspond exactly to the ordered pairs (a, b), (b, c), and (a, c) in R. Thus,

$$aRb \Longleftrightarrow f(a) > f(b).$$

A second homomorphism from $\mathfrak{A}$ into $\mathfrak{B}$ is given by $g(a) = 7, g(b) = 3, g(c) = 0$. Next, suppose $A = \{a, b, c\}$ and $R = \{(a, b), (b, c), (c, a)\}$. Then there is no homomorphism f from (A, R) into $(\Re, >)$, since aRb implies $f(a) > f(b)$, bRc implies $f(b) > f(c)$, and cRa implies $f(c) > f(a)$.

Similarly, if $A = \{0, 1, 2, \ldots\}$ and $B = \{0, 2, 4, \ldots\}$, then $f(a) = 2a$ defines a homomorphism from $\mathfrak{A} = (A, >, +)$ into $\mathfrak{B} = (B, >, +)$. For given a and b in A,

$$a > b \quad \text{iff} \quad 2a > 2b$$

and, moreover,

$$f(a + b) = 2(a + b) = 2a + 2b = f(a) + f(b).$$

A second homomorphism is given by $g(a) = 4a$. (There is no requirement that a homomorphism must be an onto function.) If $\mathfrak{A} = (\Re, >, +)$ and $\mathfrak{B} = (\Re^+, >, \times)$, where $\Re^+$ is the positive reals, then $f(a) = e^a$ defines a homomorphism from $\mathfrak{A}$ into $\mathfrak{B}$. For

$$a > b \Longleftrightarrow e^a > e^b$$

and

$$f(a + b) = e^{a+b} = e^a \times e^b = f(a) \times f(b).$$

To give a last example, suppose $A = \{a, b, c, d\}$, $B = \{\alpha, \beta, \gamma, \delta\}$, $R = \{(a, b), (b, c), (c, d)\}$, and $S = \{(\alpha, \beta), (\beta, \gamma), (\gamma, \delta)\}$. Then the function $f: A \longrightarrow B$ defined by $f(a) = \alpha$, $f(b) = \beta$, $f(c) = \gamma$, $f(d) = \delta$ gives a homomorphism from (A, R) into (B, S).

In general, we shall say that *fundamental measurement* is the assignment of a homomorphism from an observed (empirical) relational system $\mathfrak{A}$ to some (specified) numerical relational system $\mathfrak{B}$. Thus, measurement of temperature is the assignment of a homomorphism from the observed relational system (A, W) to the numerical relational system $(\mathfrak{R}, >)$, measurement of mass is the assignment of a homomorphism from the observed relational system (A, H, o) to the numerical relational system $(\mathfrak{R}, >, +)$, and so on. The homomorphism is said to give a *representation*. The triple $(\mathfrak{A}, \mathfrak{B}, f)$ will be called a *scale*, though sometimes we shall be sloppy and refer to f alone as the scale.

The first basic problem of measurement theory is the *representation problem*: Given a particular numerical relational system $\mathfrak{B}$, find conditions on an observed relational system $\mathfrak{A}$ (necessary and) sufficient for the existence of a homomorphism from $\mathfrak{A}$ into $\mathfrak{B}$. The emphasis is on finding sufficient conditions. Such conditions tell us when measurement can be performed. If all the conditions in a collection of sufficient conditions are necessary as well, that is all the better. A more important criterion is that the conditions be "testable" or empirically verifiable in some sense. The conditions are usually called *axioms* for the representation and the theorem stating their sufficiency is usually called a *representation theorem*. The axioms in a representation theorem play a different role from the axioms we have studied earlier in this book, for example in connection with trophic status (Sec. 3.5), the Shapley value (Sec. 6.7), or Arrow's Theorem (Sec. 7.2). Here, we have a representation in mind, and we want to find conditions (necessary and) sufficient to guarantee it can be accomplished. Before, we stated reasonable conditions and then sought a representation which satisfied them. If possible, the proof of a representation theorem should be constructive: It should not only show us that a representation is possible, but it should show us how actually to construct it.

A typical axiom for a representation theorem is the following. Suppose we seek a homomorphism f from (A, R) into $(\mathfrak{R}, >)$. If such a homomorphism f exists, it follows that (A, R) is transitive. For if aRb and bRc, then $f(a) > f(b)$ and $f(b) > f(c)$, from which $f(a) > f(c)$ and aRc follow. Thus, a typical measurement axiom is the requirement that (A, R) be transitive.

The second basic problem of measurement theory is the *uniqueness problem:* How unique is the homomorphism f? We shall see in the next section that a uniqueness theorem can tell us what kind of scale f is, and give rise to a theory of meaningfulness of statements involving scales.

EXERCISES

1. Give examples of relational systems of the following types:
   - (a) $(2, 2; 1)$.
   - (b) $(4, 2; 1)$.
   - (c) $(;2)$.
   - (d) $(2, 2; 2)$.

2. Suppose $A = \{a, b, c\}$ and $R = \{(a, b), (b, c), (a, c)\}$. Find a homomorphism f from $\mathfrak{A} = (A, R)$ into $\mathfrak{B} = (\mathfrak{R}, >)$ different from the two given in the text.

3. Suppose N is the set of positive integers, $2N$ is the set of even integers, Z is the set of integers, $\mathfrak{R}$ is the set of real numbers, and $\mathfrak{R}^+$ is the set of positive real numbers. In each of the following situations, give a homomorphism from $\mathfrak{A}$ into $\mathfrak{B}$.
   - (a) $\mathfrak{A} = (N, >)$, $\mathfrak{B} = (2N, >)$.
   - (b) $\mathfrak{A} = (2N, \geqq)$, $\mathfrak{B} = (N, \geqq)$.
   - (c) $\mathfrak{A} = (N, >, +)$, $\mathfrak{B} = (2N, >, +)$.
   - (d) $\mathfrak{A} = (\mathfrak{R}, +)$, $\mathfrak{B} = (\mathfrak{R}^+, \times)$.
   - (e) $\mathfrak{A} = (N, =)$, $\mathfrak{B} = (2N, =)$.
   - (f) $\mathfrak{A} = (Z, >)$, $\mathfrak{B} = (Z, <)$.
   - (g) $\mathfrak{A} = (N, >)$, $\mathfrak{B} = (Z, <)$.
   - (h) $\mathfrak{A} = (\mathfrak{R}^+, >, \times)$, $\mathfrak{B} = (\mathfrak{R}, >, +)$.
   - (i) $\mathfrak{A} = (A, R)$ and $\mathfrak{B} = (\mathfrak{R}, >)$, where
     $A = \{\alpha, \beta, \gamma, \delta\}, R = \{(\alpha, \beta), (\alpha, \gamma), (\alpha, \delta), (\beta, \delta), (\gamma, \beta), (\gamma, \delta)\}$.
   - (j) $\mathfrak{A} = (A, R)$ and $\mathfrak{B} = (\mathfrak{R}, =)$, where
     $A = \{\alpha, \beta, \gamma, \delta\}, R = \{(\alpha, \alpha), (\alpha, \beta), (\beta, \alpha), (\beta, \beta), (\gamma, \gamma), (\delta, \delta)\}$.

4. Suppose $\mathfrak{A} = (\mathfrak{R}, >, +)$ and $\mathfrak{B} = (\mathfrak{R}^+, >, \times)$. Find a homomorphism from $\mathfrak{A}$ into $\mathfrak{B}$ different from the one given in the text.

5. Is there a homomorphism from $(N, >)$ into $(N, <)$? Why?

6. In which of the following examples $\mathfrak{A} = (A, R)$ and $\mathfrak{B} = (B, S)$ is $\mathfrak{A}$ homomorphic to $\mathfrak{B}$? Why?
   - (a) $A = \{1, 2, 3\}, R = \{(1, 2), (2, 3), (1, 3)\}, B = \mathfrak{R}, S = >$.
   - (b) $A = \{1, 2, 3\}, R = \{(1, 2), (2, 3), (1, 3)\}, B = \mathfrak{R}, S = <$.
   - (c) $A = \{1, 2, 3\}$,
     $R = \{(1, 1), (1, 2), (2, 1), (2, 2), (1, 3), (3, 1), (3, 3), (2, 3), (3, 2)\}$,
     $B = \{8\}, S = \{(8, 8)\}$.
   - (d) $A = \{1, 2, 3\}, R = \{(1, 2), (2, 3), (1, 3)\}, B = \{a, b, c, d\}$,
     $S = \{(a, b), (b, c), (c, d), (a, d)\}$.

(e) $A = \{1, 2, 3\}$, $R = \{(1, 2), (2, 3), (1, 3)\}$, $B = \{a, b, c, d\}$,
      $S = \{(a, b), (b, c), (c, d), (b, d)\}$.

(f) $A = \{1, 2, 3, 4\}$, $R = \{(1, 1, 1, 1), (2, 2, 2, 2), (3, 3, 3, 3), (4, 4, 4, 4)\}$,
      $B = \{8, 9\}$, $S = \{(8, 8)\}$.

(g) $A = \{1, 2, 3, 4\}$, $R = \{(1, 1, 1, 1), (2, 2, 2, 2), (3, 3, 3, 3)\}$,
      $B = \{8, 9\}$, $S = \{(8, 8, 8, 8), (9, 9, 9, 9)\}$.

(h) $A = \{1, 2, 3, 4\}$, $R = \{(1, 1, 1, 1), (2, 2, 2, 2), (3, 3, 3, 3), (4, 4, 4, 4)\}$,
      $B = \{8, 9\}$, $S = \{(8, 8, 8, 8), (8, 9, 8, 9)\}$.

(i) $A = \{SO_2, DDT, NO_x\}$, $R = \{(SO_2, DDT), (DDT, NO_x)\}$,
      $B = \{\text{New York, Wichita, Los Angeles}\}$,
      $S = \{(\text{New York, Witchita})\}$.

7.   Let us say a homomorphism f of a relational system $\mathfrak{A}$ into a relational system $\mathfrak{B}$ is an *isomorphism* if it is one-to-one. In which of the cases in Exers. 3 and 6 is there an isomorphism from $\mathfrak{A}$ into $\mathfrak{B}$?

8.   Suppose $A = \mathfrak{R}$, $D(a, b, c, d)$ holds on A iff $a + b > c + d$, $B = \mathfrak{R}$, and $E(a, b, c, d)$ holds on B iff $a + b < c + d$. Is (A, D) homomorphic to (B, E)? Why?

9.   Suppose $\mathfrak{A}$, $\mathfrak{B}$, and $\mathfrak{C}$ are relational systems of the same type. If $\mathfrak{A}$ is homomorphic to $\mathfrak{B}$ and $\mathfrak{B}$ is homomorphic to $\mathfrak{C}$, does it follow that $\mathfrak{A}$ is homomorphic to $\mathfrak{C}$? Give a proof of your answer.

10.   Try to formulate the theory of measurement for length. (Is this similar to temperature or to mass?)

11.   Try to formulate the theory of measurement for area. (Is this similar to temperature or to mass?)

12.   Does the discussion in this section agree with your intuitive understanding of the process of measurement?

13.   What simplifications are we making in our model of the measurement process? What are we omitting?

8.4.   Scale Type and the Theory of Meaningfulness

8.4.1.   REGULAR SCALES

It makes sense to say that the number of bighorn sheep on a particular range at a particular time is 400; but it does not make sense to say that a given sheep weighs 400. It makes sense to say that one sheep weighs twice as much as another; but it does not make sense to say that today's temperature on the range at 5 P.M. was twice as much as yesterday's unless we know what temperature scale is being used (or we have an "absolute zero"). The number of sheep is specified without reference to a particular scale of measurement while the weight of a sheep is not. Similarly, the ratio of weights is the same regardless of the scale of measurement, but the ratio of temperatures is not. To be an adequate description of what we mean by measurement, a model must account for these types of observations. We shall try to account for them by introducing in our model of measurement a notion of meaningfulness. In an imprecise sense, a statement is considered meaningful if it makes sense, if it can be given a precise interpretation, if it can be tested, etc. (Meaningfulness, however, is not the same as truth; a false statement can also be given a precise interpretation, tested, etc.) We shall want to make the notion of meaningfulness more precise.

In general, what is critical in determining whether or not a statement involving a numerical assignment or homomorphism is meaningful seems to be the uniqueness of the numerical assignment. It is quite possible, given two relational systems $\mathfrak{A}$ and $\mathfrak{B}$ of the same type, for there to be two different functions f and g which map $\mathfrak{A}$ homomorphically into $\mathfrak{B}$. Since this is the case, any statement about measurement should either specify which scale (which homomorphism) is being used or be true independent of scale. We shall adopt the following notion of meaningfulness: A statement involving (numerical) scales is *meaningful* if its truth (or falsity) remains unchanged if every scale involved is replaced by another acceptable scale.

In many cases, we shall be able to check for meaningfulness by classifying admissible transformations of scale. To explain this point, let us recall that a scale is a triple $(\mathfrak{A}, \mathfrak{B}, f)$ where $\mathfrak{A}$ and $\mathfrak{B}$ are relational systems, $\mathfrak{B}$ a numerical relational system, and f is a homomorphism from $\mathfrak{A}$ into $\mathfrak{B}$. We are interested in the uniqueness of the homomorphism f. Suppose A is the set underlying $\mathfrak{A}$ and $\mathfrak{R}$ is the set underlying $\mathfrak{B}$. If ϕ is a function from $f(A)$, the range of f, into $\mathfrak{R}$, then the composition $\phi \circ f$ is a function from A into $\mathfrak{R}$. If $\phi \circ f$ is also a homomorphism from $\mathfrak{A}$ into $\mathfrak{B}$, we shall call ϕ an *admissible transformation of the scale* $(\mathfrak{A}, \mathfrak{B}, f)$. To give an example, suppose

$$A = \{a, b, c\}, \qquad R = \{(a, b), (b, c), (a, c)\},$$

$$f(a) = 10, \qquad f(b) = 8, \qquad f(c) = 4.$$

Then f is a homomorphism from $\mathfrak{A} = (A, R)$ into $\mathfrak{B} = (\mathfrak{R}, >)$. If $\phi(x)$ is defined to be $-x$, then $g = \phi \circ f$ is not a homomorphism from $\mathfrak{A}$ into $\mathfrak{B}$, for $g(a) = -10$, $g(b) = -8$, and $g(b) > g(a)$. Thus, $\phi(x) = -x$ is not an admissible transformation of the scale f. However, $\phi(x) = 2x$ is an admissible transformation, for $2f$ is a homomorphism.

If $(\mathfrak{A}, \mathfrak{B}, f)$ is a scale and $(\mathfrak{A}, \mathfrak{B}, g)$ is any other scale, it is often possible to find a function $\phi : f(A) \rightarrow \mathfrak{R}$ such that $g = \phi \circ f$. (For example, this is always true if f is one-to-one. For then we can define $\phi(x) = g(a)$ if $x = f(a)$. In our example, $g(a) = 22$, $g(b) = 16$, and $g(c) = -10$ defines a second homomorphism. We can find ϕ such that $g = \phi \circ f$. Namely, $\phi(10) = 22$, $\phi(8) = 16$, $\phi(4) = -10$.) If for every scale $(\mathfrak{A}, \mathfrak{B}, g)$, there is an admissible transformation $\phi:f(A) \rightarrow \Re$ such that $g = \phi \circ f$, we shall call the *scale* $(\mathfrak{A}, \mathfrak{B}, f)$ *regular*. A *representation* $\mathfrak{A} \rightarrow \mathfrak{B}$ is called *regular* if every homomorphism from $\mathfrak{A}$ into $\mathfrak{B}$ defines a regular scale. A representation is regular if given any two homomorphisms, we can map each into the other by an admissible transformation of scale. Almost all representations we shall encounter in this chapter are regular, though we shall give an example of an irregular scale and an irregular representation in Sec. 8.8.

If $\mathfrak{A} \rightarrow \mathfrak{B}$ is a regular representation, the class of admissible transformations of a scale $(\mathfrak{A}, \mathfrak{B}, f)$ defines the uniqueness of the scale. *If we deal only with scales arising from regular representations*, we can modify our definition of meaningfulness: A statement involving (numerical) scales is *meaningful* if and only if its truth (or falsity) remains unchanged under all admissible transformations of all numerical scales involved. Such a notion of meaningfulness is described in Suppes and Zinnes [1963] and Suppes [1959]. This definition of meaningfulness again makes precise the imprecise idea we started with. It is another example of the procedure we call *explication*, making precise an imprecise concept. We shall test our precise concept against the examples discussed at the beginning of this section. In part, this will provide a test of our entire model of the measurement process.

8.4.2. SCALE TYPE

Before testing the concept of meaningfulness, let us observe that the notion of admissible transformation allows us to define *scale type*. The idea of defining scale type in this way is due to S. S. Stevens [1946, 1951]. It is only safe to use this notion of scale type for scales which arise from regular representations. (See Exer. 31.) Thus, we shall assume that all scales discussed in this subsection come from regular representations.

We give several examples of scale types in Table 8.2. The simplest example of a scale is one where the only admissible transformation is $\phi(x) = x$. There is only one way to measure things in this situation. Such a scale is called *absolute*. Counting is an example of an absolute scale. When we say that there are x elements in a set, we mean exactly x, and there is no admissible transformation of scale (except the identity) which changes this.

To give a second example, let us suppose the admissible transformations are all the functions of the form $\phi(x) = \alpha x, \alpha > 0$. Such a function ϕ is called a *similarity transformation* and a scale with the similarity transformations as its class of admissible transformations is called a *ratio scale*. The usual scales of mass define ratio scales, as we can fix a zero point and then change the unit of mass by multiplying by a positive constant. For example, we change from grams to kilograms by multiplying by 1000. The term ratio scale arose because ratios of quantities on a ratio scale make sense, as they do in the case of mass. Temperature defines a ratio scale if we allow absolute zero, as in the Kelvin scale. Intervals of time (in minutes, hours, etc.) also define a ratio scale. According to Stevens [1959], various sensations such as loudness, brightness, etc. can also be measured on a ratio scale.

To give a third example of scale type, suppose we let the class of admissible transformations be all functions ϕ of the form $\phi(x) = \alpha x + \beta$, $\alpha > 0$. Such a function is called a *positive linear transformation*, and a corresponding scale is called an *interval scale*. The common scales of temperature are examples of interval scales. We vary the zero point (this amounts to changing β) and also the unit (this amounts to changing α). For example, when we change from Fahrenheit to Centigrade, we take $\alpha = \frac{5}{9}$ and $\beta = -\frac{160}{9}$. Time on the calendar (for example the year 1980) is another example of an interval scale. Some argue (Stevens [1959, p.25] that "standard scores" on intelligence tests define an interval scale.

Some scales are unique only up to order. For example, the scale of air quality being used in a number of cities is such a scale. It assigns a number 1 to unhealthy air, 2 to unsatisfactory air, 3 to acceptable air, 4 to good air, and 5 to excellent air. We could just as well use the numbers 1, 7, 8, 15, 23 or the numbers 1.2, 6.5, 8.7, 205.6, 750 or the numbers -10, -5, 0, 5, 10, or any numbers which preserve the order. A scale which is unique only up to order is called an *ordinal scale*. The admissible transformations are monotone increasing functions, i.e., functions $\phi(x)$ satisfying the condition that

$$x \geq y \longleftrightarrow \phi(x) \geq \phi(y).$$

Another example of an ordinal scale is the Mohs scale of hardness. Numbers are assigned to minerals reflecting their relative hardness subject to the restriction that mineral a gets a larger number than mineral b if and only if mineral a is harder than mineral b. (In practice, a is judged to be harder than b if a scratches b.) The connectedness category 0, 1, 2, or 3 of a digraph, defined in Sec. 2.2.4, is another example of an ordinal scale. So is the degree of balance in a small group defined in the exercises of Sec. 3.1. Raw scores on intelligence tests probably define only an ordinal scale (Stevens [1959, p.25]). It is possible that preference may be no more than an ordinal scale.

Finally, in some scales, all one-to-one functions $\phi : f(A) \longrightarrow \Re$ define admissible transformations. Such scales are called *nominal*. Examples of nominal scales are numbers on the uniforms of baseball players or the numbering of alternative plans as plan 1, plan 2, etc. The actual number has no significance, and any one-to-one change of numbers will contain the same information: identification of the elements of the set A.

In general, the scales listed in Table 8.2 go from "strongest" to "weakest," in the sense that absolute scales and ratio scales contain much more information than ordinal scales or nominal scales. It is often a goal of measurement to obtain as strong a scale as possible.

TABLE 8.2

Some Scale Types.

Admissible transformations	Scale type	Example
$\phi(x) = x$ (identity)	Absolute	Counting
$\phi(x) = \alpha x, \alpha > 0$ Similarity transformation	Ratio	Mass Temperature (Kelvin scale) Time intervals Loudness (sones)* Brightness (brils)*
$\phi(x) = \alpha x + \beta, \alpha > 0$ Positive linear transformation	Interval	Temperature (Fahrenheit, Centigrade) Time (calendar) Intelligence-test "standard scores"(?)*
$x \geqq y$ iff $\phi(x) \geqq \phi(y)$ Monotone increasing transformation	Ordinal	Preference? Hardness Air quality Grades of leather, lumber, wool, etc. Intelligence-test raw scores* Connectedness category Relative balance
Any one-to-one ϕ	Nominal	Numbering of uniforms Labelling of alternative plans

*According to Stevens [1959]

8.4.3. EXAMPLES OF MEANINGFUL AND MEANINGLESS STATEMENTS

Having defined several scale types, let us now test our definition of meaningfulness on the examples discussed at the beginning of this section. *We shall assume that all the scales in question come from regular representations,* and so use our second definition of meaningfulness. Let us first consider the statement

$$f(a) = 2f(b), \tag{11}$$

where $f(a)$ is some quantity assigned to a, for example its temperature or its mass. We ask under what circumstances this statement is meaningful. According to the definition, it is meaningful if and only if its truth value is preserved under all admissible transformations ϕ, i.e., if and only if under all such ϕ,

$$f(a) = 2f(b) \Longleftrightarrow (\phi \circ f)(a) = 2[(\phi \circ f)(b)].$$

If ϕ is a similarity transformation, i.e., if $\phi(x) = \alpha x$, some $\alpha > 0$, then we do indeed have

$$f(a) = 2f(b) \iff \alpha f(a) = 2\alpha f(b).$$

We conclude that the statement (11) is meaningful if the scale f is a ratio scale, as is the case in the measurement of mass. On the other hand, suppose it is only an interval scale, as is the case commonly in the measurement of temperature. Then a typical admissible transformation has the form $\phi(x) = \alpha x + \beta$, $\alpha > 0$. Certainly we can find examples where $f(a) = 2f(b)$ but $\alpha f(a) + \beta \neq 2[\alpha f(b) + \beta]$. In particular, if we choose $f(a) = 2$, $f(b) = 1$, $\alpha = 1$ and $\beta = 1$, this is the case.[4] This explains why it makes sense to say that one bighorn sheep weighs twice as much as another, but not to say that today's temperature on the range at 5 P.M. was twice as much as yesterday's.

In the same way, one can explain why it makes sense to say that the number of bighorn sheep on a range is 400, while not to say that a given sheep weighs 400. For consider the statement

$$f(a) = 400. \tag{12}$$

If f is an absolute scale, as in the case of counting, we have for every admissible transformation ϕ,

$$f(a) = 400 \iff (\phi \circ f)(a) = 400,$$

for the only admissible ϕ is the identity transformation. (Notice that $f(a)$ need not be equal to 400 for the statement $f(a) = 400$ to be meaningful. Meaningfulness is different from truth; we simply want to know whether or not it makes sense to make the assertion.) If f is a ratio scale, as in the case of weight, then the statement (12) is meaningless; for example if $f(a) = 400$ is true for some f, then taking $\alpha \neq 1$ makes $\alpha f(a) = 400$ false.

To give another example, suppose $(\mathfrak{A}, \mathfrak{B}, f)$ is a scale and a, b are in A, the underlying set of $\mathfrak{A}$. Let us consider the statement

$$f(a) + f(b) = 15. \tag{13}$$

Thus, (13) might be the statement that the sum of the weight of a and the weight of b is 15. Is this meaningful? The answer is no if f is a ratio scale. For if $f(a) + f(b) = 15$, then $\alpha f(a) + \alpha f(b) = 15\alpha \neq 15$ for $\alpha \neq 1$. However, the statement

$$f(a) + f(b) \text{ is constant for all } a, b \text{ in } A \tag{14}$$

is meaningful if f is a ratio scale. For we can rewrite this statement as

$$f(a) + f(b) = f(c) + f(d) \qquad \text{for all } a, b, c, d \text{ in } A. \tag{15}$$

---

[4]The reader should note that to demonstrate that a statement is meaningless, one simply has to give an example, whereas to demonstrate that it is meaningful requires a proof.

Then, if $\alpha > 0$, (15) holds if and only if

$$\alpha f(a) + \alpha f(b) = \alpha f(c) + \alpha f(d) \qquad \text{for all } a, b, c, d \text{ in } A.$$

To give yet another example, consider the statement

$$f(a) > f(b). \tag{16}$$

If $\phi(x) = \alpha x + \beta, \alpha > 0$, then

$$\alpha f(a) + \beta > \alpha f(b) + \beta \Longleftrightarrow \alpha f(a) > \alpha f(b)$$
$$\Longleftrightarrow f(a) > f(b).$$

Thus, the statement (16) is meaningful if f is an interval scale. Indeed, it meaningful if f is an ordinal scale. (Why?) Thus, for example, to say that the hardness of a is greater than the hardness of b is meaningful.

Continuing with additional examples, let us consider two scales, $(\mathfrak{A}, \mathfrak{B}, f)$ and $(\mathfrak{A}', \mathfrak{B}', g)$, where $\mathfrak{A}$ and $\mathfrak{A}'$ have the same underlying set, and let us consider the statement

$$f(a) = 2g(a). \tag{17}$$

For instance, this might be the statement that a certain animal's height is twice its weight. If f and g are both ratio scales, then for (17) to be meaningful, its truth or falsity should be unchanged under (possibly different) admissible transformations of each scale. That is, if $\phi(x) = \alpha x$ and $\phi'(x) = \beta x$, $\alpha, \beta > 0$, then we should have (17) holding iff

$$\alpha f(a) = 2\beta g(a).$$

But certainly this is not true: Simply take $\alpha \neq \beta$. On the other hand, certainly (17) is meaningful if f and g are both absolute scales. It is possible that in considering statements like (17), the scale g might be defined in terms of the scale f. Then, we might not want to allow all possible admissible transformations of both f and g independently, but only admissible transformations of f and the "induced" admissible transformation of g. This is a situation which arises in the theory of derived measurement, where one scale is defined in terms of another. In derived measurement, there are several versions of meaningfulness, narrow and wide, depending on whether or not we pick admissible transformations of all scales independently or not.[5] In this chapter, however, we shall allow all scales to change independently, unless otherwise mentioned.

To give a more complicated example, let us imagine that we study n animals under one kind of experimental treatment (say a special diet) and m animals under a second kind of treatment. We want to say that the average weight of the animals under the first kind of treatment is larger than the average weight of the animals under the second treatment. Specifically, if

[5]For a discussion of these concepts, see Suppes and Zinnes [1963] or Roberts [to appear].

f is the scale of weight in question, we want to consider the statement

$$\frac{1}{n} \sum_{i=1}^{n} f(a_i) > \frac{1}{m} \sum_{i=1}^{m} f(b_i). \tag{18}$$

Here, we calculate arithmetic means over two different sets, and compare them. We consider the statement (18) meaningful if for all admissible transformations ϕ, (18) holds iff

$$\frac{1}{n} \sum_{i=1}^{n} (\phi \circ f)(a_i) > \frac{1}{m} \sum_{i=1}^{m} (\phi \circ f)(b_i). \tag{19}$$

If ϕ is a similarity transformation, say $\phi(x) = \alpha x$, $\alpha > 0$, then certainly (18) holds if and only if (19) holds, for (19) is the statement

$$\frac{1}{n} \sum_{i=1}^{n} \alpha f(a_i) > \frac{1}{m} \sum_{i=1}^{m} \alpha f(b_i).$$

Statements (19) and (18) are equivalent even when ϕ is a positive linear transformation, i.e., $\phi(x) = \alpha x + \beta$, $\alpha > 0$. For then (19) becomes

$$\frac{1}{n} \sum_{i=1}^{n} [\alpha f(a_i) + \beta] > \frac{1}{m} \sum_{i=1}^{m} [\alpha f(b_i) + \beta],$$

which reduces to (18). Thus (18) is meaningful if f is a ratio scale or an interval scale. If f is an ordinal scale, then (18) is meaningless. Proof is left to the reader as Exer. 20. This result can be applied to say that the statement "group A has higher average IQ than group B" is meaningless if IQ is only an ordinal scale. ("Raw scores" on intelligence tests seem[6] to define only ordinal scales (Stevens [1959]).) The comparison of average IQ's is meaningful if IQ is an interval scale. ("Standard scores" on IQ tests seem[6] to define interval scales.) All too often in the social or biological sciences, comparisons of arithmetic means (as well as other comparisons) are made with little attention paid to whether or not these comparisons are meaningful.

Suppose next that we have several experts or individuals and each rates two alternatives, a and b. Let $f_i(a)$ be the rating of a by expert i, and $f_i(b)$ be the rating of b by expert i. We might want to consider the statement

$$\frac{1}{n} \sum_{i=1}^{n} f_i(a) > \frac{1}{n} \sum_{i=1}^{n} f_i(b). \tag{20}$$

Here, we say that the average rating of alternative a is higher than the average rating of alternative b. Even if each f_i is a ratio scale, this is now meaningless. For, we must consider simultaneously transformations $\phi_i(x) = \alpha_i x$ of each

---

[6]We say "seem" because with many scales which are constructed using empirical procedures, where there is no clear measurement representation in mind, the admissible transformations must be defined as functions preserving the empirical information captured by a scale. The judgement of what this empirical information is can be somewhat subjective. See Stevens [1968, p. 850] for a discussion of this point.

f_i, and certainly there are α_i such that (20) holds while

$$\frac{1}{n}\sum_{i=1}^{n}\alpha_i f_i(a) > \frac{1}{n}\sum_{i=1}^{n}\alpha_i f_i(b) \tag{21}$$

does not. On the other hand, comparison of geometric means[7] over individuals is meaningful, for

$$\sqrt[n]{\prod_i f_i(a)} > \sqrt[n]{\prod_i f_i(b)} \iff \sqrt[n]{\prod_i \alpha_i f_i(a)} > \sqrt[n]{\prod_i \alpha_i f_i(b)}.$$

We shall briefly mention an application of this last example. In a recent experiment (Roberts [1972, 1973]), a set of variables relevant to the growing demand for energy was presented to a panel of experts, who were asked to judge their relative importance using the method of magnitude estimation. In this method, the expert first selects that variable which seems most important and assigns it the rating 100. Then he rates the other variables in terms of the most important one, so that a variable receiving a rating of 50 is considered "half as important" as one receiving a rating of 100, etc. A typical set of variables and ratings of one of the experts is shown in Table 8.3. These importance ratings were used to choose some of the most important variables as vertices for the signed digraph of Fig. 4.8 (see Sec. 4.3). It seems

TABLE 8.3

Magnitude Estimation by One Expert of Relative Importance
for Energy Demand of Variables Related to Commuter Bus
Transportation in a Given Region.
From Roberts [1972].

Variable	Relative importance rating
1. Number of passenger miles (annually, by bus)	80
2. Number of trips (annually)	100
3. Number of miles of bus routes	50
4. Number of miles of special bus lanes	50
5. Average time home to office (or office to home, or sum)	70
6. Average distance home to office	65
7. Average speed	10
8. Average number passengers per bus	20
9. Distance to bus stop from home (or office, or sum)	50
10. Number of buses in the region	20
11. Number of stops (home to office, or vice versa, or sum)	20

[7]The *geometric mean* of a collection of numbers $x_1, x_2, \ldots, x_n$ is $\sqrt[n]{x_1 \cdot x_2 \cdot x_3 \cdots \cdots x_n} = \sqrt[n]{\prod_i x_i}$, where $\prod_i x_i$ means the product over i of x_i.

plausible that the magnitude estimation procedure leads to a ratio-scale; this is presumed by Stevens [1957, 1968] and put on a measurement-theoretic foundation by Krantz [1972] and Krantz, et al [1971, Sec. 4.6]. Thus, comparisons of geometric mean relative importance ratings (over experts) are probably meaningful, while comparisons of arithmetic means are probably not. The most important variable was chosen as the one whose geometric mean importance rating (over experts) was largest.

Remark: The analysis of this example must be done with some care. Since the scale value of the most important element was fixed to be the same for all the experts, the scales are not really independent and so it is probably not reasonable to demand that the truth value of a statement be preserved under different transformations of all the scales, but rather only under the same transformation $\phi(x) = \alpha x$ of each scale. But now it probably becomes meaningful to speak of comparisons of arithmetic means. For, if we take $\alpha_i = \alpha$, all i, then (20) holds if and only if (21) holds. Since an extra assumption is required to draw this conclusion, it is still safer to use geometric means. In any case, this reasoning points up the fact that the theory of meaningfulness has to be applied with care.

The reader interested in a more detailed discussion of the theory of meaningfulness applied to statistical analysis should consult Pfanzagl [1968].

To give one last application of the theory of meaningfulness, let us consider computation of the *consumer price index*. This index is based on the prices of n fixed commodities, which include food, clothing, fuel, etc. Let $p_i(t)$ be the price of commodity i at time t. Then $p_i(0)$ is the "base price." Bradstreet and Dutot (see Fisher [1923, p. 40]) suggest measuring consumer price index as

$$I(t) = \frac{\sum_{i=1}^{n} p_i(t)}{\sum_{i=1}^{n} p_i(0)}.$$

(Thus, $I(t)$ should be thought of as a scale *derived* from the prices.) Let us ask whether it is meaningful to say that the consumer price index has doubled in the last year. That is, is it meaningful to say that

$$I(t + 1) = 2I(t). \tag{22}$$

Prices are measured in monetary amounts, which are subject to change of unit (dollars to cents, dollars to francs, etc.) Thus, the price $p_i(t)$ defines a ratio scale. If we may use independent admissible transformations of the different prices, then the statement (22) is meaningless. Indeed, even the statement $I(t + 1) > I(t)$ is meaningless. (Why?) However, if all prices are measured in the same units, then we can only use the same admissible transformation of each price. Then, the statement (22) is meaningful. For, if $\alpha > 0$, we have

$$\frac{\sum\limits_{i=1}^{n} \alpha p_i(t+1)}{\sum\limits_{i=1}^{n} \alpha p_i(0)} = 2\left(\frac{\sum\limits_{i=1}^{n} \alpha p_i(t)}{\sum\limits_{i=1}^{n} \alpha p_i(0)}\right) \longleftrightarrow \frac{\sum\limits_{i=1}^{n} p_i(t+1)}{\sum\limits_{i=1}^{n} p_i(0)} = 2\left(\frac{\sum\limits_{i=1}^{n} p_i(t)}{\sum\limits_{i=1}^{n} p_i(0)}\right).$$

For a further discussion, see Pfanzagl [1968, p.49].

EXERCISES

Assume throughout that all scales come from regular representations, unless otherwise noted.

1. Suppose $A = \{\alpha, \beta, \gamma\}$, $R = \{(\beta, \gamma), (\gamma, \alpha), (\beta, \alpha)\}$ and f is defined as follows: $f(\alpha) = 20$, $f(\beta) = 30$, $f(\gamma) = 25$.
    (a) Show that f is a homomorphism from $\mathfrak{A} = (A, R)$ into $\mathfrak{B} = (\mathfrak{R}, >)$.
    (b) Is $\phi(x) = 2x$ an admissible transformation of scale?
    (c) What about $\phi(x) = x^2$?

2. Suppose $A = \{0, 1, 2, \ldots\}$. Then a homomorphism f from $\mathfrak{A} = (A, >)$ into $\mathfrak{B} = (\mathfrak{R}, >)$ is given by the function $f(a) = 2a$.
    (a) Is $\phi(x) = 2x$ an admissible transformation of scale?
    (b) What about $\phi(x) = x + 2$?

3. In Exer. 1, define a second homorphism g from $\mathfrak{A}$ into $\mathfrak{B}$ as follows: $g(\alpha) = 50$, $g(\beta) = 70$, $g(\gamma) = 61$. Find a function $\phi : f(A) \to \mathfrak{R}$ such that $g = \phi \circ f$.

4. Which of the following transformations ϕ is an admissible transformation of an ordinal scale whose range is all of $\mathfrak{R}$?
    (a) $\phi(x) = x^2$.
    (b) $\phi(x) = e^x$.
    (c) $\phi(x) = x + 7$.
    (d) $\phi(x) = x$.
    (e) $\phi(x) = x^2 + 1$.
    (f) $\phi(x) = 79x$.

5. Repeat Exer. 4 for ratio scale.

6. Repeat Exer. 4 for interval scale.

7. Repeat Exer. 4 for absolute scale.

8. Repeat Exer. 4 for nominal scale.

9. Assume that connectedness of a digraph is an ordinal scale. Which of the following functions $\phi \colon \{0, 1, 2, 3\} \to \mathfrak{R}$ defines an admissible transformation of scale?

    (a) $\phi(0) = 5$, $\phi(1) = 5.1$, $\phi(2) = 100$, $\phi(3) = 175$.
    (b) $\phi(0) = 10$, $\phi(1) = 9$, $\phi(2) = 8$, $\phi(3) = 7$.
    (c) $\phi(0) = 3$, $\phi(1) = 2$, $\phi(2) = 1$, $\phi(3) = 0$.
    (d) $\phi(0) = 1$, $\phi(1) = 2$, $\phi(2) = 3$, $\phi(3) = 4$.

10. Does numbering of football players' uniforms define a nominal scale? (*Hint:* Under some numbering schemes, offensive backs usually receive numbers less than 50, ends numbers in the 80's and 90's, and so on.)

11. Which of the scale types we have described applies to height? Why?

12. Show that the statement $f(a) = 2f(b)$ is meaningless for temperature by showing that it can hold under the Fahrenheit scale while failing under the Centigrade scale.

13. Suppose f is a ratio scale. Which of the following statements is meaningful?

    (a) $f(a) + f(b) > f(c)$.
    (b) $f(a) + f(b) > f(c)^2$.
    (c) $f(a)f(b) > f(c)$.
    (d) $f(a) = f(b)$.

14. Repeat Exer. 13 for f an interval scale.

15. Repeat Exer. 13 for f a nominal scale.

16. Consider the statement $f(a) + g(b) = 7$.

    (a) Is this statement meaningful if f and g are ratio scales?
    (b) If f and g are interval scales?
    (c) If f is a ratio scale and g is an absolute scale?
    (d) Suppose we modify the statement to read:

    $$f(a) + g(b) \text{ is constant for all } a, b \text{ in } A.$$

    Is this statement now meaningful if f and g are ratio scales? If f and g are interval scales?

17. Suppose f is a ratio scale and g is an absolute scale. Which of the following statements are meaningful?

    (a) $\log_{10}|f(a)| = 2\log_{10}|f(b)|$.
    (b) $f(a)$ is constant for all a in the set A.
    (c) $f(a) + g(a)$ is constant for all a in the set A.

18. A scale is a *difference scale* if the admissible transformations are functions ϕ of the form $\phi(x) = x + \beta$. (According to Suppes and Zinnes [1963, Sec. 4.2], the so-called Thurstone Case V Scale, which arises in

psychology as a measure of response strength, is an example of a difference scale.) Which of the following statements are meaningful if f is a difference scale?

(a) $f(a) > f(b)$.

(b) $f(a) = 2f(b)$.

(c) $f(a) + f(b)$ is constant for all a, b in the set A.

19.   A scale is called a *log-interval scale* if the admissible transformations are functions of the form αx^{β}, $\alpha, \beta > 0$. Log-interval scales are important in psychophysics. (See Luce [1959] and Roberts [to appear].) If f is a log-interval scale, which of the following statements are meaningful?

(a) $f(a) > f(b)$.

(b) $f(a) = 2f(b)$.

(c) $f(a) + f(b)$ is constant for all a, b in the set A.

(d) $f(a)f(b)$ is constant for all a, b in the set A.

20.   Show that the statement of Eq. (18) is meaningless if f is an ordinal scale.

21.   If arithmetic means in the statement of Eq. (18) are replaced by geometric means and if f is a ratio scale, is the resulting statement meaningful?

22.   In the situation where there are different experts, which of the following comparisons is meaningful if each f_i is an ordinal scale?

(a)   Comparison of arithmetic means.

(b)   Comparison of geometric means.

(c)   Comparison of medians if there is an odd number n of experts. (The *median* of a collection C of $n = 2k + 1$ numbers is the $(k + 1)st$ number in C if the numbers are arranged in increasing order.)

23.   Answer the questions of Exer. 22 if each f_i is a difference scale.

24.   Show that the statement $I(t + 1) > I(t)$, involving the consumer price index, is meaningless if we allow independent admissible transformations of different prices.

25.   Suppose a homomorphism $f: \mathfrak{A} \rightarrow \mathfrak{B}$ defines an interval scale. Show that ratios of intervals are preserved, by showing that if $g: \mathfrak{A} \rightarrow \mathfrak{B}$ is another homomorphism, then

$$\frac{f(a) - f(b)}{f(c) - f(d)} = \frac{g(a) - g(b)}{g(c) - g(d)}.$$

26.   Which of the following statements are meaningful. Why?

(a)   "This stick is twice as long as that one."

(b)   "This rectangle has twice the area of that one."
(c)   "John's height is twice his weight."
(d)   "John's temperature is greater than his weight."
(e)   "This flight took twice as long as the last one I took."
(f)   "Diamonds are twice as hard as coal."
(g)   "The wind today is calmer than it was yesterday." (The Beaufort wind scale classifies winds as calm, light air, light breeze, ...)
(h)   "The difference between today's temperature at 5 p.m. and yesterday's is twice the difference between yesterday's and that of the day before yesterday."
(i)   "This specimen weighs 50% more than that specimen."

27.   Discuss when it is meaningful to say that one student's grade point average is higher than a second student's.

28.   Discuss under what circumstances it is meaningful to say that one school district's average reading score is higher than another's.

29.   Exercises 29 through 34 are based on Roberts [1974]. Show that a scale $(\mathfrak{A}, \mathfrak{B}, f)$ is regular if and only if for every homomorphism $g: \mathfrak{A} \to \mathfrak{B}$, $f(a) = f(b)$ implies $g(a) = g(b)$.

30.   Show that every one-to-one homomorphism (isomorphism) defines a regular scale.

31.   This exercise shows one of the problems with irregular scales: It is hard to define scale type. Let $A = \{r, s\}$ and let $R = \{(r, r), (s, s), (r, s), (s, r)\}$. Define S on $\mathfrak{R}$ by

$$xSy \iff (x = y - 1 \quad \text{or} \quad y = x - 1 \quad \text{or} \quad x = y).$$

(a)   Show that f and g are homomorphisms of $\mathfrak{A} = (A, R)$ into $\mathfrak{B} = (\mathfrak{R}, S)$ if $f(r) = 0$, $f(s) = 0$ and $g(r) = 0$, $g(s) = 1$.
(b)   Show that $(\mathfrak{A}, \mathfrak{B}, f)$ is irregular.
(c)   Show that $(\mathfrak{A}, \mathfrak{B}, f)$ is a ratio scale (assuming the definition makes sense for irregular scales).
(d)   Show that $(\mathfrak{A}, \mathfrak{B}, g)$ is not a ratio scale.

32.   Prove that if $(\mathfrak{A}, \mathfrak{B}, f)$ is a ratio scale which is regular and $(\mathfrak{A}, \mathfrak{B}, g)$ is another scale, then $(\mathfrak{A}, \mathfrak{B}, g)$ is also a ratio scale.

33.   Repeat Exer. 32 for interval scale.

34.   Repeat Exer. 32 for ordinal scale.

35.   Discuss critically our explication of the concept of meaningfulness. Think of examples which are either explained by or not explained by the explication.

8.5.   Examples of Fundamental Measurement I: Ordinal Utility Functions

8.5.1. THE REPRESENTATION THEOREM

In studying temperature, preference, etc., we were led to the following problem. Given a relational system (A, R) with R a binary relation on A, when can we find a real-valued function f on A such that for all $a, b \in A$,

$$aRb \Longleftrightarrow f(a) > f(b). \tag{23}$$

The function f is a homomorphism from (A, R) into the numerical relational system $(\Re, >)$. In this section, we shall state necessary and sufficient conditions for the existence of such a function f. That is, we state a representation theorem for the representation (23). We also state a uniqueness theorem.

The reader will recall that a binary relation (A, R) is asymmetric if whenever aRb, then $\sim bRa$. It is negatively transitive if whenever $\sim aRb$ and $\sim bRc$, then $\sim aRc$.

Theorem 8.1.   Suppose A is a finite set and R is a binary relation on A. Then there is a real valued function f on A satisfying

$$aRb \Longleftrightarrow f(a) > f(b) \tag{23}$$

if and only if (A, R) is asymmetric and negatively transitive.

We defer the proof of Theorem 8.1.

A binary relation (A, R) which is asymmetric and negatively transitive is called a *strict weak order*. If R is (strict) preference, a function f satisfying Eq. (23) is called an (ordinal) utility function. Thus, to see if we can measure a person's preferences to the extent of producing an ordinal utility function, we simply check whether or not these preferences satisfy the conditions for a strict weak order, i.e., asymmetry and negative transitivity. In general, we could do this by doing a *pair comparison experiment*. For every pair of alternatives a and b in A, we present a and b and ask the individual to tell us which if any he prefers. We present these pairs in a random order, and use his judgements to define preference. The results might be presented in a table like Table 8.4, where an individual has been asked his preferences among vacation trips. (We have seen similar tables and pair comparison experiments in our study of tournaments in Sec. 3.2.) Using the data, we check whether the axioms of asymmetry and negative transitivity are satisfied. We check whether the individual ever said he preferred a to b and also said he preferred b to a. And we check whether he ever failed to prefer a to b, failed to prefer b to c, but said he preferred a to c.

There are two interpretations for the axioms. One is that these are testable conditions which describe what a person's preferences must be for measurement to take place. We then take the *descriptive approach*, and simply

TABLE 8.4

Preferences among Vacation Trips.
Trip i is Preferred to Trip j iff the i, j Entry is 1.

	Rome	London	Athens	Moscow	Paris	Copen-hagen	Vienna	Row sum
Rome	0	1	1	1	0	1	1	5
London	0	0	0	0	0	1	0	1
Athens	0	1	0	1	0	1	1	4
Moscow	0	1	0	0	0	1	0	2
Paris	0	1	1	1	0	1	1	5
Copenhagen	0	0	0	0	0	0	0	0
Vienna	0	1	0	0	0	1	0	2

ask whether or not a person's preferences satisfy these conditions. Alternatively, we could use these axioms to define *rationality*. We could say that an individual who violates these axioms is acting irrationally. Indeed, many would say that an individual presented with a violation of, say negative transitivity, would say: "Oh, I've made a mistake." This approach is the *prescriptive* or *normative approach* and it is usually the approach taken in economic theory. The representation theorem is used to define the class of individuals to whom the theory applies, the so-called rational individuals. It is this second approach or interpretation which we apply to this representation theorem if it is used to study measurement of temperature. If R is the relation "warmer than," we think of the conditions of asymmetry and negative transitivity as conditions of rationality, which must be satisfied before measurement can take place.

Of course, in the case of warmer than, we can also think of these conditions as testable, and we would be surprised if an individual violated them, or at least we might be tempted to think of violations more as experimental errors than "real" violations. In the case of preference, the situation is different. If we subject the asymmetry and negative transitivity conditions to an experimental test, with R taken to be strict preference, then we can find that they are violated. For example, an individual may think price is more important than quality, but choose on the basis of quality if prices are close. Thus, if a and b are close in price and b and c are close in price, he may prefer a to b because a is of higher quality than b and he may prefer b to c because b is of higher quality than c. But he may prefer c to a because c is sufficiently lower in price than a to make a difference. Then transitivity of preference is violated. Moreover, so is negative transitivity, for he does not prefer c to b and he does not prefer b to a but he prefers c to a.[8] If one of the axioms such as

[8]This argument is due to Krantz, *et al.* [1971, p. 17]. We encountered it in Sec. 3.2, as an argument against transitivity of preference. We shall present other arguments against these simple axioms for preference in Sec. 8.8.

negative transitivity is violated, then the representation (23) cannot be achieved, i.e., there is no ordinal utility function. However, measurement in some sense might still be performed. We describe how in Sec. 8.8.

To prove Theorem 8.1, let us suppose first that there is a homomorphism f satisfying Eq. (23). We show that (A, R) is a strict weak order. First, (A, R) is asymmetric. For if aRb, then $f(a) > f(b)$, whence not $f(b) > f(a)$, and $\sim bRa$. Second, (A, R) is negatively transitive. For if $\sim aRb$ and $\sim bRc$ then $\sim[f(a) > f(b)]$ and $\sim[f(b) > f(c)]$, so $f(a) \leq f(b)$ and $f(b) \leq f(c)$. It follows that $f(a) \leq f(c)$, so $\sim[f(a) > f(c)]$, so $\sim aRc$.

Conversely, suppose (A, R) is a strict weak order. The proof that a homomorphism f exists is constructive. We define $f(x)$ as follows:

$$f(x) = \text{the number of } y \text{ in } A \text{ such that } xRy^9. \tag{24}$$

Let us begin by illustrating this construction. Suppose $A = \{a, b, c, d, e\}$ and $R = \{(a, c), (a, d), (a, e), (b, c), (b, d), (b, e), (c, d), (c, e)\}$. Then it is not hard to show that (A, R) is asymmetric and negatively transitive. The function f defined by Eq. (24) is given by

$$f(a) = 3 \quad \text{(since } aRc, aRd, aRe)$$
$$f(b) = 3$$
$$f(c) = 2$$
$$f(d) = 0$$
$$f(e) = 0.$$

It is easy to check that f is a homomorphism.

To formally prove that a function f defined by Eq. (24) always satisfies Eq. (23), we first prove that (A, R) is transitive. We use the assumption that (A, R) is a strict weak order. Suppose aRb and bRc hold and suppose, by way of contradiction, that $\sim aRc$ holds. Since (A, R) is asymmetric, bRc implies $\sim cRb$. Since (A, R) is negatively transitive, $\sim aRc$ and $\sim cRb$ imply $\sim aRb$, which is a contradiction. Thus, (A, R) is transitive. We now verify Eq. (23). If aRb, then by transitivity of R, bRy implies aRy for every y. Thus, the number of y such that aRy is at least as big as the number of y such that bRy. It follows that $f(a) \geq f(b)$. Moreover, aRb but not bRb, since a strict weak order is irreflexive. (Why?) Thus, $f(a) > f(b)$. Conversely, if $\sim aRb$, then $\sim bRy$ implies $\sim aRy$, by negative transitivity. Hence, aRy implies bRy, so $f(b) \geq f(a)$, whence $\sim[f(a) > f(b)]$. This proves (23) and completes the proof of Theorem 8.1.

We should note a useful corollary of the proof.

Corollary 1. Every strict weak order is transitive.

9The reader will recognize $f(x)$ as the Borda count of Chapter 7.

This corollary gives us an additional tool for demonstrating that a homomorphism satisfying Eq. (23) does not exist. Simply demonstrate that transitivity is violated.

It should be remarked that if A is finite and f defined by (24) does not give a homomorphism, then (A, R) is not a strict weak order and so there is no homomorphism. Thus, this gives us another test for existence of a homomorphism: Simply try to build one by a fixed procedure and verify whether or not you have succeeded. This result is important enough to state as a corollary.

Corollary 2. Suppose A is a finite set and R is a binary relation on A. Then there is a real-valued function f on A satisfying Eq. (23) if and only if the following function f satisfies (23):

$$f(x) = \text{the number of } y \text{ in } A \text{ such that } xRy. \tag{24}$$

Let us apply these ideas to measurement of preference. Suppose an individual is asked his preferences among vacation trips in a pair comparison format, and gives the data shown in Table 8.4. It is easy to calculate $f(x)$ of Eq. (24) as the row sum of row x. Then $f(x)$ is

$$f(\text{Rome}) = f(\text{Paris}) = 5$$
$$f(\text{Athens}) = 4$$
$$f(\text{Vienna}) = f(\text{Moscow}) = 2$$
$$f(\text{London}) = 1$$
$$f(\text{Copenhagen}) = 0.$$

If the matrix is rearranged so that alternatives are listed in descending order of row sums, then we can easily test whether this f is a homomorphism by checking whether there are 1's in row x for all those y with $f(y) < f(x)$. This should give a block of 1's in row x from some point on to the end. In our example, the rearranged matrix is shown in Table 8.5. The blocks are also shown. We see that f is a homomorphism. Thus, f is an ordinal utility function for this individual.

On the other hand, if preferences are expressed as in Table 8.6, then $f(x)$ as defined by Eq. (24) is

$$f(\text{Paris}) = 5$$
$$f(\text{Rome}) = f(\text{Athens}) = 4$$
$$f(\text{London}) = f(\text{Moscow}) = f(\text{Vienna}) = 2$$
$$f(\text{Copenhagen}) = 0.$$

Rearranging the matrix, we see in Table 8.7 that the row corresponding to Rome has its block of 1's broken by a 0 in the Rome, London entry. Thus, f

TABLE 8.5

Table 8.4 Rearranged.

	Rome	Paris	Athens	Moscow	Vienna	London	Copen-hagen	Row sum
Rome	0	0	1	1	1	1	1	5
Paris	0	0	1	1	1	1	1	5
Athens	0	0	0	1	1	1	1	4
Moscow	0	0	0	0	0	1	1	2
Vienna	0	0	0	0	0	1	1	2
London	0	0	0	0	0	0	1	1
Copenhagen	0	0	0	0	0	0	0	0

TABLE 8.6

Preferences among Vacation Trips.
Trip i Is Preferred to Trip j iff the i, j Entry is 1.

	Rome	London	Athens	Moscow	Paris	Copen-hagen	Vienna	Row sum
Rome	0	0	1	1	0	1	1	4
London	1	0	0	0	0	1	0	2
Athens	0	1	0	1	0	1	0	4
Moscow	0	1	0	0	0	1	0	2
Paris	0	1	1	1	0	1	1	5
Copenhagen	0	0	0	0	0	0	0	0
Vienna	0	1	0	0	0	1	0	2

TABLE 8.7

Table 8.6 Rearranged.

	Paris	Rome	Athens	London	Moscow	Vienna	Copen-hagen	Row sum
Paris	0	0	1	1	1	1	1	5
Rome	0	0	1	0	1	1	1	4
Athens	0	0	0	1	1	1	1	4
London	0	1	0	0	0	0	1	2
Moscow	0	0	0	1	0	0	1	2
Vienna	0	0	0	1	0	0	1	2
Copenhagen	0	0	0	0	0	0	0	0

does not define a homomorphism and so, by Corollary 2, there can be none. (We discover easily that London is preferred to Rome, even though $f(\text{London}) < f(\text{Rome})$.) There is no ordinal utility function. This implies that one of the strict weak order axioms will have to be violated by the data, and it is not hard to discover that negative transitivity is violated: London is preferred to Rome, but Paris is not preferred to Rome and London is not preferred to Paris.

If (A, R) is a strict weak order, the function f of Eq. (24) defines a ranking of the type we have studied in Chapter 7. One element is ranked above another if it has a higher f rating, and two elements are tied if they have the same f rating. Thus, the preferences of Table 8.4 give rise to the ranking

Rome-Paris

Athens

Vienna-Moscow

London

Copenhagen

(The hyphen between two alternatives indicates that these alternatives are tied.) If (A, R) is not a strict weak order, then we cannot obtain such a ranking, for any such ranking defines a strict weak order. Rankings are a convenient way of presenting strict weak orders. As far as being useful in decisionmaking is concerned, a ranking has as much information as an ordinal utility function.

8.5.2. The Expected Utility Hypothesis

It can be a rather time-consuming process to perform a pair comparison experiment and make an individual compare every pair of elements. The set of pairs can be very large. Let us describe an alternative procedure for finding a utility function which avoids this problem. This procedure works only if we assume that a utility function exists, i.e., that preference (among what we shall call consequences) is a strict weak order.

We often face a choice of alternative actions, or gambles, or lotteries, and each action has several possible consequences or outcomes $c_1, c_2, \ldots, c_n$. Each consequence c_i has a certain probability $p(c_i)$ of occurrence and a certain utility $u(c_i)$, if we assume that utility makes sense. In what follows, we shall not assume that p and u are known, but only that they exist. The *expected utility* of the action is defined to be

$$\sum_{i=1}^{n} p(c_i)u(c_i).$$

(This is the expected value of the utility function.) The *expected utility hypothesis* says that preferences among actions or choices are the same as they would be *if* you were doing the following: For each action you calculate its expected utility and you choose (prefer) the action with the highest expected utility.[10] (This idea goes back to Bernoulli [1738].)

We describe a procedure for calculating a utility function on the set A of consequences. Let R be a binary preference relation on A.[11] Assume that there are two elements of A, a_* and a^*, such that a^*Ra_* and such that for all a in A, $\sim aRa^*$ and $\sim a_*Ra$. That is, a^* is preferred to a_*, nothing is preferred to a^*, and a_* is not preferred to anything. Among the elements of A, the consequence a^* is the "best" and the consequence a_* is the "worst." Since a^*Ra_* and u is a utility function over the set of consequences A, $u(a^*) > u(a_*)$. Thus, $u(a^*) - u(a_*) > 0$.

Given a in A, let $\lambda(a)$ be the action which has consequence a with certainty; choose a probability $f(a)$ so that you are indifferent between the action $\lambda(a)$ and the action $\theta(a)$ which has two consequences, a^* with probability $f(a)$ and a_* with probability $1 - f(a)$. Indifference means that you neither prefer $\lambda(a)$ to $\theta(a)$ nor $\theta(a)$ to $\lambda(a)$. It is easy to show that $f(a)$ defines an ordinal utility function on (A, R). For the expected utility of $\lambda(a)$ is $u(a)$ and the expected utility of $\theta(a)$ is

$$f(a)u(a^*) + [1 - f(a)]u(a_*) = f(a)[u(a^*) - u(a_*)] + u(a_*).$$

Since you are indifferent between $\lambda(a)$ and $\theta(a)$, their expected utilities may be equated, and hence

$$u(a) = f(a)[u(a^*) - u(a_*)] + u(a_*).$$

Since $u(a^*) - u(a_*) > 0$, we may divide by $u(a^*) - u(a_*)$, and we obtain

$$f(a) = \frac{u(a) - u(a_*)}{u(a^*) - u(a_*)}$$

$$= \alpha u(a) + \beta,$$

where

$$\alpha = \frac{-1}{u(a^*) - u(a_*)}$$

and

$$\beta = \frac{-u(a_*)}{u(a^*) - u(a_*)}.$$

Since $\alpha > 0$, we conclude that for all a and b in A,

$$u(a) > u(b) \Longleftrightarrow f(a) > f(b).$$

---

[10]There is no assertion that you are actually calculating expected utilities, but only that your behavior can be accounted for by saying you are behaving *as if* that is what you are doing.

[11]Note that there are two preference relations here, one on consequences and one on actions.

Thus,

$$aRb \Longleftrightarrow u(a) > u(b) \Longleftrightarrow f(a) > f(b),$$

and so f defines an ordinal utility function on (A, R).

The following sums up the procedure for finding an ordinal utility function f on A. Let a^* and a_* be chosen as the "best" and "worst" possible consequences ahead of time. Given a, ask an individual for that probability $f(a)$ so that he is indifferent between obtaining a with certainty and an uncertain situation in which he obtains a^* with probability $f(a)$ and a_* otherwise. (In practice, $f(a)$ is obtained by trying out various probabilities until a satisfactory one is reached.) The numbers $f(a)$, once computed, can then be used to help make decisions. To the best of the author's knowledge, this general procedure is due to Howard Raiffa. See Raiffa [1968, 1969] for many practical procedures for calculating utility. There is a substantial literature concerning the expected utility hypothesis. Some references are Luce and Raiffa [1957], Fishburn [1964], Savage [1954], and, from the point of view of fundamental measurement, Luce and Krantz [1971] and Krantz, *et al.* [1971, Ch. 8].

8.5.3. THE UNIQUENESS THEOREM

Turning to the uniqueness problem, let us state a uniqueness theorem for the representation (23).

Theorem 8.2. Suppose A is a finite set, R is a binary relation on A, and f is a real-valued function on A satisfying

$$aRb \Longleftrightarrow f(a) > f(b). \tag{23}$$

Then $\mathfrak{A} = (A, R) \rightarrow \mathfrak{B} = (\mathfrak{R}, >)$ is a regular representation and $(\mathfrak{A}, \mathfrak{B}, f)$ is an ordinal scale.

Proof. If $\phi: f(A) \rightarrow \mathfrak{R}$ is any monotone increasing function, then $\phi \circ f$ satisfies (23) whenever f does. For

$$(\phi \circ f)(a) > (\phi \circ f)(b) \Longleftrightarrow f(a) > f(b) \Longleftrightarrow aRb.$$

Conversely, suppose f satisfies (23) and we are given a function $\phi: f(A) \rightarrow \mathfrak{R}$ such that $\phi \circ f$ also satisfies (23). We shall show that ϕ is monotone increasing. Suppose α and β are in $f(A)$, with $\alpha = f(a)$ and $\beta = f(b)$. Then

$$\alpha > \beta \Longleftrightarrow aRb \Longleftrightarrow (\phi \circ f)(a) > (\phi \circ f)(b) \Longleftrightarrow \phi(\alpha) > \phi(\beta).$$

The conclusion is that the class of admissible transformations of f is the class of monotone increasing functions. Finally, we shall show that every function f satisfying (23) defines a regular scale, so we are dealing with a regular representation. If g is any other function satisfying (23), define $\phi: f(A) \rightarrow \mathfrak{R}$ by $\phi(\alpha) = g(a)$ if $\alpha = f(a)$. It is possible that ϕ is not well-defined. For example, if $\alpha = f(a)$ and $\alpha = f(b)$, how do we know whether to choose $\phi(\alpha)$ to be $g(a)$ or $g(b)$? Fortunately, this is not a problem. For if $f(a) = f(b)$, then since f is a homomorphism, $\sim aRb$ and $\sim bRa$, so, since g is a homomorphism, $g(a) = g(b)$. Thus, ϕ is well-defined. Moreover, $g = \phi \circ f$. Thus, f is a regular scale.   Q.E.D.

Having a uniqueness theorem, let us apply it to the case of temperature. The representation (23) arises in the measurement of temperature if R is interpreted as "warmer than." However, according to Theorem 8.2, temperature is an ordinal scale, whereas in Sec. 8.4 we suggested that temperature is an interval scale. Is there something wrong with our whole model? The answer is that we can obtain temperature as an interval scale if we observe that it is possible to make judgements of temperature difference. To make this precise in the theory, one introduces a quaternary relation $D(a, b, s, t)$ on a set A of objects whose temperatures are being compared. The relation $D(a, b, s, t)$ is interpreted to mean that the difference between the temperature of a and the temperature of b is judged to be more than the difference between the temperature of s and the temperature of t. One seeks a real valued function f on A such that for all $a, b, s, t, \in A$,

$$D(a, b, s, t) \Longleftrightarrow f(a) - f(b) > f(s) - f(t). \qquad (25)$$

Under some reasonable assumptions, this numerical assignment turns out to be regular and unique up to a positive linear transformation and hence defines an interval scale. Sufficient conditions for the representation (25) and a proof of the uniqueness theorem may be found in Suppes and Winet [1955] or Krantz, *et al.* [1971, Sec. 4.4.1].

8.5.4. REMARKS[12]

1. Theorems 8.1 and 8.2 actually hold in the more general case where A is countable, i.e., may be put in one-to-one correspondence with a set of positive integers. This result is due to Cantor [1895]. A proof may be found in Birkhoff [1948], in Krantz, *et al.* [1971] or in Roberts [to appear].

2. It is easy to show that Theorem 8.1 is false without the assumption that A be countable. To give a counterexample, suppose we let $A = \mathfrak{R} \times \mathfrak{R}$ and we define R on A by

$$(a, b)R(s, t) \Longleftrightarrow a > s \quad \text{or} \quad (a = s \ \& \ b > t).$$

R is called the *lexicographic ordering of the plane.* The lexicographic ordering of the plane (A, R) corresponds to the ordering of words in a dictionary. We order first by first letter; in case of the same first letters, by second letter; and so on. It is easy to see that (A, R) is a strict weak order. (See Exer. 20.) We shall show that there is no real-valued function f on A satisfying (23). For suppose that such an f exists. Now we know that $(a, 1)R(a, 0)$. Thus, by (23),

---

[12]The reader may omit these remarks without loss of continuity.

$f(a, 1) > f(a, 0)$. We know that between any two real numbers there is a rational. Thus, there is a rational number $g(a)$ so that

$$f(a, 1) > g(a) > f(a, 0).$$

Now the function g is defined on the set $\mathfrak{R}$ and maps it into the set of rationals. Moreover, it maps $\mathfrak{R}$ into the rationals in a one-to-one fashion. For, if $a \neq b$, then either $a > b$ or $b > a$, say $a > b$. Then we have

$$g(a) > f(a, 0) > f(b, 1) > g(b),$$

from which we conclude $g(a) > g(b)$. It is well known that there can be no one-to-one mapping from the reals into the rationals. Thus, we have reached a contradiction. We conclude that the strict weak order (A, R) cannot be represented in the form (23).

3. If A is not necessarily finite, a theorem stating conditions both necessary and sufficient on a binary relation (A, R) for the existence of a representation satisfying (23) was discovered by Milgram [1939] and by Birkhoff [1948]. For a statement and proof of this theorem, see Krantz, *et al.* [1971] or Roberts [to appear].

EXERCISES

1. Which of the following binary relations are strict weak orders?
   (a) The relation of "having more members than" on a set of organizations.
   (b) The relation of having greater average hourly nitrogen oxide output on a set of electric power generating stations.
   (c) The relation "grandson of" on the set of all people in the United States.
   (d) The relation $(\mathfrak{R}, <)$.
   (e) The relation (A, R) where $A = [0, 1]$ and R is $\leq$.
   (f) The relation (A, R) where $A = \{(1, 1), (1, 2), (2, 4)\}$ and $(a, b)R(s, t)$ iff $[a \geq s$ and $b \geq t$ and $(a > s$ or $b > t)]$.
   (g) The relation (A, R) where A is $\{(1, 1), (1, 2), (2, 1)\}$ and R is as in part (f).

2. Let $A = \{3, 4, 5, 6\}$ and let R be $<$. Use the function defined in Eq. (24) to find a homomorphism from (A, R) into $(\mathfrak{R}, >)$.

3. Repeat Exer. 2 if R is $>$.

4. Let $A = \{a, b, c, d\}$ and $R = \{(b, c), (b, a), (b, d), (d, c), (d, a)\}$.
   (a) Show that (A, R) is a strict weak order.
   (b) Find a homomorphism from (A, R) into $(\mathfrak{R}, >)$, using the function f of Eq. (24).

5. The following ranking defines a strict weak order R on the set $A = $ {Rome, Paris, Athens, Vienna, Moscow, Copenhagen, London}. Find a homomorphism from (A, R) into $(\Re, >)$.

> Rome-Paris-Athens
> Vienna
> Moscow
> Copenhagen
> London

6. Let $A = \{(a, b): a, b \in \{1, 2, 3, 4\}\}$ and suppose

$$(a, b)R(c, d) \quad \text{iff} \quad a > c.$$

(a) Show that (A, R) is a strict weak order.
(b) Use the function f defined in Eq. (24) to find a homomorphism from (A, R) into $(\Re, >)$.

7. Let $A = \{0, 1, 2, \ldots, 23\}$. Every number a in A is congruent modulo 3 to a number among $\{0, 1, 2\}$. Let this number be called a mod 3. Let aRb hold iff a mod 3 $>$ b mod 3.
(a) Show that (A, R) is a strict weak order.
(b) Use the function f defined in Eq. (24) to find a homomorphism from (A, R) into $(\Re, >)$.

8. Suppose $A = \{0, 1, 2, \ldots, 31\}$ and aRb holds iff a mod 4 $>$ b mod 4, where a mod 4 is defined analogously as in Exer. 7. Use the function f defined in Eq. (24) to find a homomorphism from (A, R) into $(\Re, >)$. Does $g(x) = -f(x)$ also define a homomorphism?

9. The preference data of Table 8.8 was obtained by performing a hypothetical pair comparison experiment. Construct an ordinal utility function or show that none exists.

TABLE 8.8

Preferences Among Cars.
Model i Is Preferred to Model j iff the i, j Entry Is 1.

	Buick	Datsun	Volkswagen	Pontiac	Cadillac	Ford
Buick	0	1	1	1	0	1
Datsun	0	0	0	0	0	0
Volkswagen	0	1	0	1	0	1
Pontiac	0	0	0	0	0	0
Cadillac	1	1	1	1	0	1
Ford	0	1	0	1	0	0

10. Repeat Exer. 9 for the data of Table 8.9.

Preferences Among Cars.
Model *i* Is Preferred to Model *j* iff the *i, j* Entry Is 1.

	Buick	Datsun	Volkswagen	Pontiac	Cadillac	Ford
Buick	0	1	1	0	0	0
Datsun	0	0	0	0	1	1
Volkswagen	1	1	0	0	0	0
Pontiac	1	1	1	0	1	1
Cadillac	1	0	1	0	0	0
Ford	1	0	1	0	1	0

11. Suppose your set of alternatives is $A = \{a, b, c, d\}$ and your utility function is defined by $u(a) = 9$, $u(b) = 12$, $u(c) = 15$, $u(d) = 18$. Calculate the expected utility of the following lotteries.

   (a) Receive a with probability $\frac{1}{3}$, b with probability $\frac{1}{3}$, and c otherwise.

   (b) Have an equal chance of receiving either b or d.

   (c) Receive a with probability $\frac{1}{4}$ and c otherwise.

12. In Exer. 11, which of the lotteries would you prefer, assuming the expected utility hypothesis is true?

13. If $1000 is the "best" alternative in a set of consequences A and $100 is the "worst," describe how you could calculate the utility of the consequence $200, assuming that a utility function exists and that the expected utility hypothesis is true. (Note that $u(\$n)$ is not necessarily $nu(\$1)$. Why?)

14. Suppose you are indifferent between the following two gambles.

   Gamble 1:   Play a game in which you are sure to win $1.

   Gamble 2:   Play a game in which you have a 50% chance of winning $20 and a 50% chance of losing $6.

   Suppose u is a utility function over consequences, and assume the expected utility hypothesis.

   (a) Derive an expression relating $u(\$1)$, $u(\$20)$, and $u(\$6)$.

   (b) Could you be indifferent if $u(\$n) = nu(\$1)$?

15. Suppose you are considering whether or not to take your umbrella, when the probability of rain is 50%. If you take the umbrella, the certain consequence is that you will have the nuisance of carrying it with you all day. If you don't take it, you have a 50% chance of having a nuisance-free day (assuming carrying the umbrella is the only nuisance

you are considering) and a 50% chance of getting soaked. How could you choose between the two actions?

16. If f is an ordinal utility function, is it meaningful to say that:
    (a) $f(a) = f(b)$.
    (b) $f(a) > 2f(b)$.

17. Suppose (A, D) is a quaternary relation and f is a real-valued function on A satisfying Eq. (25). Show that a positive linear transformation of f also satisfies (25). What else is needed to prove that f is an interval scale?

18. Suppose $\mathfrak{A} = (A, H, o)$, $\mathfrak{B} = (\mathfrak{R}, >, +)$ and $f: A \rightarrow \mathfrak{R}$ is a homomorphism from $\mathfrak{A}$ into $\mathfrak{B}$. Is it necessarily the case that (A, H) is a strict weak order? (Give proof or counterexample.)

19. Suppose A is a finite set of events and aRb is interpreted to mean that a is subjectively judged to be more probable than b. Suppose that A contains one event e which is "certain" to occur. Under what circumstances is there a real-valued function p on A, interpreted as the probability of A, such that

$$aRb \Longleftrightarrow p(a) > p(b) \quad \text{(for all } a, b \in A) \tag{26}$$

    and

$$p(e) = 1 \text{ and } p(a) \geqq 0 \quad \text{(for all } a \in A)? \tag{27}$$

    For a further discussion of the representation of subjective probability, see Exer. 13 of Sec. 8.6 and Exers. 10 and 20 of Sec. 8.7. See also Krantz, *et al.* [1971, Ch. 5] or Roberts [to appear].

20. Prove that the lexicographic ordering of the plane (Sec. 8.5.4) is a strict weak order.

21. Give an example of a binary relation which is asymmetric but not negatively transitive.

22. Give an example of a binary relation which is negatively transitive but not asymmetric.

23. Give an example of a binary relation which is negatively transitive but not transitive.

24. Give an example of a binary relation which is transitive, but not negatively transitive.

25. Consider the following "gambles":

    Gamble 1: you win $1,000,000 automatically
    Gamble 2: you win $5,000,000 with probability .10, $1,000,000 with probability .89, and $0 otherwise

Gamble 3:   you win \$5,000,000 with probability .10 and \$0 otherwise
Gamble 4:   you win \$1,000,000 with probability .11 and \$0 otherwise.

The French economist Allais [1953] has reported that most subjects prefer Gamble 1 to Gamble 2 and Gamble 3 to Gamble 4.

(a)   Does choice of Gamble 1 over Gamble 2 violate the expected utility hypothesis?
(b)   What about choice of Gamble 3 over Gamble 4?
(c)   What about both choices simultaneously?
(d)   Explain why people might be likely to have preferences like this.
(e)   How might you convince them they are making a mistake? (See Raiffa [1968, p. 80ff].)

26.   *Project:* Perform a pair comparison experiment for preference, relative loudness, or some other comparison. Determine if an order-preserving mapping into the reals exists and, if so, construct one. (Cf. Exers. 9 and 10.)

27.   Can you imagine an economic system in which people violated transitivity or negative transitivity very often? What would be some of the consequences?

8.6. Examples of Fundamental Measurement II: Extensive Measurement[13]

8.6.1. Hölder's Theorem

In studying measurement of mass and, under certain circumstances, measurement of preference, we start with the relational system (A, R, o), where R is a binary relation on A and o is a (binary) operation. We want to find a real-valued function f on A satisfying (23) and

$$f(a \circ b) = f(a) + f(b). \tag{28}$$

We shall seek conditions on (A, R, o) (necessary and) sufficient for the existence of such a function f, i.e., conditions on (A, R, o) (necessary and) sufficient for the existence of a homomorphic map into $(\Re, >, +)$.

Attributes which have additive properties, such as mass for example, have traditionally been called *extensive* in the literature of measurement and so the problem of finding conditions on (A, R, o) (necessary and) sufficient for the existence of a homomorphic map into $(\Re, >, +)$ is called the problem of *extensive measurement.*

---

[13]This section involves mathematical topics which are not discrete in the sense of most of the topics in this book. These topics may be omitted without loss of continuity. (The reader who has not previously encountered the notion of a group should consider skipping this section, though he might see how far he can get.)

The theory of extensive measurement will naturally overlap with abstract algebra, which often deals with relational systems of the form (A, o). Since algebra is not considered a prerequisite for this book, we shall try to provide the reader with a brief background. If o is an operation on the set A, the pair (A, o) is called a *group* if it satisfies the following axioms:[14]

Axiom G1 (Associativity). For all a, b, c in $A, (a \circ b) \circ c = a \circ (b \circ c)$.

Axiom G2 (Identity). There is an (identity) element e in A such that for all a in A, $a \circ e = e \circ a = a$.

Axiom G3 (Inverse). For all a in A, there is an (inverse) element b in A such that $a \circ b = b \circ a = e$.

The reader is familiar with many examples of groups. $(\mathfrak{R}, +)$ is an example, with the identity e of Axiom G2 being 0 and the inverse b of Axiom G3 being $-a$. If $\mathfrak{R}^+$ is the positive reals and $\times$ is multiplication, then $(\mathfrak{R}^+, \times)$ is a group, with the identity being 1 and the inverse of a being $1/a$. $(\mathfrak{R}, \times)$ is not a group, since the only possible identity is 1 and the element 0 has no inverse b in $\mathfrak{R}$ such that $0 \times b = b \times 0 = 1$.

If (A, o) is a group, we may define na for every positive integer n. The definition is inductive. We define $1a$ to be a. Having defined na, we define $(n + 1) a$ to be $a \circ na$. The definition makes sense since (A, o) is associative.

Often, obtaining axiomatizations for homomorphisms into the reals can involve translating properties of the real number system into an abstract relational system. Most axiomatizations in measurement theory try to capture an important property of the real numbers called the *Archimedean property*. This property can be defined as follows: If a and b are real numbers and $a > 0$, then there is a positive integer n such that $na > b$. That is, no matter how small a might be or how large b might be, if a is positive, then sufficiently many copies of a will turn out to be larger than b. This property of the real number system is what makes measurement possible: It makes it possible to roughly compare the relative magnitudes of any two quantities a and b, by seeing just how many copies of a are required to obtain a larger number than b. We shall have to translate the Archimedean property of the reals into an Archimedean axiom in (A, R, o) in order to axiomatize the representation (A, R, o) into $(\mathfrak{R}, >, +)$.

Sufficient conditions for extensive measurement, the representation (A, R, o) into $(\mathfrak{R}, >, +)$, were first given by Hölder [1901]. We state some conditions closely related to those originally given by Hölder, by giving the following definition. A relational system (A, R, o) is an *Archimedean ordered group* if it satisfies the following axioms:

Axiom A1. (A, o) is a group.

[14]Often an additional axiom, called *Closure*, is explicitly stated. This axiom asserts that for all a, b in A, $a \circ b$ is in A, which is implicit in the definition of an operation.

Axiom A2. (A, R) is a *strict simple order*, i.e., it is a strict weak order and it is *complete* in the sense that for all $a \neq b$ in A, aRb or bRa.

Axiom A3 (Monotonicity).   For all a, b, c in A,

$$aRb \text{ iff } a \circ c \, R \, b \circ c \text{ iff } c \circ a \, R \, c \circ b.$$

Axiom A4 (Archimedean).   For all a, b in A, if aRe, where e is the group identity, then there is a positive integer n such that $na \, R \, b$.

The paradigm example of an Archimedean ordered group is $(\mathfrak{R}, >, +)$. We already know that $(\mathfrak{R}, +)$ is a group. $(\mathfrak{R}, >)$ is a strict simple order because it is strict weak and for all $a \neq b$ in $\mathfrak{R}$, $a > b$ or $b > a$. $(\mathfrak{R}, >, +)$ satisfies monotonicity since $a > b$ iff $a + c > b + c$ iff $c + a > c + b$. It satisfies the Archimedean axiom, because given a and b, if $a > 0$, there is a positive integer n such that $na > b$. Another example of an Archimedean ordered group is given by $(\mathfrak{R}^+, >, \times)$.

Hölder's Theorem is the following:

Theorem 8.3 (Hölder).   Every Archimedean ordered group is homomorphic to $(\mathfrak{R}, >, +)$.

We omit the proof.[15]

The axioms for an Archimedean ordered group give a representation theorem for extensive measurement. As such, they should be tested for various examples. Let us begin with the case of mass. Here, A is a collection of objects, aRb is interpreted to mean that a is judged heavier than b, and $a \circ b$ is the combination of a and b. Let us consider first the group axioms.[16] Certainly associativity seems to make sense: Taking first the combination of a and b and then combining with c amounts to the same object (as far as mass is concerned) as first combining b and c and then combining with a. Also, at least ideally, there is an element with no mass which could serve as the identity e. Speaking of inverses, however, does not make sense. Given an object a, the axiom requires that there be another object b which, when a and b are combined, results in the identity, the object with no mass. There is no such b. Thus, to obtain a usable representation theorem for measurement of mass, it is necessary to modify this axiom.

Let us next consider the remaining axioms for an Archimedean ordered group. It might be reasonable to assume that (A, R) is a strict simple order,

[15]The reader should note that every homomorphism into $(\mathfrak{R}, >, +)$ is one-to-one. This follows since (A, R) satisfies completeness (see Axiom $A2$).

[16]To say $(A, \circ)$ is a group requires that $\circ$ be an operation, so in particular we have to make sense out of $a \circ a$, $a \circ (b \circ a)$, etc. We have already mentioned that one can make sense out of these combinations by thinking of having an infinite number of ideal copies of each element present.

at least as an idealization, although we can run into problems with completeness: Two different objects may be close enough in mass so we cannot distinguish them. Monotonicity probably is reasonable: If you think a is heavier than b, then adding the same object c to each of a and b should not change your opinion, and taking c away similarly should not. Moreover, whether you add c to a and to b or a and b to c should not make a difference. Finally, it is probably reasonable to assume the Archimedean axiom: Given an object a which is heavier than the object with no mass, if we combine enough copies of a, it seems reasonable, at least in principle, that we can create an object which is heavier than any other object b. The Archimedean axiom is not really empirically testable; no finite experiment could get enough data to refute it. In this sense, if we accept this axiom, we treat it as an idealization which seems reasonable.

Let us next consider the same axioms for the case of preference. Here, A is a set of objects or alternatives, o is again combination, and aRb means a is (strictly) preferred to b. Associativity is probably reasonable. The argument is the same as in the case of mass.[17] The axioms of identity and inverse seem to be satisfied. For certainly there is at least ideally an object with absolutely no worth at all. And given any object a, owing someone else such an object might be considered an inverse alternative. For having a and owing a amount to having nothing.[18] The strict simple order axiom is one we have previously questioned for preference; we have even questioned whether preference is a strict weak order, and we probably can question completeness: It is possible to be indifferent between two alternatives. Let us turn next to monotonicity. Even this axiom might be questioned, if objects combined can be made more useful than individual objects. For example, suppose a is black coffee, b is a candy bar, and c is sugar. You might prefer b to a (not liking black coffee), but prefer a o c to b o c. This is really an argument against the additive representation, not just the monotonicity axiom. Finally, perhaps one can even question the Archimedean axiom. For example, suppose a is a lamp and b is a long, healthy life. Will sufficiently many lamps ever be better than having a long, healthy life?

[17]The argument uses the assumption that the combination is inert and does not involve physical or chemical interaction between elements. (We tacitly assumed this about the operation of combination for mass.) If such interaction is allowed, then combining a and b first and then bringing in c might create a different object from that obtained when b and c are combined first. To give an example, if a is a flame, b is some cloth, and c is a fire retardant, then combining a and b first and then combining with c is quite different from combining b and c first and then combining with a. However, this result follows only if we allow interaction (e.g., flame lights cloth) with our combination. Similar examples may be thought of from chemistry.

[18]This is not true if we put a premium on present holdings. In this case, we would have to look for an inverse to having a among the alternatives of owing objects of more worth than a.

The discussion above suggests that it is necessary to modify Hölder's axioms to obtain a satisfactory representation theorem for extensive measurement, even if it is only mass we wish to measure. Some early attempts to improve Hölder's Theorem can be found in Huntington [1902a,b, 1917], Suppes [1951], Behrend [1953, 1956], and Hoffman [1963]. All of these improvements involve some axioms which are not necessary, as indeed does Hölder's Theorem. Sets of axioms which are necessary as well as sufficient are given by Alimov [1950], Holman [1969], and Roberts and Luce [1968]. See Krantz, *et al.* [1971, Ch. 3] for a detailed discussion and see Exers. 16 through 22.

8.6.2. Uniqueness

Before leaving the topic of extensive measurement, we ask for a uniqueness statement. Our earlier observations about measurement of mass suggest that the representation f should be unique up to a similarity transformation, i.e., measurement should be on a ratio scale. We shall prove this. This result will have significance for preference as well. For it says that if a utility function f satisfies conditions (23) and (28), then it is meaningful to say that the utility of one alternative is twice the utility of a second, or half the utility of a second, and so on. If this is the case, we can begin to use utility functions to make "quantitative" decisions, rather than just "qualitative" ones. (A utility function satisfying (23) and (28) is sometimes called a *cardinal utility function.*)

Theorem 8.4. Suppose A is a nonempty set, R is a binary relation on A, o is a (binary) operation on A, and f is a real-valued function on A satisfying the following conditions for all $a, b \in A$:

$$aRb \iff f(a) > f(b) \tag{23}$$

and

$$f(a \circ b) = f(a) + f(b). \tag{28}$$

Then $\mathfrak{A} = (A, R, \circ) \longrightarrow \mathfrak{B} = (\mathfrak{R}, >, +)$ is a regular representation and $(\mathfrak{A}, \mathfrak{B}, f)$ is a ratio scale.

Proof. To show that $\mathfrak{A} \longrightarrow \mathfrak{B}$ is a regular representation, let $(\mathfrak{A}, \mathfrak{B}, f)$ be any scale and let $(\mathfrak{A}, \mathfrak{B}, g)$ be another scale. Define $\phi: f(A) \longrightarrow \mathfrak{R}$ by $\phi(x) = g(a)$ if $x = f(a)$. That this is well-defined follows exactly as it did in the proof of Theorem 8.2. We have $g = \phi \circ f$.

To show that $(\mathfrak{A}, \mathfrak{B}, f)$ is a ratio scale, suppose first that for some $\alpha > 0$, $\phi(x) = \alpha x$ for all x in $f(A)$. Then clearly $\phi \circ f$ satisfies (23) and (28), since f does. Conversely, suppose $\phi: f(A) \longrightarrow \mathfrak{R}$ is an admissible transformation. We shall prove that ϕ is a similarity transformation, i.e., there is $\alpha > 0$ so that $\phi(x) = \alpha x$ for all x in $f(A)$. Let $g = \phi \circ f$. We show first that if $f(a) > 0$, then $g(a) > 0$. For if $g(a) \leqq 0$, then $g(a \circ a) = g(a) + g(a) \leqq g(a)$, where the equality follows since g satisfies (28). Now since g satisfies (23), $g(a \circ a) \leq g(a)$ implies $\sim[a \circ a\, R\, a]$. Since f satisfies (23), $f(a \circ a) \leq f(a)$ follows. Thus, since f satisfies (28), $f(a) + f(a) \leq f(a)$, so $f(a) \leq 0$. A similar proof shows that if $f(a) < 0$, then $g(a) < 0$ and if $f(a) = 0$, then $g(a) = 0$.

If $f(a) = 0$ for all a in A, then $g(a) = 0$ for all a in A. Then any positive α will suffice to satisfy $\phi(x) = \alpha x$ for all x in $f(A)$. Thus, we may assume that for some e in A, $f(e) \neq 0$. We shall assume that $f(e) > 0$. The proof in the case that $f(e) < 0$ is similar. Since $f(e) > 0$, $g(e)$ must be > 0. Pick α such that $g(e) = \alpha f(e)$. Since $f(e) > 0$ and $g(e) > 0$, it follows that $\alpha > 0$. We shall prove that $g(a) = \alpha f(a)$ for all a in A, which proves that $\phi(x) = \alpha x$ for all x in $f(A)$. The proof is by contradiction. We assume first that $g(a) < \alpha f(a)$. Hence,

$$\frac{g(a)}{\alpha f(e)} < \frac{f(a)}{f(e)}.$$

Since between every pair of real numbers there is a rational number, there are positive integers m and n such that

$$\frac{g(a)}{\alpha f(e)} < \frac{m}{n} < \frac{f(a)}{f(e)},$$

or

$$\frac{g(a)}{\alpha} < \frac{m}{n} f(e) < f(a).$$

It follows that $mf(e) < nf(a)$, so $f(me) < f(na)$, so $naRme$. (Why does $f(na) = nf(a)$?) But then $g(na) > g(me)$, so $ng(a) > mg(e) = m\alpha f(e)$, so

$$\frac{g(a)}{\alpha} > \frac{m}{n} f(e).$$

This is a contradiction, and shows that $g(a) < \alpha f(a)$ is impossible. A similar proof shows that $g(a) > \alpha f(a)$ is impossible. Q.E.D.

8.6.3. A COMMENT

Before closing this section, we note that, in spite of historical emphasis on additivity, there is nothing magic about the addition operation. Indeed, if f satisfies (23) and (28), then $g = e^f$ satisfies (23) and

$$g(a \circ b) = g(a) \times g(b). \tag{29}$$

Thus a multiplicative representation (23), (29) can also be obtained, with positive g. Conversely, if a multiplicative representation (23), (29) can be obtained with positive g, then $f = \ln g$ gives an additive representation (23), (28). The multiplicative representation changes the scale type, though the logarithm of the multiplicative representation gives rise to the same type of scale as the additive representation (see Exer. 15).

EXERCISES

1. Suppose $A = \{0, 1\}$, $R = >$, and o is defined by
$$0 \text{ o } 0 = 0, \quad 0 \text{ o } 1 = 1 \text{ o } 0 = 1 \text{ o } 1 = 1.$$
Which of the following conditions on (A, R, o) are satisfied?
   (a) Associativity.
   (b) Identity.
   (c) Inverse.
   (d) Strict simple order.
   (e) Monotonicity.
   (f) Archimedean Axiom. (Strictly speaking, if there is no identity e in A, we must modify the Archimedean axiom to read: If $a \text{ o } a \, R \, a$, then ... If there is no identity, is this Archimedean axiom satisfied?) axiom satisfied?)

2. Suppose $A = \{1, 2\}$, $R = >$, and o is defined by
$$1 \text{ o } 2 = 1 \text{ o } 1 = 1, \quad 2 \text{ o } 1 = 2 \text{ o } 2 = 2.$$
Repeat Exer. 1.

3. Which of the following are groups?
   (a) $(N, +)$, where N is the set of positive integers.
   (b) $(Z, +)$, where Z is the set of all integers.
   (c) $(N, \times)$.
   (d) $(Q, +)$, where Q is the set of rationals.
   (e) $(Q, \times)$.

4. Which of the binary relations of Exer. 1, Sec. 8.5, are strict simple orders?

5. Given a relational system (A, o) of type $(; 1)$, prove that if $f(a \text{ o } b) = f(a) + f(b)$ for all $a, b \in A$, then $f(na) = nf(a)$, all $a \in A$ and all positive integers n. (Use induction.)

6. Prove that $(\mathfrak{R}^+, >, \times)$ is an Archimedean ordered group.

7. Which of the following structures (A, R, o) are Archimedean ordered groups?
   (a) $(\mathfrak{R}, <, +)$.
   (b) $(\mathfrak{R}^-, <, \times)$, where $\mathfrak{R}^-$ is the negative reals.
   (c) $(N, >, +)$, where N is the positive integers.
   (d) $(Q, >, \times)$, where Q is the rationals.
   (e) $(Q^+, >, \times)$, where Q^+ is the positive rationals.

8. Suppose $A = \mathfrak{R} \times \mathfrak{R}$, suppose R is the lexicographic ordering of the plane (see Sec. 8.5.4), and suppose that $(a, b) \text{ o } (c, d)$ is defined to be $(a + c, b + d)$.

   (a)   Show that (A, R, o) is not an Archimedean ordered group, by exhibiting one of the axioms which is violated.

   (b)   Show the same thing without checking the axioms.

9.   Give an example of a relational system (A, R, o) which violates monotonicity.

10.   Suppose f is a cardinal utility function, i.e., a function satisfying (23) and (28) when R is preference. Which of the following statements are meaningful?

   (a) $f(a) = 10 f(b)$.

   (b)   The utility of a is greater than the utility of b.

   (c)   The utility of a is greater than twice the utility of b.

11.   Suppose (A, R, o) is homomorphic to $(\Re, >, +)$. Prove that $\sim [a \, o \, b \, R \, b \, o \, a]$.

12.   Which of the axioms for an Archimedean ordered group are necessary conditions for extensive measurement?

13.   In the subjective probability situation described in Exer. 19 of Sec. 8.5, it is natural to speak of the union of two events. Then, we want a function p on A which satisfies conditions (26) and (27) of Exer. 19 and also

$$p(a \cup b) = p(a) + p(b), \tag{30}$$

whenever $a \cap b = \varnothing$. (Assume that union and intersection satisfy the usual properties.[19]) Such a function p is called a *measure of subjective (or qualitative) probability*. Is the following monotonicity condition necessary for the existence of such a measure p?

Monotonicity condition: If $a \cap b = a \cap c = \varnothing$, then
$$bRc \quad \text{iff} \quad (a \cup b) \, R \, (a \cup c).$$

For a further discussion of the representation of subjective probability, see Exers. 10 and 20 of Sec. 8.7.

14.   Complete the proof of Theorem 8.4 for the case where $g(a) > \alpha f(a)$.

15.   Suppose $g : A \to \Re^+$ is a homomorphism from $\mathfrak{A} = (A, R, o)$ into $\mathfrak{B} = (\Re^+, >, \times)$. Prove that $\mathfrak{A} \to \mathfrak{B}$ is a regular representation and $\phi : g(A) \to \Re^+$ is an admissible transformation if and only if for some $\alpha > 0$, $\phi(x) = x^\alpha$ for all $x \in g(A)$.

16.   The following necessary and sufficient conditions for extensive measurement were introduced by Roberts and Luce [1968], and define an *extensive structure* (A, R, o).

---

[19]Formally, assume that the set of events A is an algebra of sets under union and intersection, i.e., it is closed under these operations and also under complementation.

Axiom E1 (Weak Associativity). For all a, b, c in A,

$$\sim[a \circ (b \circ c)\, R\, (a \circ b) \circ c] \quad \text{and} \quad \sim[(a \circ b) \circ c\, R\, a \circ (b \circ c)].$$

Axiom E2. (A, R) is a strict weak order.

Axiom E3. Monotonicity (Axiom A3).

Axiom E4 (Modified Archimedean Axiom). For all a, b, c, d in A, if aRb, then there is a positive integer n such that $na \circ c\, R\, nb \circ d$.

Show that each of the axioms for an extensive structure is a necessary condition for extensive measurement. In particular, show that Axiom E4 is indeed an Archimedean axiom in the sense that it follows from the Archimedean property of the real number system.

17. Which of the systems of Exer. 7 satisfy all of the axioms for an extensive structure (Exer 16)?

18. Comment on the axioms for an extensive structure (Exer. 16) as axioms for mass.

19. Comment on the axioms for an extensive structure as axioms for preference.

20. Suppose (A, R) is a strict weak order and $(A, \circ)$ is associative (or weakly associative; see Exer. 16). If $2a\, R\, a$, then a is called *positive*. Assume that every a in A is positive. Is it possible for A to be finite? (Give proof or counterexample.)

21. If $(A, R, \circ)$ is an extensive structure (Exer. 16), is it possible for A to be finite? If so, describe what $(A, R, \circ)$ would have to look like. (*Hint:* Use the notion of positivity defined in Exer. 20. Can there be any positive elements?)

22. Suppose $(A, R, \circ)$ satisfies the first three axioms for an extensive structure (Exer. 16). (You may assume associativity in place of weak associativity.) A *pair* of elements a and b in A is called *anomalous* if either aRb or bRa and either for all positive integers n,

$$na\, R\, (n+1)b \quad \text{and} \quad nb\, R\, (n+1)a,$$

or for all positive integers n,

$$(n+1)b\, R\, na \quad \text{and} \quad (n+1)a\, R\, nb.$$

Show that if $(A, R, \circ)$ is homomorphic to $(\Re, >, +)$, then there is no anomalous pair. (*Note:* Alimov [1950] proved that the first three axioms for an extensive structure plus the assumption that there are no anomalous pairs provide necessary and sufficient conditions for extensive measurement.)

23. *Project:* Perform the following experiment. Use four simple objects, of different weights, and let A consist of all simple objects plus all objects a o b for a, b simple objects. Perform a pair comparison experiment for judgement of relative heaviness among all pairs of objects in A. Test to see which of the axioms for Archimedean ordered groups or extensive structures (Exer. 16) are satisfied. Which axioms cannot be tested by this experiment?

24. Discuss the axioms of Hölder's Theorem as axioms for measurement of
    (a) Length.
    (b) Area.

25. Discuss the axioms for an extensive structure (Exer. 16) as axioms for
    (a) Length.
    (b) Area.

8.7. Examples of Fundamental Measurement III: Conjoint Measurement

In this section, we consider a third example of fundamental measurement which is a bit different in character from our first two examples.

Very often when we are making choices among alternatives, we consider these alternatives from several points of view. For example, we might choose an automobile on the basis of price, appearance, handling, economy of operation, etc. In designing a rapid transit system, we might consider power source, vehicle design, right-of-way design, and so on. In such cases, each alternative a in the set A of alternatives can often be looked at as an n-tuple $(a_1, a_2, \ldots, a_n)$. The component a_i represents some measure of alternative a on the ith *dimension* or attribute. Thus, in the case of cars, a_1 might be price, a_2 might be some measure of appearance, and so on. In economics, a_i is often thought of as the amount of the ith commodity and $a=(a_1, a_2, \ldots, a_n)$ is then thought of as a "market basket." Consumers are asked to express their preferences among alternative market baskets.

We let A_i be the set of all possible a_i and think of A as a Cartesian product $A_1 \times A_2 \times \cdots \times A_n$. We say that A has a *product structure*. There is an alternative interpretation possible for the meaning of a_i and A_i: Each A_i might be a set of possibilities for the ith attribute, and a_i would then be an element of the set A_i. For example, A_1 might be a set of power sources, A_2 a set of vehicle designs, and so on. To choose an alternative, we pick one member a_i from each set A_i, for example, one power source, one vehicle design, etc.

In applications of utility theory to decisionmaking, it is a very important problem to construct a product structure on an unstructured set of alternatives. The most natural procedure builds up this structure hierarchically. The reader is referred to Manheim and Hall [1968], Miller [1969], or Raiffa [1968, 1969] for a discussion.

We shall assume that we start with a set of alternatives A which has been given a product structure. In order to calculate utility of elements of A, it would be considerably easier if we could calculate utility on each attribute separately and then add. That is, if $u:A \longrightarrow \Re$ is an ordinal utility function, we would like to find real-valued functions $u_1, u_2, \ldots, u_n$ on $A_1, A_2, \ldots, A_n$ respectively so that for all $a = (a_1, a_2, \ldots, a_n)$ in A,

$$u(a) = u_1(a_1) + u_2(a_2) + \cdots + u_n(a_n). \tag{31}$$

A utility function satisfying (31) is called *additive*.

Suppose R is a binary relation on the set A with a product structure. If we interpret R as (strict) preference, then we seek real-valued functions u_i on A_i such that for all $a = (a_1, a_2, \ldots, a_n)$ and $b = (b_1, b_2, \ldots, b_n)$ in A,

$$aRb \Longleftrightarrow$$
$$u_1(a_1) + u_2(a_2) + \cdots + u_n(a_n) > u_1(b_1) + u_2(b_2) + \cdots + u_n(b_n). \tag{32}$$

Strictly speaking, the representation (32) does not fit our general framework for measurement. We cannot speak of a homomorphism of a relational system into another relational system. However, we shall treat (32) in the same spirit as the measurement representations of the previous two sections, and seek (necessary and) sufficient conditions on (A, R) for the existence of functions u_i satisfying (32). The representation (32) is often called *(additive) conjoint measurement*, because different components are being measured conjointly.

To illustrate, suppose $n = 2$, $A_1 = A_2 = \{0, 1\}$, and strict preference gives rise to the following ranking:

$$(1, 1)$$
$$(1, 0)$$
$$(0, 1)$$
$$(0, 0)$$

An ordinal utility function u exists: Take

$$u(1, 1) = 3, \qquad u(1, 0) = 2, \qquad u(0, 1) = 1, \qquad u(0, 0) = 0.$$

Moreover, we may find an additive conjoint representation (32). We pick

$$u_1(1) = 2, \qquad u_1(0) = 0, \qquad u_2(1) = 1, \qquad \text{and} \qquad u_2(0) = 0.$$

Then

$$u_1(1) + u_2(1) = 3, u_1(1) + u_2(0) = 2,$$
$$u_1(0) + u_2(1) = 1, u_1(0) + u_2(0) = 0,$$

and so

$$u_1(1) + u_2(1) > u_1(1) + u_2(0) > u_1(0) + u_2(1) > u_1(0) + u_2(0).$$

Using the same A_1 and A_2, if $(0, 1)$ is strictly preferred to $(0, 0)$ and $(1, 0)$ is strictly preferred to $(1, 1)$, then no additive conjoint representation exists. For suppose one did. Then $(0, 1) \, R \, (0, 0)$ implies

$$u_1(0) + u_2(1) > u_1(0) + u_2(0),$$

so

$$u_2(1) > u_2(0),$$

so

$$u_1(1) + u_2(1) > u_1(1) + u_2(0),$$

which implies $(1, 1) \, R \, (1, 0)$, a contradiction.

Conjoint measurement has potential applications in areas other than utility theory. For example, in studying response strength, psychologists often consider two factors, drive and incentive. They sometimes want to measure response strength conjointly over these two factors. That is, if R is a binary relation "responds more strongly than," and d and k are drive and incentive respectively, they want functions δ and κ such that for all d_1, d_2 in a set of drives A_1 and k_1, k_2 in a set of incentives A_2,

$$(d_1, k_1)R(d_2, k_2) \longleftrightarrow \delta(d_1) + \kappa(k_1) > \delta(d_2) + \kappa(k_2). \tag{33}$$

Similarly, in studying binaural loudness in auditory perception, psychologists present sounds of different intensity to each ear and study the effects conjointly. If the relation R is "louder than," the first component l measures sound intensity presented to the left ear, and the second component r measures sound intensity presented to the right ear, they seek functions $\mathcal{L}$ and $\mathcal{R}$ such that for all l_1, l_2 in a set A_1 and r_1, r_2 in a set A_2,

$$(l_1, r_1)R(l_2, r_2) \longleftrightarrow \mathcal{L}(l_1) + \mathcal{R}(r_1) > \mathcal{L}(l_2) + \mathcal{R}(r_2). \tag{34}$$

The conjoint measurement representation also arises in the theory of mental testing. Let A_1 be a set of subjects in such a test and A_2 be a set of items on the test. Then $(s_1, t_1)R(s_2, t_2)$ is interpreted to mean that subject s_1 scores better on item t_1 than subject s_2 does on item t_2. It is tempting to assume that subjects and items are independent, and that the score of a subject depends only on his ability and the difficulty of the item, and not on how hard the item is for him personally. Then, we would like ability and difficulty functions α and δ such that for all $s_1, s_2 \in A_1$ and $t_1, t_2 \in A_2$,

$$(s_1, t_1)R(s_2, t_2) \longleftrightarrow \alpha(s_1) + \delta(t_1) > \alpha(s_2) + \delta(t_2). \tag{35}$$

We shall study the conjoint measurement representation (32) in the special case where $n = 2$, i.e., we shall study the representation

$$(a_1, a_2)R(b_1, b_2) \longleftrightarrow u_1(a_1) + u_2(a_2) > u_1(b_1) + u_2(b_2). \tag{36}$$

Conditions on (A, R) sufficient for conjoint measurement were first presented by Debreu [1960], though some of his conditions were topological. An algebraic representation theorem in the spirit of those in the previous two sections was presented by Luce and Tukey [1964]. More refined conditions can be found in Krantz, *et al.* [1971, Ch. 6]. We present them in the exercises.

We shall restrict ourselves to the case where each A_i is finite. In this case, conditions have been found which are necessary as well as sufficient. These conditions are due to Scott [1964].

It is convenient to state Scott's conditions in terms of the binary relation S on A which is defined as follows:

$$aSb \iff \sim bRa.$$

If R is strict preference, then S is weak preference: aSb if and only if we think a is at least as good as b.

The first Scott condition is the following:

Axiom C1.   For all a_1, b_1 in A_1 and a_2, b_2 in A_2,

$$(a_1, a_2)S(b_1, b_2) \quad \text{or} \quad (b_1, b_2)S(a_1, a_2).$$

This condition obviously follows from the representation (36). (Why?) The second Scott condition is the following:

Axiom C2.   Suppose $x_0, x_1, \ldots, x_{n-1}$ are in A_1 and $y_0, y_1, \ldots, y_{n-1}$ are in A_2 and suppose π and σ are permutations of $\{0, 1, \ldots, n-1\}$.[20] If $(x_i, y_i)S(x_{\pi(i)}, y_{\sigma(i)})$ for $i = 1, 2, \ldots, n-1$, then $(x_{\pi(0)}, y_{\sigma(0)})S(x_0, y_0)$.

To illustrate Axiom C2, let us take $A_1 = A_2 = \{0, 1\}$, $n = 2$, $x_0 = 1$, $x_1 = 0$, $y_0 = 0$, and $y_1 = 1$. Suppose π is the identity permutation, the permutation which takes 0 into 0 and 1 into 1. Suppose σ is the permutation which takes 0 into 1 and 1 into 0. Axiom C2 says that if $(x_1, y_1)S(x_{\pi(1)}, y_{\sigma(1)})$, then $(x_{\pi(0)}, y_{\sigma(0)})S(x_0, y_0)$. Thus, if $(x_1, y_1)S(x_1, y_0)$, then $(x_0, y_1)S(x_0, y_0)$. In our example, this says that if $(0, 1)S(0, 0)$, then $(1, 1)S(1, 0)$. If this condition (which we have already encountered) is violated, then there are no functions u_1 and u_2 satisfying (36).

To show that Axiom C2 follows from the representation (36), note that since π and σ are permutations,

$$\sum_{i=0}^{n-1} [u_1(x_i) + u_2(y_i)] = \sum_{i=0}^{n-1} u_1(x_i) + \sum_{i=0}^{n-1} u_2(y_i)$$

$$= \sum_{i=0}^{n-1} u_1(x_{\pi(i)}) + \sum_{i=0}^{n-1} u_2(y_{\sigma(i)})$$

$$= \sum_{i=0}^{n-1} [u_1(x_{\pi(i)}) + u_2(y_{\sigma(i)})],$$

for each x_i and y_i is listed once and only once in the next-to-last expression. Note also that $(x_i, y_i)S(x_{\pi(i)}, y_{\sigma(i)})$ for $i = 1, 2, \ldots, n-1$ implies that

$$\sum_{i=1}^{n-1} [u_1(x_i) + u_2(y_i)] \geq \sum_{i=1}^{n-1} [u_1(x_{\pi(i)}) + u_2(y_{\sigma(i)})].$$

[20]The reader will recall that a permutation is just a one-to-one function of the set onto itself. It amounts to a relabelling or rearranging of the elements of the set.

Thus,

$$u_1(x_0) + u_2(y_0) \leqq u_1(x_{\pi(0)}) + u_2(y_{\sigma(0)}),$$

so

$$(x_{\pi(0)}, y_{\sigma(0)})S(x_0, y_0).$$

Theorem 8.5 (Scott). Suppose A_1 and A_2 are finite sets and R is a binary relation on $A = A_1 \times A_2$. Then Axioms C1 and C2 are necessary and sufficient for the existence of functions $u_1 : A_1 \longrightarrow \Re$ and $u_2 : A_2 \longrightarrow \Re$ such that for all $a_1, b_1 \in A_1$ and $a_2, b_2 \in A_2$,

$$(a_1, a_2)R(b_1, b_2) \Longleftrightarrow u_1(a_1) + u_2(a_2) > u_1(b_1) + u_2(b_2). \tag{36}$$

If u_1' and u_2' also satisfy (36), then there are real numbers α, β, and γ, with $\alpha > 0$, such that

$$u_1' = \alpha u_1 + \beta$$
$$u_2' = \alpha u_2 + \gamma.^{21}$$

We shall skip the sufficiency proof for Theorem 8.5. (The proof uses a clever variant of a famous theorem, the separating hyperplane theorem.)

Axiom C1 seems quite reasonable for the case of preference and utility. Axiom C2 also seems reasonable here. But it is impossible to completely verify it empirically. For really, Axiom C2 is a whole infinite bundle of axioms, one for each n.[22]

Even though Scott's axioms are not completely testable, the additive representation can be tested directly. In one such test, computer programs are used to "fit" the best possible additive representation, and then statistical tests are used to see if the data is accounted for by this additive representation. Using this procedure in the binaural loudness situation, Levelt, Riemersma, and Bunt [1972] discovered that an additive representation fit their data very well.

Another approach to testing whether conjoint measurement is a good model for a particular situation is to test various necessary conditions which follow from it. If these necessary conditions are violated, then of course the conjoint measurement representation is not possible. One such necessary condition is called independence. We say that $(A, R) = (A_1 \times A_2, R)$ satisfies *independence* if whenever $(a, x)R(b, x)$ then $(a, y)R(b, y)$ for all y, and whenever $(a, x)R(a, y)$ then $(b, x)R(b, y)$ for all b. This axiom seems reasonable in the preference and utility situation. For example, the axiom says that if you prefer ($10,000, 1 house) to ($10,000, 0 houses), you will also prefer

---

[21] Thus in particular, each u_i is a (regular) interval scale, but admissible transformations of u_1 and u_2 cannot be taken independently.

[22] We do not get away with using only a finite number of n, since the x_i and y_i are not necessarily distinct. That is, $x_0, x_1, \ldots, x_{n-1}$ might all be the same element.

($1 million, 1 house) to ($1 million, 0 houses). Independence is violated if there is some interaction between alternatives. To repeat an example used in Sec. 8.6, if you don't like black coffee, you might prefer (no coffee, no sugar) to (1 coffee, no sugar), but not prefer (no coffee, 1 sugar) to (1 coffee, 1 sugar). Independence is often violated in the mental testing situation also. For example, if subject a is good in arithmetic and subject b is good in vocabulary, and if item x is an arithmetic item and item y is a vocabulary item, we might very well get $(a, x)R(b, x)$ but not $(a, y)R(b, y)$.

If the additive representation of conjoint measurement does not apply to certain situations, perhaps because of interactions between terms, it is possible to modify the representation. For example, we could seek real-valued functions u_1 and λ_1 on A_1 and u_2 and λ_2 on A_2 such that for all a_1, b_1 in A_1 and a_2, b_2 in A_2,

$$(a_1, a_2)R(b_1, b_2) \Longleftrightarrow$$
$$u_1(a_1) + u_2(a_2) + \lambda_1(a_1)\lambda_2(a_2) > u_1(b_1) + u_2(b_2) + \lambda_1(b_1)\lambda_2(b_2). \quad (37)$$

The term $\lambda_1(a_1)\lambda_2(a_2)$ represents the interaction effect. Representation (37) is called the *quasi-additive representation*. Other related representations are studied under the heading *polynomial conjoint measurement*. For a discussion of such representations, the reader is referred to Krantz, *et al.* [1971, Ch. 7].

EXERCISES

1. A shopper is asked to express preferences among the following market baskets:

    a: 3 loaves of bread, 6 dozen eggs
    b: 3 loaves of bread, 2 dozen eggs
    c: 1 loaf of bread, 6 dozen eggs
    d: 1 loaf of bread, 2 dozen eggs.

    Suppose the shopper defines the following preference relation R:

    $$R = \{(a, b), (a, c), (a, d), (b, c), (b, d), (c, d)\}.$$

    Find an additive conjoint representation for utility.

2. Suppose a psychologist tests two drives d_1 and d_2 and two incentives k_1 and k_2, and he measures response strength for each combination of drive and incentive. Suppose judgements of response strength give rise to the following ranking:

    $$(d_1, k_1)$$
    $$(d_1, k_2)$$
    $$(d_2, k_1)$$
    $$(d_2, k_2)$$

Find functions δ and κ such that for all i, j, s, t,

(d_i, k_j) shows stronger response than $(d_s, k_t) \Longleftrightarrow$
$$\delta(d_i) + \kappa(k_j) > \delta(d_s) + \kappa(k_t).$$

3. State Scott's Axiom C2 if $x_0 = a$, $x_1 = b$, $x_2 = c$, $y_0 = 0$, $y_1 = 1$, $y_2 = 2$, $\pi(0) = 1$, $\pi(1) = 2$, $\pi(2) = 0$, $\sigma(0) = 0$, $\sigma(1) = 1$, and $\sigma(2) = 2$.

4. Suppose in Exer. 2 that the judgements of response strength give rise to the following ranking:

$$(d_1, k_2)$$
$$(d_1, k_1)$$
$$(d_2, k_1)$$
$$(d_2, k_2)$$

   (a) Show that there are no functions δ and κ as in Exer. 2 by assuming that such functions exist and deriving a contradiction.
   (b) Illustrate an axiom of Scott's which is violated.

5. In a binaural loudness experiment, we present stimuli (l, r) consisting of sounds of intensity l to the left ear and sounds of intensity r to the right ear. Both l and r may be either 20 dB, 21 dB, or 22 dB. (dB stands for decibels and is a measure of sound intensity.) Suppose that the following ranking in terms of loudness has been obtained. Is there an additive conjoint representation?

$$(22, 22)$$
$$(22, 21) - (21, 22)$$
$$(22, 20)$$
$$(21, 20)$$
$$(21, 21)$$
$$(20, 21)$$
$$(20, 22)$$
$$(20, 20)$$

6. Let $A_1 = A_2 = \{0, 1\}$ and let R on $A = A_1 \times A_2$ be lexicographic preference (see Sec. 8.5.4). Does (A, R) satisfy

   (a) Scott's Axiom C1?
   (b) Scott's Axiom C2?
   (c) Independence?
   (d) Additive conjoint measurement? (If so, what are u_1 and u_2?)

7. Suppose u_1 and u_2 define an additive conjoint representation (36) and $u(a_1, a_2) = u_1(a_1) + u_2(a_2)$. Which of the following statements is meaningful?

   (a) $u(a_1, a_2) > u(b_1, b_2)$
   (b) $u(a_1, a_2) = 2u(b_1, b_2)$.
   (c) $u(a_1, a_2)$ is constant for all a_1 in A_1 and a_2 in A_2.

8. Suppose $A = A_1 \times A_2$ and (A, R) is a strict weak order and satisfies independence. Define R_i on A_i as follows:

$$aR_1b \quad \text{iff} \quad \text{for some } x \text{ in } A_2, (a, x)R(b, x)$$
$$xR_2y \quad \text{iff} \quad \text{for some } a \text{ in } A_1, (a, x)R(a, y).$$

   What are R_1 and R_2 in Exer. 2?

9. In Exer. 8, show that each (A_i, R_i) always defines a strict weak order.

10. In the subjective probability situation of Exers. 19 of Sec. 8.5 and Exer. 13 of Sec. 8.6, suppose there is a real-valued function p on A satisfying Eqs. (26), (27), and (30). Suppose $a_0, a_1, \ldots, a_{n-1}, b_0, b_1, \ldots, b_{n-1}$ are in A and suppose a_iSb_i for $i = 1, 2, \ldots, n - 1$, where S is defined from R as in conjoint measurement. Show that if $a_0, a_1, \ldots, a_{n-1}$ are pairwise disjoint and $b_0, b_1, \ldots, b_{n-1}$ are pairwise disjoint, and if
$$a_0 \cup a_1 \cup \cdots \cup a_{n-1} = b_0 \cup b_1 \cup \cdots \cup b_{n-1},$$
   then b_0Sa_0. (A generalization of this condition is used by Scott [1964] to give necessary and sufficient conditions for the existence of a measure of subjective probability.)

11. Show that if $(A_1 \times A_2, R)$ satisfies additive conjoint measurement, then it satisfies *multiplicative conjoint measurement*, i.e., there are functions $v_1:A_1 \rightarrow \Re^+$ and $v_2:A_2 \rightarrow \Re^+$ such that for all a_1, b_1 in A_1 and a_2, b_2 in A_2,
$$(a_1, a_2)R(b_1, b_2) \Longleftrightarrow v_1(a_1)v_2(a_2) > v_1(b_1)v_2(b_2).$$

12. Assume that the expected utility hypothesis discussed in Sec. 8.5.2 holds. Suppose the set A of consequences has a product structure $A = A_1 \times A_2$. Suppose λ_x and θ_x are actions all of whose consequences are of the form (a, x) for a fixed x in A_2. Suppose λ_y and θ_y are obtained from λ_x and θ_x respectively by substituting for every (a, x) the consequence (a, y). Assume that the probabilities $p(a, x)$ and $p(a, y)$ are equal. We say that *strong independence* holds *on the first component* if for all x and y in A_2, λ_x is preferred to θ_x if and only if λ_y is preferred to θ_y. (A similar definition holds for *strong independence on the second component*.) If the utility function on the set of consequences is additive, does strong independence on the first component necessarily hold?

13. Under the expected utility hypothesis, does the quasi-additive representation of Eq. (37) imply strong independence on the first component (Exer. 12)? (Give proof or counterexample.)

14. Exercises 14 through 18 develop an alternative axiomatization for additive conjoint measurement. If $A = A_1 \times A_2$, define E on A by
$$(x, y)E(u, v) \Longleftrightarrow \sim[(x, y)R(u, v)] \;\&\; \sim[(u, v)R(x, y)].$$

(If R is preference, then E corresponds to indifference: You are indifferent between two alternatives if and only if you prefer neither.) We say that (A, R) satisfies the *Thomsen condition* if for all x, y, z in A_1 and q, r, s in A_2,

$$(x, s)E(z, r) \ \& \ (z, q)E(y, s) \Rightarrow (x, q)E(y, r).$$

Is the Thomsen condition a necessary condition for conjoint measurement?

15. If $A = A_1 \times A_2$, we say that (A, R) satisfies *restricted solvability on the first component* if whenever $x, \bar{y}, y$ are in A_1 and q, r are in A_2 and $(\bar{y}, r)R(x, q)$ and $(x, q)R(\underbar{y}, r)$, then there exists y in A_1 such that $(y, r)E(x, q)$, where E is defined as in Exer. 14. A similar definition holds on the second component. (The point is that under some restricted assumptions, we can "solve" for y.) Is restricted solvability on each component a necessary condition for conjoint measurement?

16. If $A = A_1 \times A_2$ and (A, R) is a strict weak order satisfying independence, we say that the ith component ($i = 1, 2$) is *essential* if there are x, y in A_i such that xR_iy or yR_ix, where R_i is defined in Exer. 8. Is the axiom that each component is essential a necessary condition for conjoint measurement?

17. Suppose $A = A_1 \times A_2$, suppose R is a binary relation on A, and let E be defined as in Exer. 14. Suppose that p, q are in A_2 and suppose that $a, a_1, a_2, \ldots, a_n, \ldots$ are in A_1.

    (i) Suppose that $(a, p)R(a, q)$.
    (ii) Suppose that for all i, $(a_i, p)E(a_{i+1}, q)$.
    (iii) Suppose that there are b in A_1 and c in A_2 such that $(b, c)R(a_i, c)$ holds for all i.

    If (A, R) satisfies the conjoint measurement representation of Eq. (36), show that the sequence $a_1, a_2, \ldots, a_n, \ldots$ is finite. (*Note: A* sequence $a_1, a_2, \ldots, a_n, \ldots$ satisfying assumptions (i) and (ii) is called a *standard sequence*. In a standard sequence, the differences between successive elements are constant. A standard sequence is called *strictly bounded* if it satisfies assumption (iii). The axiom that every strictly bounded standard sequence is finite is a common Archimedean axiom in axiomatizations of conjoint measurement.)

18. The following conditions define an *additive conjoint structure* (A, R), where $A = A_1 \times A_2$:

    Axiom CS1. (A, R) is a strict weak order.
    Axiom CS2. (A, R) satisfies independence.
    Axiom CS3. (A, R) satisfies the Thomsen condition (Exer. 14).
    Axiom CS4 (Archimedean Axiom). Every strictly bounded standard sequence is finite (Exer. 17).

Axiom CS5. (A, R) satisfies restricted solvability on each component (Exer. 15).

Axiom CS6. Each component of (A, R) is essential (Exer. 16).

Krantz, *et al.* [1971, Ch. 6] prove the following theorem: If (A, R) is an additive conjoint structure, then (A, R) satisfies additive conjoint measurement.

(a)   Is (A, R) of Exer. 2 an additive conjoint structure?

(b)   What about (A, R) of Exer. 6?

19. Discuss the axioms of Exer. 18 as axioms for preference.

20. Consider the subjective probability situation of Exer. 19 of Sec. 8.5, Exer. 13 of Sec. 8.6, and Exer. 10 above. Suppose that there is a measure of subjective probability (Exer. 13, Sec. 8.6). Define E on A by

$$aEb \iff \sim aRb \quad \text{and} \quad \sim bRa.$$

(E corresponds to the relation "judged equally probable.") Suppose that $a_1, a_2, \ldots, a_n, \ldots$ is a sequence from A and suppose that there exist a, b_i, c_i in A such that the following conditions hold:

$$a_1 = b_1, b_1 Ea$$
$$b_i \cap c_i = \varnothing$$
$$b_i Ea_i$$
$$c_i Ea_i$$
$$a_{i+1} = b_i \cup c_i.$$

Show that if $aR\varnothing$, where $\varnothing$ is the empty set, then the a_i must form a finite sequence. (As in Exer. 17, the a_i form what is called a *strictly bounded standard sequence*.)

21. Think of other potential applications of additive conjoint measurement.

22. Think of potential applications of multiplicative conjoint measurement (Exer. 11).

23. Think of potential applications of the quasi-additive representation (Eq. (37)).

8.8.   Semiorders

8.8.1.   THE SCOTT-SUPPES THEOREM

In Sec. 8.5 we gave an example where (strict) preference was not transitive. In such a situation, if R is preference on a set A, there is no homomorphism from (A, R) into $(\Re, >)$. In this section, we consider whether or not measurement is still possible. (Analogous considerations might apply for the other measurement representations we have studied.)

If R is preference, then indifference corresponds to the binary relation E on A defined by

$$aEb \quad \text{iff} \quad {\sim}aRb \quad \text{and} \quad {\sim}bRa. \tag{38}$$

That is, you are indifferent between a and b if and only if you neither prefer a to b nor prefer b to a. Suppose there is an ordinal utility function u, that is, a function $u{:}A \longrightarrow \Re$ satisfying

$$aRb \longleftrightarrow u(a) > u(b). \tag{39}$$

(By Theorem 8.1, the existence of an ordinal utility function implies that (A, R) is a strict weak order.) If u exists, then

$$aEb \longleftrightarrow u(a) = u(b). \tag{40}$$

Equation (40) implies that (A, E) is transitive, for aEb and bEc imply that $u(a) = u(b)$ and $u(b) = u(c)$, so $u(a) = u(c)$, and aEc.

The economist Armstrong [1939, 1948, 1950, 1951] was one of the first to argue that indifference is not necessarily transitive. (Menger [1951] claims that such arguments go back as far as Poincaré in the the nineteenth century.) Luce [1956] suggests as one argument against the transitivity of indifference the following. Most people would prefer a cup of coffee with one spoon of sugar to a cup with five spoons. But if sugar were added to the first cup at the rate of $\frac{1}{100}$ of a gram, they would almost certainly be indifferent between successive cups. If indifference were transitive, they would have to be indifferent between the cup with one spoon and the cup with five spoons. Similarly, if preference between air environments is determined on the basis of eye irritation, then you probably prefer an air environment with .05 parts per million (ppm) of ozone to one with .5 ppm. But you remain indifferent if ozone is added in amounts of 10^{-10} ppm at a time.

A considerably different example is the following. Suppose you are indifferent between two alternative plans for water pollution abatement, plans a and b, where plan a would allocate a budget of 2 billion dollars to the federal Environmental Protection Agency (EPA) and plan b would allocate 2 billion dollars to various state agencies. It seems likely that you would still be indifferent between plan a and plan b' which would allocate 2 billion and one dollars to the state agencies. For probably if you have any preference when budgets are so close it will be based on a choice of a particular approach to water pollution abatement (federal vs. state). On the other hand, if you want to see the government spend money on water pollution abatement, you would certainly prefer b' to b, which violates transitivity of indifference.

Other arguments against the transitivity of indifference and many references to the literature of this issue can be found in Fishburn [1970a], Krantz, et al [to appear, Chapter 15], and Roberts [to appear].

The problem of nontransitivity of indifference led Luce [1956] to slightly modify the demands in the measurement of preference. Motivated by examples like the first two, and by the notion of threshold in psychology, he suggested that we ask for a real-valued function u on A so that for all $a, b \in A$, a is preferred to b if and only if $u(a)$ is not only larger than $u(b)$ but "sufficiently larger" so that we can tell a and b apart. To formalize this representation problem, we fix a positive number δ, the *threshold* or *just noticeable difference*, and ask for conditions on the relational system (A, R) necessary and sufficient for the existence of a real-valued function u on A such that for all $a, b \in A$,

$$aRb \Longleftrightarrow u(a) > u(b) + \delta. \tag{41}$$

(To formulate this as a homomorphism problem, we would define a binary relation $>_\delta$ on $\Re$ by

$$x >_\delta y \Longleftrightarrow x > y + \delta.$$

Then we would ask for a homomorphism from (A, R) into $(\Re, >_\delta)$.)

If A is a finite set, then conditions on (A, R) necessary and sufficient for the representation (41) can be explicitly stated. We should note that this representation, which is designed to take account of examples like the cups of coffee and the comparison of air environments, does not account for examples like the alternative budgets. For if a' is the plan of budgeting 2 billion and one dollars to EPA, you probably prefer a' to a, but are indifferent between a' and b and between a' and b'. Thus, your preference relation R on the set of plans $\{a, a', b, b'\}$ is probably the relation $\{(a', a), (b', b)\}$. But this relation cannot be represented in the form (41). For if it could, then we would have

$$u(a') > u(a) + \delta \geq u(b') > u(b) + \delta.$$

(The second inequality follows since b' is not preferred to a.) Now $u(a') > u(b) + \delta$ implies $a'Rb$, which is contrary to assumption.

In studying the representation (41), Luce [1956] introduced the concept of a semiorder. Our definition of semiorder is formulated following that of Scott and Suppes [1958]. We say that a binary relation (A, R) is a *semiorder* if for all $a, b, c, d \in A$, the following axioms are satisfied.

Axiom S1. $\sim aRa$.
Axiom S2. $(aRb$ and $cRd) \Longrightarrow (aRd$ or $cRb)$.
Axiom S3. $(aRb$ and $bRc) \Longrightarrow (aRd$ or $dRc)$.

To explain the axioms, and to see that they follow from the representation (41), let us first note that Axiom S1 says that (A, R) is irreflexive, which follows since $u(a)$ can never be larger than $u(a) + \delta$. To see that Axiom S3 holds, suppose aRb and bRc. Then $u(a)$, $u(b)$, and $u(c)$ have positions like those in Fig. 8.3. Now $u(d)$ can have any one of the positions $u(d_1)$, $u(d_2)$, $u(d_3)$, $u(d_4)$. We have d_4Rc, d_3Rc, aRd_2, aRd_1. To see that Axiom S2 holds, consider two cases: $u(a) \geq u(c)$ and $u(c) \geq u(a)$. In the first case, we have

$$u(a) \geq u(c) > u(d) + \delta,$$

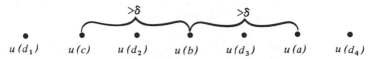

FIGURE 8.3. Axiom $S3$ of the definition of a semiorder is a necessary condition.

so aRd. In the second case, we have

$$u(c) \geqq u(a) > u(b) + \delta,$$

so cRb. If A is finite, then the axioms for a semiorder are sufficient as well as necessary.

Theorem 8.6 (Scott and Suppes).   Suppose R is a binary relation on a finite set A and δ is a positive number. Then (A, R) is a semiorder if and only if there is a real-valued function u on A such that for all $a, b \in A$,

$$aRb \Longleftrightarrow u(a) > u(b) + \delta. \tag{41}$$

For a proof of this theorem, which is constructive, the reader is referred to Scott and Suppes [1958], Suppes and Zinnes [1963], or Rabinovitch [1976]. It is a corollary that if (A, R) is representable in the form (41) for *some* positive number δ, then it is representable in the form (41) for *any* positive number δ, in particular, for $\delta = 1$. (Why?)

At this point, let us consider an example. Let $A = \{1, 2, 3, 4, 5\}$ and let

$$R = \{(1, 3), (1, 4), (1, 5), (2, 4), (2, 5), (3, 5)\}. \tag{42}$$

Then (A, R) is not a strict weak order. For $\sim 1R2$, $\sim 2R3$, but $1R3$, so (A, R) is not negatively transitive. Thus there is no function u on A satisfying Eq. (39). But there is a function u on A satisfying Eq. (41), for (A, R) is a semiorder. To verify that, note that Axiom S1 for semiorders is straightforward. Axiom S2 is rather tedious to check. In particular, since $1R3$ and $2R4$, Axiom S2 implies $1R4$ or $2R3$. We have $1R4$. Similarly, since $3R5$ and $2R4$, Axiom S2 implies that $3R4$ or $2R5$. We have $2R5$. And so on. This axiom is sometimes a little easier to check if one observes what it means graph-theoretically: Whenever there are two independent arcs as shown by solid lines in Fig. 8.4, then one of the two dotted diagonal arcs must also be present. The reader is urged to draw a digraph corresponding to this semiorder. To check Axiom

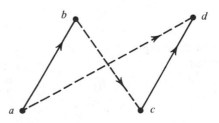

FIGURE 8.4. A graphical test of Axiom $S2$ for semiorders.

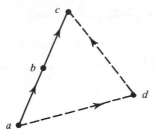

FIGURE 8.5. A graphical test of Axiom
S3 for semiorders.

S3, again it is sometimes easier to observe what it means graph-theoretically:
Whenever there are two arcs as shown by solid lines in Fig. 8.5, and d is any
vertex (including possibly, a, b, or c), then one of the two dotted arcs shown
must be present. In our example, we see that $1R3$ and $3R5$, so using $d = 4$, we
conclude that $1R4$ or $4R5$ must hold. We have $1R4$. The rest of the check
proceeds similarly.

Sometimes an easier way to check that (A, R) is a semiorder is to find a
function u satisfying Eq. (41). If $\delta = 1$, such a function is given for our
example (42) by

$$u(1) = 2.6, u(2) = 1.8, u(3) = 1.5, u(4) = .7, u(5) = 0.$$

It is left to the reader to check that u satisfies Eq. (41).

8.8.2. UNIQUENESS

For the representation (41), there can be no uniqueness theorem which
specifies the class of admissible transformations. For a function satisfying
Eq. (41) does not always define a regular scale. To see this, let $A = \{1, 2, 3\}$
and let $R = \{(1, 3), (1, 2)\}$. Then two functions satisfying Eq. (41) with $\delta = 1$
are given by

$$u(1) = 2, \qquad u(2) = 0, \qquad u(3) = 0$$

and

$$v(1) = 2, \qquad v(2) = .1, \qquad v(3) = 0.$$

There is no function $\phi : u(A) \rightarrow \Re$ such that $v(a) = (\phi \circ u)(a)$, for all a in A.
For this would imply that $v(2) = v(3)$. Thus, $[(A, R), (\Re, >_\delta), u]$ is not a
regular scale and the representation $(A, R) \rightarrow (\Re, >_\delta)$ is not a regular
representation.

8.8.3. INTERVAL ORDERS AND MEASUREMENT WITHOUT NUMBERS

For another insight into the representation (41), let us consider the
interval

$$J(a) = \left[u(a) - \frac{\delta}{2}, u(a) + \frac{\delta}{2} \right].$$

If J and J' are two real intervals, we shall say that

$$J \succ J' \quad \text{iff} \quad (a > b \text{ for all } a \in J \text{ and } b \in J').$$

If u satisfies (41), then

$$aRb \iff J(a) \succ J(b). \tag{43}$$

We may think of $J(a)$ as a collection of alternatives not noticeably different from a, or alternatives within threshold of a. $J(a)$ defines a range of "fuzziness" about a, or a range of possible values. (In Sec. 3.4.2 we discussed the idea of estimating the monetary values of different fine wines. $J(a)$ could be a range of estimates.) We prefer a to b if and only if we are sure that every possible value of a is larger than every possible value of b. If (A, R) is a semiorder, then all of the intervals $J(a)$ have the same length. But it is interesting to think of the possibility of letting them have different lengths. (Certainly in the case of estimating monetary values of wines, this makes sense.) We ask: When does there exist an assignment to each $a \in A$ of an interval $J(a)$ so that for all $a, b \in A$, (43) is satisfied? If we take a more general point of view than we did in Sec. 8.3, then the assignment of intervals satisfying Eq. (43) is as legitimate a form of measurement as the assignment of numbers satisfying the representation

$$aRb \iff u(a) > u(b). \tag{39}$$

For, one of the goals of measurement is to reflect empirical relations by well-known relations on mathematical objects. Having translated an empirical relational system into what we shall loosely call a mathematical relational system, we can apply the whole collection of mathematical tools at our disposal to better understand the mathematical system and hence the empirical one. In particular, we can apply our mathematical tools to help in decisionmaking. In this broad sense, assignment of vectors, sets, intervals, etc. is a perfectly legitimate form of measurement if a representation theorem stating a homomorphism from an empirical relational system to a mathematical relational system can be proved. A similar point of view is expressed in Krantz [1968].

Having expressed this point of view, let us state a representation theorem for the representation (43). A binary relation (A, R) is called an *interval order* if it satisfies the first two axioms in the definition of a semiorder. Thus, every semiorder is an interval order. But it is not too hard to give an

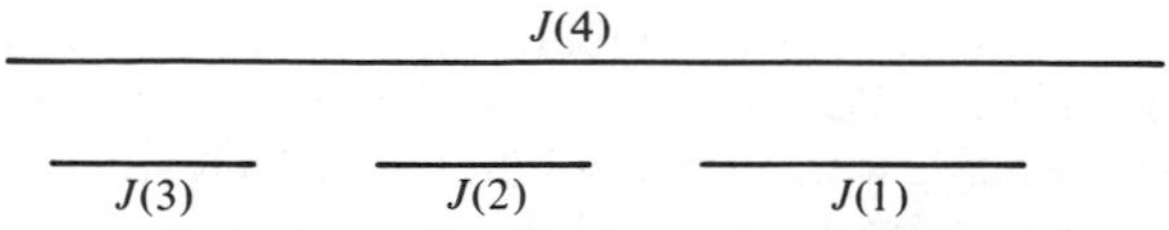

$J(4)$

$J(3)$ $J(2)$ $J(1)$

FIGURE 8.6. An interval representation for the interval order (A, R), where $A = \{1, 2, 3, 4\}$ and $R = \{(1, 2), (2, 3), (1, 3)\}$. Intervals are displaced vertically for ease of comparison.

example of an interval order which is not a semiorder. Let $A = \{1, 2, 3, 4\}$ and define R on A by $R = \{(1, 2), (2, 3), (1, 3)\}$. An interval representation satisfying Eq. (43) for (A, R) is shown in Fig. 8.6. Proof that (A, R) is not a semiorder is left to the reader (Exer. 6). We now have the following representation theorem.

Theorem 8.7 (Fishburn [1970b, 1970c]). Suppose (A, R) is a binary relation on a finite set A. Then (A, R) is an interval order if and only if there is an assignment of an interval $J(a)$ to each $a \in A$ so that for all $a, b \in A$,

$$aRb \Longleftrightarrow J(a) \succ J(b). \tag{43}$$

In closing, we consider the relation E of indifference defined from an interval order (A, R) by Eq. (38). It is easy to see, using Fishburn's Theorem, that

$$aEb \Longleftrightarrow J(a) \cap J(b) \neq \varnothing.$$

We have encountered this representation for indifference in Sec. 3.4, where we called (A, E) an *interval graph* if there was such an assignment of intervals. In Sec. 3.4.3, we gave conditions on (A, E) necessary and sufficient for the existence of such an assignment. We also saw that Eq. (43) defined a transitive orientation of the complementary graph of (A, E). It is not hard to prove (Exer. 19) that if (A, R) is a binary relation on a finite set A, then (A, R) is an interval order if and only if the binary relation (A, E) defined by Eq. (38) is an interval graph and (A, R) is a transitive orientation of the complementary graph of (A, E).

EXERCISES

1. Suppose A is a set of sounds and aRb holds if and only if a is judged louder than b. Give an argument, analogous to Luce's argument for preference, which suggests that there is no real-valued function u on A satisfying Eq. (39).

2. For the binary relation defined in Eq. (42), find a function u on A satisfying Eq. (41) if $\delta = 2$.

3. A well-known example in utility theory, due to Armstrong [1939], is the following. Suppose a boy is indifferent between receiving as a gift a pony or a bicycle. He will undoubtedly prefer the bicycle if a bell is added to the bicyle without the bell. But he is likely still to be indifferent between the bicycle with bell and the pony.

    (a) Show that in this case, indifference is not transitive.

(b)   Add a pony with bridle and argue that there is no function u satisfying Eq. (41).

(c)   Do you agree with this example?

4.   Suppose (A, R) is defined as follows: $A = \{1, 2, 3, 4\}$ and $R = \{(1, 4)\}$.

 (a)   Show from the definition that (A, R) is a semiorder.

 (b)   Find a function u on A satisfying Eq. (41).

5.   Suppose $A = \{0, 1, 2\}$ and R is $\{(2, 1), (2, 0), (1, 0)\}$. Show from the definition that (A, R) is a semiorder. (*Note:* (A, R) is also a strict weak order.)

6.   Prove that if $A = \{1, 2, 3, 4\}$ and $R = \{(1, 2), (2, 3), (1, 3)\}$, then (A, R) is not a semiorder.

7.   Give an example of an irreflexive binary relation (A, R) such that A has 3 elements and such that (A, R) is not a semiorder.

8.   Suppose (A, R) is defined as follows. Find a function u on A satisfying Eq. (41), and conclude that (A, R) is a semiorder.

$$A = \{1, 2, 3, 4, 5, 6, 7\}$$

$$R = \{(1, 4), (1, 5), (1, 6), (1, 7), (2, 5), (2, 6), (2, 7), (3, 5),$$
$$(3, 6), (3, 7), (4, 5), (4, 6), (4, 7), (5, 7)\}.$$

9.   An individual tastes five wines and rates their possible dollar values as follows: w_1 is worth between \$1 and \$4, w_2 between \$2 and \$3, w_3 between \$10 and \$15, w_4 between \$5 and \$6, and w_5 between \$1 and \$1.50. Assuming Eq. (43), what are his preferences?

10.   Suppose (A, R) is defined as follows:

$$A = \{1, 2, 3, 4, 5\}$$

$$R = \{(5, 3), (5, 1), (5, 2), (4, 1), (4, 2), (1, 2)\}.$$

 (a)   Show that (A, R) is an interval order.

 (b)   Show that (A, R) is not a semiorder.

 (c)   Find an interval assignment $J(a)$ satisfying Eq. (43).

 (d)   Draw the interval graph (A, E) and show that (A, R) is a transitive orientation of the complementary graph of (A, E).

11.   Suppose an individual's preferences are defined as in Table 8.10. Is there a function u on A satisfying Eq. (41)?

12.   Prove by verifying the axioms that every strict weak order is a semiorder.

13.   Is every semiorder a strict weak order? (Give proof or counterexample.)

14.   Consider the following semiorder: $A = \{0, 1, 2, 3\}$, $R = \{(3, 0)\}$. Show that the representation $(A, R) \longrightarrow (\mathfrak{R}, >_\delta)$, with $\delta = 1$, is not a regular

TABLE 8.10

Preferences among Cars.

Model *i* Is Preferred to Model *j* iff the *i, j* Entry Is 1.

	Pontiac	Volkswagen	Buick	Cadillac	Datsun
Pontiac	0	0	0	0	0
Volkswagen	0	0	0	0	0
Buick	1	1	0	1	0
Cadillac	0	0	0	0	0
Datsun	1	0	0	1	0

representation, by finding two functions u and v satisfying Eq. (41) such that v is not $\phi \circ u$ for any $\phi : u(A) \rightarrow \mathfrak{R}$.

15. This exercise is designed to prove that Theorem 8.6 is false if the hypothesis that A be finite is removed. Let N be the set of positive integers and let x be any element not in N. Let $A = N \cup \{x\}$ and define R on A as follows: For all a, b in N,

$$aRb \Longleftrightarrow a > b + 1$$

holds, and so does

$$xRa.$$

   (a) Show that (A, R) is a semiorder.
   (b) Assume that there is a function $u : A \rightarrow \mathfrak{R}$ satisfying (41) and show that u is one-to-one.
   (c) Show that in addition, $u(2n + 1) > u(1) + n\delta$.
   (d) Conclude that no such u exists.

16. (Roberts [1969].) Suppose A is finite and (A, R) is a semiorder. Define E on A by Eq. (38). Prove that there is an x in A with the following property: Whenever xEa and xEb, then aEb. (*Hint:* Use the representation theorem.)

17. (Roberts [1971].) Suppose A is finite and there is a representation $u : A \rightarrow \mathfrak{R}$ satisfying Eq. (41). Define W on A by

$$aWb \Longleftrightarrow u(a) > u(b).$$

Define E on A by Eq. (38). Show that if

$$aWu \quad \text{and} \quad uWv \quad \text{and} \quad vWb$$

then

$$aEb \Longrightarrow uEv.$$

In words, intervals of preference cannot be contained within intervals of indifference. This result is called, by Goodman [1951], the *weak mapping rule*.

18. Using the assumptions of Exer. 17, Sec. 8.7, suppose that instead of the conjoint measurement representation of Eq. (36), we assume the following representation with δ a positive constant:

$$(a_1, a_2)R(b_1, b_2) \Longleftrightarrow u_1(a_1) + u_2(a_2) > u_1(b_1) + u_2(b_2) + \delta.$$

    Does the conclusion that the sequence $a_1, a_2, \ldots, a_n, \ldots$ is finite still hold? Why?

19. Prove that if (A, R) is a binary relation on a finite set A, then (A, R) is an interval order if and only if the binary relation (A, E) defined by Eq. (38) is an interval graph and (A, R) is a transitive orientation of the complementary graph of (A, E).

20. Using the data from Exer. 26 of Sec. 8.5, determine if it defines a semi-order or an interval order, and if so, find appropriate mappings into the reals or real intervals.

21. Your threshold in hearing depends on the intensity of the sounds. If the same is true of preference, how would you have to modify the semi-order representation to take account of this? Can you prove some theorems about this representation? (See Roberts [1971] or Krantz, et al. [to appear, Ch. 15].)

22. Formulate the discussion of the representation of ecological phase space of Sec. 3.6.1 in terms of measurement without numbers.

23. Is measurement without numbers really a useful idea? Can you think of potential applications?

24. Discuss critically Luce's argument about coffee and sugar.

References

ADAMS, E. W., "Elements of a Theory of Inexact Measurement," *Phil. Sci.*, **32** (1965), 205–228.

ALIMOV, N. G., "On Ordered Semigroups," *Izv. Akad. Nauk. SSSR. Ser. Mat.*, **14** (1950), 569–576.

ALLAIS, M., "Le comportement de l'homme rationnel devant le risque: Critique des postulats de l'école Americaine," *Econometrica*, **21** (1953), 503–546.

ARMSTRONG, W. E., "The Determinateness of the Utility Function," *Econ. J.*, **49** (1939), 453–467.

ARMSTRONG, W. E., "Uncertainty and the Utility Function," *Econ. J.*, **58** (1948), 1–10.

ARMSTRONG, W. E., "A Note on the Theory of Consumer's Behavior," *Oxford Economic Papers*, **2** (1950), 119–122.

ARMSTRONG, W. E., "Utility and the Theory of Welfare," *Oxford Economic Papers*, **3** (1951), 259–271.

BEHREND, F. A., "A System of Independent Axioms for Magnitudes," *J. Proc. Roy.*

Soc. New South Wales, **87** (1953), 27–30.

BEHREND, F. A., "A Contribution to the Theory of Magnitudes and the Foundations of Analysis," *Math. Z.,* **63** (1956), 345–362.

BERNOULLI, D., "Specimen Theoriae Novae de Mensura Sortis," *Comentarii Academiae Scientiarum Imperiales Petropolitanae,* **5** (1738), 175–192; translated by L. Sommer in *Econometrica,* **22** (1954), 23–36.

BIRKHOFF, G., *Lattice Theory,* American Mathematical Society Colloquium Publication No. XXV, New York, 1948, 1967.

CAMPBELL, N. R., *Physics: The Elements,* Cambridge University Press, Cambridge, England, 1920; reprinted as *Foundations of Science: The Philosophy of Theory and Experiment,* Dover Publications, Inc., New York 1957.

CAMPBELL, N. R., *An Account of the Principles of Measurement and Calculation,* Longmans, Green & Co., Inc., London, 1928.

CANTOR, G., "Beiträge zur Begründung der Transfiniten Mengenlehre," *Math. Ann.,* **46** (1895), 481–512.

COHEN, M. R. and NAGEL, E., *An Introduction to Logic and Scientific Method,* Harcourt, Brace & World, Inc., New York, 1934.

DEBREU, G., "Topological Methods in Cardinal Utility Theory," in *Mathematical Methods in the Social Sciences,* K. J. Arrow, S. Karlin, and P. Suppes (eds.), Stanford University Press, Stanford, Calif., 1960, pp. 16–26.

ELLIS, B., *Basic Concepts of Measurement,* Cambridge University Press, London, 1966.

FISHBURN, P. C., *Decision and Value Theory,* John Wiley & Sons, Inc., New York, 1964.

FISHBURN, P. C., "Utility Theory," *Management Science,* **14** (1968), 335–378.

FISHBURN, P. C., "Intransitive Indifference in Preference Theory: A Survey," *Operations Research,* **18** (1970), 207–228, a.

FISHBURN, P. C., "Intransitive Indifference with Unequal Indifference Intervals," *J. Math. Psychol.,* **7** (1970), 144–149, b.

FISHBURN, P. C., *Utility Theory for Decisionmaking,* John Wiley & Sons, Inc., New York, 1970, c.

FISHER, I., *The Making of Index Numbers,* Houghton Mifflin Co., Boston, Mass., 1923.

GOODMAN, N., *Structure of Appearance,* Harvard University Press, Cambridge, Mass., 1951.

HELMHOLTZ, H. v., "Zählen und Messen, Erkenntnistheoretisch Betrachtet," in *Philosophische Aufsätze Edward Zeller gewidmet,* Leipzig, 1887; Engl. translation by C. L. Bryan, *Counting and Measuring,* D. van Nostrand Co. Inc., Princeton, N.J., 1930.

HOFFMAN, K. H., "Sur Mathematischen Theorie des Messens," *Rozprawy Mat. (Warsaw),* **32** (1963), 1–31.

HÖLDER, O., "Die Axiome der Quantität und die Lehre vom Mass," *Ber. verh. Kgl. Sächsis. Ges. Wiss. Leipzig, Math.-Phys. Klasse,* **53** (1901), 1–64.

HOLMAN, E. W., "Strong and Weak Extensive Measurement," *J. Math. Psychol.,* **6** (1969), 286–293.

HUNTINGTON, E. V., "A Complete Set of Postulates for the Theory of Absolute Continuous Magnitude," *Trans. Amer. Math. Soc.,* **3** (1902), 264–279, a.

HUNTINGTON, E. V., "Complete Sets of Postulates for the Theories of Positive Integral and of Positive Rational Numbers," *Trans. Amer. Math. Soc.*, **3** (1902), 280–284, b.

HUNTINGTON, E. V., *The Continuum and Other Types of Serial Order*, Harvard University Press, Cambridge, Mass., 1917.

KRANTZ, D. H., "A Survey of Measurement Theory," in *Mathematics of the Decision Sciences*, Part 2, G. B. Dantzig & A. F. Veinott, Jr. (eds.), Vol. 12 of Lectures in Applied Mathematics, American Mathematical Society, Providence, R. I., 1968, pp. 314–350.

KRANTZ, D. H., "A Theory of Magnitude Estimation and Cross-Modality Matching," *J. Math. Psychol.*, **9** (1972), 168–199.

KRANTZ, D. H., LUCE, R. D., SUPPES, P., and TVERSKY, A., *Foundations of Measurement*, Vol. I, Academic Press, Inc., New York, 1971.

KRANTZ, D. H., LUCE, R. D. SUPPES, P., and TVERSKY, A., *Foundations of Measurement*, Vol. II, Academic Press, Inc., New York, to appear.

LEVELT, W. J. M., RIEMERSMA, J. B., and BUNT, A. A., "Binaural Additivity of Loudness," *British J. of Math. and Stat. Psychol.*, **25** (1972), 51–68.

LUCE, R. D., "Semiorders and a Theory of Utility Discrimination," *Econometrica*, **24** (1956), 178–191.

LUCE, R. D., "On the Possible Psychophysical Laws," *Psych. Rev.*, **66** (1959), 81–95.

LUCE, R. D., and KRANTZ, D. H., "Conditional Expected Utility," *Econometrica*, **39** (1971), 253–271.

LUCE, R. D., and RAIFFA, H., *Games and Decisions*, John Wiley & Sons, Inc., New York, 1957.

LUCE, R. D., and SUPPES, P., "Preference, Utility, and Subjective Probability," in *Handbook of Mathematical Psychology*, Vol. 3, R. D. Luce, R. R. Bush, and E. Galanter (eds.), John Wiley & Sons, Inc., New York, 1965, pp. 249–410.

LUCE, R. D., and TUKEY, J. W., "Simultaneous Conjoint Measurement: A New Type of Fundamental Measurement," *J. Math. Psychol.*, **1** (1964), 1–27.

MANHEIM, M. L. and HALL, F. L., "Abstract Representation of Goals," Paper P-67-24, Department of Civil Engineering, Massachusetts Institute of Technology, January 1968.

MENGER, K., "Probabilistic Theories of Relations," *Proc. Nat. Acad. Sci.*, **37** (1951), 178–180.

MILGRAM, A. N., "Partially Ordered Sets, Separating Systems, and Inductiveness," in *Reports of a Mathematical Colloquium*, Second Series No. 1, K. Menger (ed.), University of Notre Dame, 1939.

MILLER, J. R., "Assessing Alternative Transportation Systems," Rand Corporation Memorandum RM-5865-DOT, 1969.

PFANZAGL, J., *Theory of Measurement*, John Wiley & Sons, Inc., New York, 1968.

RABINOVITCH, I., "The Scott-Suppes Theorem on Semiorders," *J. of Math. Psychol.*, to appear, 1976.

RAIFFA, H., *Decision Analysis*, Addison-Wesley, Reading, Mass., 1968.

RAIFFA, H., "Preferences for Multi-attributed Consequences," Rand Corporation Memorandum RM-5868-DOT, 1969.

REESE, T. W., "The Application of the Theory of Physical Measurement to the

Measurement of Psychological Magnitudes, with Three Experimental Examples," *Psychol. Monogr.*, **55** (1943), 1–89.

ROBERTS, F. S., "Indifference Graphs," in *Proof Techniques in Graph Theory*, F. Harary (ed.), Academic Press, New York, 1969, pp. 139–146.

ROBERTS, F. S., "On the Compatibility between a Graph and a Simple Order," *J. Comb. Theory*, **11** (1971), 28–38.

ROBERTS, F. S., "Building an Energy Demand Signed Digraph I: Choosing the Nodes," Rand Corporation Report R-927/1-NSF, April 1972.

ROBERTS, F. S., "Building and Analyzing an Energy Demand Signed Digraph," *Environment and Planning*, **5** (1973), 199–221.

ROBERTS, F. S., "On the Theory of Uniqueness in Measurement," Mimeographed, Department of Mathematics, Rutgers University, New Brunswick, N.J., 1974.

ROBERTS, F. S., *Measurement Theory*, a volume in the Encyclopedia of Mathematics and its Applications, Addison-Wesley, Reading, Mass., to appear.

ROBERTS, F. S., and LUCE, R. D., "Axiomatic Thermodynamics and Extensive Measurement," *Synthese*, **18** (1968), 311–326.

SAVAGE, L. J., *The Foundations of Statistics*, John Wiley & Sons, Inc., New York, 1954.

SCOTT, D., "Measurement Models and Linear Inequalities," *J. Math. Psychol.*, **1** (1964), 233–247.

SCOTT, D. and SUPPES, P., "Foundational Aspects of Theories of Measurement," *J. Symbolic Logic*, **23** (1958), 113–128.

STEVENS, S. S., "On the Theory of Scales of Measurement," *Science*, **103** (1946), 677–680.

STEVENS, S. S., "Mathematics, Measurement and Psychophysics," in *Handbook of Experimental Psychology*, S. S. Stevens (ed.), John Wiley & Sons, Inc., New York, 1951, pp. 1–49.

STEVENS, S. S., "On the Psychophysical Law," *Psych. Rev.*, **64** (1957), 153–181.

STEVENS, S. S., "Measurement, Psychophysics, and Utility," in *Measurement: Definitions and Theories*, C. W. Churchman and P. Ratoosh (eds.), John Wiley & Sons, Inc., New York, 1959, pp. 18–63.

STEVENS, S. S., "Measurement, Statistics, and the Schemapiric View," *Science*, **161** (1968), 849–856.

SUPPES, P., "A Set of Independent Axioms for Extensive Quantities," *Portugal. Math.*, **10** (1951), 163–172.

SUPPES, P., *Introduction to Logic*, D. Van Nostrand Co., Inc., Princeton, N.J., 1957.

SUPPES, P. "Measurement, Empirical Meaningfulness and Three-Valued Logic," in *Measurement: Definitions and Theories*, C. W. Churchman and P. Ratoosh (eds.), John Wiley & Sons, Inc., New York, 1959, pp. 129–143.

SUPPES, P. and WINET, M., "An Axiomatization of Utility Based on the Notion of Utility Differences," *Management Science*, **1** (1955), 259–270.

SUPPES, P. and ZINNES, J., "Basic Measurement Theory," in *Handbook of Mathematical Psychology*, Vol. I, R. D. Luce, R. R. Bush and E. Galanter (eds.), John Wiley & Sons, Inc., New York, 1963, pp. 1–76.

THRALL, R. M., MORTIMER, J. A., REBMAN, K. R., and BAUM, R. F. (eds.), *Some Mathematical Models in Biology*, Revised Edition, Report No. 40241-R-7, University of Michigan, 1967.

Author Index

Subject Index